M N V Prasad

Heavy Metal Stress in Plants

Springer

Berlin
Heidelberg
New York
Hong Kong
London
Milan
Paris
Tokyo

M N V Prasad (Ed.)

Heavy Metal Stress in Plants

From Biomolecules to Ecosystems

Second Edition

With 76 Figures, two in Color, and 56 Tables

 Springer

M N V Prasad
Department of Plant Sciences
University of Hyderabad
Hyderabad, Andhra Pradesh
India

First Edition 1999

© Springer-Verlag Berlin Heidelberg 2004
Printed in India

Springer-Verlag is a part of Springer Science+Business Media
springeronline.com

ISBN 3-540-40131-8

31/3150 – 5 4 3 2 1

Preface

Biogeochemical cycling of essential and non-essential elements in ecosystem is a complex problem. Al, Ca, Fe, Na, P, K, S, Si, Ti, Mg; bioelements, viz. C, H, N and O, constitute about 99% of the elemental composition of the environment. Elements such as As, Cd, Co, Cu, Cr, Hg, Mo, Mn, Ni, Pb, Se, Zn etc constitute about 1% of the total elemental content of the soil and hence are called trace elements. We use the term 'heavy metal' for those weighing more than 5 g cm^{-3} e.g. Zn (7.1), Cr (7.2), Cd (8.6), Ni (8.7), Co (8.9), Cu (8.9), Mo (10.2), Pb (11.4), and Hg (13.5); Al (2.7) is a light metal; 'As and Sn' half-metals and 'Se' non-metal. Some metals can occur in different valence states rendering more or less toxic in different states. One example is less toxic Cr(III) and the more toxic Cr(VI). A few metals are precious, like gold or platinum. Cd and As are poisonous and others like Cs, Hg and Ga are liquids at room temperature. Elements viz., As, B, Cd, Cr, Cu, Hg, Ni, Pb, Se, U, V, and Zn are present naturally in soils in low concentrations but may be elevated because of human activities, fossil fuel combustion, mining, smelting, sludge amendment to soil, fertilizer and agrochemical application. At low concentrations some trace elements (e.g. Cu, Cr, Mo, Ni, Se and Zn etc.) are essential for healthy functioning and reproduction of microorganisms, plants and animals including man. However, at high concentrations, the same essential elements may pose toxicity. Some trace elements are also non-essential e.g. As, Cd, Pb and Hg etc. and even low concentrations of these elements in the environment can cause toxicity to both plants and animals

Beneficial as well as detrimental effects of mining, use and disposal of various metals are described in reports dating back to ancient history. However, it is astonishing that the interest in plant metal interaction aroused only recently, mostly in the last couple of decades. This volume brings a state-of-the-art review focussing the toxic symptoms, resulting biomolecules in plants. In this treatise topics of contemporary importance on the chosen area are covered in various chapters, i.e the molecular level, cellular level, organismal level (plant) and ecosystem level. Additionally two chapters, one on experimental characterization of heavy metal tolerance in plants and the other on species-selective analysis for metals and metalloids in plants are also included.

This book was prepared for all those who are interested in heavy metals and their interaction with plants, particularly students and teachers of plant physiology and biochemistry, molecular biology, ecophysiology, agronomy, forestry, chemistry and geography. This herculean task could be completed with the cooperation of several colleagues from the diverse scientific fields of chemistry, plant physiology, ecology, forestry and agriculture. I would like to thank them all for their competent contributions.

I sincerely thank the editorial team of Springer-Verlag, particularly for their excellent technical help which hastened the production of this second revised version Thanks are due to Dr Jürgen Hagemeyer, who inspired me to undertake revision though he could not join due to reasons beyond his control. I am indebted to all contributors and peers for supporting me to develop this facinating field of research.

I am extremely thankful to Dr Kota Harinarayana, Vice-Chancellor, University of Hyderabad; Professor T. Suryanarayana, Dean, School of Life Sciences for supporting my academic pursuits. I deeply appreciate the inputs of my students and several collaborators from whom I learnt a lot. I am pleased to record the excellent cooperation of my wife, Savithri.

Certainly this book may not be free from typographical errors and mistakes, as the future will show, but 'the point is not to be right'. 'The point is to make progress, and one cannot make progress if afraid of to be wrong' (D.Hohanson and J.Shreeve, *Lucy's Child*)!

M N V Prasad

List of Contributors

Adriano, Domy C. Savannah River Ecology Laboratory, University of Georgia, Aiken, SC 29802, USA

Arduini, I. Dipartimento di Agronomia e Gestione dell'Agro-Ecosistema, Università degli Studi Pisa, Via S. Michele degli Scalzi 2, 56124 Pisa, Italy

Barceló, J. Laboratorio de Fisiología Vegetal, Facultad de Ciencias, Universidad Autónoma de Barcelona, 08193 Bellaterra, Spain

Dirk Schaumlöffel, CNRS UMR 5034, Group of Bio-Inorganic Analytical Chemistry, 2, av. Pr. Angot, 64000 Pau, France

Enzo Lombi, Adelaide Laboratory, PMB2 Glen Osmond, SA 5064 Australia

Godbold, D.L. School of Agricultural and Forest Sciences, University of Wales, Bangor, Gwynedd LL57 2UW, UK

Hüttermann, A. Forstbotanisches Institut, Abteilung Technische Mykologie, Universität Göttingen, 37077 Göttingen, Germany

Joanna Szpunar, CNRS UMR 5034, Group of Bio-Inorganic Analytical Chemistry, 2, av. Pr. Angot, 64000 Pau, France

Jürgen Hagemeyer, Im Bergsiek 43, 33739 Bielefeld, Germany

Köhl, Karin I. Max Planck Institut für Molekulare Pflanzenphysiologie, Am Mühlenberg 1, 14476 Golm, Germany

Łobiński, Ryszard. CNRS UMR 5034, Group of Bio-Inorganic Analytical Chemistry, 2, av. Pr. Angot, 64000 Pau, France

Lösch, R. Abt. Geobotanik, H. Heine-Universität, Universitätsstr. 1, 40225 Düsseldorf, Germany

Maria Greger, Department of Botany, Stockholm University, 106 91 Stockholm, Sweden

Mishra, R.K. Institute of Life Sciences, Nalco Square, Bhubaneswar 751023, Orissa, India

Myśliwa -Kurdziel, B. Department of Plant Physiology and Biochemistry, Faculty of Bio-technology, Jagiellonian University, ul. Gronostajowa 7, 30-387 Kraków, Poland

Pohlmeier, A. Institut für Chemie und Dynamik der Geosphäre 7, Forschungszentrum, 52425 Jülich, Germany

Poschenrieder, Ch. Laboratorio de Fisiología Vegetal, Facultad de Ciencias, Universidad Autónoma de Barcelona, 08193 Bellaterra, Spain

Prasad, M.N.V. Department of Plant Sciences, University of Hyderabad, Hyderabad 500046, Andhra Pradesh, India

Rainer Lösch, Abt. Geobotanik, H. Heine-Universität, Universitätsstr. 1, 40225 Düsseldorf, Germany

Rama Devi, S. Department of Plant Sciences, University of Hyderabad, Hyderabad 500046, Andhra Pradesh, India

Sahu, S.K. Institute of Life Sciences, Nalco Square, Bhubaneswar 751023, Orissa, India

Shaw, B.P. Institute of Life Sciences, Nalco Square, Bhubaneswar 751023, Orissa, India

Sreekrishnan, T.R. Department of Biochemical Engineering and Biotechnology, Indian Institute of Technology, Hauz Khas, New Delhi 110 016, India

Strzałka, K. Department of Plant Physiology and Biochemistry, Faculty of Biotechnology, Jagiellonian University, ul. Gronostajowa 7, 30-387 Kraków, Poland

Tyagi, R.D. Institut national de la recherche scientifique (INRS-Eau), Université du Québec, Complexe Scientifique, 2700 rue Einstein, C.P. 7500, Sainte-Foy, Québec, G1V4C7 Canada

Wenzel, Walter W. BOKU - University of Natural Resources and Applied Life Sciences, Vienna, Institute of Soil Science, Gregor Mendel Strasse 33, 1180 Vienna, Austria

Zdenko Rengel, Soil Science and Plant Nutrition, Faculty of Natural and Agricultural Sciences, The University of Western Australia, 35 Stirling Highway, Crawley WA 6009, Australia

Contents

Chapter 1

Metal Availability, Uptake, Transport and Accumulation in Plants

Maria Greger

Department of Botany, Stockholm University, 106 91 Stockholm, Sweden

1.1 HEAVY METAL ABUNDANCE

Heavy metals are natural elements that are found at various high background levels (Table 1.1) at different places throughout the world, due to various concentrations in the bedrock. Thus, for example, Ni, Cr and Co are abundant in serpentine soils, whereas Zn, Pb and Cd are high in calamine soils. Heavy metals are persistent and cannot be deleted from the environment. Thus, a problem arises when their availability is high due to high background levels or to human activity.

Table 1.1 Background levels in natural water and sediment (Förstner and Wittmann 1979) and the upper limit of non-polluted soil (Temmerman et al. 1984)

Metal	Natural water, $\mu g\ l^{-1}$		Soil, $\mu g\ g^{-1}$		Sediment, $\mu g\ g^{-1}$	
	Seawater	Freshwater	Sandy soil	Loam	Lake	Sea
Cd	0.01-0.07	0.07	1	1	0.14-2.5	0.02-0.43
Cr	0.08-0.15	0.5	15	30	7-77	11-90
Co	0.04	0.05	5	15		0.1-74
Cu	0.04-0.1	1.8	15	25	16-44	4-250
Hg	0.01	0.01	0.15	0.15	0.004-0.2	0.001-0.4
Mn	0.2	<5	500	800		390-6700
Mo	10	1	5	5		0.2-27
Ni	0.2-0.7	0.3	1	1	34-55	2-225
Pb	0.001-0.015	0.2	50	50	14-40	7-80
Zn	0.01-0.62	10	100	150	7-124	16-165

Heavy metals are enriched in the environment by human activities of different kinds (Dean et al. 1972). Results of these activities end up in outlets and wastes where heavy metals are transported to the environment by air, water or deposits, thereby increasing the metal concentrations in the environment. For example, the metal concentration in river waters has been shown to be increased several thousand fold by effluents from mining wastes (Föstner and Wittmann 1979). Metals supplied to the environment are transported by water and air, ultimately reaching the soil and sediment where they become bound. However, the time taken for them to become bound may be fairly long and it has been shown that the bioavailable fraction of metals in soil is high at the beginning of the binding period, but decreases with time (Martin and Kaplan 1998). Thus, there is probably a bigger problem with anthropogenically supplied metals, with high levels of bioavailable metals, than with high background levels originating from bedrock, with slow weathering.

1.2 FACTORS INFLUENCING METAL AVAILABILITY TO PLANTS

Metals have to be in an available form for plants to take up or plants must have mechanisms to make the metals available. Soil and sediment colloids roughly consist of inorganic clay minerals and organic substances. Due to the hydroxyl groups and electron pairs of oxygen in the structure of clay minerals, and to the carboxyl and phenolic groups of organic substances, the soil and sediment colloids are negatively charged (Mengel and Kirkby 1982). Positive metal ions are attracted to these charges. Anion adsorption occurs when anions are attracted to positive charges on soil colloids, and hydrous oxides of Fe and Al are usually positively charged and so tend to be the main sites for anion exchange. In water, metals are also bound to negatively charged small particles and cells, and metals are also found in complexes with anions or humic substances. Factors affecting metal release from colloids and complexes influence the bioavailable metal concentration, and many different abiotic factors influence the availability of metals to plants.

1.2.1 Soil

Partitioning of metals between solid and liquid phases in the soil is strongly affected by soil properties such as soil pH, organic matter content, soil solution ionic strength, Mn and Fe oxides, redox potential and the nature of sorbing soil surfaces (Salomons and Förstner 1984; König et al. 1986; Buffle 1988; King 1988; McGrath et al. 1988; Brown et al. 1989). Factors known to affect the solubility and plant availability of metals include their chemical characteristics, loading rate, pH, cation exchange capacity (CEC), redox potential, soil texture, clay content and organic matter content (Lagerwerff 1972; Haghiri 1974; Williams et al. 1980; Logan and Chaney 1983; Verloo and Eeckhout 1990). Chang et al. (1987) found soil temperature to be one of the major factors accounting for variations in metal accumulation by crops.

The most discussed factors influencing the bioavailability of metals in soil are pH, CEC, clay content and organic matter content. Low pH increases the metal availability since the hydrogen ion has a higher affinity for negative charges on the colloids, thus competing with the metal ions of these sites, therefore releasing metals. High organic matter content also means immobilization of the metals that bind to, e.g., fulvic or humic

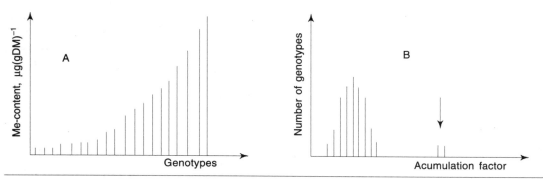

Fig. 1.7A,B Schematic graphs of **A** metal uptake in different genotypes of one and the same species and **B** distribution of genotypes in relation to their accumulation factor of the metal. The *arrow* denotes number of genotypes with very high accumulation factors, so-called hyperaccumulators

plants differed 43, 23, and 130 times between the highest and lowest accumulating clones of Cd, Cu and Zn, respectively. The differences in accumulation of a metal were not correlated with tolerance to that metal. Instead, it has been suggested that *Salix* spp. growing on highly contaminated areas have low translocation of the metal to the shoot, thus phytosynthesis is protected and the plant becomes tolerant (Landberg and Greger 1996). The distribution to the shoot of what has been taken up by the roots was between 1-72, 1-32 and 9-86% for Cd, Cu and Zn, respectively, depending on clone (Greger and Landberg 1999). However, metal distribution did not correlate with tolerance. Thus, low transport of metal to the shoot is probably only valid as a tolerance property for genotypes growing on highly contaminated areas and not for a population growing on uncontaminated soil. The property of having low or high uptake was specific, and not general, for each of the three metals studied.

A common strategy to grow on metal-contaminated sites is to immobilize the metals in the roots. This strategy was found for plants growing on mine tailings (Stoltz and Greger 2002a) and near metal smelters (Dahmani-Muller et al. 2000). Recent results showed that, at least when dealing with mechanisms such as organic acid release and pH changes of the rhizosphere to regulate the uptake of Zn, Cu and Cd, the mechanisms behind the differences between low and high accumulating clones of Salix were found in the low accumulating ones (Greger, unpubl.).

Low plant metal accumulation could also be obtained in some plant species by accumulating a high level of the metals in the leaves, which the plants get rid of by leaf fall. This is likely the case in Salix (Riddle-Black 1994) and *Armeria maritime* ssp. *halleri* (Dahmani-Muller et al. 2000).

The reason for very high accumulation in some genotypes or species is not yet known. High biomass production has been given as a cause of high uptake and accumulation. There was, however, no correlation between biomass production rate and metal accumulation when different metal-accumulating Salix clones were investigated (Greger and Landberg 2001). Furthermore, a relationship between high biomass and high metal uptake is not valid either for hyperaccumulators, which often have low biomass production. High biomass production may also involve dilution of internal metals, causing a lowering of the metal concentration, which weakens this idea. Different species and different genotypes of a species also vary in accumulating organs, which may influence the total accumulation by the plant. This means that a plant may be both a high and a low accumulator depending on which plant organ is analyzed. One thing, however, is clear, the big differences in accumulation of metals within a species show that the use of a species for mapping metal-contaminated areas may be questionable.

1.8.2 Hyperaccumulators

The variation in metal accumulation within a species often has a normal distribution; however, sometimes there is a second "bump" to the right of the first one (Fig. 1.7B). The genotypes in this second bump are called hyperaccumulators, and they have a very high accumulation factor. The concentration of accumulated metals may be very high, and, to be called a hyperaccumulator, the concentration of the metal has to be as high as shown in Table 1.6. Hyperaccumulators have a low biomass production since they have to use their energy in the mechanisms they have evolved to cope with the high metal concentrations in the tissues. The mechanisms behind the high accumulation as well as the tolerance are today poorly understood, and more information can be found in the review written by Baker et al. (1998).

Table 1.6 Definitions of hyperaccumulators (lowest metal concentration in leaves), numbers of taxa and families which are hyperaccumulators, and examples of hyperaccumulators (Baker *et al.* 1998; Reeves and Baker 1998)

Metal	Concentration in leaves $mg(g\ DW)^{-1}$	Number of Taxa	Number of Families	Example of species
Cd	>0.1	1	1	*Thlaspi caerulescens*
Pb	>1	14	6	*Minuartia verna*
Co	>1	28	11	*Aeollanthus biformifolius*
Cu	>1	37	15	*Aeollanthus biformifolius*
Ni	>1	317	37	*Berkheya coddi*
Mn	>10	9	5	*Macadamia neurophylla*
Zn	>10	11	5	*Thlaspi caerulescens*

Hyperaccumulators are populations of species found in soils rich in heavy metals, and natural mineral deposits containing particularly large quantities of metals are present in many regions around the globe. These are, e.g., calamine soils rich in Zn, Pb, Cd, and serpentine soils with high levels of Ni, Cr and Co. Hyperaccumulators have been found in areas of New Caledonia, Australia, southern Europe/Mediterranean, southeast Asia, Cuba, the Dominican Republic, California in the USA, Zimbabwe, Transvaal in South Africa, Goiás in Brazil, Hokkaido in Japan, and Newfoundland in Canada (Baker et al. 1998). There are reports on accumulators of Co and Cu (Brooks et al. 1978), Ni (Brooks et al. 1979), Mn (Brooks et al. 1981), and Pb and Zn (Reeves and Brooks 1983). Recently, ladder brake (*Pteris vittata* L.) was found to hyperaccumulate As (Ma et al. 2001b). Some species of hyperaccumulators are mentioned in Table 1.6.

REFERENCES

Arduini I, Godbold DL, Onnis A (1996) Cadmium and copper uptake and distribution in Mediterranean tree seedlings. Phys Plant 97:111-117

Arazi T, Sunkar R, Kaplan B, Fromm H (1999) A tobacco plasma membrane calmodulin-binding transporter confers Ni^{2+} tolerance and Pb^{2+} hypersensitivity in transgenic plants. Plant J Cell Mol Biol 20:171-182

Baker AJM (1981) Accumulators and excluders - strategies in the response of plants to heavy metals. J Plant Nutr 3:643-654

Baker AJM, McGrath SP, Reeves RD, Smith JAC (1998) Metal hyperaccumulator plants: a review of the ecology and physiology of a biological resource for phytoremediation of metal-polluted soils. In: Terry N, Banuelos GS (eds) Phytoremediation. Ann Arbor Press, Ann Arbor, MI

Beauford W, Barber J, Barringer AR (1977) Uptake and distribution of mercury within higher plants. Physiol Plant 39:261-265

Beckett RP, Brown DH (1984) The control of cadmium uptake in the lichen genus *Peltigra*. J Exp Bot 35:1071-1082

Blinda A, Koch B, Ramanjulu S, Dietz K-J (1997) De novo synthesis and accumulation of apoplastic proteins in leaves of heavy metal-exposed barley seedlings. Plant Cell Environ 20:969-981

Bowen JE (1987) Physiology of genotyping differences in zinc and copper uptake in rice and tomato. Plant Soil 99:115-125

Brix H (1993) Macrophyte-mediated oxygen transfer in wetlands: Transport mechanisms and rates. In: Moshire GA (ed) Contructed wetland for water quality improvement. Lewis Publ, Boca Raton, pp 391-398

Brooks RR, Morrison RS, Reeves RD, Malaisse F (1978) Copper and cobalt in African species of *Aeolanthus* Mart. (Plectranthinae, Labiatae). Plant Soil 50:503-507

Brooks RR, Morrison RS, Reeves RD, Dudley TR, Akman Y (1979) Hyperaccumulation of nickel by *Alyssum Linnaeus* (Cruciferae). Proc R Soc Lond B Biol Sci 203:387-403

Brooks RR, Thow JM, Veillon J, Jaffre T (1981) Studies on manganese-accumulating *Alyxia* from New Caledonia. Taxon 30:420-423

Brown PH, Dunemann L, Schulz R, Marschner H (1989) Influence of redox potential and plant species on the uptake of nickel and cadmium from soil. Z Pflanzenernaehr Bodenkd 152:85-91

Buffle J (1988) Complexation reactions in aquatic systems, an analytical approach. John Wiley, Chichester

Cataldo DA, Garland TR, Wildung RE (1978) Nickel in plants. II. Distribution and chemical form in soybean plants. Plant Physiol 62:566-570

Cataldo DA, Garland TR, Wildung RE (1983) Cadmium uptake kinetics in intact soybean plants. Plant Physiol 73:844-848

Cavallini A, Natali L, Durante M, Maserti B (1999) Mercury uptake, distribution and DNA affinity in durum wheat (*Triticum durum* Desf.) plants. Sci Tot Environ 243/244:119-127

Chang AC, Page AL, Warneke JE (1987) Long-term sludge application on cadmium and zinc accumulation in Swiss chard and radish. J Environ Qual 16:217-221

Chardonnens AN, Koevoets PLM, van Zanten A, Schat H, Verkleij JAC (1999a) Properties of enhanced tonoplast zinc transport in naturally selected zinc-tolerant *Silene vulgaris*. Plant Phys 120:779-785

Chardonnens AN, Ten Bookum WM, Vellinga S, Schat H, Verkleij JAC, Ernst WHO (1999b) Allocation patterns of zinc and cadmium in heavy metal tolerant and sensitive *Silene vulgaris*. J Plant Phys 155:778-787

Chawla G, Singh J, Viswanathan PN (1991) Effect of pH and temperature on the uptake of cadmium by *Lemna minor* L. Bull Environ Contam Toxicol 47:84-90

Clarkson DT (1966) Effect of aluminum and some other trivalent metal cations on cell division in the root apices of *Allium cepa*. Ann Bot NS 29:309-315

Cohen CK, Fox TC, Garvin DF, Kochian LV (1998) The role of iron-deficiency stress responses in stimulating heavy-metal transport in plants. Plant Physiol 116:1063-1072

Costa G, Morel JL (1994) Water relations, gas exchange and amino acid content in Cd-treated lettuce. Plant Physiol Biochem 32:561-570

Cutler JM, Rains DW (1974) Characterization of cadmium uptake by plant tissue. Plant Physiol 54: 67-71

Dahmani-Muller H, van Oort F, Gélie B, Balabane M (2000) Strategies of heavy metal uptake by three plant species growing near a metal smelter. Environ Pollut 109:231-2380

Dean JG, Bosqui FL, Lanouette VH (1972) Removing heavy metals from waste water. Environ Sci Technol 6:518-522

DeKock PC, Mitchell RL (1957) Uptake of chelated metals by plants. Soil Sci 84:55-62

Du ShH, Fang ShC (1982) Uptake of elemental mercury vapour by C3 and C4 species. Environ Exp Bot 22:437-443

Ekvall L, Greger M (2003) Effects of environmental biomass-producing factors on Cd uptake in two Swedish ecotypes of *Pinus sylvestris* (L.). Environ Qual 121:401-411

Fan TWM, Lane AN, Shenker M, Bartley JP, Crowley D, Higashi RM (2001) Comprehensive chemical profiling of gramineous plant root exudates using high-resolution NMR and MS. Phytochemistry 57:209-221

Förstner U (1979) Metal transfer between solid and aqueous phases. In: Förstner U, Wittmann GTW (eds) Metal pollution in the aquatic environment. Springer, Berlin Heidelberg New York, pp 197-270

Förstner U, Wittmann GTW (eds) (1979) Metal pollution in the aquatic environment. Springer, Berlin Heidelberg New York

Fortin C, Campbell PGC (2001) Thiosulphate enhances silver uptake by a green alga: role of anion transporters in metal uptake. Adv Sci Technol 35:2214-2218

Franke W (1967) Mechanism of foliar penetration of solutions. Annu Rev Plant Physiol 18:281-300

Galli U, Schüepp H, Brunold C (1994) Heavy metal binding by mycorrhizal fungi. Physiol Plant 92:364-368

Gobran GR, Clegg S, Courchesne F (1999) The rhixosphere and trace element acquisition. In: Selim HM, Iskander A (eds) Fate and transport of heavy metals in the vadouse zone. CRC Press, Boca Raton, pp 225-250

Greger M (1997) Willow as phytoremediator of heavy metal contaminated soil. Proceedings of the 2nd international conference on element cycling in the environment, Warsaw, pp 167-172

Greger M, Bertell G (1992) Effects of Ca^{2+} and Cd^{2+} on the carbohydrate metabolism in sugar beet (*Beta vulgaris*). J Exp Bot 43:167-173

Greger M, Johansson M (1992) Cadmium effects on leaf transpiration of sugar beet (*Beta vulgaris*). Physiol Plant 86:465-473

Greger M, Kautsky L (1993) Use of macrophytes for mapping bioavailable heavy metals in shallow coastal areas, Stockholm, Sweden. Appl Geochem Suppl 2:37-43

Greger M, Landberg T (1995) Cadmium accumulation in *Salix* in relation to cadmium concentration in the soil. Report from Vattenfall Utveckling AB 1995/9 (in Swedish)

Greger M, Landberg T (1999) Use of willow in phytoextraction. Int J Phytorem 1:115-124

Greger M, Landberg T (2001) Investigations on the relation between biomass production and uptake of Cd, Cu and Zn in Salix viminalis. In: Greger M, Landberg T, Berg B (eds) *Salix* clones with different properties to accumulate heavy metals for production of biomass. Academitryck AB, Edsbruk, ISBN 91-631-1493-3, pp 19-27

Greger M, Lindberg S (1986) Effects of Cd^{2+} and EDTA on young sugar beets (*Beta vulgaris*). I. Cd^{2+} uptake and sugar accumulation. Physiol Plant 66:69-74

Greger M, Brammer E, Lindberg S, Larsson G, Idestam-Almquist J (1991) Uptake and physiological effects of cadmium in sugar beet (*Beta vulgaris*) related to mineral provision. J Exp Bot 42:729-737

Greger M, Tillberg J-E, Johansson M (1992) Aluminum effects on *Scenedesmus obtusiusculus* with different phosphorus status. I. Mineral uptake. Physiol Plant 84:193-201

Greger M, Johansson M, Stihl A, Hamza K (1993) Foliar uptake of Cd by pea (*Pisum sativum*) and sugar beet (*Beta vulgaris*). Physiol Plant 88:563-570

Greger M, Kautsky L, Sandberg T (1995) A tentative model of Cd uptake in *Potamogeton pectinatus* in relation to salinity. Environ Exp Bot 35:215-225

Hagemeyer J, Lohrie K (1995) Distribution of Cd and Zn in annual xylem rings of young spruce trees (*Picea abies* (L.) Karst.) grown in contaminated soi

l. Trees 9:195-199

Haghiri FE (1974) Plant uptake of cadmium as influenced by cation exchange capacity, organic matter, zinc, and soil temperature. J Environ Qual 3:180-182

Hardiman RT, Jacoby B (1984) Absorption and translocation of Cd in bush beans (*Phaseolus vulgaris*). Physiol Plant 61:670-674

Hardiman RT, Jacoby B, Banin A (1984) Factors affecting the distribution of cadmium, copper and lead and their effect upon yield and zinc content in bush beans (*Phaseolus vulgaris* L.). Plant Soil 81:17-27

Harmens H, Koevoets PLM, Verkleij JAC, Ernst WHO (1994) The role of low molecular weight organic acids in mechanisms of increased zinc tolerance in *Silene vulgaris* (Moench) Garcke. New Phytol 126:615-621

Hemphill DD, Rule J (1978) Foliar uptake and translocation of ^{210}Pb and ^{109}Cd. Int Conf Heavy Metals Environ, Symp. Proc. II(I), Toronto, Ontario, 1975, pp 77-86

Herren T, Feller U (1994) Transfer of zinc from xylem to phloem in the penduncle of wheat. J Plant Nutr 17:1587-1598

Herren T, Feller U (1996) Effect of locally increased zinc contents on zinc transport from the flag leaf lamina to the maturing grains of wheat. J Plant Nutr 19:379-387

Holloway PJ (1982) Structure and histochemistry of plant cuticular membranes: an overview. In: Cutler DF, Alvin KL, Price CE (eds) The plant cuticle. Academic Press, London, pp 1-32

Hooda PS, Alloway BJ (1993) Effects of time and temperature on the bioavailability of Cd and Pb from sludge-amended soils. J Soil Sci 44:97-110

Hu S, Tang CH, Wu M (1996) Cadmium accumulation by several seaweeds. Sci Total Environ 187:65-71

Huang JW, Chen J, Berti, WR, Cunningham SD (1997) Phytoremediation of lead-contaminated soils: role of synthetic chelates in lead phytoextraction. Environ Sci Technol 31:800-805

Hunt GM, Baker EA (1982) Developmental and environmental variations in plant epicuticular vaxes: some effects on the penetration of naphthylacetic acid. In: Cutler DF, Alvin KL, Price CE (eds) The plant cuticle. Academic Press, London, pp 279-292

Jackson PJ, Unkefer PJ, Delhaize E, Robinson NJ (1990) Mechanisms of trace metal tolerance in plants. In: Katterman F (ed) Environmental injury to plants. Academic Press, San Diego, pp 231-258

Jarvis SC, Jones LHP, Hopper MJ (1976) Cadmium uptake from solution by plants and its transport from roots to shoots. Plant Soil 44:179-191

Johansson L-Å (1985) Chromatographic analysis of epicuticular plant waxes. Sv UtsädesförenTidskr 95:129-136

Joner EJ, Leyval C (2001) Time-course of heavy metal uptake in maize and clover as affected by root density and different mycorrhizal inocluuation regimes. Biol Fertil Soils 33:351-357

Kabata-Pendias A, Pendias H (1992) Trace elements in soils and plants, 2nd edn. CRC Press, Boca Raton

Keller P, Deuel H (1957) Kationenaustauschkapazität und Pektingehalt von Pflanzenwurzeln. Z Pflansenerngehr Dueng Bodenkd 79:119-131

King LD (1988) Effect of selected soil properties on cadmium content in tobacco. J Environ Qual 17:251-255

Knauer K, Behra R, Sigg L (1997) Adsorption and uptake of copper by the green alga *Scenedesmus subspicatus* (Chlorophyta). J Phycol 33:596-601

Kocjan G, Samardakiewicz S, Wozny A (1996) Regions of lead uptake in *Lemna minor* plants and localization of the metal within selected parts of the root. Biol Plant 38:107-117

König N, Baccini P, Ulrich B (1986) The influence of natural organic matter on the transport of metals in soils and soil solutions in an acidic forest soil (in German). Z Pflanzenernaehr Bodenkd 149:68-82

Kozuchowski J, Johnson DL (1978) Gaseous emissions of mercury from an aquatic vascular plant. Nature 274:468-469

Krämer U, Cotter-Howells JD, Charnock JM, Baker AJM, Smith JAC (1996) Free histidine as a metal chelator in plants that accumulate nickel. Nature 379:635-639

Lagerwerff JV (1971) Uptake of cadmium, lead and zinc by radish from soil and air. Soil Sci 111:129-133

Lagerwerff JV (1972) Pb, Hg, and Cd as contaminants. In: Mortvedt JJ, Giordano PM, Lindsay WL (eds) Micronutrients in agriculture. Soil Sci Soc Am, Madison, pp 593-636

Landberg T, Greger M (1994a) Influence of selenium on uptake and toxicity of copper and cadmium in pea (*Pisum sativum*) and wheat (*Triticum aestivum*). Physiol Plant 90:637-644

Landberg T, Greger M (1994b) Can heavy metal tolerant clones of *Salix* be used as vegetation filters on heavy metal contaminated land? In: Aronsson P, Perttu K (eds) Willow vegetation filters for municipal wastewaters and sludges. A biological purification system. Swed Univ Agric Sci, Report 50. SLU Info/Repro, Uppsala, pp 145-152

Landberg T, Greger M (1996) Differences in uptake and tolerance to heavy metals in *Salix* from unpolluted and polluted areas. Appl Geochem 11:175-180

Lasat MM, Pence NS, Garvin DF, Ebbs SD, Kochian LV (2000) Molecular physiology of zinc transport in the Zn hyperaccumulator *Thlaspi caerulescens*. J Exp Bot 51:71-79

Lindberg SE, Meyers TP, Taylor GE Jr, Turner RR, Schroeder WH (1992) Atmosphere-surface exchange of mercury in a forest: results of modeling and gradient approaches. J Geophys Res 97:2519-2528

Little P (1973) A study of heavy metal contamination of leaf surfaces. Environ Pollut 5:159-172

Little P, Martin MH (1972) A survey of zinc, lead and cadmium in soil and natural vegetation around a smelting complex. Environ Pollut 3:241-254

Lodenius M, Kuusi T, Laaksovirta K, Liukkonen-Lilja H, Piepponen S (1981) Lead, cadmium and mercury contents of fungi in Mikkeli, SE Finland. Ann Bot Fenn 18:183-186

Logan TJ, Chaney RL (1983) Metals. In: Page AL (ed) Utilization of municipal wastewater and sludge on land. University of California, Riverside, CA, pp 235-326

Ma JF, Ryan PR, Delhaize E (2001a) Aluminium tolerance in plants and the complexing role of organic acids. Trends Plant Sci 6:273-278

Ma LQ, Komar KM, Tu C, Zhang W, Cai Y, Kennelley ED (2001b) A fern that hyperaccumulates arsenic. Nature 409:579

Macek T, Kotrba P, Suchova M, Skacel F, Demnerova K, Ruml T (1994) Accumulation of cadmium by hairy-root cultures of *Solanum nigrum*. Biotechnol Lett 16:621-624

Maier-Maercker U (1979) "Peristomatal transpiration" and stomatal movements: a controversial view. I. Additional proof of peristomatal transpiration by photography and a comprehensive discussion in the light of recent results. Z Pflanzenphysiol 91:25-43

Markert B (1994) Plants as biomonitors - potential advantages and problems. In: Adriano DC, Chen ZS, Yang SS (eds) Biogeochemistry of trace elements. Science and Technology Letters, Northwood, NY, pp 601-613

Marschner H (1995) Mineral nutrition of higher plants. Academic Press, Cambridge

Martin HW, Kaplan DI (1998) Temporal changes in cadmium, thallium, and vanadium mobility in soil and phytoavailability under field conditions. Water Air Soil Pollut 101:399-410

Martin TJ, Juniper EB (1970) The cuticles of plants. Arnold, Edinburgh

Mathys W (1977) The role of malate, oxalate and mustard oil glucosides in the evolution of zinc-resistance in herbage plants. Physiol Plant 40:130-136

Mautsoe PJ, Beckett RP (1996) A preliminary study of the factors affecting the kinetics of cadmium uptake by the liverwort *Dumortiera hirsuta*. S Afr J Bot 62:332-336

McGrath SP, Sanders JR, Shalaby MH (1988) The effect of soil organic matter levels on soil solution concentrations and extractabilities of manganese, zinc and copper. Geoderma 42:177-188

Meharg AA (1994) Integrated tolerance mechanisms: constitutive and adaptive plant responses to elevated metal concentrations in the environment. Plant Cell Environ 17:989-939

Mench M, Martin E (1991) Mobilization of cadmium and other metals from two soild by root exudates of *Zea mays* L., *Nicotiana tabacum* L. and *Nicotiana rustica* L. Plant Soil 132:187-196

Mench M, Morel JL, Guckert A (1987) Metal binding properties of high molecular weight soluble exudates from maize (*Zea mays* L.) roots. Biol Fertil Soils 3:165-169

Mench M, Morel JL, Cuckert A, Guillet B (1988) Metal binding with root exudates of low molecular weight. J Soil Sci 33:521-527

Mengel K, Kirkby EA (1982) Principles of plant nutrition. International Potash Institute Bern, Switzerland

Mérida T, Schönherr J, Schmidt HW (1981) Fine structure of plant cuticles in relation to water permeability: the fine structure of the cuticle of *Clivia miniata* Reg. leaves. Planta 152:259-267

Momoshima N, Bondietti EA (1990) Cation binding in wood: applications to understanding historical changes in divalent cation availability to red spruce. Can J For Res 20:1840-1849

Morel F, McDuff RE, Morgan JJ (1973) Interactions and chemostasis in aquatic chemical systems: role of pH, pE, solubility, and complexation. In: Singer PC (ed) Trace metals and metal-organic interactions in natural waters. Ann Arbor Press, Ann Arbor, MI, pp 157-200

Morel JL, Mench M, Guckert A (1986) Measurement of Pb^{2+}, Cu^{2+} and Cd^{2+} binding with mucilage exudates from maize (*Zea mays* L.) roots. Biol Fertil Soils 2:29-34

Munda IM, Hudnik V (1988) The effects of Zn, Mn and Co accumulation on growth and chemical composition of *Fucus vesiculosus* L. under different temperature and salinity conditions. Mar Ecol 9:213-225

Muranyi A, Seeling B, Ladewig E, Jungk A (1994) Acidification in the rhizosphere of rape seedlings and in bulk soil by nitrification and ammonium uptake. Z Pflanzenernähr Bodenkd 157:61-65

Nieboer E, Richardson DHS (1980) The replacement of the nondescript term "heavy metals" by biologically and chemically significant classification of metal ions. Environ Pollut (Ser B) 1:2-26

Österås AH, Ekvall L, Greger M (200) Sensitivity to, and accumulation of, cadmium in Betula *pendula*, *Picea abies*, and *Pinus sylvestris* seedlings from different regions in Sweden. Can J Bot 78:1440-1449

Pearson R (1968) Hard and soft acids and bases. HSAB, part I. Fundamental principles. J Chem Educ 45:581-587

Puthotá V, Cruz-Ortega R, Johnson J, Ownby J (1991) An ultrastructural study of the inhibition of mucilage reaction in the wheat root cap by aluminium. In: Wright RJ, Baligar VC, Murrmann RP (eds) Plant-soil interactions at low pH. Kluwer, Dordrecht, pp 779-787

Ray TC, Callow JA, Kennedy JF (1988) Composition of root-mucilage polysaccharides from *Lepidium sativum*. J Exp Bot 39:1249-1261

Reeves RD, Baker AJM (1998) Metal-accumulating plants. In: Ensley BD, Raskin I (eds) Phytoremediation of toxic metals: using plants to clean the environment. Wiley, New York, pp 193-230

Reeves RD, Brooks RR (1983) Hyperaccumulation of lead and zinc by two metallophytes from mining areas of central Europe. Environ Pollut 31:277-285

Ribeyre F, Boudou A (1994) Experimental study of inorganic and methylmercury bioaccumulation by four species of freshwater rooted macrophytes from water and sediment contamination sources. Ecotox Environ Safety 28:270-286

Riddell-Black D (1994) Heavy metal uptake by fast growing willow species. In: Aronsson P, Perttu K (eds) Willow vegetation filters for municipal wastewaters and sludges. A biological purification system. Swed Univ Agric Aci, Report 50. SLU Info/Repro, Uppsala, pp 145-152

Römheld V (1991) The role of phytosiderophores in acquisition of iron and other micronutrients in graminaceous species: an ecological approach. Plant Soil 130:127-134

Ross S (1994) Retention, transformation and mobility of toxic metals in soils. In: Ross S (ed) Toxic metals in soil-plant systems. Wiley, Chichester

Rugh CL, Bizily SP, Meagher RB (2000) Phytoreduction of environmental mercury pollution. In: Ensley BD, Raskin I (eds) Phytoremediation of toxic metals: using plants to clean the environment. Wiley, New York, pp 151-169

Salim R, Al-Subu MM, Douleh A, Khalaf S (1992) Effects on growth and uptake of broad beans (*Vicia fabae* L) by root and foliar treatments of plants with lead and cadmium. J Environ Sci Health A 27:1619-1642

Salomons W, Förstner U (1984) Metals in the hydrocycle. Springer, Berlin Heidelberg New York

Santa-María GE, Cogliatti D (1998) The regulation of zinc in wheat plants. Plant Sci 137:1-12

Sauvé S, Cook N, Hendershot WH, McBride MB (1996) Linking plant tissue concentrations and soil copper pools in urban contaminated soils. Environ Pollut 94:153-157

Schönherr H, Bukovac MJ (1970) Preferential polar pathways in the cuticle and their relationship to ectodesmata. Planta 92:189-201

Siegel AM, Puerner NJ, Speitel TW (1974) Release of volatile mercury from vascular plants. Physiol Plant 32:174-176

Sposito G (1989) The chemistry of soils. Oxford University Press, Oxford

Steffens JC (1990) The heavy metal-binding peptides of plants. Annu Rev Plant Physiol Plant Mol Biol 41:553-575

Stephan UW, Scholz G (1993) Nicotinamine: mediator of transport of iron and heavy metals in phloem? Physiol Plant 88:522-529

Stoltz E, Greger M (2002a) Accumulation properties of As, Cd, Cu, Pb and Zn by four wetland plant species growing on submerged mine tailings. Environ Exp Bot 47:271-280

Stoltz E, Greger M (2002b) Cottongrass effects on trace elements in submersed mine tailings. J Environ Qual 31:1477-1483

Svenningsson M, Liljenberg C (1986) Changes in cuticular transpiration rate and cuticular lipids of oat (*Avena sativa*) seedlings induced by water stress. Physiol Plant 66:9-14

Temmerman LO, Hoenig M, Scokart PO (1984) Determination of "normal" levels and upper limit values of trace elements in soils. Z Pflanzen Bodenkd 147:687-694

Thursby GB (1984) Root-exudated oxygen in the aquatic angiosperm *Ruppia maritima*. Mar Ecol Prog Ser 16:303-305

Van de Geijn SC, Petit CM (1979) Transport of divalent cations. Cation exchange capacity of intact xylem vessels. Plant Physiol 64:954-958

Van Hoof NALM, Koevoets PLM, Hakvoont HWJ, Ten Bookum WM, Schat H, Verkleij JAC, Ernst WHO (2001) Enhanced ATP-dependent copper efflux across the root cell plasma membrane in copper-tolerant *Silene vulgaris*. Physiol Plant 113:225-232

Verloo M, Eeckhout M (1990) Metal species transformations in soil: an analytical approach. Int J Environ Anal Chem 39:179-186

Vesely J, Majer V (1994) The effect of pH and atmospheric deposition on concentrations of trace elements in acidified freshwaters: a statistical approach. Water Air Soil Pollut 88:227-246

White MC, Decker AM, Chaney RL (1981a) Metal complexation in xylem fluid; I: Chemical composition of tomato and soyabean stem exudate. Plant Physiol 67:292-300

White MC, Baker FD, Chaney RL, Decker AM (1981b) Metal complexation in xylem fluid; II: Theoretical equilibrium model and computational computer program. Plant Physiol 67:301-310

Williams DE, Vlamis J, Purkite AH, Corey JE (1980) Trace element accumulation movement, and distribution in the soil profile from massive applications of sewage sludge. Soil Sci 1292:119-132

Wierzbicka M (1998) Lead in the apoplast of *Allium cepa* L. root tips - ultrastructural studies. Plant Sci 133:105-119

Wolterbeek HT (1987) Cation exchange in isolated xylem cell walls of tomato. I. Cd^{2+} and Rb^{2+} exchange in adsorption experiments. Plant Cell Environ 10:30-44

Wood T, Bormann FH (1975) Increases in foliar leaching caused by acidification of an artificial mist. Ambio 4:169-171

Wright D, Otte ML (1999) Wetland plant effects on the biogeochemistry of metals beyond the rhizosphere. Biology and environment. Proc Royal Irish Acad 99B:3-10

Yamada Y, Bukovac MJ, Wittwer SH (1964) Ion binding by surfaces of isolated cuticular membranes. Plant Physiol 39:978-982

Chapter 2

Metal Speciation, Chelation and Complexing Ligands in Plants

A. Pohlmeier

Institut für Chemie und Dynamik der Geosphäre 7, Forschungszentrum, 52425 Jülich, Germany

2.1 INTRODUCTION

Most transition metal ions and some main group metal ions form complexes with small (Cl⁻, carboxylic acids, amino acids) and macromolecular organic substances (proteins, DNA, polysaccharides) present in the xylem and phloem of plants (Lobinski and Potin-Gautier 1998). In the first section of this chapter some remarks and definitions concerning the aqueous chemistry of metal ions will be given to facilitate further reading. The following sections present a brief outline of theories describing the nature of the coordination bond, the basic thermodynamics of complexation, the hydrolysis and chelation of metal ions, and reactions with important groups of ligands present in plants. The chapter will close with an overview of some special aspects relevant for the understanding of complexation reactions at macromolecules and the reaction mechanism of complex formation.

2.2 TERMINOLOGY AND DEFINITIONS

Most of the *metal ions* forming complexes may exist in aqueous solution in different oxidation states. Under environmental conditions, i.e. at moderate pH-values and under slightly oxidizing conditions, +1 to +3 are the most usual formal oxidation states for the metal ions concerned in the context of this book. Some important exceptions are Mo^{6+}, which exists in the natural environment as stable MoO_4^{2-} ions, forming no complexes with anionic ligands present in plants, and V^{4+}, present in most cases as VO^{2+}, forming complexes with ligands. The most stable oxidation state of chromium and iron is +3. However, due to their strong tendency to hydrolyze at moderate pH-values (see below), these ions exist as (poly)hydroxo species or as insoluble minerals.

In the following, when no special metal ions are considered, complex-forming metal ions will be termed as M^{x+}. *Metal ion complexes*, often also termed as *coordination compounds,* are formed by at least two partners: the *central metal ion,* which is generally M^{x+}, and the *ligand,* abbreviated in the following as L^{y-}. In most cases one metal ion binds at least four ligands, which can be identical, as in the case of the hexaquo-complexes $M(H_2O)_6^{x+}$, or different like $M(H_2O)_4Cl_2$. The *coordination number Z* is the total number of ligand ions or molecules directly associated with the metal ion. Its value depends on several factors like the size of M^{x+} and L and the specific electron configuration of M^+. Table 2.1 gives a survey of the most usual coordination numbers and geometries of some important metal ions. The *charge* of the complexes can be negative, zero or positive depending on the sum of charges of M^{x+} and all L. The stability of metal ion complexes is expressed by means of the *equilibrium constant K,* often also termed the *stability constant* or *complexation constant.* This describes the thermodynamic equilibrium state, keeping in mind that, at equilibrium, the rates of complex formation and dissociation are equal. The definition is given in Section 2.4.

Table 2.1 Coordination numbers and geometry of some important metal ions in their most frequent oxidation state in aqueous solution. *o* Octaheder; *t* tetraheder; *s* square planar

	K^+	Al^{3+}	Cr^{3+}	Mn^{2+}	Co^{2+}	Ni^{2+}	Zn^{2+}	Cd^{2+}	Cu^+	Cu^{2+}	Pb^{2+}
Z	6	6	6	6	4,6	6	4	6	4,6	6	
Geometry	o	o	o	o	t,o	t	t	o,t	t,s,	o	o

2.3 CHEMICAL STRUCTURE OF SIMPLE COMPLEXES

Most metal ions (especially the transition metals) form coordination bonds with ligands, which should not be confused with the weaker, mainly electrostatically controlled ion bond, or the stronger covalent bond, typical for organic chemistry, although the borders are not rigid. The binding strength depends on several factors, such as the electron configuration of M^{x+} and L and the relative energy of the atom and molecule orbitals of the free and coordinated partners. Several theories exist describing the nature of the chemical bonds between metal ions and ligands on different levels of abstraction. Here only two theories will be outlined; more detailed treatments can be found in textbooks of physical and theoretical chemistry, e.g. Gerloch and Constable (1994).

2.3.1 Crystal Field Theory

The crystal field theory, developed in the 1930s, is based on the distribution of electrons in the d-shell of metal ions. For example, Ni^{2+} has eight electrons in the 3d-shell and the higher shells are unoccupied. In the absence of ligands, all five d-orbitals possess the same energy. The ligands are placed in a certain symmetry around the central metal ion and so exert a varying influence on d-orbitals of different symmetry. For example, imagine the Ni^{2+} ion octahedrally surrounded by six Cl^- or six ligands. This leads to a splitting of the five d-orbitals into a set of three orbitals of low energy and a set of two orbitals of higher energy, each of which may be occupied maximally by two electrons (Fig. 2.1). A typical example for this orbital splitting are complexes like $Ni(H_2O)_6^{2+}$, where six of the eight electrons in the outer shell of Ni^{2+} occupy the t2 g set and two electrons occupy the

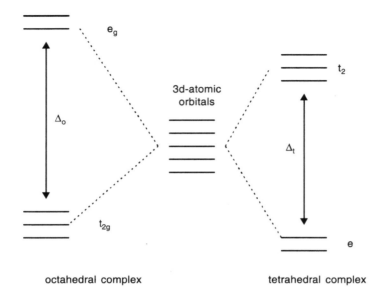

Fig. 2.1 Schematic energy level diagram for complexes in the crystal field theory. t2g, eg, t2, and e denote the sets of d-orbitals in different energy levels, Δo and Δt refer to the energy difference of the sets in octahedral and tetrahedral symmetry, respectively

eg-orbitals. On the other hand, a tetrahedral crystal field leads to a low-energy set of two orbitals and a high-energy set of three d-orbitals. Other symmetries, mixed ligands and especially chelates give more complicated splittings. This theory does not explain energies, but is a useful tool for the understanding of magnetic and spectroscopic properties like color and the intensity of the absorption bands of metal complexes.

2.3.2 Molecular Orbital Theory (*Ab Initio* Calculations)

The molecular orbital (MO) theory calculates a molecular orbital scheme for the complex by linear combination of all atomic orbitals (AOs) of the metal ions and the ligand. It minimizes free energy and thus optimizes the geometry of the complex. This requires a considerable effort of numerical mathematics, but has the great advantage that it yields all parameters defining the stabilities of bonds between the metal ion and the ligands: force constants, vibration frequencies, energy difference of the molecular orbitals, magnetic and spectroscopic properties. An example is given in Fig. 2.2a for the $[\text{Ni-formate}]^+$ complex. A comparison with Fig. 2.1 shows that within this theory both metal AOs and ligand AOs participate in the resulting MOs. The optimized geometry shows that both oxygen atoms of the formate ion participate in the binding (Fig. 2.2b). In many cases the MO theory is too complicated for practical applications like those in the context of this book, but is very useful for understanding the chemical nature of the bond.

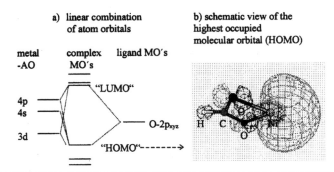

Fig. 2.2 **a** Simplified schematic molecular orbital diagram for the Ni-formate$^+$ complex. **b** Shape of the highest occupied molecular orbital (HOMO) of the Ni-formate$^+$ complex. *AO* Atomic orbital; *LUMO* lowest unoccupied molecular orbital; *MO* molecular orbital

2.4 THERMODYNAMIC STABILITY OF SIMPLE COMPLEXES

For practical purposes absolute bond strength is of minor interest, since all reactions occurring in aqueous complex chemistry are exchange reactions of ligands in the inner coordination sphere of M^{x+}. Therefore, it is convenient to define the *equilibrium constant of complexation* which describes the exchange of one ligand by another. This is illustrated in Eq. (2.1) for the formation of a simple 1:1 complex:

$$M^{2+} (H_2O)_6 + L^- \rightleftarrows ML(H_2O)_5^+ + H_2O \tag{2.1}$$

For reasons of lucidity the solvation molecules are often omitted and the reaction is written as follows:

$$M^{2+} + L^- \underset{k_b}{\overset{k_f}{\rightleftarrows}} ML^+ \tag{2.2}$$

where k_f and k_b are the *rate constants* for the association and the dissociation reaction, respectively. The *equilibrium constant K* for the reaction is defined as:

$$K - \frac{k_f}{k_b} = \frac{ML^+ \, f_{ML}}{M^{2+} \, [L] \, f_M \, f_L} \tag{2.3}$$

$$\log f_i = -0.5 z_i \left(\frac{\sqrt{I}}{1 + \sqrt{I}} - 0.3I \right) \tag{2.4}$$

where z_i is the charge number of the respective ion and I is the ionic strength in mol dm^{-3}. The equilibrium constant K defined above is one of the most important parameters describing the complexation process. However, since all complexation reactions are exchange reactions between three or more partners, their values depend on the exact stoichiometry of the reaction, and only values for identical stoichiometries may be compared. Data tables must contain the definition of the reaction process, otherwise the data are meaningless. For example, one of the most popular texts on complexation constants (Martell and Smith 1972) presents the data in the form:

Metal ion	Equibrium	Log K 20°C, 0.1
Ni^{2+}	ML/M.L	5.4

This means that Ni^{2+} forms a complex with, e.g., $citrate^{3-}$ at 20 °C in a medium of ionic strength of 0.1 mol dm^{-3}, and the equilibrium constant is:

$$K = [\text{Ni-citrate}]/[Ni^{2+}][citrate^{3-}] = 10^{5.4}\ dm^3\ mol^{-1}$$

Often in aqueous solution a complexation reaction may proceed in the following way:

$$M^{2+} + LH \underset{k_{b2}}{\overset{k_{f2}}{\rightleftharpoons}} ML^+ + H^+ \tag{2.5}$$

which means that the ligand releases a proton [in aqueous solution only stable as hydronium ion, $H_3O^+·(H_2O)_x$]. This type of reaction often proceeds with amino acids or anthropogenic complexants like EDTA. The equilibrium constant $K_2 = k_{f2}/k_{b2}$ is not identical to K in Eq. (2.3), since it incorporates the acidity of the ligand:

$$K_a = [L^-][H^+]f_L f_H/[LH] \tag{2.6}$$

K_a generally denotes the equilibrium constant of acid dissociation (often also termed *acidity constant*). By multiplication of Eqs. (2.3) and (2.6) it can easily be seen that for the above example the following relation is valid:

$$K = K_2/K_a \tag{2.7}$$

Whether L^- or LH reacts with M^{x+} depends mainly on the pH of the solution. As a *rule of thumb* one may say that the ligand reacts predominantly in the deprotonated form if pH > pK_a, and vice versa. However, in natural systems, the situation is more complex, since most of the ligands occurring in plant tissues may release more than one proton. This is especially true for chelates and polymeric ligands like pectins or proteins (see below).

2.4.1 Hydrolysis of Metal Ions

Many metal ions tend to hydrolyze in aqueous solutions (Baes and Messmer 1976). This means the following reactions take place:

$$M^{x+}(H_2O)_n \xrightarrow{K_{H1}} MOH^{(x-1)+}(H_2O)_{n-1} + H_3O^+, \text{ and} \tag{2.8}$$

$$M^{x+}(H_2O)_n \xrightarrow{K_{H2}} M(OH)_2^{(x-2)+}(H_2O)_{n-2} + 2H_3O^+ \tag{2.9}$$

In this reaction scheme, K_{H1} is defined analogously to an acidity constant, which allows an easy estimation of the fraction of hydrolyzed metal ions: one has to take into consideration hydrolysis if pH >(pK_{H1}-2), e.g. for Pb^{2+} at a pH above 5.7 (Table 2.2). It should be noted that in many cases one has to consider the formation of further hydrolytic products, as shown in Eq. (2.9), and also the precipitation of insoluble metal hydroxides. Details exceed the scope of this text and should be taken from the literature (Martell and Smith 1972; Baes and Messmer 1976). However, it should be noted that often the hydrolysis is formally treated as complexation with OH^- ligands, but the definition by Eqs. (2.8) and (2.9) has the advantage of an easy comparison with weakly acidic ligands. Table 2.2 presents some hydrolysis constants of ecologically important heavy metal ions. It is easy

Table 2.2 First and second hydrolysis constants of some important metal ions in the notation of Eqs. (2.8) and (2.9), respectively

	Fe^{3+}	Cr^{3+}	Al^{3+}	Cu^{2+}	Pb^{2+}	Zn^{2+}	Co^{2+}	Ni^{2+}	Cd^{2+}
pK_{H1}	2.19	3.96	4.95	7.7	7.7	9.0	9.7	9.9	10.12
pK_{H2}	4.7	10.8	9.3	13.5	17.1	17.8	19.6	19.0	20.3

to recognize that Fe^{3+}, Cr^{3+} and Al^{3+} exist only as hydrolyzed species at pH values of physiological interest. Often the hydrolysis of metal ions, which is indeed a complex formation with OH^- ligands, is a competitive reaction with other ligands and must be taken into account for an estimation of metal ion transport in plants (see also Fig. 2.4).

2.4.2 Chelates

In nature many ligands exist which are able to form more than one bond to a metal ion (*bi-* or *multidentate* ligands, *chelating agents*). The complex formed is termed a *chelate* or *multidentate complex.* Examples are citric acid, all amino acids, dicarboxylic acids like succinic acid, and several anthropogenic ligands like EDTA, NTA or ethylenediamine. Figure 2.3 shows the schematic structures of two simple chelate complexes. Since these ligands carry two or more donor atoms coordinated to the same metal ion, their stability is greater than that of the bis-complex with two monodentate ligands. For example, the stability constant for $Ni(ethylenediamine)^{2+}$ is

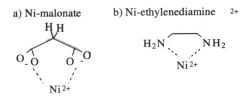

Fig. 2.3 Schematic structure of two simple chelate complexes

$10^{7.45}$ dm^3 mol^{-1}, which is greater than for the $Ni(NH_3)_2^{2+}$-complex ($10^{5.04}$ dm^3 mol^{-1}; Gerloch and Constable 1994). This means that if equal amounts of both ligands are present, the metal ion prefers the coordination of the chelating ligand. For 4- or 6-dentate ligands like NTA or EDTA, the stability constants increase up to $10^{11.5}$ dm^3 mol^{-1} and $10^{18.4}$ dm^3 mol^{-1} for Ni^{2+}.

2.4.3 Speciation of Metal Complexes

The term speciation means the distribution of several metal complexes when one or more ligands are present in solution. The relative contribution of each component as well as the pH-value determines which species of complex exists and whether one species is dominant. Figure 2.4 gives an example of the speciation of Pb^{2+} in the presence of citrate as a function of pH. The calculation of such diagrams requires the knowledge of all equilibrium constants involved and mass balances leading to a multidimensional, nonlinear, inhomogeneous system of equations, which must be solved numerically. For this purpose several computer programs exist, like MINTEQ of the US environmental administration EPA (US-EPA 1991) or GEOCHEM (Parker et al. 1995). These programs are equipped with thermodynamic databases and so they allow easy calculation of such speciation diagrams. Generally, it is useful to estimate the most relevant ligands and metal ions in a system from analytical data and so reduce the number of interesting components.

In natural waters trace metals often exist in several different forms (either as different oxidation states or forms complexed or bound to inorganic and organic matter). The

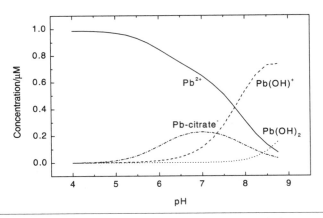

Fig. 2.4 Speciation diagram of 1 µM Pb(NO$_3$)$_2$ in 20 µM citrate solution

distribution between these forms is often referred to as the metal's "speciation". It has been shown that a metal's toxicity is related to its speciation; some forms of the metal are more bioavailable and toxic than others. This fact has been recognized in recent environmental legislation relating to surface water quality, e.g. for copper, aluminum and silver. As a consequence, industrial and environmental regulators require an increasing amount of information on metal speciation (Fig. 2.5b).

2.5 RELEVANT LIGANDS IN PLANTS

In the tissue of plants a great variety of possible ligands for metal ions exist (Zimmermann and Milburn 1975). Therefore, in this section, the most relevant groups will be presented, together with some examples of metal ion complexes. For a given ligand the reader should first check the number and nature of functional groups (e.g. R-COOH, -R-NH$_2$, etc.). Then the pK_a value of each group must be identified and compared with the pH value of the medium of interest. Table 2.3 gives a survey of pK_a ranges. If only one functional group is present, a simple complex can be formed at pH values higher than pK_a-2 (keeping in mind that there is no rule without an exception!). The extent of complex formation depends strongly on the value of the individual equilibrium constant of the complex. Next, the formation of chelators must be checked according to the following rule of thumb: the optimal ring size should be between 5 and 7, due to minimal bond angle stress. If chelates are possible, the pK_a values of the second, third, etc., coordinating functional groups are generally lower than those of the single functional group. This means if a ligand with different types of functional groups contains at least one with a small pK_a (e.g. carboxylic groups), and the resulting ring size is around 6 or 7, the formation of a chelate is most probable.

Table 2.3 pK_a ranges of functional groups of organic ligands defined according to Eq. (2.6)

	Aliphatic	Aromatic	Aliphatic	Aromatic	Aliphatic	Aromatic	Aliphatic
	RCOOH	–RCOOH	R-OH	R-OH	R-NH$_3$	R-NH$_3$	R-SH
pK_a range	2-5	2-5	>14	7-14	9-11	5-8	9-12

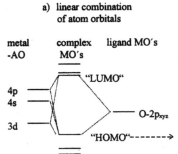
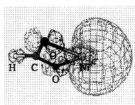

2.5.1 Carboxylates

The most important simple carboxylic acids are present in plants in very different concentrations. Table 2.4 gives a survey of the equilibrium constants of some metal ions for the formation of complexes with the predominant species at pH »7.0. Only 1:1 complexes are considered, since at low metal ion and ligand concentrations the formation of higher complexes is less probable. However, at higher concentrations of the ligand, the formation of complexes M/L=1:2 must be considered (compare with Fig. 2.5). It should be noted that the coordination number Z of the metal ion is higher than the number of bonds to the ligand's functional groups, so that the remaining valencies are occupied by water. Furthermore, in mixed ligand environments, the formation of mixed ligand complexes is obviously possible, so the number of complexes present may increase greatly, and the speciation (Fig. 2.5b) must be calculated by the computer programs mentioned in Section 2.4.3. General rules for the prediction of complex stability are different, but three tendencies should be kept in mind:

1. The higher the metal's tendency for hydrolysis, the greater are the complex equilibrium constants (compare with Ca^{2+}, Mg^{2+}, Ni^{2+}, Cu^{2+}, Al^{3+}, Fe^{3+}). Please note that the equilibrium constants of K^+-Mg^{2+}-, and Ca^{2+}-complexes are rather small compared with those of the transition metal complexes.

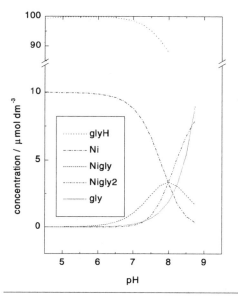

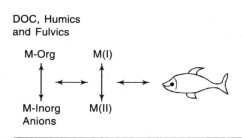

Fig. 2.5 Speciation diagram of 100 µM glycine, 10 µM Ni^{2+} for ionic strength of 0.1 M

Fig. 2.5b Schematic representation of metal speciation in the environment

Table 2.4 Equilibrium constants of complexes with mono-, di- and tricarboxylic acids (Martell and Smith 1972) for ionic strength of 0.1 M, if not otherwise indicated. pK_{an} are the negative logarithms of the acidity constants for the reaction: $LH_n \Leftrightarrow LH_{n-1}^{x-} + H^+$. Log K are the logarithms of the equilibrium constants for complex formation with the completely deprotonated species of the ligand L^{n-}: $M^{x+} + L^{n-} \Leftrightarrow ML^{(n-x)-}$

	pK_a values (see Eq. 2.6)	Ligand species present at pH 7	Complex	log K, $I=0.1$ mol dm^{-3} (see Eq. 2.3)
Citrate	pK_{a1}=5.69 pK_{a2}=4.35 pK_{a3}=2.87	L^{3-}	KL^{2-}	0.6
			MgL^-	3.37
			CaL^-	3.5
			NiL^-	5.4
			CuL^-	5.9
			CoL^-	5.0
			PbL^-	4.34 ($I=0$)
			MnL^-	4.15
Oxalate	pK_{a1}=3.82 pK_{a2}=1.04	L^{2-}	$KL-$	-0.8
			CaL	1.66
			MgL	2.76
			MnL	3.19
			CoL	3.8
			NiL	5.2 ($I=0$)
			CuL	4.8
			FeL^+	7.5 ($I=0$)
			AlL^+	6.1 ($I=1.0$)
Malate	pK_{a1}=4.71 pK_{a2}=3.24	L^{2-}	$KL-$	0.18
			MgL	1.7
			NiL	3.17
			CuL	3.42
			FeL^+	7.1
Succinate	pK_{a1}=5.24 pK_{a2}=4.00	L^{2-}	NiL	1.6
			CuL	2.6
Tartrate	pK_{a1}=3.95 pK_{a2}=2.82	L^{2-}	KL^-	0.0
			NiL	2.06 ($I=1.0$)
			CuL	3.39
			FeL^+	6.49
Phthalate	pK_{a1}=4.93 pK_{a2}=2.75	L^{2-}	NiL	2.95
			CuL	4.04
			CoL	2.83
			MnL	2.74
			AlL^+	3.18
Salicylate	pK_{a1}=13.4 pK_{a2}=2.81	LH^-	NiL	6.95
			CuL	10.62
			FeL^+	16.3
			AlL^+	12.9
Acetate	pK_{a1}=4.56	L^-	NiL^+	1.43 ($I=0$)
			CuL^{+1}	1.83
			FeL^2	3.38

2. The higher the charge and chelation ability, the greater are the stability constants (compare citrate, malate, acetate)
3. Five- and six-membered rings are most favored (compare Ni-oxalate, malate, succinate)

2.5.2 Amino Acids and Mercaptic Acids

Amino acids may be present in concentrations of up to several millimoles per liter and therefore are major candidates for metal ion binding ligands. They possess at least one carboxylic group and one amino group and can form at least two bonds to a metal ion. The pK_a values are in the range 2-4 for the carboxylic group and around 9 to 10 for the amino group. This means that at pH 7, simple amino acids exist as zwitterions with a negative charge at the carboxylic group and a positive charge at the amino group. If the amino group is deprotonated at higher pH values, a strong coordination bond can be formed via the lone electron pair at the N-atom, leading to high values of the equilibrium constant K. Table 2.5 gives a survey of equilibrium constants for some important amino acids. For example, the equilibrium constant for the complex formation of Ni-glycine$^+$ is presented as the reaction of Ni^{2+} with a completely deprotonated glycinate ion L$^-$ as $K=[ML]/[M][L]=10^{5.78}$ dm^3 mol^{-1}. Since at, e.g., pH 7, only the species LH is present (see pK_{a1} in Table 2.5), the acidity of this species must be included in the calculation according to Eq. (2.6).

At high concentrations of the ligand, the formation of complexes M/L=1:2 is probable. Figure 2.5 shows a speciation diagram for a solution of 100 μmol dm^{-3} glycine and 10 μmol dm^{-3} Ni^{2+} at an ionic strength of 0.1 mol dm^{-3}. Due to the high surplus of the ligand, the formation of the 1:2 complex is dominant at pH values >8.0. Two points are striking. First, the lack of a monodentate complex MLH^{2+} which may be formed at low pH values when the amino group is protonated. However, its equilibrium constant may be estimated to be in the range compared to that of acetate, and therefore it is too small to contribute significantly to the whole speciation. Second, the relatively small contributions of complexes at pH <7, which is a consequence of the strong competition of H$^+$ with respect to Ni^{2+} for the binding site at the amino group. Due to its high affinity to the N-atom (high value of $pK_{a2}=9.78$), it prevents the formation of the chelate complex with Ni^{2+}, although the complexation constant is $K=10^{6.18}$ dm^3 mol^{-1}, which is higher than those of the dicarboxylic acids forming chelate complexes of similar ring size (see Table 2.5: Ni-oxalate: $K=10^{5.2}$). This means that in a mixed ligand environment at lower pH values the carboxylate complexes will dominate over the amino acid complexes. Figure 2.6 shows the structure of the Ni^{2+}-glycine complex. The situation is different for complexes with amino acids carrying more than one carboxylic group (e.g. aspartate). These obviously can form chelate complexes even at lower pH values. A special case is *histidine*. Figure 2.6 shows that histidine is capable of forming three bonds with metal ions, including the ring N-atom, whose pK_a value (6.02) is relatively low, so that at moderate pH values H$^+$ is not a strong competitor for Ni^{2+}. This leads to the importance of histidine as a chelating agent for metal ions in the natural environment (Krämer et al. 1996).

Table 2.5 Equilibrium constants of complexes with amino acids (Martell and Smith 1972) for ionic strength of 0.1 M, if not otherwise indicated. pK_{an} are the negative logarithms of the acidity constants for the reaction: $LH_n \Leftrightarrow LH_{n-1}^{x-} + H^+$. Log K are the logarithms of the equilibrium constants for complex formation with the completely deprotonated species of the ligand L^{n-}: $M^{x+} L^{n-} \Leftrightarrow ML^{(n-x)-}$

	pK_a values (see Eq. 2.6)	Ligand species present at pH 7	Complex	log K, I=0.1 mol dm^{-3} (see Eq. 2.3)
Glycine	pK_{a1}=9.57	LH	MgL^+	2.22
	pK_{a2}=2.36		CaL^{+-}	1.39 (I=0)
			NiL^+	5.78
			NiL2	10.58
			CuL^+	8.15
			CuL2	15.3
			Fe2+1	10.0 (I=0)
Glutamic acid	pK_{a1}=9.59	LH^{2-}	MgL	1.9
	pK_{a2}=4.20		NiL	5.6
	pK_{a3}=2.18		NiL_2	9.76
			CuL	7.87
			CuL_2	14.16
			FeL^+1	12.1 (I=1)
Histidine	pK_{a1}=7.99	LH, LH_2^+	NiL^+	8.76
	pK_{a2}=6.73		CuL^+	10.2
	pK_{a3}=2.90		ZnL^+	6.55
			CdL^+	5.39
Cysteine	pK_{a1}=10.92	LH_2^+, LH	NiL	9.82
	pK_{a2}=8.15		ZnL	9.17
	pK_{a3}=1.88		PbL	11.6
Gly-gly-gly-his	pK_{a1}=7.99	LH^-, LH	NiL	8.38 (I=0.16)
	pK_{a2}=6.73		CuL	9.17 (I=0.16)
	pK_{a3}=2.90			
Val-lys-ser Glu-gly	pK_{a1}=7.74	L^-, LH	CuL	5.4
	pK_{a2}=4.48			
	pK_{a3}=3.28			
NTA	pK_{a1}=9.65	LH^{2-}	KL^{2-}	0.6
	pK_{a2}=2.48		MgL^-	5.47
	pK_{a2}=1.8		NiL^-	11.5
			CuL^-	12.94
			FeL	15.9
EDTA	pK_{a1}=10.19	LH^{3-}, LH_2^{2-}	KL^{3-}	0.8
	pK_{a2}=6.13		MgL^{2-}	8.85
	pK_{a3}=2.69		CaL^{2-}	10.65
	pK_{a4}=2.5		NiL^{2-}	18.4
	pK_{a5}=1.5		CuL^{2-}	18.78
			FeL^-	25.1

Anthropogenic N-containing complexants like EDTA, NTA or IDA are strong chelates due to the high number of carboxylic groups, and the additional complexation via the N-atom yields six, four, and three five-membered rings, respectively, resulting in a high stability of the chelate complexes (Table 2.5).

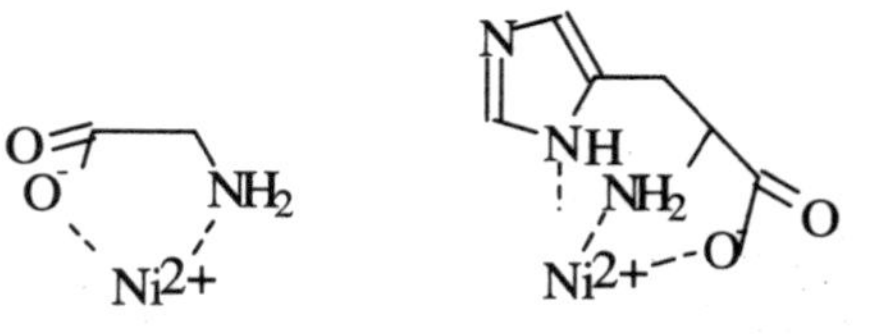

Fig. 2.6 Structures of **a** the Ni-glycine$^+$ complex and **b** the Ni-histidine$^+$ complex

Polypeptides (proteins) are polymers of amino acids where the carboxylic group of one amino acid forms an amide bond (also termed peptide bond) with the α-amino group of the next one. These functional groups may coordinate metal ions via the free electron pair at the N-atom (e.g. Hay et al. 1993). Also, many amino acids possess carboxylic amino or mercapto groups in their side chains. These may form complexes with metal ions. However, this complexation is a complicated process, which may also be influenced by the polyelectrolytic properties of these polymers (see next section).

Mercaptic acids contain R-SH functional groups. Due to the high affinity of the sulfur atom for metal ions, these compounds are also strong chelates, if small ring sizes are possible. As for the amino acids, H$^+$ is a strong competitor for metal ions (pK_a 9—11) so that the complexes become important only at higher pH values. For example, the equilibrium constant of the Zn-mercaptoacetate complex ($K=10^{7.86}$ dm^3 mol^{-1}) is rather high compared with that of simple amino acids. The same is valid for M^{x+}-cysteine complexes.

2.5.3 Polymers

Generally polymers are complexants for metal ions that must be taken into account since they exist in large amounts in plants. Examples are proteins (as mentioned in Sect. 2.5.2), polysaccharidic acids like pectins, DNA, RNA, lignins, and polysaccharides. They are composed of a linear or branched chain of monomers, which may possess functional groups like the small carboxylic acids and amino acids discussed above. At these functional groups the complexation takes place. Principally, the same reactions as described above (i.e. competition between M^{x+} and H$^+$) are expected and the stability of the complexes is controlled by the resulting ring size, the pK_a values of the functional groups, and the tendency of M^{x+} to hydrolyze.

However, there is a significant difference between complexation at polymers and at small ions: at polymers the binding sites can exist in close proximity to each other, so that the complexation at one site influences the complexation at one or more neighboring sites. This effect is often encountered at proteins containing amino acids with functional groups for the coordination of metal ions on the side chains. These proteins contain amino acids in special steric arrangements, so that well-suited binding sites for a special metal ion are created. For example, metallothioneins possess a high affinity for metal ions due to their high cysteine content which is able to form strong bonds with metal ions via the sulfur atom leading to high values of the equilibrium constant (e.g. 10^{16} dm^{-3} mol^{-1} for Cd^{2+}; Hunziker and Kägi 1985). Therefore, the chemistry of protein-metal ion interaction has huge importance for life (see also Chap. 3, this Vol.)

Other important polymeric complexants in plants belong to the group of polysaccharidic acids like, e.g., pectins and alginates. These polymers have a polysaccharide backbone, carrying free carboxylate functional groups which may be deprotonated, resulting in the formation of *polyions* or *polyelectrolytes*. These deprotonated groups are binding sites for metal ions (Buffle 1988).

Figure 2.7 shows a section of polygalacturonic acid. The most important characteristic is the high charge density with the consequence that the complex formation at one binding site is influenced by the state of the neighboring sites. The total interaction free energy (DG) is composed of a constant term, DG_{chem}, describing the chemical affinity of H^+ or M^{x+} to the functional group, and the electrical work term $w_{electr.}=zF\psi$, describing the work of bringing one mole of charge (zF) to binding sites with the electrical potential ψ.

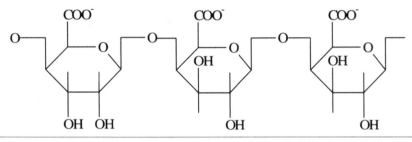

Fig. 2.7 Section of a non-esterified polygalacturonic acid in completely deprotonated state

$$\Delta G = \Delta G_{chem} + w_{electr.} \tag{2.10}$$

Identifying ΔG with $-RT\ln K$ and ΔG_{chem} with $-RT\ln K_{chem}$ leads to the following expression:

$$-RT \ln K = -RT \ln K_{chem} + zF \tag{2.11}$$

or

$$K \equiv K_{cond} = \frac{L_{polymer} \, M^{(x-1)+}}{L_{polymer}^- \, M^{x+}} = K_{chem} \, \exp\left(-\frac{zF\psi}{RT}\right) \tag{2.12}$$

where K_{cond} is termed the *conditional equilibrium constant* (depending on ionic strength, degree of charging, and conformation of the polyelectrolyte), which is experimentally available. Equations (2.11) and (2.12) are also valid for deprotonation reactions, see Eq. (2.6), in which case "M" must be replaced by "H^+", and K corresponds to the reciprocal acidity constant. The electrical potential y depends on the charge density of the polymer, the ionic strength and the conformation, and in principle it can be calculated by solving the Poisson-Boltzmann equation for the given problem. However, this requires knowledge of the polymer size and conformation, which is often unknown. There are two ways to overcome this problem.

In the first approach, one may assume that the potential approaches zero at zero charge, i.e. the potential term in Eq. (2.12) approaches unity. This means, one has to determine K_{cond} at various degrees of charging and plot K_{cond} vs charge. The extrapolation then gives K_{chem}. Figure 2.8 shows schematically such a plot for the deprotonation of the technical polyelectrolyte polymethacrylic acid resulting in $pK_{a,chem}=5.2$, which is well in the range of conventional carboxylic acids. The second approach is to introduce a distribution function P(K) which takes into account that the value of K_{cond} varies with different carboxylic acids. The second approach is to introduce a distribution function P(K) which takes into

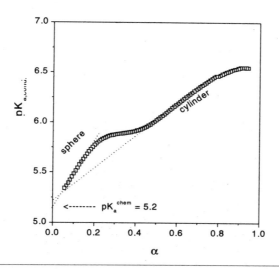

Fig. 2.8 Conditional acidity constant $K_{a,cond}$ for polymethacrylic acid as a function of the degree of charging $\alpha=[L^-]/([L^-]+[LH])$

account that the value of K_{cond} varies with different degrees of bound H^+ or M^{x+}. This approach means that the polyelectrolyte effect, which is caused by the varying potential, is not calculated explicitly, but is described by the distribution function. Figure 2.9 shows schematically such a distribution function for the binding of Ni^{2+} at a polymethacrylic acid at a constant pH value. Such a plot is also termed the *affinity spectrum*. The maximum of the curve represents the mean equilibrium constant defined according to Eq. (2.3). In such a diagram, complexation with a non-polymeric ligand (e.g. dicarboxylation) is described by a d-function, as indicated by the very narrow peak for the complexation of Ni^{2+} with malate.

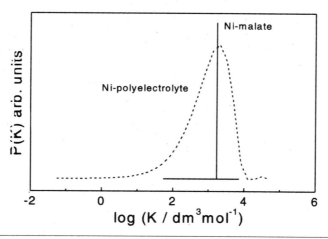

Fig. 2.9 Distribution function of log K_{cond} for the binding of Ni^{2+} at a polyelectrolyte (schematically), and for comparison that of Ni-malate, where $K_{cond} = K$ (Table 2.4)

These two approximations are satisfactory descriptions of the processes taking place, but they cannot replace exact modelling, taking into account the electrostatic interactions explicitly. Although these approaches give some insight into the behavior of polyelectrolytes, the description of complexation at such polyelectrolytes and its modelling are challenging tasks for the future, since these polymers are of great importance in understanding reactions in living organisms

2.6 INORGANIC, HARD AND SOFT LIGANDS

Inorganic ligands like chloride, fluoride, sulfate, etc., are also very important complexants in plants. The equilibrium constants for the formation of the 1:1 halide complexes vary over a great range, due to the different charges and structures of the atomic orbitals. Also, in many cases, the formation of higher coordinated complexes is observed at higher ligand concentrations. For example, log K values of the 1:1, 1:2, and 1:3 complexes of Cd^{2+} follow the order 1.98, 2.6, and 2.4, where the decrease for the 1:3 complex is due to the reduced charge. The stability constants of sulfate complexes are also in the same order of magnitude. The data tables (e.g. Martell and Smith 1972) mentioned above as well as the speciation programs (e.g. US-EPA 1991) contain the data for many inorganic complexes. One popular model for the explanation of stability data of inorganic as well as of simple organic complexes is the *hard and soft acids and bases (HSAB) concept* (Gerloch and Constable 1994). It classifies metal ions into hard and soft ones. The former ones are small ions characterized by a high positive charge density, e.g. Fe^{3+}, Zn^{2+}, Mn^{2+}, and the latter ones have a low charge density due to their greater radius: Cu^+, Pb^{2+}, Cd^{2+}. Many metal ions are also classified as 'intermediates'. The concept also defines hard ligands, whose electron shell is less distortable, e.g. F^-, Cl^-, SO_4^{2-}, $RCOO^-$, and soft ligands like I^-, CN^- and RS^-. Now, the rule states that *hard-hard* and *soft-soft* interactions are preferred over *hard-soft* or *soft-hard*.

This explains, for example, why the stability of 1:1 halide complexes increases for a soft metal ion like Pb^{2+} from log K=1.23 to 1.94 for F^- to I^-, respectively, whereas it decreases for a hard metal ion like Zn^{2+} from 1.15 to -2.91. Finally, it should be mentioned that the hard-hard interaction is more electrostatic in nature, and the soft-soft interaction is more covalent.

2.7 KINETICS AND MECHANISM OF COMPLEXATION

So far, complexation reactions have been treated thermodynamically, i.e. only the difference between initial (uncomplexed metal ion) and final state (equilibrium) has been of interest; the pathways by which complexes are formed are not covered in this approach. Investigations of the *chemical kinetics* of complexation processes not only determine reaction rates, but also allow statements about how complexes are formed, i.e. about the *reaction mechanisms*.

The generally accepted mechanism for the formation of simple metal ion complexes follows the theory of Eigen and Tamm developed in the 1950s and 1960s (Hewkin and Prince 1970; Strehlow and Knoche 1977), and is illustrated in Fig. 2.10. The mechanistic approach starts with the diffusion of the hydrated metal ion and the hydrated ligand to each other, step (a) in Fig. 2.10. The relative diffusion rate is very high, and depends

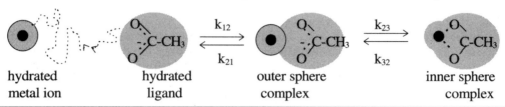

a) diffusion of metal ion to ligand   b) H_2O exchange

hydrated metal ion — hydrated ligand — outer sphere complex — inner sphere complex

Fig. 2.10 Eigen-Tamm mechanism of the formation of simple complexes (simplified) where a metal ion forms a complex with a carboxylic acid. The *black circle* symbolizes the metal ion, the *grey circles* and *ellipses* symbolize the hydration sphere of the metal ion, the ligand and the complexes. k_{12} and k_{21} are the rate constants of the outer sphere complex formation and dissociation, respectively, k_{23} and k_{32} are the rate constants of the formation and the dissociation, respectively, of the inner sphere complex

mainly on the diffusion coefficients in the medium (aqueous solution) and on their charges. For example, for the reaction of M^{2+} (di-cation) with a ligand L^- (mono-anion), k_{12} is $10^{10.4}$ dm^{-3} mol^{-1} s^{-1} (Strehlow and Knoche 1977). Coming into contact, they form an *outer sphere* complex, in which the ions are separated by the inner hydration shell of the metal ion. (It should be noted that two types of outer sphere complexes exist: one where both ions are hydrated and the above mentioned case where the ions are separated by the inner hydration shell of the metal ion. However, in most cases, the hydration shell of the ligand is very weakly bound, so that it can be neglected.)

The rate-determining step is the exchange of one or more water molecules of the inner hydration shell of the metal ion by the ligand, combined with the formation of the coordination bond, step (b) in Fig. 2.10. This step, as described by the rate constant k_{23}, depends mainly on the nature of the metal ion, so that rate constants for the formation of different complexes of the same metal ion are in the same range.

The approach to obtain the rate constants is firstly to set up the rate equation for the problem. For the example shown in Fig. 2.10, this reads:

$$\frac{d[\text{inner} - \text{sphere c.}]}{dt} = k_{23}[\text{outer} - \text{sphere c.}] - k_{32}[\text{inner} - \text{sphere c.}] \tag{2.13}$$

where the brackets represent the concentrations of the outer and inner sphere complex. Assuming that the formation of the outer sphere complex is very rapid compared with that of the inner sphere complex, and by introducing a reaction variable $x=[\text{inner-sph.}]-[\text{inner-sph.}]_{eq}$ (difference of the actual concentration from the equilibrium concentration), one can derive the rate law for pseudo first-order conditions (Strehlow and Knoche 1977):

$$x = x°\exp(-t/\tau)$$

with

$$\frac{1}{\tau} = k_{32} + k_{23}\frac{K_{12}([M]+[L])}{1+K_{12}([M]+[L])} \tag{2.14}$$

where τ is termed the relaxation time and $K_{12}=k_{12}/k_{21}$ is the equilibrium constant for the formation of the outer sphere complex. The rate constants k_{32} and k_{23} as well as K_{12} are obtainable by measuring the relaxation times (which are reciprocal pseudo first-order rate constants) for different concentrations of metal ions and ligands, and fitting Eq. (2.14) to the data. It should be noted that Eq. (2.14) is valid only for the simple reaction scheme shown in Fig. 2.10, and in many cases additional deprotonation and hydrolysis reactions have to be taken into account, so that the rate equations may look more complicated.

Furthermore, it should be mentioned that most complexation reactions are quite rapid (exceptions: Al^{3+}, Cr^{3+}), so that classical experimental techniques are not convenient for the measurement. So, in the last 40 years, several relaxation techniques have been developed to access the time range of seconds to micro-seconds. Table 2.6 presents some rate constants k_{23}. It can be seen that k_{23} is in most cases in the order of magnitude of k_{ex} (with the exception of Fe^{3+}) and confirms the model presented above. The model also explains the complexation of Ni^{2+} with the polycarboxylate poly(maleic, acrylic) acid, suggesting that also the complexation at polyelectrolytes follows the same basic model, although the physicochemical properties of polyelectrolytes are very different to those of simple ions (Hermeier et al. 1996). This consequence is important for the understanding of metal ion behavior in plants.

Table 2.6 Kinetic parameters for the formation of complexes for some metal ions (taken from Hewkin and Prince 1970, unless otherwise indicated)[a]

Metal ion	Log k_{ex}/s^{-1}	Ligand	Log k_{23}/s^{-1}	Remarks
Ni^{2+}	4.4	Alanine	4.3	
		Bi-cysteine	4.2	
		H-oxalate	3.7	
		Oxalate^{2-}	4.8	
		EDTA	5.2	
		Acetate$^-$	4.1	Bonsen et al. (1976)
		Polycarboxylate	3.8	Hermeier et al. (1996)
Cu^{2+}	9.9	Glycine$^-$	9.6	
Fe^{3+}	6.5	H-oxalate$^-$	2.9	
Al^{3+}	0.2	Formate$^-$	1.0	Pohlmeier et al. (1993)
		NTAH^{2-}	0.3	Pohlmeier and Knoche (1996)

[a]k_{ex} is the water exchange rate constant.

2.8 MECHANISM OF CHELATION

Figure 2.10 describes the formation of simple monodentate complexes. If the ligand is a chelate, further steps must be considered (Wilkins 1991):

$$M^{2+} + L \xrightarrow[]{k_O} o.s. \xrightleftharpoons[k_{32}]{k_{23}} M-L \xrightleftharpoons[k_{43}]{k_{34}} M=L \qquad (2.15)$$

The kinetics of the formation of further coordination bonds may be either faster than the inner sphere complexation, since the exchange rate of H_2O at the metal ion is enhanced by the presence of the first bond, or slower, if steric factors like internal H-bonds dominate. An example for the first possibility is the formation of the glycine complex of Cr^{3+} (Abdullah et al. 1985), where the overall rate is determined by the rate of water exchange at the Cr^{3+} ion. However, the opposite may also be true if the formation of the chelate bond is either sterically hindered or internal H-bonds exist, which must be broken. This is the case with the formation of NTA complexes of Al^{3+}, where the rate constant, $k_{34}=0.27$ s^{-1}, is nearly ten times smaller than k_{23} (Pohlmeier and Knoche 1996). These two examples show that the mechanism of chelation can become complicated, and no general rules may be stated since the mechanism of chelation depends strongly on steric conditions within the complex.

Kinetic investigations supply further information about reaction mechanisms by the investigations of the temperature and pressure dependence of the reaction rate constants k. The scope of this book does not allow us to go into details, and only the main aspects can be given. Temperature dependence is often described by the Arrhenius equation:

$$k = A \exp(-E_A/RT) \tag{2.16}$$

where E_A is the activation energy, R is the gas constant, T is the absolute temperature, and $A=k_B T/h \exp(\Delta S^{\#}/R)$ where $\Delta S^{\#}$ is the activation entropy, k_B is Botzmann's constant, and h is Planck's constant. From E_A, the activation enthalpy $\Delta H^{\#}$ is obtainable ($\Delta H^{\#}$ $E_A=RT$). A further quantity often encountered in the original literature about metal ion complexation is the activation volume $\Delta V^{\#}$.

$$\left(\frac{d \ln k}{dp}\right)_T = \frac{\Delta V^{\#}}{RT} \tag{2.17}$$

Equations (2.16) and (2.17) imply that the reaction proceeds via a transition state, where the inner energy of the complex is enhanced by E_A. This energy barrier must be overcome before the product is formed. A discussion of the values of E_A, $\Delta S^{\#}$, and $\Delta V^{\#}$ leads to a deeper insight into the detailed mechanisms of the reaction (Wilkins 1991).

ACKNOWLEDGEMENTS

I am very grateful to Wilhelm Knoche, University of Bielefeld, Germany, for many helpful discussions about metal ion complexation. I am also thankful to Anke Schauer and Jörg Kleimann, University of Bielefeld, Germany, for data on the binding of Ni^{2+} and H^+ at polymethacrylic acid, which are the bases for Figs. 2.8 and 2.9, and Ekkehard Koglin, ICG-7, Research Centre Jülich, Germany, for the MO calculations presented in Fig. 2.2. Many thanks also to my wife, Sabina Haber-Pohlmeier, for discussion and critical reading of this manuscript.

REFERENCES

Abdullah MA, Barret J, O'Brien P (1985) The kinetics of the reaction of chromium(III) with L-cysteine. Inorg Chim Acta 96 pp

Baes CF, Messmer RE (1976) The hydrolysis of cations. Wiley-Interscience, New York

Bonsen A, Eggers F, Knoche W (1976) Formation of Ni-acetate complexes. Inorg Chem 15:1212–1215

Buffle J (1988) Complexation reactions in aquatic systems: an analytical approach. Horwood, Chichester

Cotton FA, Wilkinson G (1972) Advanced inorganic chemistry. Wiley, New York

Davies CW (1962) Ion association. Butterworths, London

Gerloch M, Constable EC (1994) Transition metal chemistry. VCH, Weinheim

Hay RW, Hassan MM, You-Quan C (1993) Kinetic and thermodynamic study of the copper(II) and nickel(II) complexes of glycyl glycyl-L-histidine. J Inorg Biochem 52:17-25

Hermeier I, Herzig M, Knoche W, Narres HD, Pohlmeier A (1996) Kinetics of complexation of heavy metal ions with polyelectrolytes in aqueous solution. Ber Bunsenges Phys Chem 100:788-795

Hewkin DJ, Prince RH (1970) The mechanism of octahedral complex formation by labile metal ions. Coord Chem Rev 5:45-73

Hunziker PE, Kagi JHR (1985) Metallothioneins. In: Harrison P (ed) Metalloproteins, part 2. VCH, Weinheim

Krämer U, Cotter-Howells JD, Baker AJM, Smith JAC (1996) Free histidine as a metal chelator in plants that accumulate nickel. Nature 379:635-637

Lobinski R, Potin-Gautier M (1998) Metals and biomolecules—bioinorganic analytical chemistry. Analusis 26:21-24

Martell AE, Smith RM (1972) Critical stability constants. Plenum, New York

Parker DR, Norvell WA, Chaney RL (1995) Chemical equilibrium and reaction models. Soil Sci Soc Am Spec Publ 42. Madison, Wisconsin, pp 253-269

Pohlmeier A, Knoche W (1996) Kinetics of the complexation of Al^{3+} with amino acids, IDA and NTA. Int J Chem Kin 28:125-136

Pohlmeier A, Thesing U, Knoche W (1993) Formation of aluminium(III) mono-carboxylates in aqueous solution. Ber Bunsenges Phys Chem 97:10-15

Strehlow H, Knoche W (1977) Fundamentals of chemical relaxation. VCH, Weinheim

US-EPA (1991) Minteq version 3.11, US Environmental Protection Agency, CEAM, College Station Road, Athens, GA 30613-0801

Wilkins RG (1991) Kinetics and mechanism of reactions of transition metal complexes. VCH, Weinheim

Zimmermann MH, Milburn JA (1975) Encyclopedia of plant physiology, vols 1 and 2. Springer, Berlin Heidelberg New York

Chapter 3

Metallothioneins, Metal Binding Complexes and Metal Sequestration in Plants

M.N.V. Prasad

Department of Plant Sciences, University of Hyderabad, Hyderabad 500046, India

3.1 INTRODUCTION

A heavy metal (HM) contaminated atmosphere, geosphere and hydrosphere pose a serious threat to plants. Plants growing in these environments acquire a wide range of adaptive strategies, the most prominent mechanisms being the synthesis of phytochelatins (PCs) and metallothioneins (MTs; Rauser 1990b, 1995; Reddy and Prasad 1990; Steffens 1990; Prasad 1997; Rengel 1997; Sanità Di Toppi et al. 2002). The HM deposition pattern has been correlated with forest decline and the concentration of PCs (Gawel et al. 1996). It has also been reported that certain plants function as hyperaccumulators of specific heavy metals owing to their efficient metal complexation processes (Reeves et al. 1995; Krämer et al. 1996; see Chaps. 1, 8, 13 and 14, this Vol.). Thus, metal biomolecular complexes are of considerable interest not only in the context of ecotoxicology, but also from a social point of view (Lobinski and Potin-Gautier 1998; Prasad 1998). The main mechanism of heavy metal tolerance in plant cells is chelatin through the induction of metal-binding peptides and the formation of metal complexes. The introduction or overexpression of metal-binding proteins has been used to increase the metal-binding capacity, tolerance or accumulation ability of bacteria and plants. Modification in the biosynthesis of PCs in plants has recently been utilized to enhance the metal accumulation in bacteria. In addition, various peptides consisting of metal-binding amino acids have been studied for increased heavy metal accumulation by bacteria. MTs and phytochelatins in plants contain a high percentage of cysteine sulfhydryl groups, which bind and sequester heavy metal ions in very stable complexes (Figs. 3.1-3.5)

Phytochelatin PC2

Peptide bond

$$PC_2 = \gamma E\text{-}C\text{-}\gamma E\text{-}C\text{-}G \qquad n = 2$$

Structure of phytochelatin

232

n $n = 2 - 4 (11)$

| Glu | Cys | Gly |

Glutathione $(n = 1)$

Fig. 3.1a,b Structure of phytochelatin showing peptide bonds

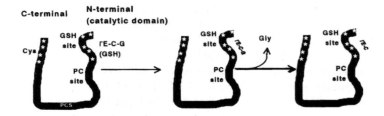

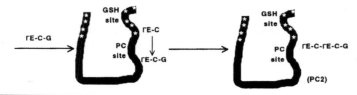

Fig. 3.2 Schematic model showing synthesis of phytochelatin

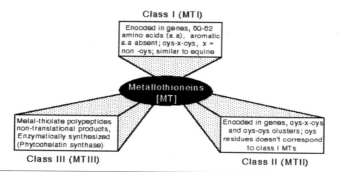

Fig. 3.3 Classification of metallothioneins (MTs)

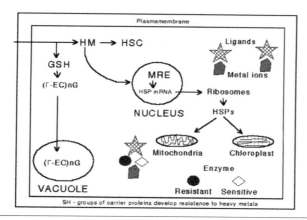

Fig. 3.4 Cellular responses of HM toxicity. Metal ions are complexed in the cytosol with *GSH* and the derived PCs are transported into the vacuole. The following functions have been implicated in Cd complexation and transport (see Rauser 1995 and references therein): (1) ATP-binding cassette-type transport activity at the tonoplast, (2) Cd^{2+}/H^+ antiport, and (3) vacuolar-type ATPase, generating a proton gradient. Another notable cellular response depicted is that some metals interact with genes having metal-regulating elements (*MRE*) at the promotor region as found for MT genes of animals. For example, in the soybean, similar sequences have been found in the upstream region of the heat-shock gene coded for 17.5 kDa HSP (Czarnecka et al. 1984). Some heat-shock genes and small HSPs have been induced by Cd ions in soybean (Czarnecka et al. 1984). These HSPs thus may function as molecular chaperones (Reddy and Prasad 1993; Prasad 1997)

3.2   METALLOTHIONEIN CLASSIFICATION AND OCCURRENCE

Polypeptides are designated "metallothioneins" when they show several of the features characteristic of equine renal metal-binding proteins, like high metal content, high cysteine content with absence of aromatic acids and histidine, an abundance of Cys-x-Cys sequences, where x is an amino acid other than Cys, spectroscopic features characteristic of metal thiolates, and metal thiolate clusters (see Rauser 1990a; Reddy and Prasad 1990; Steffens 1990). Considering the structural aspects, MTs have been classified into three groups (Fig. 3.3; Kagi and Schaffer 1988).

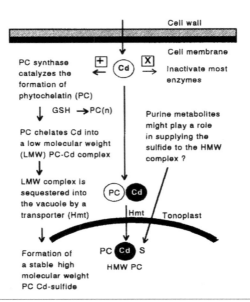

Fig. 3.5 Reductive sulfate assimilation and GSH synthesis, LMW complex formation and vacuolar internalization, S^{2-} production and incorporation into HMW complexes is required for an optimal exploitation of PC-mediated HM detoxification. Adequate sulfur nutrition (either natural or with fertilizer supplementation) and a high capacity for sulfur assimilation through the multienzyme pathways leading from sulfate to sulfide and then to Cys and GSH appear to be particularly important. In fact, natural HM accumulator plants such as *Brassica juncea* have an especially strong sulfur assimilation capacity and an *Arabidopsis* mutant, *man1*, with the characteristics of a natural polymetallic accumulator has a sulfur content in leaves about 3-fold higher than the wild type (Delhaize et al. 1989). On the other hand, the multiple connections between sulfur metabolism, PC biosynthesis, and downstream processing of PC-metal complexes leave room for a number of potentially negative, cellular and environmental interferences that may impair the effectiveness of the PC-mediated response

The class III metallothioneins (MTs) are only known from the plant kingdom. They occur in organisms as metal-binding complexes of various sizes, *Mr* (molecular mass) 3000 to 10,000, depending on the ionic strength of the eluants, suggesting that these are accretions of multiple peptides of various lengths with the metal atoms. The incorporation of varying amounts of sulfide or sulfite ions also contribute to the size heterogeneity of the class III MTs. Nevertheless, two broad categories of the complexes are generally recognized, low molecular weight (LMW) and high molecular weight (HMW). The grouping is based on the good resolution of the metal-binding complexes obtained in extracts from fission yeast exposed to Cd in some of the earliest known experiments in the line (Murasugi et al. 1981a, 1983). Currently, several situations are known to exist in plants, ranging from good resolution of low molecular weight (LMW) and high molecular weight (HMW) complexes (Kneer and Zenk 1992), partial resolution of LMW complexes on the trailing shoulder of an abundant HMW complex (Howden et al. 1995b), to no evidence for LMW complexes (Rauser and Meuwly 1995). Such variations are attributed to differences between organisms, type of nutrient medium for growth, concentration of the gel-filtration matrix, and column bed dimensions (Rauser 1995).

An indication of the existence of class III MTs in plants was first provided by Rauser and Curvetto (1980) in roots of *Argostis* tolerant to Cu, and by Weigel and Jager (1980) in bean exposed to Cd. However, the metal-binding complexes were only characterized to any extent by Murasugi et al. (1981a,b) in an extract from fission yeast, *Schizosaccharomyces pombe*, grown on Cd solution. Subsequently, these have been reported to be produced in cultured plant cells (Grill et al. 1985), algae (Gekeler et al. 1988) and virtually all higher plants tested (Grill et al. 1985; Gekeler et al. 1989). They have also been reported to be induced by a variety of metals, including Cd, Cu, Zn, Pb, Hg, Bi, Ag, and Au (Grill et al. 1987; Ahner and Morel 1995; Maitani et al. 1996), some of which are soft metals on Pearson's scale of softness. In addition, the complexes are also induced by multi-atomic anions like SeO_4^{2-}, SeO_3^{2-} and AsO_4^{3-} (Grill et al. 1987). Their induction is not only dependent on the type of the metals, but also on the plant species (Ahner et al. 1995).

The information on the composition and structure of phytochelatins (PCs) comes from the pioneering work of two groups, Kondo et al. (1983, 1984) on the fission yeast *Schizosacchromyces pombe*, and Grill et al. (1985) on cultured cells of *Rauvolfia serpentina*. The complexes produced by *S. pombe* have a structure identical to those of plants, consisting of heterogeneous populations of polypeptides, each representing repeating units of γ glutamyl-cysteine (γ-Glu-Cys) followed by a single C-terminal glycine [(γ-Glu-Cys)$_n$-Gly] with the number of repeating units (n) ranging from 2-11 (Kondo et al. 1983, 1984; Grill et al. 1985). Before their classification as class III MTs, various trivial names were given depending on one or more features associated with them, such as cadystin, as were found to be induced by cadmium in fission yeast (Murasugi et al. 1981a,b, 1983; Kondo et al. 1983, 1984), phytochelatin (PC), representing the metal-binding peptides of kingdom phyta (Grill et al. 1985), γ-glutamyl peptides, after the presence of γ-glutamyl (Reese and Winge 1988), poly(γ-glutamylcysteinyl)glycine, after their basic constituents (Jackson et al. 1987). However, none of these trivial names are appropriate, as these polypeptides are also induced by metals other than Cd, fungi is not considered to belong to the kingdom phyta, and diverse γ-glutamyl di- and tripeptides are known (Rauser 1990b). Nevertheless, the term phytochelatin is popularly used, as a fungus is always considered to be a plant, and also the term is meaningful suggesting the chelating function of the molecule, and hence will continued to be used in this chapter.

3.3 PHYTOCHELATINS

Phytochelatins (PCs) are simple γ-glutamyl peptides containing glutamate, cysteine and glycine [(γEC)$_n$G] in ratios of 2:2:1 to 11:11:1 (Grill and Zenk 1985). These peptides are synthesized from glutathione in the presence of HMs by an enzyme γ-glutamyl cysteine dipeptidyl transpeptidase (phytochelatin synthase; Grill et al. 1989). These peptides bind free metals like MTs in animal cells. PCs are induced by many metals and multi-atomic anions and have been named differently (Tables 3.1 and 3.2). Vast amounts of information are available on Cd-induced PCs and other metal-binding complexes (Prasad 1995). However, this is not the situation with Cu. One of the reasons could be due to Cu-induced depletion of glutathione on account of oxidative stress. Löffler et al. (1989) purified the PC synthase (γ-Glu-Cys dipeptidyl transpeptidase) to homogeneity in cell cultures of *Silene cucubalus* (Caryophyllaceae), *Podophyllum peltatum* (Berberidaceae), *Eschscholtzia californica* (Papaveraceae), *Beta vulgaris* (Chenopodiaceae) and *Equisetum giganteum* (Pteridophyte, Equisetaceae). This enzyme catalyzes the formation of metal-chelating peptides

Table 3.1 Inducers and non-inducers of PCs (Rauser 1995)

Potential inducers	Non-inducers
Ag, Bi, Cd, Cu, Hg, Ni, Sn, Sb, Te, W, Zn, SeO_4^{2-}, SeO_3^{2-} and AsO_4^{3-}	Na, Mg, Al, Ca, V, Cr, Mn, Fe, Co and Cs

Table 3.2 Terminology used for PCs

Metallocompound	Reference
Cadystin	Murasugi et al. (1981a)
Poly(γ-glutamylcysteinyl) glycine	Robinson and Jackson (1986)
Phytometallothioneins	Tripathi et al. (1996)
γ-Glutamyl metal-binding peptide	Reese et al. (1988)
γ-Glutamyl cysteinyl isopeptides	Stillman (1995)
Metallothiopeptide	Verkleij et al. (1990)
Metallopeptides	Ernst et al. (1992)
Des glycyl peptides (g-glu-cys)	Meuwly et al. (1995)
Phytochelatins	Grill and Zenk (1985); Rauser (1990b); Reddy and Prasad (1990); Steffens (1990)

(PCs) from glutathione in the presence of HM ions. Incubation of PC synthase under standard conditions in the absence of HM ions does not lead to the formation of PCs (Löffler et al. 1989). However, addition of Cd to the incubation mixture instantaneously reactivated the enzymes (Kneer and Zenk 1992).

The primary structure of the polypeptides described above is related to glutathione (GSH), γ-Glu-Cys-Gly. However, in some plants such as soybean (*Glycine max*), glutathione is replaced by homo-glutathione having a non-protein amino acid b-alanine (γ-Glu-Cys-β-Ala). In such plants, glycine at the carboxy terminal in the polypeptide contains β-Ala, and hence has been named homo-phytochelatin (h-PC; Grill et al. 1986). This has the general structure (γ-Glu-Cys)$_n$-β-Ala, and has been observed in 36 species of legume, with 13 species producing only h-PC, and 23 species both h-PC and PC depending on whether the plant produces only h-GSH or both GSH and h-GSH (Grill et al. 1986).

The third family of the PC polypeptides was observed by Klapheck et al. (1994) in certain species of Poaceae (rice, wheat and oats) in which the terminal Gly is replaced by serine (Ser) showing the primary structure (γ-Glu-Cys)$_n$-Ser. As these peptides are related to the tripeptide hydroxymethyl-glutathione (γ-Glu-Cys-Ser), the polymer is termed hydrooxymethyl-phytochelatin (hm-phytochelatin). In addition, the species studied also produced (γ-Glu-Cys)$_n$-Gly and (γ-Glu-Cys)$_n$ (desglycyl or desGly peptide).

The most recent addition to the family of phytochelatin polypeptides is that related to the novel tripeptide γ-Glu-Cys-Glu, with the structure (γ-Glu-Cys)$_n$-Glu. This was identified in maize exposed to Cd (Meuwly et al. 1995). Maize also produces in abundance another family of polypeptides, the desGly-peptides [(γ-Glu-Cys)$_n$], which were first noticed as minor constituents of Cd-binding complexes in the fission yeast (Meuwly et al. 1995). The production of (γ-Glu-Cys)$_n$-Gly also occurred.

Thus, altogether, there are five families of polypeptides in class III MTs. They have the common features (1) Glu occupies the amino-terminal position, (2) Cys forms the next residue forming peptide bond with the γ-carboxyl group of Glu, and (3) γGlu-Cys pairs are repeated two or more times with the subscript (n) specifying the exact number of repeats. The division of the polypeptides into five classes is only on the basis of the variation in the carboxy terminal amino acid. All five classes of phytochelatins thus belong to one specific family of dipeptide γGlu-Cys, and the complement of $(\gamma$Glu-Cys$)_n$ peptides from the five families largely varies according to the species.

The γ-glutamyl linkages present in PCs suggest that these polypeptides are not primary gene products, i.e. they are not the translational products of mRNA, and must be formed by ribosome-independent enzyme reactions. The similarity of PCs to GSH in containing a γGlu-Cys moiety suggested that this could be involved in the synthesis of polypeptides. A number of other observations also support the function of GSH as a precursor of PCs. First, the metal-induced synthesis of PCs is accompanied by a depletion of the GSH pool in cell cultures (Grill et al. 1987; Scheller et al. 1987; Delhaize et al. 1989) and in plant tissue, root (Ruegsegger et al. 1990; Tukendorf and Rauser 1990). Second, GSH is synthesized by the action of γ-glutamylcysteine (γGlu-Cys) synthetase (EC 6.3.2.2), which joins Glu with Cys followed by addition of Gly by GSH synthetase (EC 6.3.2.3). The activity of one or both the enzymes increases upon exposure of the plants to Cd (Ruegsegger et al. 1990; Ruegsegger and Brunold 1992; Chen and Goldsbrough 1993). Furthermore, plant cells incubated with buthionine sulfoximine (BSO), a potent inhibitor of γGlu-Cys synthetase, are unable to synthesize PCs, and addition of GSH reestablishes PC synthesis (Scheller et al. 1987; Ruegsegger et al. 1990). Third, the mutants of *S. pombe* that lack either γGlu-Cys synthetase or GSH synthetase do not synthesize PCs in response to Cd (Mutoh and Hayashi 1988). It has been demonstrated that in cells of *Datura innoxia* [35S]GSH is rapidly incorporated into PCs after exposure to cadmium (Berger et al. 1989). In the presence of BSO and GSH, PCs produced upon exposure of tomato cells to cadmium incorporate little [35S]cysteine, indicating that these peptides are not synthesized by sequential addition of cysteine and glutamate to GSH (Mendum et al. 1990).

At least three possible pathways of biosynthesis of PCs could be visualized: (1) transpeptidation, with GSH or the oligomeric PC peptide acting as an acceptor for the successive addition of γGlu-Cys moieties from GSH by transpeptidation reaction; (2) dipeptide addition in which γGlu-Cys units synthesized by the action of γGlu-Cys synthetase are transferred to GSH and/or the oligomeric PC peptides; and (3) polymerization of the γGlu-Cys units to $(\gamma$Glu-Cys$)_n$ oligomeric molecules that are transferred subsequently to Gly in a similar fashion known for GSH synthetase. Grill et al. (1989) identified one activity in an extract from *Silene cucubalus* (=*vulgaris*) that conformed to the first possible pathway. In vitro experiments with a 15-fold purified enzyme activity fraction with GSH as substrate showed induction of PC synthesis immediately after the addition of 0.1 mM Cd^{2+}. $(\gamma$Glu-Cys$)_2$-Gly appeared without a noticeable lag phase. Fifteen minutes after the addition of Cd^{2+}, heptapeptide (n=3) was detected, and after a further 20 min nanopeptide (n=4) was detected. The reaction came to a halt after 100 min, which resumed after a further addition of Cd^{2+}. When $(\gamma$Glu-Cys$)_2$-Gly was present along with GSH, heptapeptide formation occurred immediately after the addition of Cd^{2+}. In addition, in the presence of only $(\gamma$Glu-Cys$)_2$-Gly as substrate, the formation of first heptapeptides and then nanopeptides occurred. Simultaneously, the concentration of the pentapeptide (the substrate) decreased with concomitant release of GSH. The authors concluded that the enzyme, which was named γ-glutamylcysteine dipetidyl transpeptidase (trivial name phytochelatin synthetase or PC synthetase), in addition to adding a γ-Glu-Cys unit to GSH could also add this to PC molecules. And, secondly, the source of the γ-Glu-Cys moiety can be both GSH and the PC molecules.

Chen et al. (1997), in addition to confirming the conclusion of Grill et al. (1989), also showed that no PCs could be synthesized by the enzyme (PC synthetase) extract from tomato cells in the presence of either γ-Glu-Cys alone or γ-Glu-Cys and Gly. They concluded that the enzyme identifies only two substrates, GSH and PCs, and probably has two binding sites, one specific for GSH and the other, less specific, for GSH and PCs. This conclusion was also in context of observations of Klapheck et al. (1995) who working on pea (*Pisum sativum* L.), which produced both GSH and h-GSH, showed that the crude enzyme preparation from root produced (γ-Glu-Cys)$_2$-Gly in the presence of GSH, but, given only γ-Glu-Cys-β-Ala (h-GSH) or γ-Glu-Cys-Ser (hm-GSH), the rate of production of their n_2 oligomers was much less. However, in the presence of both GSH and h-GSH or hm-GSH, the synthesis of the respective β-alanyl or seryl n_2 oligomers was greatly increased. This led them to conclude that the enzyme has a γ-Glu-Cys donor-binding site specific for GSH and a less specific γ-Glu-Cys acceptor-binding site, being able to bind several tripeptides, namely GSH, h-GSH, hm-GSH, and of course the PCs, and thus the synthesis of not only PCs, but also of H-PCs and hm-PCs is possible only by a single enzyme. However, it has not been investigated whether the enzyme preparation of both Klapheck et al. (1995) and Chen et al. (1997) could carry out PCs synthesis even in the absence of GSH, with only added (γ-Glu-Cys)$_2$-Gly. If so, then the possibility of the existence of a GSH-specific binding site does not exist. Of course, it may be possible that the specificity of the γ-Glu-Cys donor site of the enzyme lies in the recognition of Gly residue at the carboxy terminal end of the tripeptide, or the PCs.

The other two pathways of (γ-Glu-Cys)$_n$-Gly synthesis are indicated in the observation of Hayashi et al. (1991). The crude preparation of the enzyme differs from that described above in two respects: (1) Cd is not necessary for its catalysis, and (2) some (γ-Glu-Cys)$_2$ appears in the reaction mixture along with PCs with GSH as substrate. Incubation of GSH with γ-Glu-Cys, (γ-Glu-Cys)$_2$ or (γ-Glu-Cys)$_3$ in the presence of the enzyme produces n+1 oligomers of the (γ-Glu-Cys)$_n$ provided, i.e. PCs are produced by dipeptide addition, the second pathway. The preparation also polymerized γ-Glu-Cys into (γ-Glu-Cys)$_{2,3}$, suggesting a dipeptide transfer function of the enzyme. This is the only work to suggest a biosynthetic origin for (γ-Glu-Cys)$_n$. Furthermore, GSH synthetase added Gly to (γ-Glu-Cys)$_{2,3}$ giving n=2 and n=3 oligomers of the PCs, respectively, allowing Hayashi et al. (1991) to propose that polymerization of γ-Glu-Cys to (γ-Glu-Cys)$_n$ followed by GSH synthetase adding Gly could be a third pathway for PC biosynthesis.

Klapheck et al. (1995), however, believe the production of (γ-Glu-Cys)$_n$ to be a result of catabolic processes, by the action of carboxypetidase on (γ-Glu-Cys)$_n$-Gly removing the Gly moiety. By analogy to the reactions catalyzed by carboxypeptidase C, a proteolytic enzyme that also acts as dipeptidyl transpeptidase (Wurz et al. 1962), desGly-PCs may also arise from a hydrolytic activity of PC synthetase, i.e. cleavage of the Gly after binding of a PC molecule at the donor binding site and transfer to water instead of to a γ-Glu-Cys acceptor (Klapheck et al. 1995). The assumption of a catabolic process in the formation of $(\gamma\text{-Glu-Cys})_n$ is strengthened by the observation of production of $(\gamma\text{-Glu-Cys})_n\text{-Glu}$ in maize (Meuwly et al. 1995). The tripeptide $\gamma\text{-Glu-Cys-Glu}$ is found in maize only after the Cd-induced appearance of $(\gamma\text{-Glu-Cys})_n\text{-Gly}$ and $(\gamma\text{-Glu-Cys})_n$, offering the possibility that the family of $\gamma\text{-Glu-Cys}$ peptides with amino-terminal Glu are degradation products of other thiol peptides. Just action of a γ-glutamyl transpeptidase cleaving intramolecular γ-Glu linkages would be required (Rauser 1995). Study of the Cd-sensitive mutant of *Arabidopsis thaliana* also supports a simple transpeptidation of a $\gamma\text{-Glu-Cys}$ moiety from GSH to GSH or $(\gamma\text{-Glu-Cys})_n\text{-Gly}$ as the probable pathway for PC synthesis; the mutant is deficient in GSH synthesis, and produces significantly less PCs despite having PC synthetase activity similar to the wild type (Howden et al. 1995a,b). Nevertheless, the presence of desGly-PCs at the beginning of Cd incubation and at low Cd concentrations does indicate that de novo synthesis of these peptides is possible (Klapheck et al. 1994)

3.4 INDUCTION OF PHYTOCHELATINS BY HEAVY METALS

Voluminous literature exists on the induction of PCs by heavy metals, and their possible involvement in metal tolerance, which have also been reviewed many times (Rauser 1990a,b, 1995; Steffens 1990; Ernst et al. 1992). In fact the phytochelatin response, or the synthesis of heavy metal binding polypeptides, is one of the few examples in plant stress biology in which it can be readily demonstrated that the stress response (PC synthesis) is truly an adaptive stress response. Nevertheless, there are also several exceptions. During the course of the stress-response studies, attention has been focused not only on the rate of synthesis of the polypeptides, but also on the role of the precursors and the enzymes involved in their synthesis. Most of the information, however, mainly comes from the work involving the heavy metal cadmium, in response to which the induction of PC synthesis was first detected.

The argument in favor of the possible involvement of PCs in heavy metal tolerance mainly comes from its induction and accumulation by a wide range of plant species, including algae, in response to Cd, and also in response to a range of heavy metals (see Rauser 1990a; Reddy and Prasad 1990; Steffens 1990; Ernst et al. 1992; de Knecht et al. 1994; Klapheck et al. 1994; Ahner et al. 1995; Howden et al. 1995a,b; Vögeli-Lange and Wagner 1996). In cell suspension cultures (of *Rauvolfia serpentina*) it has been observed that the tendency of metals to induce PCs decreases in the order Hg>>Cd, As, Fe>Cu, Ni>Sb, Au>Sn, Se, B>Pb, Zn (Grill et al. 1987). In the root culture of *Rubia tinctorum*, the PC induction by various heavy metals was in the order

Hg>>Ag>Cd>>As>Cu>Pd>Se>Ni>Pb>Zn>In>Ga (Maitani et al. 1996). For the metals common to both the cases, the order of induction is more or less similar, except of Ni and Se. The order, however, is based on the total metal concentration in the culture medium. For free ionic metals, the order may be different. This is evident from the work of Huang et al. (1987) who applied the metal concentrations to the cell suspension culture depending on their toxicity. Furthermore, Grill et al. (1989) showed that the activation of the purified enzyme from cell suspension cultures of *Silene cucubalus* by Hg was only 27% of activation produced by an equimolar concentration of Cd.

Further support in favor of a possible role of PCs in providing plant resistance to heavy metals comes from a study on metal (mostly Cd) tolerant culture cell lines and strains of algae and mutants. The uptake of Cd by Cd-tolerant plant cell lines is somewhat greater than by the non-tolerant ones prior to Cd becoming toxic (Jackson et al. 1984; Huang et al. 1987; Huang and Goldsborough 1988). In addition, the Cd-tolerant cells bind more than 80% of the cellular Cd as Cd-PC complexes, whereas little binding of Cd occurs in the non-tolerant cells, which grow poorly and die prematurely (Jackson et al. 1984; Huang et al. 1987; Huang and Goldsborough 1988; Delhaize et al. 1989). Furthermore, Gupta and Goldsbrough (1991) observed that the tomato cell lines selected for resistance to various concentrations of Cd showed increased Cd and PC accumulation concomitant with increase in their tolerance level, and at least 90% of the Cd in the most tolerant cell line was associated with Cd-PC complexes.

The evidence of PCs having a protective function against heavy metal toxicity also comes from the studies of the influence of the precursors of PCs, and the enzyme(s) involved in their synthesis, on the resistance of cell cultures or intact plants to the metals. Upon exposure of Cu-sensitive and Cu-tolerant *Silene cucubalus* (L.) to Cu, the loss of the GSH pool was only observed in the former (de Vos et al. 1992), suggesting that the maintenance of the GSH pool for continued synthesis of phytochelatin is necessary for survival under metal stress. In a similar study on tomato cells it was observed that the tolerance of CdR6-0 cells (cells selected for Cd tolerance) was associated with their enhanced capacity to synthesize GSH, nearly 2-fold higher than the unselected CdS cells, to maintain the production of PC (Chen and Goldsbrough 1994). It has also been demonstrated that the transgenic Indian mustard (*Brassica juncea*) overexpressing GSH synthetase contains greater amounts of GSH and phytochelatin, accumulates more Cd and shows greater tolerance to Cd than the wild type (Zhu et al. 1999). Furthermore, the growth of the Cd-tolerant cells (Steffens et al. 1986; Huang et al. 1987) or that of the non-tolerant cells (Mendum et al. 1990) remains unaffected in the presence of Buthionine-S-sulfoximine (BSO) alone, but is greatly inhibited in the presence of BSO together with Cd. The cell growth is, however, restored in the presence of exogenous GSH and is accompanied by PC synthesis (Scheller et al. 1987; Mendum et al. 1990).

Howden et al. (1995a,b) used a genetic approach to establish the relationship between Cd tolerance and phytochelatin synthesis and its accumulation. They isolated an allelic series of Cd-sensitive mutants, *cad1* (*cad1-1*, *cad1-2*...), and a second Cd-hypersensitive mutant, *cad2*, affected at a different locus. They observed that the hypersensitivity of *cad1* mutants to Cd was associated with deficiency in their ability to accumulate PCs due to deficient PC synthetase activity, while that of *cad2* was associated with deficient GSH levels resulting in deficient PC synthesis. Genetic studies using *S. pombe* have also shown that GSH-deficient mutants are also PC-deficient and Cd-hypersensitive (Mutoh and Hayashi 1988; Glaeser et al. 1991).

In contrast to the above, there are many studies on naturally evolved heavy metal tolerant varieties of plants, as well as on laboratory-selected tolerant cell lines that do not demonstrate a clear relationship between heavy metal resistance and PC production creating doubts about the involvement of PCs in metal tolerance. First, tolerant plants often do not produce more PCs than non-tolerant ones (de Vos et al. 1992; Schat and Kalff 1992; de Knecht et al. 1994). Second, while the level of PCs in Cu-sensitive ecotypes of *Silene cucubalus* increases significantly at 0.5 μM concentration of Cu, which is a non-toxic concentration for those plants, a significant increase in the level of PCs in the Cu-tolerant plants occurs only at 40 μM or higher Cu concentrations, which are toxic for the ecotype (de Vos et al. 1992). Third, distinctly Cu-tolerant (Marsberg) and non-tolerant (Amsterdam) ecotypes of *Silene vulgaris* produce equal amounts of PCs if they are grown at Cu concentrations which cause equal degrees of root growth inhibition, but such concentrations of the metal for tolerant plants are always greater than that for the non-tolerant plants (Schat and Kalff 1992). Fourth, the roots of Cd-tolerant plants of *S. vulgaris* exposed to a range of Cd concentrations accumulated greater amounts of the metal than the roots of Cd-sensitive plants, but contained significantly less PCs than the latter, particularly at the higher exposure concentrations (de Knecht et al. 1994).

3.5 HIGH MOLECULAR WEIGHT PHYTOCHELATINS AND THEIR ROLE IN METAL TOLERANCE

Although the accumulation of PCs could be a major component of the heavy metal detoxification process, the increased tolerance to metals may involve other aspects of PC function. The first argument in favor of this came from Delhaize et al. (1989) who observed that, although both Cd-sensitive and -tolerant cells of *Datura innoxia* synthesized the same amount of PCs during the initial 24 h exposure to 250 μM Cd, the concentration was toxic to the Cd-sensitive cells only, and not for the Cd-tolerant cells, as revealed by cell viability study. However, they differed in their ability to form PC-Cd complexes with the sensitive cells forming complexes later than the tolerant cells. In addition, the complexes formed by the sensitive cells were of lower molecular weight than those of tolerant cells, and did not bind all the Cd unlike in the tolerant cells. Thus, the rapid formation of PC-Cd complexes that sequester most of the Cd within a short period could be a necessity for plants or cells showing tolerance to heavy metals. Evidence in support of this also comes from work on Cd-sensitive mutants of *Arabidopsis thaliana*, *CAD1*, which is deficient in its ability to sequester Cd (Howden and Cobbett 1992). Furthermore, Gupta and Goldsbrough (1991) observed that the cell lines of tomato selected for their tolerance to various concentrations of Cd showed a trend towards accumulation of higher molecular weight (HMW) PCs in addition to showing their enhanced synthesis, and at least 90% of the Cd in the most tolerant cells was associated with PC complexes containing large amounts of SH. Thus the size of PCs may be a determining factor in tolerance to heavy metals. It has been shown that PC_7 is more efficient than PC_2 in complexing Cd per mole of γ-Glu-Cys (Löffler et al. 1989). Moreover, exposure of maize seedlings to increasing concentrations of Cd results in the accumulation of longer PCs with PC_4 being the largest peptide accumulating (Tukendorf and Rauser 1990).

Yet another way by which the metal-binding capacity of PCs (per mol PC-SH) can be increased is upon their association with acid-labile sulfur (S^{2-}), which has been reported to increase the stability of Cd-PC complexes as well in *S. pombe* (Reese and Winge 1988). The relevance of the presence of S^{2-} in PC-metal complexes to metal detoxification is substantiated by the observation that mutants of *S. pombe* that produce PC-Cd complexes without sulfide are hypersensitive to Cd (Mutoh and Hayashi 1988). In addition, Cd-tolerant *Silene vulgaris* plants exhibit a higher S:Cd ratio in the PC complexes than the Cd-sensitive plants (Verkleij et al. 1990). It has also been observed that the HMW PC-metal complexes contain greater amounts of S^{2-} than the LMW PC-metal complexes.

Two distinct peaks for HMW and LMW have not generally been observed in plants, and have only been described for tomato (Reese et al. 1992) and an Se-tolerant variety of *Brassica juncea*. Nevertheless, the two forms do exist even if they may not be distinctly separated. The acid-labile sulfur associated with the two forms varies from species to species. HMW Cd-binding complexes in maize seedlings exposed to 3 μM Cd show a S^{2-}:Cd molar ratio of 0.18, while no acid-labile sulfur occurred in the LMW complexes (Rauser and Meuwly 1995). *Brassica juncea* grown in synthetic medium with 100 μM Cd for 7 days produced HMW complexes with a S^{2-}:Cd molar ratio of 1.0 and LMW complexes with ratio 0.42 (Speiser et al. 1992a). Incompletely resolved complexes from roots of tomato exposed to 100 μM Cd for 4 weeks had continuous S^{2-}:Cd molar ratios ranging from 0.15 to 0.41 for the HMW complexes and from 0.04 to 0.13 for the LMW complexes (Reese et al. 1992). The yeasts *S. pombe* and *C. glabrata* grown in different media for 16-48 h and exposed to 500 or 1000 μM Cd showed complexes with S^{2-}:Cd molar ratios of 0.11 to 0.55 (Reese et al. 1988; Dameron et al. 1989). How sulfide, Cd and PC peptides interact within the complex is unclear for the cases in which the ratio is low. Cd-PC complexes with S^{2-}:Cd ratios greater than 0.4 appear as dense aggregates of 2 nm diameter particles called CdS crystallites. In yeast each crystallite contains about 80 CdS units stabilized by a coating of about 30 peptides of glycyl [$(\gamma$-Glu-Cys$)_n$Gly] and desGly [$(\gamma$-Glu-Cys$)_n$] forms (Dameron et al. 1989). Reese et al. (1992) showed the presence of such crystallites in plants (tomato) in PCs-Cd formation with a S^{2-}:Cd ratio of 0.41, but the crystallites were of less than 2 nm diameter coated with only $(\gamma$-Glu-Cys$)_n$Gly peptides.

The number of γ-Glu-Cys dipeptide repeats influences the stability of the complexes. Complexes formed with shorter peptides (n=1, 2) are more labile, and accretion of the crystallite to larger particles is more facile (Reese et al. 1992). In the yeasts *S. pombe* and *C. glabrata*, although $(\gamma$-Glu-Cys$)_{2,3}$ peptides are present in Cd-binding complexes, $(\gamma$-Glu-Cys$)_{2-4}$-Gly peptides are usually more concentrated (Dameron et al. 1989; Barbas et al. 1992). In tomato the number of γ-Glu-Cys units varies from 3 to 6, with n=4 being the predominant peptides (Reese et al. 1992). The Cd-binding complexes from several other sources are composed of $(\gamma$-Glu-Cys$)_n$-Gly, with n=3 and n=4 oligomers being the most abundant (Gupta and Goldsbrough 1991; Strasdeit et al. 1991). In soybean (*Glycine max*), n=1, 2, 3, 4 oligomers of $(\gamma$-Glu-Cys$)_n$-β-Ala form the Cd-binding complexes (Grill et al. 1986). The HMW complexes in maize are formed by the peptides from three families, $(\gamma$-Glu-Cys$)_n$-Gly, $(\gamma$-Glu-Cys$)_n$-Glu and $(\gamma$-Glu-Cys$)_n$, of which $(\gamma$-Glu-Cys$)_n$ peptides remain present in highest concentrations, followed by $(\gamma$-Glu-Cys$)_n$-Gly and $(\gamma$-Glu-Cys$)_n$-Glu, and the n=3 oligomers of all the three families form the highest constituents, followed by an equally dominating concentration of n=4 oligomers (Rauser and Meuwly 1995).

The preponderance of n=3 and n=4 oligomers in Cd-binding complexes from maize corroborates the increasing affinity of Cd for longer peptides (Hayashi et al. 1988), and their presence in HMW complexes together with acid-labile sulfur testifies to the importance of HMW Cd complexes in metal (Cd) detoxification. In their study, Rauser and Meuwly (1995) showed that the concentration of n=3 and n=4 oligomers increased in maize with the increase in the number of days of its exposure to Cd, and the HMW complexes sequestered 59% of the Cd after day 1, which increased to 88-92% by days 4 to 7.

3.6 CELLULAR COMPARTMENTATION OF PHYTOCHELATINS

Another important aspect of PC-mediated tolerance of plants to heavy metals is probably the effective transportation of the metal to vacuoles for storage in which they could play an important role. Arguments in favor of this come from several observations. Vögeli-Lange and Wagner (1990) isolated mesophyll protoplast from tobacco exposed to Cd and showed that the vacuoles contained 110±8% of the protoplast Cd and 104±8% of the protoplast PCs. These workers envisioned the synthesis of PCs in the cytosol, and transfer of Cd and the peptides, perhaps as a complex, across the tonoplast into the vacuole where the metal is chelated by the peptides and organic acids. Gupta and Goldsbrough (1990) working on tomato cells observed the highest level of PCs after 4 days of their exposure to Cd, which coincided with the peak of cellular Cd concentration (0.6 mM), and at this time there was a 8-fold molar excess of PC over Cd. However, the PCs could not be detected after 12 days while the cellular concentration of Cd was still 0.2 mM (the intracellular concentration of Cd decreased as a result of increase in the cell mass). This led them to suggest that PCs possibly function as transport carriers for Cd into the vacuole where the acidic pH favors dissociation of the Cd-PC complexes, followed by breakdown of the PCs and possible sequestration of the metal in some other form, in agreement with the model proposed by Vögeli-Lange and Wagner (1990). Later, de Knecht et al. (1994), while working on Cd-tolerant and -sensitive plants of *Silene vulgaris*, observed that in response to a range of Cd concentrations the root tips of Cd-tolerant plants exhibited a lower rate of PC production accompanied by a lower rate of larger chain PC synthesis than those of Cd-sensitive plants, although both the plants (root tips) accumulated nearly similar levels of Cd at a particular metal-exposure concentration. Secondly, the tolerant plants reached the same PC concentration as the sensitive plants only after exposure to high Cd concentrations, and at an equal PC concentration the composition of PC and the amount of sulfide incorporated per unit PC-thiol were the same in both the populations. The authors concluded that the lower concentration of PCs in the Cd-tolerant plants than in the Cd-sensitive plants could be because of greater transport of Cd-PC complexes in vacuoles in the former, and, as suggested by Vögeli-Lange and Wagner (1990) and Gupta and Goldsbrough (1990), the PC-Cd complexes might be getting dissociated in the vacuole because of its acidic pH, followed by breakdown of the PCs or their reshuttling into the cytoplasm. Thus, the observed lower PC concentration in Cd-tolerant plants might be a result of a lower Cd concentration in the cytoplasm caused by a faster transport of the metal into the vacuole when compared with that in the Cd-sensitive plants, and secondly by return of the dissociated PCs (in the vacuole) into the cytoplasm obviating the need of their fresh synthesis for the additional Cd uptake.

As the enzymes involved in PC synthesis are present in the cytoplasm but PCs are also found in the vacuole, there must be a transport mechanism involved, and an insight into this comes from the work on *S. pombe*. A Cd-hypersensitive mutant deficient in producing HMW complexes was observed (Mutoh and Hayashi 1988). This was found to be as the result of mutation within the *hmt1* (heavy metal tolerance 1) gene encoding an ATP-binding cassette (ABC)-type protein associated with vacuolar membrane (Ortiz et al. 1995). ABC-type proteins represent one of the largest known families of membrane transporters. They can mediate tolerance to a wide diversity of cytosolic agents. The presence of HMT1 protein in the vacuolar membrane suggests the possibility of an ABC-type transporter mediated resistance to Cd, by its sequestration in the vacuole (Ortiz et al. 1995). The yeast *hmt1⁻* mutant harboring *hmt1*-expressing multicopy plasmid (pDH35) exhibited enhanced resistance to Cd compared with the wild strain (*hmt1⁻* mutant) and accumulated more Cd with HMW complex formation (Ortiz et al. 1995). The vacuolar vesicle derived from the *hmt1⁻* mutant complemented with *hmt1* cDNA (*hmt1⁻*/pDH35), i.e. HMT1 hyper-producer, exhibited ATP-dependent uptake of LMW apophytochelatin and LMW-Cd complexes, whereas that from the *hmt1⁻* mutant did not show any such activity. The HMW-Cd complex was not an effective substrate for the transporter proteins. The vacuolar uptake of Cd^{2+}, which was ATP-dependent, was also observed, but was not attributable to HMT1. The electrochemical potential generated by vacuolar ATPase did not drive transport of peptides or complexes.

The observation of Ortiz et al. (1995) is also supported by work on oat tonoplast vesicles (Salt and Wagner 1993; Salt and Rauser 1995). Tonoplast vesicles from oat roots have a Cd^{2+}/H^+ antiporter (Salt and Wagner 1993). The vesicles also show MgATP-dependent transport of PCs and Cd-PC complexes (Salt and Rauser 1995), and the peptide transport is not driven by electrochemical potential generated by the vacuolar ATPase. Based on the information available, Rauser (1995) proposed a model, somewhat similar to that proposed by Ortiz et al. (1995), for the transport of Cd and Cd-binding complexes across the tonoplast. PCs synthesized in the cytosol combine with Cd to form a LMW complex that is moved across the tonoplast by ABC-type transporters. Apo-PCs are also transported by them. The energy required for both transports is derived from ATP.

Once inside the vacuole more Cd, transported by the Cd^{2+}/H^+ antiporter, is added along with apo-PCs and sulfide to the LMW complexes to produce HMW complexes. Genetic and biochemical analyses suggest that the formation of a sulfide moiety in the HMW PC-Cd-S^{2-} complex involves purine metabolism, which serves as a source of sulfide (Speiser et al. 1992b; Juang et al. 1993). The sulfide-rich HMW complex is more stable in the acidic environment of the vacuole and has a higher Cd-binding capacity than the LMW complex. The LMW complex functions as a cytosolic carrier and the vacuolar HMW complex is the major storage form of cellular Cd. Whether LMW and HMW complexes in plants are compartmentalized as depicted in the model and are of the same peptide composition, however, awaits direct evaluation. Nevertheless, the studies (Ortiz et al. 1995; Salt and Rauser 1995) do indicate a central role of the vacuole in sequestration and detoxification of Cd, and perhaps heavy metals in general, and that tolerance to metals could also be due to increased ability of plants to transport them into the vacuole.

3.7 ROLE OF PHYTOCHELATINS IN HEAVY METAL TOLERANCE

As stated earlier, synthesis of PCs is induced by most of the heavy metals including the multi-atomic anions (Grill et al. 1987; Maitani et al. 1996) in most of the higher plants (Gekeler et al. 1989; Grill et al. 1986). It has also been observed that the enzyme involved in its synthesis, PC synthetase, needs the presence of heavy metals for its activation; a crude preparation of the enzyme from *S. vulgaris* was activated best by Cd^{2+}, and by Ag^+, Bi^{3+}, Pb^{2+}, Zn^{2+}, Cu^{2+} and Au^+ in decreasing order (Grill et al. 1989). No activation of the enzyme was detected by metals of the hard acceptor category including Al^{3+}, Ca^{2+}, Fe^{3+},

Mg^{2+}, Mn^{2+}, Na^+ and K^+. The trend of activation observed by Grill et al. (1989) was, however, not observed for PC synthetase from tobacco cells, except that Cd was the most effective activator, followed by Ag+. The activation by Cu^{2+} was next to Ag^+, while Pb^{2+}, Zn^{2+} and Hg^{2+} produced only weak stimulation of the enzyme activity (Chen et al. 1997). Thus, although the enzyme has a rather non-selective domain for binding with metals, it is mostly activated by heavy metals. This strongly suggests that intracellular metabolism of heavy metals, other than Cd also, might be largely mediated through PCs. The view also stems from the fact that the heavy metal ions that activate PC synthetase in vitro are also able to induce PC synthesis in vivo (Chen et al. 1997), of course with one exception; Ni^{2+} induced PC synthesis in vivo, but did not activate PC synthetase activity in vitro. Furthermore, the indication of possible involvement of PCs in heavy metal tolerance also comes from genetic evidence; phytochelatin-deficient *cad1* mutants of *Arabidopsis* besides being sensitive to Cd^{2+} are also sensitive to Hg^{2+} (Howden and Cobbett 1992), and GSH-deficient strains of *S. pombe* show reduced tolerance to Pb^{2+} (Glaeser et al. 1991) besides showing no tolerance to Cd^{2+}.

PC synthetase activity is detected mostly in roots, but not in leaves or fruits (Rauser 1990a; Steffens 1990). The constitutive presence of PC synthetase in roots suggests an important role of PCs in metal detoxification. Since plants assimilate various metal ions from soil, the first organ exposed to these ions is the root. Localization of PC synthetase to roots and stems probably provides an effective means of restricting the heavy metals to these organs itself by chelation in the form of Cd-PC complexes. It has been demonstrated that PCs are able to protect enzymes from heavy metal poisoning in vitro (Thumann et al. 1991; Kneer and Zenk 1992); many metal-sensitive plant enzymes, Rubisco, nitrate reductase, alcohol dehydrogenase, glycerol-3-phosphate dehydrogenase and urease, were more tolerant to Cd in the form of a Cd-PC complex compared with the free metal ion, and free PCs reactivated the metal-poisoned enzymes (nitrate reductase poisoned by Cd-acetate) in vitro more effectively than other chelators such as GSH or citrate (Kneer and Zenk 1992).

Recognition of PCs as the chelators of heavy metals in general, and protectors of plants against their toxic effect, however, requires careful consideration. For instance, in tobacco cells not selected for metal tolerance, BSO increased the toxicity of Cd but not of Zn or Cu, as if the control of sequestration differed between the metals (Reese and Wagner 1987). This may of course be true, as we will see below, but has not been properly demonstrated. Secondly, while most of the heavy metals are able to induce synthesis of PCs in plants, only a few of them (Cd, Cu and Ag) form complexes with the peptides (Maitani et al. 1996). Recently, As has been reported to form complexes with PCs in arsenate-tolerant *Holcus lanatus* (Hartley-Whitaker et al. 2001). In fact PC-metal complex formation has been reported mostly for Cd. A few reports, other than that of Maitani et al. (1996), of PCs forming complexes with Cu are also available (Rauser 1984a,b; Grill et al. 1987; Reese et al. 1988), while formation of PC-Zn complexes has been observed in cells (of *Rauvolfia*) grown in micronutrient concentrations of Zn (Kneer and Zenk 1992). PCs have also been reported to form complexes with Hg and Pb, but in vitro. Nevertheless, there is genetic evidence that PCs are involved in tolerance to these metals; PC-deficient *cad1* mutants of *Arabidopsis* are also sensitive to Hg^{2+} (Howden and Cobbett 1992), and GSH-deficient strains of *S. pombe* show reduced tolerance to Pb^{2+} (Glaeser et al. 1991).

Thus, while the involvement of PCs in making plants resistant to heavy metals other than Cd cannot be overlooked, more information is required on their induction by individual heavy metals in different plant species, and also information is required on the formation of PC-metal complexes and cellular localization of the metals (individual), apo-PCs and metal-PC complexes before the functional significance of PCs known for Cd can be generalized for all heavy metals.

PCs are considered to have an important role in cellular metal-ion homeostasis (Steffens 1990). The enzymatically inactive, metal-requiring apoforms of diamine oxidase and of carbonic anhydrase were reactivated by Cu- and Zn-PC complexes, respectively (Thumann et al. 1991). When the metal chelator diethyl-dithiocarbamate was added to the purified diamine oxidase from pea (*Pisum sativum*) seedlings, it removed Cu ions from the enzyme. However, addition of various concentrations of $CuSO_4$ to the apoform of the enzyme resulted in maximal restoration of its activity to 80% (Kneer and Zenk 1992). PCs (poly γ-glutamyl cysteinyl glycines) are also implicated in HM detoxification (Jackson et al. 1987; Kneer et al. 1992).

PCs have been shown to bind Cd and Cu directly (Murasugi et al. 1981a,b; Reese et al. 1988) and are believed to bind Pb and Hg by competition with Cd (Abrahamson et al. 1992). HMs cause cell death in plants by inactivating enzymes and structural proteins by acting on metal-sensitive groups such as sulfhydryl or histidyl groups (Van Assche and Clijsters 1990; Kneer and Zenk 1992).

PCs reduce cytoplasmic toxicity of certain concentrations of HMs by complexation or binding. PC-metal complexes are less toxic to cellular plant metabolism than free metal ions (Figs. 3.3 and 3.4). The best indirect evidence for such an assumption is available from the experiments of Steffens et al. (1986) who demonstrated that, in tomato cells selected for Cd tolerance, PCs are accumulated at a considerably higher level than in normal, sensitive cells.

More direct evidence for the role of PCs in protecting plant enzymes was reported by Kneer and Zenk (1992) using suspension cell cultures of *Rauvolfia serpentina* and treating them with 100 μM $CdCl_2$ containing 100 μGi $^{109}CdCl_2$. They reported that a series of metal-sensitive plant enzymes, such as alcohol dehydrogenase, glyceraldehyde-3-phosphate dehydrogenase, nitrate reductase, ribulose-1,5-bisphosphate carboxylase and urease, tolerate Cd in the form of a PC complex from 10- to 1000-fold the amount as compared with the free metal ion. PCs reactivated metal-poisoned nitrate reductase in vitro up to 1000-fold better than chelators such as glutathione and citrate, thereby proving the extraordinary sequestering potential of PCs (Kneer and Zenk 1992).

The high molecular weight (HMW) complex has not been a common feature in the extracts of plant cells exposed to metals. However, acid-labile S^{2-} and sulfite in PC-metal complexes of tomato cells were isolated (Steffens et al. 1986; Eanetta and Steffens 1989). Verkleij et al. (1990) reported increased levels of S^{2-} in PC-Cd complexes from metal-resistant isolates of *Silene vulgaris*. HMW PC-Cd complexes from tomato (*Lycopersicon esculentum*; Reese et al. 1992) and Indian mustard (*Brassica juncea*; Speiser et al. 1992a) have also been reported.

Schat and Kalff (1992), during their investigations on Cu-sensitive and Cu-tolerant populations of *S. vulgaris*, found that both populations produce PCs when exposed to Cu. However, the total non-protein sulfhydryl content of the roots responding to the external

Cu concentration was population-dependent. Further, they also found that the dose-response curves for non-protein sulfhydryl accumulation and root growth were very consistent. When populations of *S. vulgaris* are exposed to Cu (no effect concentration or 50% effect concentration), tolerant and non-tolerant plants synthesize PCs equally in the root apex, which is the primary target for Cu. Therefore, the authors concluded that PCs are not decisively involved in differential Cu tolerance in *S. vulgaris*. Variations in PC-SH production seems to be a mere consequence of variation in tolerance. Differential Cu tolerance in the above plant does not appear to rely on differential PC production. Induction and loss of Cd tolerance was reported in *Holcus lanatus* and several other grasses (Baker et al. 1986).

3.8 SYMPLASMIC COMPLEXATION OF METALS

Glutathione is a precursor of PC synthesis. Metal-induced PC production decreases cellular levels of glutathione. Glutathione and its homologues, viz., homoglutathione and hydroxymethyl-glutathione, are the abundant LMW thiols in plants. Glutathione has been implicated in playing a major role in plants subjected to metal stress (Prasad 1997). Sulfur (10μmM) reduced the toxicity of Cd to the chloroplast pigments in leaves of young sugar beet plants, probably by influencing the glutathione (Rauser 2000). Alteration of thiol pools in maize exposed to Cd could be regulated by intermediates and effectors of glutathione synthesis. Direct evidence of glutathione involvement in PC synthesis was demonstrated in tomato cells. Glutathione was depleted due to Cu-induced PCs, causing oxidative stress in *Silene cucubalus*. Oxidative damage in plants is also caused either by an excess or deficiency of metals like Zn, Fe, Cu, B, Mg and K (Rengel 1997 and references therein).

In Fabales, homo-glutathione is the initial molecule and synthesizes homo-PCs (Grill et al. 1986; Tomsett and Thurman 1988). The key role of glutathione in detoxification or conferring tolerance to HMs is the crux of the metal resistance mechanism. The functions of the enzymes involved under metal stress and the altered glutamyl cycle are of considerable interest in the area of heavy metal resistance to plants (Agrawal et al. 1992; Reddy and Prasad 1992a; Prasad 1995, 1997). Buthionine-S-sulfoximine (BSO) is an inhibitor of γ glutamylcysteine synthetase (EC 6.3.2.2). Thus, its inhibition prevents the formation of metal-chelating peptides, i.e. PCs. However, *Chlamydomonas reinhardtii* treated with BSO bound three times more Cd than untreated cells, suggesting that some component other than a PC may bind Cd. Further, the Cd-binding component induced by BSO appears to be localized in the cell membrane or on the cell wall and may not be a PC (Cai et al. 1995).

Glutathione (0.01-1 mM) stimulated the transcription of defense genes including those that encode cell wall hydroxyproline-rich glycoproteins and the phenylpropanoid biosynthetic enzymes, viz., phenylalanine ammonia lyase and chalcone synthase, involved in lignin and phytoalexin production (Wingate et al. 1988). Likewise, $(\gamma EC)_nG$ may also cause expression of some genes resulting in many metal-inducible proteins.

3.9 METAL EXCLUSION/SEQUESTRATION BY ORGANIC ACIDS

Hue et al. (1986) demonstrated that addition of citric, oxalic and tartaric acid to a hydroponic solution alleviated the inhibitory effect of Al on root elongation in cotton. This antagonistic effect of chelating agents on Al toxicity has also been demonstrated for corn

(Berlett and Riego 1972), ryegrass (Muchovej et al. 1988), and sorghum (Shuman et al. 1991). It has been known for some time that several plant species excrete organic acids (citric acid and other) from their roots in response to P deficiency (Gardner et al. 1983; Lipton et al. 1987; Dinkelaker et al. 1989). Later cell cultures of carrot (*Daccus carota* L.) and tobacco (*Nicotiana tabacum* L.) selected for Al tolerance were also shown to exhibit enhanced ability to excrete citric acid upon Al treatment (Ojima et al. 1984, 1989; Ojima and Ohira 1988; Koyama et al. 1990). These studies led to the development of the hypothesis that organic acid secretion might be involved in the Al exclusion mechanism (Matsumoto 2002).

Organic acids are the other group of biomolecules that can function as chelators of heavy metals inside the cell, converting the metals to almost inactive and non-toxic forms. With regard to Al, at least two organic acids are known to function as chelators. One is citric acid (Ma et al. 1997c): nearly two-thirds of the Al in hydrangea leaves remains present in the cell sap in soluble form, as a Al-citrate complex at a 1:1 molar ratio of Al to citrate, a non-toxic form of Al. Another acid that has been reported to form intracellular complexes with Al is oxalic acid (Ma et al. 1998). About 90% of Al in buckwheat remains present as a soluble Al-oxalate complex in the symplasm, and the intracellular concentration of Al detected is as high as 2 mM. The complex occurs in a molar ratio of 1:3, Al:oxalate. Oxalic acid can form three species of complexes with Al at Al to oxalic acid molar ratios of 1:1, 1:2 and 1:3, but the 1:3 Al-oxalate complex is the most stable, with a stability constant of 12.4 (Nordstrom and May 1996). This stability constant is much higher than that of Al-citrate (8.1) or Al:ATP (10.9), meaning that formation of the 1:3 Al-oxalate complex can prevent binding of Al to cellular components, thereby detoxifying Al very effectively. The report is in contrast to the order of stability constant for Al-organic acid complexes: Al-citrate>Al-oxalate>Al-malate (Zheng et al. 1998b). It is, however, not known whether the Al complexes of citrate or oxalate remain located in the cytoplasm or in the vacuole.

Among the heavy metals that have been reported to be chelated by organic acids inside cells are Zn and Ni. Vacuoles of Zn- and Ni-tolerant plants, and also those of the non-tolerant plants, after exposure to high concentrations of various heavy metals often contain high concentrations of zinc and nickel (Ernst 1972; Brookes et al. 1981), and also to a lesser degree of Cu and Pb (Mullins et al. 1985) and Cd (Heuillet et al. 1986; Rauser and Ackerley 1987; Krotz et al. 1989). The results of the studies on Zn- and Ni-tolerant plants suggested that organic acids could be involved in their sequestration in the vacuole; while the Zn-tolerant plants, including *Silene vulgaris*, exhibited enhanced accumulation of malate (Brooks et al. 1981; Godbold et al. 1984), the Ni-tolerant plants showed accumulation of malate, malonate or citrate (Pelosi et al. 1976; Brooks et al. 1981) upon their exposure to Zn and Ni, respectively. Details of their transportation and sequestration inside the vacuole and the roles of the organic acids in the process are, however, not available.

In one of the models for the transport of Zn into the vacuole (see Ernst et al. 1992) it has been postulated that malic acid would bind Zn in the cytosol, thereby detoxifying it, and the Zn-malate complex would be transported over the tonoplast into the vacuole where it would dissociate. After this, malate would be retransported into the cytosol. Vacuolar Zn would remain bound to stronger chelators, such as citrate, oxalate, etc., when present. Brune et al. (1994) reported that barley mesophyll cell vacuoles contain appreciable concentrations of phosphate (30-100 mol m^{-3}), malate (>10 mol m^{-3}), sulfate (>4 mol m^{-3}), citrate (~1 mol m^{-3}) and amino acids (>10 mol m^{-3}) when grown in hydroponic culture. They hypothesized that these organic and inorganic salts interact with the divalent cations, thereby buffering the vacuolar free Zn concentration to low values even in the presence of high Zn levels (292 mmol m^{-3}) in the vacuolar space. According to Wang et al. (1992), citrate is the most efficient ligand for metal complexation in the vacuole at vacuolar pH-values of 6-6.5. The results of Brune et al. (1994) demonstrate the importance of compartmentation and transport as homeostatic mechanisms within leaves to handle, possibly toxic, zinc levels in shoot. The dependence of plants on organic acids for detoxification of Zn could be the reason for the poor induction of PCs by the metal (Grill et al. 1987; Maitani et al. 1996).

The mechanism of detoxification adapted by plants probably varies from metal to metal, and for a metal from species to species, and it is difficult to reconcile the idea of tolerance by means of any single mechanism. For example, Zn-tolerant *Agrostis capillaries* and *Silene vulgaris*, which exhibit increases in malate levels (Ernst 1976), are only slightly Ni-tolerant (Schat and Ten Bookum 1992), whereas Ni-tolerant *Alyssum bertolonii*, which is very rich in malate (Pelosi et al. 1976), is non-tolerant to Zn (Ernst et al. 1992). Similarly, as stated earlier, BSO increases the toxicity of Cd to the tobacco cells not selected for Cd tolerance, but not of Zn or Cu.

An increased organic acid exudation as a metal resistance mechanism has been shown in a number of plant species (Rengel 1997). Aluminum stress in buckwheat (*Fagopyrum esculentum*) caused secretion of oxalic acid into the rhizosphere, and the detoxified Al-oxalate complex was taken up by roots and translocated into the leaves. Aluminum is excluded from the symplasm by complexation with organic anions into the rhizosphere (Rengel 1997). High turnover of organic acids, viz. phytate, malate, citrate, oxalate, succinate, aconitate, a-ketoglutarate and malonic, has been reported as a mechanism of metal resistance. Cysteine (41.3 μM), glutamate (34 μM) and glycine (66.6 μM) separately reduced Cd toxicity in terms of dry weight and chlorophyll a (Chl a) in *S. quadricauda*. By doubling cysteine, glutamate and glycine concentrations, the protective effect in terms of growth recovery (dry wt, Chl a) was further enhanced (Reddy and Prasad 1992b). An increase in the Chl a value was observed in the Cd-treated cultures when they were preincubated with cysteine, glutamate or citrate. A reversal of Cu, Cd and Hg toxicity by amino acids was also reported in *Chlorella vulgaris* and *Anabaena variabilis* (Kosakowasaka et al. 1988).

Citrate preincubation also increased the dry weight and Chl a content of Cd-treated *S. quadricauda* cultures (Reddy and Prasad 1992b). Citrate has been reported to reverse the inhibition of wheat root growth caused by aluminum (Ownby and Popham 1989). Complexation of organic acids (citrate, malate, phytate, oxalate, etc.) with metals (Cd and Zn) in vacuoles was reported in tobacco suspension cultures (Kortz et al. 1989). Whether a similar situation is ubiquitous under metal stress needs to be verified.

3.9.1 Citric Acid

Miyasaka et al. (1991) using differentially Al-resistant cultivars of snapbean (*Phaseolus vulgaris*) demonstrated in whole plant experiments that Al-resistant cultivars excreted a higher level of citric acid into the rhizosphere, 70 times more than the Al-sensitive cultivar in response to Al stress. In addition, the tolerant cultivar secreted 10 times more citric acid than the sensitive cultivar even in the absence of Al. This led them to suggest that the resistance to Al in the Al-tolerant cultivar could be due to a decrease in the active form (Al^{3+}) of Al around the rhizosphere as a result of its complex formation with the acid. However, they also noticed the formation of Al-phosphate precipitates, which could have caused P deficiency, and the latter is known to trigger organic acid secretion (Ojima et al. 1989). Thus, the relationship between citric acid secretion and Al exclusion remained unclear. Nevertheless, enhanced secretion of citric acid has also been observed in an Al-resistant maize line (Pellet et al. 1995; Kollmeier et al. 2001) and an Al-resistant species, *Cassi tora* L. (Ma et al. 1997a), in response to Al, which further supports the existence of a relationship between citric acid secretion and Al exclusion. A possible role of citric acid in Al resistance further stems from the observation that tobacco and papaya plants genetically engineered for over-production of citric acid by introducing a citrate synthase gene from *Pseudomonas aeruginosa* show increased Al resistance (Fuente et al. 1997). Also, the Al-resistant mutant, *alr*-108, *alr*-128 and *alr*-131, of *A. thaliana* shows enhanced cellular exudation of citrate, malate and pyruvate than the wild type, although the enhanced exudation of citric acid is not sustained upon exposure to Al^{3+} (Larsen et al. 1998).

3.9.2 Malic Acid

Delhaize et al. (1993b) used a genetic approach to prove the relationship between Al tolerance and organic acid secretion. They used near-isogenic wheat (*Triticum aestivum* L.) lines, which showed 5- to 10-fold difference in Al tolerance, and differed in Al tolerance at a single locus (*Alt1*). The test species, however, excreted malic acid and succinic acid instead of citric acid, and the malic acid excretion was 5- to 10-fold greater in the Al-tolerant (ET3) seedlings than in the Al-sensitive (ES3) seedlings, despite the cellular content of the acid remaining nearly unchanged and similar. A significant correlation between Al-triggered malate release, Al resistance, and Al exclusion from the root apex was observed (Delhaize et al. 1993a). It was proposed that the release of malic acid from roots exposed to Al could be the Al-tolerance mechanism encoded by the *Alt1* locus. This is because: (1) a consistent correlation of the *Alt1* locus occurred with malic acid excretion in the population of seedlings segregating for Al tolerance; (2) Al stimulated malic acid excretion within 15 min, consistent with the observation that Al tolerance is apparent after short exposure to Al; (3) malic acid excretion was localized at root apices, the primary site of Al toxicity; and (4) malic acid added to nutrient solution was shown to ameliorate Al toxicity. These authors also demonstrated that the low external Pi conditions did not stimulate malic acid excretion over 24 h, and high external Pi concentration did not prevent Al from stimulating malic acid secretion. Later, Basu et al. (1994) observed a similar difference in malate efflux from roots of several cultivars differing in Al tolerance, and Ryan et al. (1995b) after screening 36 different wheat cultivars for Al resistance proposed that Al-stimulated malate efflux might be a general mechanism for Al tolerance in wheat. This is further substantiated by the observation that the inhibition of malate exudation results in enhanced accumulation of Al in an Al-resistant wheat (cv. Atlas) exposed to the metal (Osawa and Matsumoto 2001).

Concomitant with malate excretion, Basu et al. (1994) also observed enhanced de novo synthesis of the organic acid, which is consistent with data that the malate content of

Al-tolerant root apices is replenished over five times during the initial 2 h of Al exposure (Delhaize et al. 1993b). Furthermore, it was observed that, although the root apices of Al-tolerant seedlings synthesized more malate in response to Al than the root apices from the Al-sensitive seedlings, the root apices of both the genotypes showed similar activities of phosphoenolpyruvate carboxylase and malate dehydrogenase, the two enzymes important in malate synthesis (Ryan et al. 1995a). Since the root apices of Al-sensitive and Al-tolerant genotypes showed nearly similar malic acid contents, whether exposed to Al or not (Delhaize et al. 1993b), and they had the same capacity to synthesize the acid (Ryan et al. 1995a), it was hypothesized that the difference in efflux probably lay in their relative ability to transport malate across the plasma membrane in response to Al (Delhaize et al. 1993b; Ryan et al. 1995a), the cytoplasm pool being replenished by fresh synthesis (Basu et al. 1994).

Taking into consideration all these observations, Delhaize and Ryan (1995) proposed a working model for the transport of malic acid across the membrane. Malate exists primarily as a divalent anion (malate^{2-}) in the cytoplasm, and, if transported out of the cell in this form, electroneutrality must be maintained either by an equivalent uptake of anions or by an equivalent efflux of cations. Ryan et al. (1995a) and Kollmeier et al. (2001) showed that excretion of malate is in fact accompanied by efflux of K^+. Zhang et al. (2001) further observed that the efflux of K^+ in the Al-tolerant line of wheat (ET8) is not maintained because of insensitivity of the K^+ outward rectifying channel to Al, as suggested by Kollmeier et al. (2001). Rather, Al inhibits the K^+ outward rectifying channel in ET8 strongly. Later, however, the inhibited channel, or additional K^+ outward rectifying channel, is activated, a process in which cAMP is involved (Zhang et al. 2001). The movement of malate^{2-} could be mediated by anion channels in the plasma membrane. The evidence for this was provided by Ryan et al. (1995a); the rapid release of malate in response to Al was inhibited by anion channel antagonists, anthracene-6-carboxylic acid (A-9-C) and niflumic acid (NIF). The existence of a malate-permeable channel and its activation by Al in wheat has also been confirmed by Zhang et al. (2001) using anion channel antagonists. Furthermore, it has been observed that in an Al-tolerant maize cultivar (cv. ATP-Y) the malate channel is permeable to citrate as well (Kollmeier et al. 2001). Delhaize and Ryan (1995) proposed three ways in which Al, probably as Al^{3+}, could trigger the opening of the putative malate^{2-}-permeable channel: (1) Al may interact directly with the channel protein causing a change in conformation, increasing its mean open time or conductance; (2) Al may interact with a specific receptor on the membrane surface or with the membrane itself, which through a series of secondary messages in the cytoplasm could change the channel activity; and (3) Al^{3+} may enter the cytoplasm and could alter the channel activity either directly by binding with the channel or indirectly through a signal transduction pathway. They further suggested that the *Alt1* locus could code for a malate^{2-}-permeable channel that is responsive to Al or for a component of the pathway that regulates the activity of the putative channel leading to enhanced excretion of malate^{2-} in the Al-tolerant cultivar, but not in the Al-sensitive one. Recently, it has been reported that the exudation of malate in the roots of an Al-tolerant cultivar of wheat (cv. Atlas) is inhibited by K-252a, a broad-range inhibitor of protein kinases, suggesting that the opening of the channel is preceded by protein phosphorylation (Osawa and Matsumoto 2001). Treatment of the root apices by K-252a prior to exposure to Al also led to enhanced accumulation of the metal. The interaction of Al with the malate channel is thus likely through the second or third pathway proposed by Delhaize and Ryan (1995).

3.9.3 Oxalic Acid

Ma et al. (1997a-c) and Zheng et al. (1998a) observed Al tolerance in buckwheat to be much greater than in the Al-tolerant cultivar of wheat (cv. Atlas 66), and found this to be a result of secretion of oxalic acid, the simplest dicarboxylic acid, from the root apex, the Al-sensitive region. The secretion was specific to Al stress, as neither exposure to La^{3+} nor P deficiency resulted in any enhanced secretion of the acid. They also observed that the secretion of oxalic acid in response to Al was inhibited in the presence of an anion channel inhibitor, phenylglyxol (PG), with subsequent inhibition of root elongation by as much as 40%, suggesting that the secretion of oxalic acid might be contributing to high Al resistance in buckwheat. The secretion was, however, not inhibited by NIF or A-9-C, which inhibited the secretion of malic acid in wheat; the secretion of oxalic acid probably occurs through the anion channel, which differs in characteristics from the $malate^{2-}$ anion channel in wheat, and hence the tolerance mechanism in wheat and buckwheat could be mediated through different gene functions.

Although the secretion of organic acid as a mechanism of Al resistance is well established in many plants, it is still not clear why there is a difference in the requirement of the type of organic acid to be secreted by the plants to achieve the resistance. As far as the Al-detoxifying capacities of organic acids are concerned, for a plant species these can be grouped into strong (citric, oxalic and tartaric), moderate (malic, malonic and salicyclic) and weak (succinic, lactic, formic, acetic and phthalic; Hue et al. 1986). Using a 1:1 ratio of organic acid to Al experimentally it has been proved that for a species (corn) the detoxifying capacity is in the order citric>oxalic>malic (Zheng et al. 1998b). The difference in capacity of organic acids to ameliorate Al toxicity is attributable to their different stability constants with Al (stability constant: Al-citrate>Al-oxalate>Al-malate), which probably results in different activity of free Al^{3+}.

3.9.4 Necessity of Continuous Secretion of Organic Acids

Irrespective of the type of organic acid secretion by a species for detoxification of Al, it is, however, necessary that continuous secretion of the acid at a high level is maintained for Al resistance (Zheng et al. 1998b). According to the total amount of organic acids secreted, three patterns are observed with different cultivars differing in Al-resistance/-sensitivity (Zheng et al. 1998b): (1) the amount secreted is very low during the treatment (wheat cv. Scout 66, oat) - sensitive; (2) the amount of secretion is high at the initial phase of exposure, but gradually decreases with duration of treatment (wheat cv. Atlas 66, rape oilseeds) - moderately tolerant to tolerant; (3) the amount of secretion is maintained at a high level during the whole period of Al treatment (buckwheat and radish) - highly tolerant. The categorization, however, may not be strict, particularly for the sensitive category, as the tolerance mechanism other than organic acid secretion may provide tolerance to the species against the metal (Pellet et al. 1996). Furthermore, as already known, it is not necessary that all of the Al molecules in solution need to be detoxified, rather it is the concentration of Al around the root apex, possibly just at the cell plasma membrane, that needs to be reduced. In this context, the mucilage exuded by the root cap may be of much importance as it will increase the unstirred layer around the root apex helping the root to maintain the organic acid concentration necessary to protect the root cap (Henderson and Ownby 1991). So the Al tolerance of a plant may also be determined by its ability to exude mucilage around the root cap. This view is substantiated further by the observation that the root border cells (the living cells surrounding the root apices) of an Al-tolerant cultivar (cv. Dade) of *Phaseolus vulgaris* produce a thicker mucilage layer than the Al-sensitive cultivar (cv. Romano) in response to Al treatment (Miyasaka and Hawes 2001).

3.9.5 Extracellular Detoxification of Metals Other than Aluminum

The indication of the possible involvement of plant exudates in ameliorating the toxic effect of metals other than Al mainly comes from work on algal systems, and that too only on cyanobacteria. The indication mainly stems from the fact that many amino acids and organic acids supplied along with the heavy metals (Cu, Cd, Ag) alleviated their toxicity (Reddy and Prasad 1992b; Mallick and Rai 2002). However, the relationship between the excretion of organic compounds and their detoxification role has never been experimentally demonstrated (Vymazal 1990). There is no report of any protective effect of plant exudates against heavy metal toxicity. However, Arduini et al. (1996) working on Cd and Cu distributions in various Mediterranean tree seedlings suggested the well-developed root cap in plants to have a protective role against metal uptake.

3.10 RESISTANCE MEDIATED BY INTRACELLULAR SEQUESTRATION

Plant resistance to heavy metals, unlike that to Al^{3+}, is achieved through their uptake and proper sequestration inside the cell, rather than by their exclusion that works for Al^{3+}. The adaptation of a totally different mechanism by plants for tolerance to heavy metals than for tolerance to Al^{3+} is probably because of the fact that there are many essential metals in the heavy metal category, which must be taken up from the environment for various metabolic functions to continue, but their chemical properties match greatly with many of the non-essential heavy metals making it difficult for plants to go for their selective uptake, and hence the non-essential heavy metals are also taken up along with the essential ones when present in the environment. Besides, the essential heavy metals are also more or less equally toxic to the organisms as the non-essential ones at a similar concentration. Thus, probably the only way left for plants to counter the presence of elevated levels of heavy metals in the environment, whether essential or non-essential, is to sequester them properly out of the cytoplasm in an inactive form. The problems associated with the presence of elevated levels of heavy metals in the environment may rather be viewed under a broad perspective: it is essential for plants to have mechanisms that (1) maintain the internal concentrations of essential metals between deficient and toxic limits, and (2) keep the non-essential metals below their toxicity threshold (Rauser 1995).

At present approximately 400 plant species belonging at least to 45 families are hyperaccumulators of heavy metals to various degrees (Kramer et al. 1997; Guerinot and Salt 2001). Field-collected samples of plants from metal-rich soils have been found to contain metals like Cu, Co, Cd, Mg, Ni, Se or Zn up to levels that are 100 to 1000 times those normally accumulated by plants (Kramer et al. 1997; Guerinot and Salt 2001), and their concentration for some metals may reach as high as 1000 ppm or more on a dry mass basis (Baker and Brooks 1989; Reeves 1992). The exact mechanism involved in such hyperaccumulation is, however, still debatable. Although there are a few reports of involvement of organic acids in the hyperaccumulation process, there is a general agreement in the scientific community that the hyperaccumulation, and so also the resistance to the heavy metals being accumulated, is facilitated by a thiol-rich polypeptide, similar in function to that of metallothioneins rich in the amino acid cysteine, which was first discovered in horse kidney (see Hamer 1986).

3.11 ENZYMES

Chromium(IV) altered the activities of oxido-reductases in *Chlorella pyrenoidosa* (superoxide dismutase, catalase and peroxidases). Lead treatment to maize seedings reduced the activity of nitrate reductase, piling up the nitrate amount. Lead treatment also increased aspartate amino transferase activity, while alanine amino transferase activity decreased. When *Zea mays* seedlings were grown in 50 μM Cd for 5 days, they incorporated more label from $^{35}SO_4^{2-}$, and Cd increased two enzymes involved in sulfate reduction, viz. ATP-sulfurylase and adenosine 5'-phosphosulfate sulfotransferase (Nussbaum et al. 1988). *Scenedesmus quadricauda* cultured with sulfate salts increased the uptake of sulfate, resulting in a high turnover of GSH (glutathione, reduced) and, consequently, PCs that help in detoxification of metal ions (Reddy and Prasad 1992b). Refer to Rauser (1995) for additional information on the enzymology of PCs and related peptides.

3.12 PHYTOFERRITINS

Iron is an essential micronutrient. It plays a pivotal role in many important redox reactions. Iron is present in cytochromes, FeS proteins like ferredoxin. It is a constituent of catalytic enzymes, viz., catalase, peroxidase, nitrogenase, etc. With the advent of an aerobic atmosphere on earth, free and insoluble forms of iron posed a serious oxidative threat to life. Hence, plants developed extracellular binders, siderophores, storage systems and ferritins. Phytoferritins are usually abundant in non-photosynthesizing tissues such as roots, root nodules, senescing cells and seeds, but present in smaller amounts in photosynthesizing tissues. The protein moiety of ferritin is composed of 24 subunits of spherical shells with a central cavity. The subunits form channels with deposition of Fe. Iron is stored as ferric hydroxyphosphate. Legume seeds are a rich source of ferritin (Table 3.3). The functions of ferritin in plants are depicted in Fig. 3.6.

3.13 METALLOTHIONEIN GENES IN PLANTS

A pea gene (designated *PsMT-A*) showed homology (with predicted amino acid sequence of *PsMT-A*) to class I and II MT genes from different organisms (Evans et al. 1990; Table 3.4). A cDNA sequence from *Mimulus guttatus* with a single open reading frame (ORF) encoding a putative 72 amino acid polypeptide sequence, 19 of 23 matches were MTs when compared with a protein database, the top 8 matches being from *Neurospora crassa*, sea urchin, Chinese hamster and *Drosophila* (de Miranda et al. 1990). This strong similarity

Table 3.3 Iron content of the ferritin from some legume seeds

Taxon	Fe atoms/molecule	Reference
Pea (*Pisum sativum*)	1800-2100	Crichton et al. (1978)
Soybean (*Glycine max*)	2500	Laulhere et al. (1988)
Lentil (*Lens esculenta*)	2100	Laulhere et al. (1988)
Jackbean (*Canavalia ensiformis*)	900	Briat et al. (1990)
Black gram (*Vigna mungo*)	1100	Kumar and Prasad (2000)

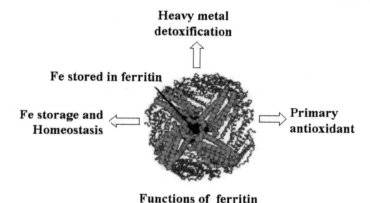

Fig. 3.6 Functions of ferritins

Table 3.4 Metallothionein genes in plants

Taxa	Reference
Pea (*Pisum sativum*)	Evans et al. (1990)
Soybean (*Glycine max*)	Kawashima et al. (1991)
Thale cress (*Arabidopsis*)	Zhou and Goldsbrough (1994, 1995)
Monkey flower (*Mimulus guttatus*)	Macnair (1977); de Miranda et al. (1990)
Maize (*Zea mays*)	Rauser (1984a,b); Leblova et al. (1986); White and Rivin (1995)
Barley (*Avena sativa*)	Rauser (1999)
Wheat (*Triticum aestivum*)	Kawashima et al. (1992)
Castor bean (*Ricinus communis*)	Weig and Komor (1992)
Rape seed (*Brassica napus*)	Misra and Gedamu (1989); Buchanan-Wallston (1994)

with class I and II MTs is due to two domains of 14/15 amino acids, each of which contain six cysteine residues arranged exclusively as cys-x-cys clusters. Similar domains are also present in a 75 amino acid ORF in a cDNA isolated from *Pisum sativum* (Tommey et al. 1991). The pea and *M. guttatus* nucleotide sequences are strongly homologous. Within the coding regions, 66% of the bases are identical and, in the regions encoding domains 1 and 2, 74% are identical.

In earlier reports where MT-like proteins were purified, some percentage of metal ions also bound to low M_r peptides, probably PCs. In *Datura innoxia* 80% of the Cd was bound to low molecular weight $(\gamma EC)_nG$ and about 15% to MT-like proteins (Robinson et al. 1987). The amino acid composition of this MT-like protein is similar in some respects to that of mammalian and microbial MTs, but contains more glutamine/glutamate than these. In maize, about 72% of the Cd is bound to low M_r peptides, possibly $(\gamma EC)_nG$, and about 18% is bound to an MT-like protein of 11.35 kDa (Leblova et al. 1986). For Cu, about 60% was bound to an MT-like protein and only about 25% to low M_r peptides, possibly $(\gamma EC)_nG$ (Leblova et al. 1986).

Kawashima et al. (1991) deduced the amino acid sequence of a soybean MT-like protein and compared it with MT genes of rat, *N. crassa*, *M. guttatus* and *P. sativum*. The N-terminal domain exhibited significant homology to *N. crassa* MT (9/13, 70%) and the b-domain of rat MT (9/14, 64%). However, by contrast, the middle domain contains several aromatic and hydrophobic amino acids and it showed little homology to other MTs. No cysteine was detected in the middle domain. The deduced amino acid sequence of the soybean protein is 51% and 59% homologous to corresponding sequences from *M. guttatus* and pea, respectively. These proteins from *M. guttatus*, *P. sativum* and soybean show a close resemblance to each other in the following three respects: (1) the number of amino acid residues range from 72 (*M. guttatus*) to 79 (soybean); (2) the N- and C-terminals of these three proteins contain cysteine clusters and show a high degree of homology in terms of amino acid sequence; and (3) the middle domains of all the three proteins are relatively rich in hydrophobic amino acid residues and contain no cysteine (Kawashima et al. 1991).

Using a synthetic oligonucleotide which corresponds to the consensus nucleoside sequence of the N-terminal region of a mammalian MT gene as a probe to screen a soybean cDNA library for a gene for an MT-like protein, a cDNA clone 21-1-A was isolated (Kawashima et al. 1991). This clone had an ORF that encoded a protein of 79 amino acids; the deduced amino acid sequence of this protein is 51% and 59% homologous to corresponding sequences from *M. guttatus* (de Miranda et al. 1990) and pea (Tommey et al. 1991), respectively. A probe prepared from this clone in Northern blots showed a transcript of approximately 700 bp. However, the level of transcript in soybean roots did not increase in the presence of Cu (Kawashima et al. 1991). These results indicate that the expression of this gene is constitutive and not increased by Cu treatment.

PsMT-A showed homology to some extent with monkey MT-I, fruit fly MT, horse MT-IA and B, Chinese hamster MT-II, green monkey MT-II, human and monkey MT-II, rat MT-II, mouse MT-II and *N. crassa* MT (Evans et al. 1990). By contrast, the wheat Ec protein, a higher plant class II MT, did not show significant homology to *PsMT-A* sequences. *PsMT-A* expression was studied in *Escherichia coli* using a carboxy terminal extension of glutathione-S-transferase (GST). This fusion protein was found to bind greater amounts of metal ions (Zn, Cd and Cu) than GST alone, suggesting that metal ions bind to the *PsMT-A* portion of the protein. It was also found that a cleaved *PsMT-A* portion from the fusion protein was able to bind Zn. The pH values of half dissociation of Zn, Cd and Cu from *PsMT-A*-GST, purified from *E. coli* and grown in media supplemented with the respective metal ions, were estimated to be 5.25, 3.95 and 1.45, respectively (Tommey et al. 1991). Since GST also binds to metals, the role of metal dissociation from the GST part of *PsMT-A*-GST to half dissociation pH is to be evaluated (Reddy 1992).

Pea roots also synthesize $(\gamma\text{-EC})_n G$ upon exposure to Cd salts, and *PsMT-A* transcript is abundant in roots that have not been exposed to elevated concentrations of trace metals (de Miranda et al. 1990). De Miranda et al. (1990) suggested (-EC)$_n$G may detoxify Cd and possibly excesses of other metals in pea roots; the putative *PsMT-A* product may have a role in essential trace metal metabolism (de Miranda et al. 1990). However, Grill et al. (1985) reported $(\gamma\text{-EC})_n G$ induction in cell cultures in standard media and proposed a role in homeostasis of essential metals, the microelements.

Mammalian MTs also function in plants (Lefévbre et al. 1987). Chinese hamster MT-II is expressed and functional in *Brassica campestris* leaf tissue when they are infected with a cloned CMV (cabbage mosaic virus), pCa-BB1 containing cDNA of the Chinese hamster MT-II gene (Lefévbre et al. 1987). HM-tolerant transgenic *Brassica napus* and *Nicotiana tabacum* plants were obtained by infecting with *Agrobacterium tumefaciens* containing a human MT-II processed gene on a disarmed *Ti* plasmid (Misra and Gedamu 1989). Mouse MT gene expression was studied in transgenic tobacco (Maiti et al. 1989). However, in such transgenic plants, synthesis of metal-inducible proteins including PCs and their role in metal tolerance is to be studied.

Copper and Cd are reported to replace Zn in zinc-fingers (Sarkar 1997). Transcription-activating factors, TFIIIA from *Xenopus* and GAL4 from yeast contain zinc-fingers. Displacement of Cd and Cu resulted in dissociation of TFIIIA from RNA (Miller et al. 1985). Hence, presence of transcription-activating factors, Cd- and Cu-substituted zinc-fingers in transcription-activating factors, and functions in higher plants require critical investigation.

Various MT genes - mouse *MTI*, human *MTIA*, human *MTII*, Chinese hamster *MTII*, yeast *CUPI*, pea *PsMTA* - have been transferred to *Nicotiana* sp., *Brassica* sp. or *Arabisopsis thaliana*. Varying degrees of constitutively enhanced Cd tolerance have been achieved, being maximally 20-fold. Metal uptake was not markedly altered, in some cases there was no difference, but 70% less or 60% more Cd maximally was taken up by the shoots or leaves in other cases. Karenlampi et al. (2000) isolated an MT gene from metal-tolerant *Silene vulgaris* and transferred it into several metal-sensitive yeast species. It showed increases in both Cd and Cu tolerance in the modified yeasts. This suggests that an MT gene may be useful in improving metal tolerance of plants, but it did not seem to have significant effects on uptake or translocation of metals.

3.14 CO-STRESS MANIFESTATIONS AND INVOLVEMENT OF OTHER PROTEINS AND METAL TRANSPORTERS IN HEAVY METAL TOLERANCE

Plants often exhibit multiple stress resistance mechanisms. When pretreated with HMs, seedlings of different plants acquired tolerance to lethal and sub-lethal temperatures. Conversely, when exposed to heat, the heat-shock proteins synthesized prevented cellular damage against HM toxicity (Prasad 1997). Thermoprotection by heavy metals by heat-shock cognates, the role of heat-shock proteins in protecting membranes from toxic HMs in plants, is a manifestation of co-stress which evokes coping mechanisms for vice versa stresses, resulting in adaptation at the cellular and molecular level (Lesham and Kuiper 1996; Prasad 1997).

It has been shown that some nuclear proteins bind in the promoter region of a small *Hsp* gene, causing its expression (Czarnecka et al. 1984). Cadmium induction of several proteins and similar induction by other stresses show that they share a common induction mechanism, i.e. these stresses must activate DNA binding nuclear protein(s) synthesis and cause stress-related gene(s) expression, or activate the existing factors to bind DNA. Cadmium induction of *Hsp* may be due to activation of the heat-shock transcription factors. Presence of any transcription activation factors and their induction/activation by Cd needs critical investigation.

Eide et al. (1996) and Grotz et al. (1998) pointed out that several zinc and iron transporters such as ZIP1, ZIP3 and IRT1 are expressed in response to metal deficiency, and IRT1 may also play a role in the uptake of metals such as Cd, which could inhibit Fe uptake of IRT1. Changing the regulation of the expression of these transporters may modify the uptake of metals to the cells or organelles in a useful way.

Overexpression of proteins involved in intracellular metal sequestration (MTs, phytochelatin synthase, vacuolar transporters) may significantly increase metal tolerance, but may not be useful for metal accumulation. These proteins presumably improve accumulation only by delaying the metal-responsive transcriptional down-regulation of plasma membrane transporter expression. Substantially enhanced accumulation may only be achieved by overexpression of plasma membrane transporters put under the control of nonmetal-responsive promoters (Karenlampi et al. 2000).

Schat et al. (2000) concluded that the tolerance, uptake and root-to-shoot transport of metals in the hyperaccumulator *Thlaspi caerulescens* are subject to uncorrelated variation, even with respect to the patterns of metal specificity. It would allow for the selection and breeding of varieties with useful combinations of properties. Tolerance and accumulation are largely independent properties in *Thlaspi caerulescens*; they should both be engineered to produce a suitable plant for phytoextraction.

Some proteins involved in heavy metal tolerance including MTs, PCs and IRT1 may be good sources for engineering. Overexpression of proteins, however, involved in intracellular metal sequestration (MTs, phytochelatin synthase, vacuolar transporters) may significantly increase metal tolerance, but may not be useful for metal accumulation. These proteins presumably improve accumulation only by delaying the metal-responsive transcriptional down-regulation of plasma membrane transporter expression. Substantially enhanced accumulation may only be achieved by overexpression of plasma membrane transporters put under the control of nonmetal-responsive promoters. The tolerance, uptake and root-to-shoot transport of metals in the hyperaccumulator *Thlaspi caerulescens* are subject to uncorrelated variation.

Different combinations of genes involved in heavy metal tolerance will be transferred into *Thlaspi caerulescens*. First, these combinations will be transferred into yeasts for screening possibly effective gene combinations. Then, those genes will be transferred into the target plant, *Thlaspi caerulescens*, and be tested in the laboratory and at field scale.

ACKNOWLEDGEMENTS

Thanks are due to funding agencies, viz., the Ministry of Environment and Forests, Department of Science and Technology, Council of Scientific and Industrial Research, and the New Delhi, Government of India, that supported trace metal research in the author's laboratory.

REFERENCES

Abrahamson SL, Speiser DM, Ow DW (1992) A gel electrophoresis assay for phytochelatins. Anal Biochem 20: 239-243

Agrawal SB, Agrawal M, Lee EH, Kramer GF, Pillai P (1992) Changes in polyamines and gluta-thione content of green alga, *Chlorella elongatum* (Dang) France exposed to mercury. Environ Exp Bot 32: 145-151

Ahner BA, Morel FMM (1995) Phytochelatin production in marine algae. 2. Induction by various metals. Limnol Oceanogr 40: 658-665

Ahner BA, Kong S, Morel FMM (1995) Phytochelatin production in marine algae. 1. An interspecies comparison. Limnol Oceanogr 40: 649-657

Arduini I, Godbold DL, Onnis A (1996) Cadmium and copper uptake and distribution in Mediter-ranean tree seedlings. Physiol Plant 97: 111-117

Baker AJM, Brooks RR (1989) Terrestrial higher plants which accumulate metallic elements - a review of their distribution, ecology and phytochemistry. Biorecovery 1: 81-126

Baker AJM, Grant CJ, Martin MH, Shaw SC, Whitebrook J (1986) Induction and loss of cadmium tolerance in *Holcus lanatus* L. and other grasses. New Phytol 102: 575-587

Barbas J, Santhanagopalan V, Blaszczynski M, Ellis WR Jr, Winge DR (1992) Conversion in the peptide coating cadmium: sulfide crystallites in *Candida glabrata*. J Inorg Biochem 48: 95-105

Basu U, Godbold D, Taylor GJ (1994) Aluminum resistance in *Triticum aestivum* associated with enhanced exudation of malate. J Plant Physiol 144: 747-753

Berger JM, Jackson PJ, Robinson NJ, Lujan LD, Delhaize E (1989) Precursor-product relationships of poly(g-glutamylcysteinyl) glycine biosynthesis in *Datura innoxia*. Plant Cell Rept 7: 632-635

Berlett RJ, Riego DC (1972) Effect of chelation on the toxicity of aluminum. Plant Soil 37: 419-423

Briat JF, Massenet O, Laulhere JP (1990) Purification and characterization of an iron-induced fer-ritin from soybean (*Glycine max*) cell suspensions. J Biol Chem 264: 11550-11553

Brookes A, Collins JC, Thurman DA (1981) The mechanism of zinc tolerance in grasses. J Plant Nutr 3: 695-705

Brooks RR, Malaisse F (1989) Metal-enriched sites in south central Africa. In: Shaw AJ (ed) Heavy metal tolerance in plants: evolutionary aspects. CRC Press, Boca Raton, pp. 53-73

Brooks RR, Shaw S, Asensi Marfil A (1981) The chemical form and physiological function of nickel in some Iberian *Alyssum* species. Physiol Plant 51: 167-17

Brune A, Urback W, Dietz K-J (1994) Compartmentation and transport of zinc in barley primary leaves as basic mechanisms involved in zinc tolerance. Plant Cell Environ 17: 153-162

Buchanan-Wallston V (1994) Isolation of cDNA clones for genes that are expressed during leaf senescence in *Brassica napus*. Identification of a gene encoding a senescence-specific metallothionein-like protein. Plant Physiol 105: 839-846

Cai XH, Traina SJ, Logan TJ, Gustafson T, Sayre R (1995) Application of eukaryotic algae for the removal of heavy metals from water. Mol Mar Biol Biotechol 4: 338-344

Chen J, Goldsbrough PB (1993) Characterization of cadmium tolerant tomato cell line (abstract no 933). Plant Physiol 102: S162

Chen J, Goldsbrough PB (1994) Increased activity of γ-glutamylcysteine synthetase in tomato cells selected for cadmium tolerance. Plant Physiol 106: 233-239

Chen J, Zhou J, Goldsbrough P (1997) Characterization of phytochelatin synthase from tomato. Physiol Plant 101: 165-172

Crichton RR, Ortiz YP, Koch MHJ, Parfait R, Sturmann BH (1978) Isolation and characterization of phytoferritin from pea (*Pisum sativum*) and lentil (*Lens esculenta*). Biochem J 171: 349-356

Czarenecka E, Edelman L, Schoffl F, Key JL (1984) Comparative analysis of physical stress responses in soybean seedlings using cloned heat-shock cDNAs. Plant Mol Biol 3: 45-58

Dameron CT, Smith BR, Winge DR (1989) Glutathione-coated cadmium-sulfide crystallites in *Candida glabrata*. J Biol Chem 264:17355-17360

De Knecht JA, van Dillen M, Koevoets PLM, Schat H, Verkleij JAC, Ernst WHO (1994) Phytochelatins in cadmium-sensitive and cadmium-tolerant *Silenevulgaris*. Plant Physiol 104: 255-261

Delhaize E, Ryan PR (1995) Aluminum toxicity and tolerance in plants. Plant Physiol 107: 315-321

Delhaize E, Jackson PJ, Lujan LD, Robinson NJ (1989) Poly(γ-glutamylcysteinyl) glycine synthesis in *Datura innoxia* and binding with cadmium. Plant Physiol 89: 700-706

Delhaize E, Craig S, Beaton CD, Bennet RJ, Jagadish VC, Randall PJ (1993a) Aluminum tolerance in wheat (*Triticum aestivum* L.). I. Uptake and distribution of aluminum in root apices. Plant Physiol 103: 685-693

Delhaize E, Ryan PR, Randall PJ (1993b) Aluminum tolerance in wheat (*Triticum aestivum* L.). II. Aluminum-stimulated excretion of malic acid from root apices. Plant Physiol 103: 695-702

De Miranda JR, Thomas MA, Thurman DA, Tomsett AB (1990) Metallothionein genes from the flowering plant *Mimulus guttatus*. FEBS Lett 260: 277-280

De Vos CHR, Vonk MJ, Vooijs R, Schat H (1992) Glutathione depletion due to copper-induced phytochelatin synthesis causes oxidative stress in *Silene cucubalus*. Plant Physiol 98: 853-858

Dinkelaker B, Romheld V, Marschner H (1989) Citric acid excretion and precipitation of calcium citrate in the rhizosphere of white lupin (*Lupinus albus* L.). Plant Cell Environ 12: 285-292

Eanetta NT, Steffens JC (1989) Labile sulfide and sulfite in phytochelatin complexes. Plant Physiol 89: 76

Eide D, Broderius M, Fett J, Guerinot ML (1996) A novel iron-regulated metal transporter from plants identified by functional expression in yeast. Proc Natl Acad Sci USA 93: 5624-5628

Ernst WHO (1972) Ecophysiological studies on heavy metal plants in south central Africa. Kirkia 8: 125-145

Ernst WHO (1976) Physiological and biochemical aspects of metal tolerance. In: Mansfield TA (ed) Effects of air pollutants in plants. Cambridge University Press, Cambridge, pp 115-133

Ernst WHO (1990) Poly g-glutamylcysteinyl glycines or phytochelatins and their role in cadmium tolerance of *Silene vulgaris*. Plant Cell Physiol 13: 913-921

Ernst WHO, Verkleij JAC, Schat H (1992) Metal tolerance in plants. Acta Bot Neerl 41: 229-248

Evans IM, Gatehouse LN, Gatehouse JA, Robinson NJ, Coy RRD (1990) A gene from pea (*Pisum sativum*) and its homology to metallothionein genes. FEBS Lett 262: 29-32

Fuente JM, Ramirez-Rodriguez V, Cabrera-Ponce JL, Herrera-Estrella L (1997) Aluminum tolerance in transgenic plants by alteration of citrate synthesis. Science 276: 1566-1568

Gardner WK, Barber DA, Parbery DG (1983) The acquisition of phosphorus by *Lupinus albus* L. III. The probable mechanism by which phosphorus movement in the soil/root interface is enhanced. Plant Soil 70: 107-124

Gawel JE, Ahner BA, Friedland AJ, Morel FMM (1996) Role for heavy metals in forest decline indicated by phytochelatin measurements. Nature 381: 64-65

Gekeler W, Grill WE, Winnacker E-L, Zenk MH (1988) Algae sequester heavy metals via synthesis of phytochelatin complexes. Arch Microbiol 150: 197-202

Gekeler W, Grill E, Winnacker E-L, Zenk MH (1989) Survey of the plant kingdom for the ability to bind heavy metals through phytochelatins. Z Naturforsch 44c: 361-369

Glaeser H, Coblenz A, Kruczek R, Ruttke I, Ebert-Jung A, Wolf K (1991) Glutathione metabolism and heavy metal detoxification in *Schizosccharomyces pombe*: isolation and characterization of glutathione-deficient, cadmium-sensitive mutants. Curr Genet 19: 207-213

Godbold DL, Horst WJ, Collins JC, Thurman DA, Marschner H (1984) Accumulation of zinc and organic acids in roots of zinc tolerant and non-tolerant ecotypes of *Deschampsia caespitosa*. J Plant Physiol 116: 59-69

Grotz N, Fox T, Connolly E, Park W, Gcuerinot ML, Eide D (1998) Identification of a family of zinc transporter genes from *Arabidopsis thaliana* that respond to zinc deficiency. Proc Natl Acad Sci USA 95: 7220-7225

Grill E, Zenk MH (1985) Induction of heavy metal sequestering phytochelatin by cadmium in cell cultures of *Rauvolfia serpentina*. Naturwissenschaften 72: 432-433

Grill E, Winnacker EL, Zenk MH (1985) Phytochelatins, the principal heavy-metal complexing peptides of higher plants. Science 230: 674-676

Grill E, Gekeler W, Winnacker EL, Zenk MH (1986) Homo-phyto-chelatins are heavy metal binding peptides of homo-glutathione containing fabales. FEBS Lett 205: 47-50

Grill E, Winnacker EL, Zenk MH (1987) Phytochelatins, a class of heavy-metal binding peptides from plants are functionally analogous to metallothioneins. Proc Nat Acad Sci USA 84: 439-443

Grill E, Löffler S, Winnacker EL, Zenk MH (1989) Phytochelatins, the heavy-metal binding peptides of plants, are synthesised from glutathione by a specific γ-glutamyl-cysteine dipeptidyl transpeptidase phytochelatin synthase. Proc Natl Acad Sci USA 86: 6838-6842

Gruen LC (1975) Interaction of amino acids with silver(I) ions. Biochem Biophys Acta 386:270-274

Guerinot ML, Salt DE (2001) Fortified foods and phytoremediation. Two sides of the same coin. Plant Physiol 125: 164-167

Gupta SC, Goldsbrough PB (1990) Phytochelatin accumulation and stress tolerance in tomato cells exposed to cadmium. Plant Cell Rep 9: 466-469

Gupta SC, Goldsbrough PB (1991) Phytochelatin accumulation and cadmium tolerance in selected tomato cell lines. Plant Physiol 97: 306-312

Hamer DH (1986) Metallothionein. Annu Rev Biochem 55: 913-951

Hartley-Whitaker J, Ainsworth G, Vooijs R, Ten Bookum W, Schat H, Meharg AA (2001) Phytochelatins are involved in differential arsenate tolerance in *Holcus lanatus*. Plant Physiol 126: 299-306

Hayashi Y, Nakagawa CW, Uyakul D, Imai K, Isobe M, Goto T (1988) The change of cadystin components in Cd-binding peptides from the fission yeast during their induction by cadmium. Biochem Cell Biol 66: 288-295

Hayashi Y, Nakagawa CW, Mutoh N, Isobe M, Goto T (1991) Two pathways in the biosynthesis of cadystins (γ-EC)ₙG in the cell free system of the fission yeast. Biochem Cell Biol 69: 115-121

Henderson M, Ownby JD (1991) The role of root cap mucilage secretion in aluminum tolerance in wheat. Curr Top Plant Biochem Physiol 10: 134-141

Heuillet E, Moreau A, Halpern S, Jeanne N, Puiseux-Dao S (1986) Cadmium binding to a thiol-molecular in vacuoles of *Dunaliella bioculata* contaminated with $CdCl_2$: electron probe analysis. Biol Cell 58: 79-86

Howden R, Cobbett CS (1992) Cadmium-sensitive mutants of *Arabidopsis thaliana*. Plant Physiol 99: 100-107

Howden R, Anderson CR, Goldsbrough PB, Cobbett CS (1995a) A cadmium-sensitive, glutathione-deficient mutant of *Arabidopsis thaliana*. Plant Physiol 107: 1067-1073

Howden R, Goldsbrough PB, Andersen CR, Cobbet CS (1995b) Cadmium-sensitive, cad1 mutants of Arabidopsis thaliana are phytochelatin deficient. Plant Physiol 107: 1059-1066

Huang B, Goldsborough PB (1988) Cadmium tolerance in tobacco cell culture and its relevance to temperature stress. Plant Cell Rep 7: 119-122

Huang B, Hatch E, Goldsbrough PB (1987) Selection and characterization of cadmium tolerant cells in tomato. Plant Sci 52: 211-221

Hue NV, Craddock GR, Adams F (1986) Effect of organic acids on aluminum toxicity in subsoils. Soil Sci Soc Am J 50: 28-34

Jackson PJ, Roth EJ, McClure PR, Naranjo CM (1984) Selection, isolation, and characterization of cadmium resistant *Datura innoxia* suspension cultures. Plant Physiol 75: 914-918

Jackson PJ, Unkefer CJ, Doolen JA, Watt K, Robinson NJ (1987) Poly γ-glutamyl cysteinyl glycine: its role in cadmium resistance in plant cells. Proc Natl Acad Sci USA 84: 6619-6623

Juang R-H, McCue KF, Ow DW (1993) Two purine biosynthetic enzymes that are required for Cd tolerance in Schizosacromyces pombe utilize cysteine sulfinate in vitro. Arch Biochem Biophys 304: 392-401

Kagi JHR, Schaffer A (1988) Biochemistry of metallothionein. Biochemistry 27: 8509-8515

Karenlampi S, Schat H, Vangronsveld J, Verkleij JAC, van der Lelie D, Mergeay M, Tervahauta AI (2000) Genetic engineering in the improvement of plants for phytoremediation of metal polluted soils. Environ Pollut 107: 225-231

Kawashima I, Inokuchi Y, Chino M, Kimura M, Shimizu N (1991) Isolation of a gene for a metallothionein protein from soybean. Plant Cell Physiol 32: 913-916

Kawashima I, Kennedy TD, Chino M, Lane BG (1992) Wheat Ec metallothionein genes like mammalian Zn^{2+} metallothionein genes, wheat Zn^{2+} metallothionein genes are conspicuously expressed during embryogenesis. Eur J Biochem 209: 971-976

Klapheck S, Fliegner W, Zimmer I (1994) Hydroxymethyl-phytochelatins [(g-glutamylcysteine)$_n$-serine] are metal-induced peptides of the Poaceae. Plant Physiol 104: 1325-1332

Klapheck S, Schlunz S, Bergmann L (1995) Synthesis of phytochelatins and homo-phytochelatins in Pisum sativum L. Plant Physiol 107: 515-521

Kneer R, Zenk MH (1992) Phytochelatins protect plant enzymes from heavy-metal poisoning. Phytochemistry 31: 2663

Kneer R, Kutchan TM, Hochberger A, Zenk MH (1992) Saccharomyces cerevisiae and Neurospora crassa contain heavy metal sequestering phytochelatin. Arch Microbiol 157: 305-310

Kollmeier M, Dietrich P, Bauer CS, Horst WJ, Hedrich R (2001) Aluminum activates a citrate-permeable anion channel in the aluminum-sensitive zone of the maize root apex. A comparison between an aluminum-sensitive and an aluminum-resistant cultivar. Plant Physiol 126: 397-410

Kondo N, Isobe M, Imai K, Goto T, Murasugi A, Hayashi Y (1983) Structure of cadystin, the unit-peptide of cadmium-binding peptides induces in a fission yeast, Schizosaccharomyces pombe. Tetrahedron Lett 24: 925-928

Kondo N, Imai K, Isobe M, Goto T, Murasugi A, Wada-Nakagawa C, Hayashi Y (1984) Cadystin A and B, major unit peptides comprising cadmium binding peptides induced in a fission yeast - separation, revision of structures and synthesis. Tetrahedron Lett 25: 3869-3872

Kortz RM, Evangelou BP, Wagner GJ (1989) Relationships between cadmium, zinc, cd-peptide and organic acid in tobacco suspension cells. Plant Physiol 91: 780-787

Kosakowasaka A, Falkowski L, Lewandowska J (1988) Effect of amino acids on the toxicity of heavy metals to phytoplankton. Bull Environ Contam Toxicol 40: 532-538

Koyama H, Ojima K, Yamaya T (1990) Utilization of anhydrous aluminum phosphate as a sole source of phosphorus by a selected carrot cell line. Plant Cell Physiol 31: 173-177

Krämer U, Smith RD, Wenzel WW, Raskin I, Salt DE (1997) The role of metal transport and tolerance in nickel hyperaccumulation by Thlaspi goesingense Halacsy. Plant Physiol 115: 1641-1650

Krämer U, Cotter-Howells JD, Charnock JM, Baker AJM, Smith JAC (1996) Free histidine as a metal chelator in plants that accumulate nickel. Nature 379: 635-638

Kumar TR, Prasad MNV (2000) Partial purification and characterization of ferritin from Vigna mungo (Black gram) seeds. J Plant Biol 27: 241-246

Larsen PB, Degenhardt J, Tai C-Y, Stenzler LM, Howell SH, Kochian LV (1998) Aluminum-resistant Arabidopsis mutants that exhibit altered patterns of aluminum accumulation and organic acid release from roots. Plant Physiol 117: 9-18

Laulhere JP, Laboure AM, Briat JF (1988) Purification and characterization of ferritin from maize, pea and soybean seeds. J Biol Chem 263: 10289-10294

Leblova S, Mucha A, Spirhanzova E (1986) Compartmentation of cadmium, copper, lead and zinc in seedlings of maize (*Zea mays* L.) and induction of metallothionein. Biologia 41: 777-785

Lefévbre DD, Miki BL, Laliberte JF (1987) Mammalian metallothionein functions in plants. Biotechnology 5: 1053-1056

Leshem YY, Kuiper PJC (1996) Is there a GAS (general adaptation syndrome) response to various types of environmental stress? Biol Plant 36: 1-18

Lipton DS, Blanchar RW, Blevins DG (1987) Citrate, malate, and succinate concentration in exudates from P-sufficient and P-stressed *Medicago sativa* L. seedlings. Plant Physiol 85: 315-317

Lobinski R, Potin-Gautier M (1998) Metals and biomolecules - bioinorganic analytical chemistry Analusis 26: 21-24

Löffler S, Hochberger A, Grill E, Winnacker EL, Zenk MH (1989) Termination of the phytochelatin synthase reaction through sequestration of heavy metals by the reaction product. FEBS Lett 258: 42-46

Ma JF, Hiradate S, Matsumoto H (1998) High aluminum resistance in buckwheat. II. Oxalic acid detoxifies aluminum internally. Plant Physiol 117: 753-759

Ma JF, Zheng SJ, Matsumoto H (1997a) Specific secretion of citric acid induced by Al stress in Cassia tora L. Plant Cell Physiol 38: 1019-1025

Ma JF, Zheng SJ, Hiradate S, Matsumoto H (1997b) Detoxifying aluminum with buckwheat. Nature 390: 569-570

Ma JF, Kiradate S, Nomoto K, Iwashita T, Matsumoto H (1997c) Internal detoxification mechanism of Al in hydrangea. Identification of Al form in the leaves. Plant Physiol 113: 1033-1039

Macnair MR (1977) Major genes for copper tolerance in *Mimulus guttatus*. Nature 268:428-430

Maitani T, Kubota H, Sato K, Yamada T (1996) The composition of metals bound to class II metallothionein (phytochelatin and its desglycyl peptide) induced by various metals in root cultures of *Rubia tinctorum*. Plant Physiol 110: 1145-1150

Maiti IB, Wagner GJ, Yeargen R, Hunt AG (1989). Inheritance and expression of the mouse metallothionein gene in tobacco. Impact on Cd tolerance and tissue Cd distribution in seedlings. Plant Physiol 91: 1020-1024

Mallick N, Rai LC (2002) *Physiological responses of non-vascular plants to heavy metals. In: Prasad MNV, Strzałka K (eds) Physiology and biochemistry of metal toxicity and tolerance in plants.* Kluwer Academic, Dordrecht, pp 111-147

Matsumoto H (2002) Metabolism of organic acids and metal tolerance in plants exposed to aluminum. In: Prasad MNV, Strzałka K *(eds) Physiology and biochemistry of metal toxicity and* tolerance in plants. Kluwer Academic, Dorrdrecht, pp 95-109

Mendum LM, Gupta SC, Goldsbrough PB (1990) Effect of glutathione on phytochelatin synthesis in tomato cells. Plant Physiol 93: 484-488

Meuwly P, Thibault P, Schwan AL, Rauser WE (1995) Three families of thiol peptides are induced by cadmium in maize. Plant J 7: 391-400

Miller J, McLachlan AD, Klug A (1985) Repetitive zinc-binding domains in the protein transcription factor IIA from *Xenopus* oocytes. EMBO J 4: 1609-1614

Misra S, Gedamu L (1989) Heavy metal tolerant transgenic *Brassica napus* L. and *Nicotiana tabacum* L. plants. Theoret Appl Genet 78: 161-168

Miyasaka SC, Hawes MC (2001) Possible role of root border cells in detection and avoidance of aluminum toxicity. Plant Physiol 125: 1978-1987

Miyasaka SC, Kochian LV, Shaff JE, Foy CD (1989) Mechanism of aluminum tolerance in wheat. An investigation of genotypic differences in rhizosphere pH, K^+, and H^+ transport, and root-cell membrane potentials. Plant Physiol 91: 1188-1196

Miyasaka SC, Buta JG, Howell RK, Foy CD (1991) Mechanism of aluminum tolerance in snapbeans. Root exudation of citric acid. Plant Physiol 96: 737-743

Moffett JW, Brand LE (1996) Production of strong, extracellular Cu chelators by marine cyanobacteria in response to Cu stress. Limnol Oceanogr 41: 388-395

Muchovej RCM, Allen VG, Martens DC, Muchovej JJ (1988) Effects of aluminum chelates in nutrient solution on the growth and composition of ryegrass. J Plant Nutr 11: 117-129

Mullins M, Hardwick K, Thurman DA (1985) Heavy metal location by analytical electron microscopy in conventionally fixed and freeze-substituted roots of metal tolerant and non-tolerant ecotypes. In Lekkas TD (ed) Heavy metals in the environment (Athens 1985). CEP Consultants, Edinburgh, pp 43-50

Murasugi A, Wada C, Hayashi Y (1981a) Cadmium-binding peptide induced in fission yeast *Schizosaccharomyces pombe*. J Biochem 90: 1561-1564

Murasugi A, Wada C, Hayashi Y (1981b) Cadmium-binding peptide induced in fission yeast *Schizosacharomyces pombe*. J Biochem 103: 1021-1028

Murasugi A, Wada C, Hayashi Y (1983) Occurrence of acid labile sulfide in cadmium-binding peptide 1 from fission yeast. J Biochem 93: 661-664

Mutoh N, Hayashi Y (1988) Isolation of mutants of *Schizosaccharomyces pombe* unable to synthesize cadystin, small cadmium-binding peptides. Biochem Biophys Res Commun 151: 32-39

Nordstrom DK, May HM (1996) Aqueous equilibrium data for mononuclear aluminum species. In: Sposito (ed) Environment chemistry of aluminum. CRC Press, Boca Raton, pp 39-80

Nussbaum S, Schmutz D, Brunold C (1988) Regulation of assimilatory sulfate reduction by cadmium in *Zea mays* L. Plant Physiol 88: 1407-1410

Odum HT (2000) Heavy metals in the environment using wetlands for their removal. CRC Press, Boca Raton

Ojima K, Ohira K (1988) Aluminum-tolerance and citric acid release from a stress-selected cell line of carrot. Commun Soil Sci Plant Anal 19: 1229-1238

Ojima K, Abe H, Ohira K (1984) Release of citric acid into the medium by aluminum-tolerant carrot cells. Plant Cell Physiol 25: 855-858

Ojima K, Koyama H, Suzuki R, Yamaya T (1989) Characterization of two tobacco cell lines selected to grow in the presence of either ionic Al or insoluble Al-phosphate. Soil Sci Plant Nutr 35: 545-551

Oritz DF, Ruscitti T, McCue KF, Ow DW (1995). Transport of metal-binding peptides by HMT1, a fission yeast ABC-type B vacuolar membrane protein. J Biol Chem 270: 4721-4728

Osawa H, Matsumoto H (2001) Possible involvement of protein phosphorylation in aluminum-responsive malate efflux from wheat root apex. Plant Physiol 126: 411-420

Ownby JD, Popham HP (1989) Citrate reverses the inhibition of wheat root growth caused by aluminium. J Plant Physiol 135: 588-591

Pellet DM, Grunes DL, Kochian LV (1995) Organic acid exudation as an aluminum-tolerance mechanism in maize (Zea mays L.). Planta 196: 788-795

Pellet DM, Papernik LA, Kochian LV (1996) Multiple aluminum-resistance mechanisms in wheat: roles of root apical phosphate and malate exudation. Plant Physiol 112: 591-597

Pelosi P, Fiorentini R, Galoppini C (1976) On the naure of nickel compounds in *Alyssum bertolonii* Desv. Agric Biol Chem 40: 1641-1642

Prasad MNV (1995) Cadmium toxicity and tolerance in vascular plants. Environ Exp Bot 35: 525-540

Prasad MNV (1997) Trace metals. In: Prasad MNV (ed) Plant ecophysiology. Wiley, New York, pp 207-249

Prasad MNV (1998) Metal-biomolecule complexes in plants: occurrence, functions and applications. Analusis 26: 25-28

Rauser WE (1999) Structure and function of metal chelators produced by plants: the case for organic acids, amino acids, phytin and metallothioneins. Cell Biochem Biophys 31: 19-46

Rauser WE (1984a) Isolation and partial purification of cadmium-binding protein from roots of the grass *Agrostis gigantea*. Plant Physiol 74: 1025-1029

Rauser WE (1984b) Estimating metallothionein in small root samples of *Agrostis gigantea* and *Zea mays* exposed to cadmium. J Plant Physiol 116: 253-260

Rauser WE (1986) The amount of cadmium associated with Cd-binding protein in roots of *Agrostis gigantea*, maize and tomato. Plant Sci 43: 85-91

Rauser WE (1990a) Changes in glutathione and phytochelatins in roots of maize seedlings exposed to cadmium. Plant Sci 70: 155-166

Rauser WE (1990b) Phytochelatins. Annu Rev Biochem 59: 61-86

Rauser WE (1995) Phytochelatins and related peptides. Structure, biosynthesis, and function. Plant Physiol 109: 1141-1149

Rauser WE (2000) The role of thiols in plants under metal stress. In: Brunold C, Rennenberg H, De Kok LJ (eds) Sulfur nutrition and sulfur assimilation in higher plants. Haupt, Bern, pp 169-183

Rauser WE, Ackerley CA (1987) Localization of cadmium in granules within differentiating and mature root cells. Can J Bot 65: 643-646

Rauser WE, Curvetto NR (1980) Metallothionein occurs in roots of *Agrostis* tolerant to excess copper. Nature 287: 563-564

Rauser WE, Meuwly P (1995) Retention of cadmium roots of maize seedlings. Role of complexation by phytochelatins and related thiol peptides. Plant Physiol 109: 195-202

Reddy GN (1992) Heavy metal inducible proteins and metallothionein genes in higher plants. Biochem Arch 8: 87-93

Reddy GN, Prasad MNV (1990) Heavy metal binding proteins peptides, occurrence, structure, synthesis and functions review. Environ Exp Bot 30: 252-264

Reddy GN, Prasad MNV (1992a) Cadmium induced peroxidase activity and isozymes in *Oryza sativa*. Biochem Arch 8: 101-106

Reddy GN, Prasad MNV (1992b) Characterization of cadmium binding protein from *Scenedesmus quadricauda* and Cd toxicity reversal by phytochelatin constituting amino acids and citrate. J Plant Physiol 140: 156-162

Reddy GN, Prasad MNV (1993) Tyrosine is not phosphorylated in cadmium induced hsp70 cognate in maize (*Zea mays* L.) seedlings. Role in chaperone function? Biochem Arch 9: 25-32

Reese NR, White CA, Winge DR (1992) Cadmium-sulfide crystallites in Cd-$(\gamma EC)_nG$ peptide complexes from tomato. Plant Physiol 98: 225-229

Reese RN, Wagner GJ (1987) Effects of buthionine sulfoximine on Cd-binding peptide levels in suspension-cultured tobacoo cells treated with Cd, Zn, or Cu. Plant Physiol 84: 574-577

Reese RN, Winge DR (1988) Sulfide stabilization of the cadmium γ-glutamyl peptide complex of *Schizosaccharomyces pombe*. J Biol Chem 262: 112832-112835

Reese RN, Mehra RK, Tarbet EB, Winge DR (1988) Studies on the γ-glutamyl Cu-binding peptide from *Schizosaccharomyces pombe*. J Biol Chem 263: 4186-4192

Reeves RD (1992) The hyperaccumulation of nickel by serpentine plants. In: Baker AJM, Proctor J, Reeves RD (eds) The vegetation of ultramafic (serpentine) soils. Intercept, Andover, Hampshire, UK, pp 253-277

Reeves RD, Baker AJM, Brooks RR (1995) Abnormal accumulation of trace metals by plants. Mining Environ Manage 3: 4-8

Rengel Z (1997) Mechanisms of plant resistance to aluminium and heavy metals. In: Basra AS, Basra RK (eds) Mechanisms of environmental stress resistance in plants. Harwood Academic Publ, Amsterdam, pp 241-276

Robinson NJ, Jackson PJ (1986) «Metallothionein-like» metal complexes in angiosperms, their structure and function. Physiol Planta 67: 499-506

Robinson NJ, Barton K, Naranjo CM, Sillerud LO, Trewhella J, Watt K, Jackson PJ (1987) Characterization of metal peptides from cadmium resistant plant cells. Experentia 52: 323-327

Ruegsegger A, Brunold C (1992) Effect of cadmium on γ-glutamylcysteine synthesis in maize seedlings. Plant Physiol 99: 428-433

Ruegsegger A, Schmutz D, Brunold C (1990) Regulation of glutathione synthesis by cadmium in *Pisum sativum* L. Plant Physiol 93: 1579-1584

Ryan PR, Delhaize E, Randall PJ (1995a) Characterization of Al-stimulated efflux of malate from the apices of Al-tolerant wheat roots. Planta 196: 103-110

Ryan PR, Delhaize E, Randall PJ (1995b) Malate efflux from root apices: evidence for a general mechanism of Al-tolerance in wheat. Aust J Plant Physiol 22: 531-536

Salt DE, Rauser WE (1995) MgATP-dependent transport of phytochelatins across the tonoplast of oat roots. Plant Physiol 107: 1293-1301

Salt DE, Wagner GJ (1993) Cadmium transport across tonoplast of vesicles from oat roots. Evidence for Cd^{2+}/H^{+} antiport activity. J Biol Chem 268: 12297-12302

Sanità Di Toppi L, Prasad MNV, Ottonello S (2002) Metal chelating peptides and proteins in plants. In: Prasad MNV, Strzalka K (eds) Physiology and biochemistry of metal toxicity and tolerance in plants. Kluwer Academic Publ, Dordrecht, pp 59-93

Sarkar B (1997) Zinc finger-DNA interactions: effect on metal replacement, free radical generation and DNA damage and its relevance to carcinogenesis. In: Hadjiliadis ND (ed) Cytotoxic, mutagenic and carcinogenic potential of heavy metals related to human environment. Kluwer Academic Publ, Dordrecht. NATO ASI ser 2. Environ 26: 1-14

Schat H, Kalff MMA (1992) Are phytochelatins involved in differential metal tolerance or do they merely reflect metal imposed strain? Plant Physiol 99: 1475-1480

Schat H, Ten Bookum WM (1992) Genetic control of copper tolerance in *Silene vulgaris*. Heredity 68:219-229

Schat H, Kalff MMA (1992) Are phytochelatins involved in differential metal tolerance or do they merely reflect metal-imposed strain? Plant Physiol 99: 1475-1480

Schat M, Llugany M, Bernhard R (2000) Metal-specific patterns of tolerance, uptake and transport of heavy metals in hyperaccumulating and nonhyperaccumulating metallophytes. In: Norman T, Banuelos G (eds) Phytoremediation of contaminated soil and water, CRC Press, Boca Raton, 184 pp

Schmöger MEV, Oven M, Grill E (2000) Detoxification of arsenic by phytochelatins in plants. Plant Physiol 122: 793-801

Scheller HV, Huang B, Hatch E, Goldsbrough PB (1987) Phytochelatin synthesis and glutathione levels in response to heavy metals in tomato cells. Plant Physiol 85: 1031-1035

Shuman LM, Wilson DO, Ramseur EL (1991) Amelioration of aluminum toxicity to sorghum seedlings by chelating agents. J Plant Nutr 14: 119-128

Speiser DM, Abrahamsom SL, Banuelos G, Ow DW (1992a) *Brassica juncea* produces a phytochelatin-cadmium-sulfide complex. Plant Physiol 99: 817-821

Speiser DM, Ortiz DF, Kreppel L, Scheel G, McDonald G, Ow DW (1992b) Purine biosynthetic genes are required for cadmium tolerance in *Schizosaccharomyces pombe*. Mol Cell Biol 12: 5301-5310

Steffens JC (1990) The heavy metal-binding peptides of plants. Annu Rev Plant Physiol Plant Mol Biol 41: 553-575

Steffens JC, Hunt DF, Williams BG (1986) Accumulation of non-protein metal-binding poly-peptides γ-glutamyl-cysteinyl glycine in selected cadmium-resistant tomato cells. J Biol Chem 261: 13879-13882

Stillman MJ (1995) Metallothioneins. Co-ordination. Chem Rev 144: 461-511

Strasdeit H, Duhme A-K, Kneer R, Zenk MH, Hermes C, Nolting H-F (1991) Evidence for discrete $Cd(Scys)_4$ units in cadmium phytochelatin complexes from EXAFS spectroscopy. J Chem Soc Chem Commun 16: 1129-1130

Thumann J, Grill E, Einnackder E-L, Zenk MH (1991) Reactivation of metal-requiring apoenzymes by phytochelatin-metal complexes. FEBS Lett 282: 66-69

Tommey AM, Shi J, Lindsay WP, Urwin PE, Robinson NJ (1991) Expression of the pea gene PsMT$_A$ in *E. coli* meta-binding properties of the expressed protein. FEBS Lett 292: 48-52

Tomsett AB, Thurman DA (1988) Molecular biology of metal tolerance of plants. Plant Cell Environ 11: 388-394

Tripathi RD, Yunus M, Mehra RK (1996) Phytochelatins and phytometallothioneins: the potential of these unique metal detoxifying systems in plants. Physiol Mol Biol 2: 101-104

Tukendorf A, Rauser WE (1990) Changes in glutathione and phytochelatins in roots of maize seedlings exposed to cadmium. Plant Sci 70: 155-166

Van Assche F, Clijsters H (1990) Effects of metals on enzyme activity in plants. Plant Cell Environ 13: 195-206

Verkleij JAC, Koevoets P, Riet JV, Bank R, Nijdam Y, Ernst WHO (1990) Poly-(γ-glutamyl cysteinyl) glycines or phytochelatins and their role in cadmium tolerance of *Silene vulgaris*. Plant Cell Environ 13: 913-921

Vögeli-Lange R, Wagner GJ (1990) Subcellular localization of cadmium-binding peptides in tobacco leaves. Implications of a transport function for cadmium-binding peptides. Plant Physiol 92: 1086-1093

Vögeli-Lange R, Wagner GJ (1996) Relationship between cadmium, glutathione and cadmium-binding peptides (phytochelatins) in leaves of intact tobacco seedlings. Plant Sci 114: 11-18

Vymazal J (1990) Toxicity and accumulation of lead with respect to algae and cyanobacteria: a review. Acta Hydrochim Hydrobiol 18: 531-535

Wang J, Evangelou BP, Nielsen MT, Wagner GJ (1992) Computer-simulated evaluation of possible mechanisms for quenching heavy metal ion activity in plant vacuoles. II. Zinc. Plant Physiol 99: 621-626

Weig A, Komor E (1992) Isolation of a class II metallothionein cDNA from *Ricinus communis* L. GenBank Accession no L02306

Weigel HJ, Jager HJ (1980) Subcellular distribution and chemical form of cadmium in bean plants. Plant Physiol 65: 480-482

White CN, Rivin CJ (1995) Characterization and expression of a cDNA encoding a seed-specific metallothionein in maize. Plant Physiol 108: 831-832

Wingate VPM, Lawton MA, Lamb CJ (1988) Glutathione uses a massive and selective induction of plant defense genes. Plant Physiol 87: 207-210

Wurz H, Tanaka A, Fruton JS (1962) Polymerizaion of dipetide amides by cathepsin C. Biochemistry 1: 19-29

Zhang W-H, Ryan PR, Tyerman SD (2001) Malate-permeable channels and cation channels activated by aluminum in the apical cells of wheat roots. Plant Physiol 125: 1459-1472

Zheng SJ, Ma JF, Matsumoto H (1998a) High aluminum resistance in buckwheat. I. Al-induced specific secretion of oxalic acid from root tips. Plant Physiol 117: 745-751

Zheng SJ, Ma JF, Matsumoto H (1998b) Continuous secretion of organic acids is related to aluminum resistance during relatively long-term exposure to aluminum stress. Physiol Plant 103: 209-214

Zhou J, Goldsbrough PB (1994) Functional homologs of fungal metallothionein genes from *Arabidopsis*. Plant Cell 6: 875-884

Zhou J, Goldsbrough PB (1995) Structure, organization and expression of the metallothionein gene family in *Arabidopsis*. Mol Gen Genet 248: 318-328

Zhu YL, Pilon-Smits AH, Jouanin L, Terry N (1999) Overexpression of glutathione synthetase in Indian mustard enhances cadmium accumulation and tolerance. Plant Physiol 119:73-79

Chapter 4

Heavy Metal Induced Oxidative Damage in Terrestrial Plants

B.P. Shaw, S.K. Sahu, R.K. Mishra
Institute of Life Sciences, Nalco Square, Bhubaneswar 751023, Orissa, India

4.1 METALS

There are 110 elements in the periodic table with the elements from 104 to 110 being of somewhat recent discovery (www.sdfine.com). Of these chemical elements, metals make up the largest group; some 69 of the currently known elements, excluding the trans-uranium series, are metallic in character (Fig. 4.1). Also, out of the 10 most abundant elements in the earth's crust, seven are metals (Table 4.1); aluminum occupies the third place, followed by iron, calcium, sodium, potassium, magnesium and titanium (Mason 1958). Their characteristics, however, differ greatly within the biosphere.

Table 4.1 Common elements in the earth's crust (after Mason 1958)

Elements	Relative abundance (weight per cent)
Oxygen	46.60
Silicon	27.72
Aluminum	8.13
Iron	5.00
Calcium	3.63
Sodium	2.83
Potassium	2.59
Magnesium	2.09
Titanium	0.44
Hydrogen	0.14
Phosphorus	0.12
Manganese	0.1
All other elements	0.61
Total	100.00

11 Na Sodium 0.97	
Element symbol	
Element symbol	
Element name	
Density, 20°C	

s-block | *d*-block | *p*-block

1	2	3	4	5	6	7	8	9	10	11	12	13	14	15	16	17	18
Ia	IIa	III b	IV b	V b	VI b	VII b	VIII			I b	II b	IIIa	IVa	Va	VIa	VIIa	O
1 H Hydrogen 0.084																	2 He Helium 0.17
3 Li Lithium 0.53	4 Be Beryllium 1.85											5 B Boron 2.46	6 C Carbon 2.267	7 N Nitrogen 1.17	8 O Oxygen 1.33	9 F Fluorine 1.58	10 Ne Neon 0.84
11 Na Sodium 0.97	12 Mg Magnesium 1.74											13 Al Aluminium 2.7	14 Si Slicon 2.33	15 P Phosphorus 1.82	16 S Sulphur 2.06	17 Cl Chlorine 2.95	18 Ar Argon 1.66
19 K Potassium 0.86	20 Ca Calcium 1.54	21 Sc Scandium 2.99	22 Ti Titanium 4.51	23 V Vanadium 6.09	24 Cr Chromium 7.14	25 Mn Manganese 7.44	26 Fe Iron 7.87	27 Co Cobalt 8.89	28 Ni Nickle 8.91	29 Cu Copper 8.92	30 Zn Zinc 7.14	31 Ga Gallium 5.91	32 Ge Germanium 5.32	33 As Arsenic 5.72	34 Se Selenium 4.82	35 Br Bromine 3.14	36 Kr Krypton 3.48
37 Rb Rubidium 1.53	38 Sr Strontium 2.63	39 Y Yttrium 4.47	40 Zr Zirconium 6.51	41 Nb Niobium 8.58	42 Mo Molybdenum 10.28	43 Tc Technetium 11.49	44 Ru Ruthenium 12.45	45 Rh Rhodium 12.41	46 Pd Palladium 12.02	47 Ag Silver 10.49	48 Cd Cadmium 8.64	49 In Indium 7.31	50 Sn Tin 7.29	51 Sb Antimony 6.69	52 Te Tellurium 6.25	53 I Iodine 4.94	54 Xe Xenon 5.49
55 Cs Cesium 1.9	56 Ba Barium 3.65	Lanthanides 57-71	72 Hf Hafnium 13.31	73 Ta Tantalum 16.68	74 W Tungsten 19.26	75 Re Rhenium 21.03	76 Os Osmium 22.61	77 Ir Iridium 22.65	78 Pt Platinum 22.65	79 Au Gold 19.32	80 Hg Mercury 13.55	81 Tl Thallium 11.85	82 Pb Lead 11.34	83 Bi Bismuth 9.80	84 Po Polonium 9.20	85 At Astatine	86 Rn Radon 9.23
87 Fr Francium	88 Ra Radium 5.00	Actinides 89-103	104 Rf Rutherfordium	105 Ha Hahnium	106 Unh Unnilhexium	107 Ns Neilsbohrium	108 Hs Hassium	109 Mt Meitnerium	110 Uun Ununnilium								

Fig. 4.1 Periodic table of elements depicting their atomic number and density. The elements of the f-block (lanthanides and actinides not shown) are not shown. The elements generally considered heavy metals, the metals of environmental concern, are *shaded*

The term "metal" designates an element which is a good conductor of electricity and whose electric resistance is directly proportional to absolute temperature. In addition to this distinctive characteristic, metals share several other typical physical properties, such as high thermal conductivity, density, malleability and ductility (Forstner and Wittmann 1979). Several non-metallic elements exhibit one or more of these properties. And, hence, the only feature that defines a metal unambiguously is electric conductivity, which decreases with increase in temperature. There are, of course, elements in the periodic table, like boron, silicon, germanium, arsenic and tellurium, which show electric conductivity, but their electric conductivity is low, and it increases with a rise in temperature. These are termed metalloids (or half-metals) situated between metals and non-metals in the periodic table (Forstner and Wittmann 1979).

4.2 CLASSIFICATION OF METALS

4.2.1 The HSAB Principle

A metal in a chemical reaction reacts as an electron pair acceptor (Lewis acid) with an electron pair donor (Lewis base) to form various chemical groups, such as an ion pair, a metal complex, a coordination compound, or a donor-acceptor complex. The reaction may be generalized as follows:

$$M + L \rightarrow M{:}L$$

where M represents the metal ion, L the ligand, and M:L the product (complex). The stability of the complex will depend on the magnitude of the equilibrium constant, K_{ML}, also called the stability constant.

$$K_{ML} = [ML]/[M][L]$$

The larger the magnitude of K_{ML}, the more stable will be the product (ML) in solution.

Pearson (1968a,b) classified the metal acceptors and the ligand donors into "hard" and "soft" categories to explain the stability of the product complex (see also, http://chemistry.uttyler.edu/~coe/lectures/num16.ppt). The chief criteria for such classification are electron mobility or polarizability (the degree to which the electron cloud is distorted by interaction with a charge or electric field), electron negativity (a measure of the power of an atom to attract an electron to itself in a covalent bonding), and ionic charge density. A hard acceptor is characterized by low polarizability, low electronegativity and large positive charge density (high oxidation state and small radius), and the opposite is true for a soft acceptor. A hard donor, on the other hand, is characterized by low electron mobility or polarizability, but high electronegativity and a high negative charge density, and the reverse constitutes the characteristics of a soft donor. In between lies the intermediate donors and acceptors (Table 4.2).

Experimental evidence suggests that hard acceptors prefer to bind hard donors and soft acceptors prefer to bind soft donors to form stable compounds (Ahrland 1968; Pearson 1968a,b). This is called the HSAB (hard soft acids and bases) principle. The HSAB principle is very much at work in nature: some metals occur in the earth's crust as ores of oxide and carbonate, whereas other metals occur as sulfides. This is because hard acids, like Mg^{2+}, Ca^{2+} and Al^{3+}, form strong bonds with hard bases, like O_2^{2-} or CO_3^{2-}, and, conversely, softer acids, like Hg_2^{2+} or Hg^{2+} and Pb^{2+}, prefer soft bases like S^{2-}.

Table 4.2 Different metal/ligand acceptors and donors (after Pearson 1968a)

	Hard	Intermediate	Soft
Acceptors	H^+, Na^+, K^+, Be^{2+}, Mg^{2+}, Ca^{2+}, Mn^{2+}, Al^{3+}, Cr^{3+}, Co^{3+}, Fe^{3+}, As^{3+}	Fe^{2+}, Co^{2+}, Ni^{2+}, Cu^{2+}, Zn^{2+}, Pb^{2+}	Cu^+, Ag^+, Au^+, Tl^+, Hg_2^{2+}, Pd^{2+}, Cd^{2+}, Pt^{2+}, Hg^{2+}, CH_3Hg^+
Donors	H_2O, OH^-, F^-, Cl^-, PO_4^{3-}, SO_4^{2-}, CO_3^{2-}, O_2^{2-}	Br^-, NO_2^-, SO_3^{2-}	SH^-, S^{2-}, RS^-, CN^-, SCN^-, CO, R_2S, RSH, RS^-

R= alkyl or aryl group

4.2.2 Covalent and Ionic Indices

Nieboer and Richardson (1980) drew up a relationship between the ionic index (Z^2r^{-1}) and the covalent index (X^2_mr) of a metal and metalloid ions and grouped them into three categories, class A, class B and borderline (Fig. 4.2). The result of the classification is more or less similar to that obtained using the HSAB principle, with some exceptions; the class

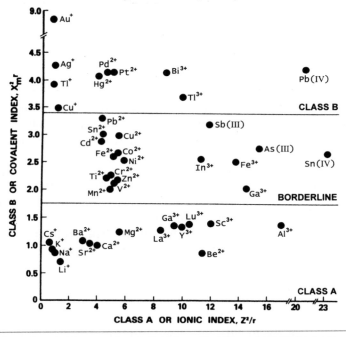

Fig. 4.2 Separation of metal ions and metalloid ions [As(III) and Sb(III)] into three categories: class A, borderline and class B. The class B index X^2_mr is plotted for each ion against the class A index Z^2/r. In these expressions, X_m is the metal ion electronegativity, r its ionic radius and Z its formal charge. (Reproduced from Nieboer and Richardson 1980, with permission of the senior author)

A metals correspond to hard acceptors, class B to soft acceptors and the borderline to that of the intermediate category. A few metals of soft and hard acceptor categories find places in the borderline group instead of in the class B and class A group, respectively (Table 4.2, Fig. 4.2). Nieboer and Richardson (1980) also observed that the three categories of metal ions showed different binding patterns for a ligand, or, in other words, the metals of a group showed similar binding patterns to a ligand, similar to that shown by hard, soft and the intermediate acceptors. In general, the metals of class A show preference for ligands containing oxygen (hard donors), the metals of class B show preference for binding with ligands containing sulfur and nitrogen (soft donors), and the borderline metals show characteristics intermediate to the characteristics of class A and class B metals. Nieboer and Richardson (1980) emphasized that the binding preference of metals for different natural groups of ligands could be the reason for the similar pattern of their toxicity even for dissimilar organisms.

4.2.3 Toxicity and Availability

Viewed from the standpoint of environmental pollution, Wood (1974) classified metals according to three criteria, non-critical, toxic but very insoluble or very rare, and very toxic and relatively accessible (Table 4.3).

Table 4.3 Non-critical and toxic metals (after Wood 1974)

Non-critical	Toxic but rare	Toxic and accessible
Na, K, Mg, Ca, Fe, Li, Sr and Al	Ti, Hf, Zr, W, Nb, Ta, Re, Ga, La, Os, Rh, Ir, Ru, Ba	Be, Co, Ni, Cu, Zn, Sn, As, Se, Te, Pd, Ag, Cd, Pt, Au, Hg, Tl, Pb, Sb, Bi

From a comparison of the classification of metals according to their toxicity and availability with the classifications given by Pearson (1968a) and Nieboer and Richardson (1980), it is possible to establish a relationship which reveals that non-critical elements such as Na, K, Mg and Ca belong to the hard acceptors (Lewis acids) or class A group of metals, which, in accordance with the HSAB theory, form stable bonds with hard donors (Lewis bases) or ligands containing oxygen, such as H_2O, OH^-, PO_4^{3-}, Cl^-, etc. And, secondly, most of the very toxic and relatively accessible metals belong to the soft acceptors or class B group of metals, which form stable bonds with soft donors or ligands containing S or N, such as the SH group, imidazole group, etc. The preference of toxic metals for binding with SH groups is of much toxicological significance, as this forms the active sites of many proteins and enzymes.

4.2.4 Heavy Metals

One of the commonly used terms in the scientific literature on metal toxicity or metal pollution is the term "heavy metal". In fact, the term "heavy metal" has become entrenched in the literature of environmental pollution. Nevertheless, there is a considerable difference of opinion in the use of the term. Going by the definition given in encyclopedias and dictionaries of the scientific term, heavy metal refers to all metals having a specific gravity greater than 4 (Anonymous 1964; see Nieboer and Richardson 1980), or 5 or more than this (Lapedes 1974; see Nieboer and Richardson 1980). Taking this into consideration the term heavy metal would include all the lower members (metals) of the periodic table (Fig. 4.1). However, in fact, this is not so. The term is always used in context with environmental pollution. Only those lower (metal) members of the periodic table which are (1) relatively abundant in the earth's crust, (2) extracted and used in reasonable amounts, (3) used in places where the public may come into contact with them, and (4) toxic to human beings are generally referred to as heavy metals. On this basis all the heavy metals fall in the category of class B (or soft acceptors) and borderline (intermediate acceptor) metals. Elements like rhodium, titanium, manganese, gallium, and indium in the class B and borderline metals rarely constitute environmental hazards, and thus are not referred to as heavy metals. Also, the elements of the lanthanide and actinide series, and metals like zirconium, niobium, technetium, tungsten, rhenium and osmium, are excluded from the heavy metal category, although they have specific gravities greater than 5 (Martin and Coughtrey 1982).

Thus the term heavy metal should not be used taking into consideration merely the density of a metal. A great deal of confusion appears in the literature where authors have been uncritical in their use of the term, and have even used the term "light metal" to draw a differentiation with heavy metals. Nieboer and Richardson (1980) have reviewed this problem, and emphasized the lack of agreement between different authors. They also proposed avoiding the use of the term heavy metal, and emphasized the need to address a metal by its biologically or chemically significant classes, like oxygen-seeking (class A), nitrogen- or sulfur-seeking (class B) or border-line metals, as described above. Nevertheless, for environmentalists and pollution control officers, the more descriptive terminology of the metals of environmental concern, i.e. heavy metals, may well continue to be more acceptable than class A, class B and borderline metals.

4.3 HEAVY METAL POLLUTION: SOME FACTS

It is important to realize that heavy metal 'pollution' represents a subtly different form of pollution than do many other forms of contamination. The primary source of heavy metals in the environment is from naturally occurring geochemical materials; all heavy metals occur to a varying extent within all components of the environment. Although this occurrence may be enhanced by a human activity, this activity is not itself the source of the heavy metal, rather it is the cause of an elevated occurrence. Hence, heavy metal 'pollution' of the environment does not represent a unique occurrence of the metal within the ecosystem, rather it represents an increase in concentration of the heavy metal relative to the natural occurrence of the element.

Literature on the contamination of the environment by heavy metals is enormous (Martin and Coughtrey 1982; Shaw et al. 1985, 1986; Denton and Burdon-Jones 1986; Shaw and Panigrahi 1986, 1987; Shaw et al. 1988; Barak and Mason 1990; Fowler 1990; Thompson et al. 1990; Baker et al. 1994; Cruz et al. 1994; Fabris et al. 1994; Francesconi et al. 1994; Muller et al. 1994; Prudente et al. 1994; Yan et al. 1994; Yurukova and Kochev 1994; Babiarz and Andren 1995; Hamasaki et al. 1995; Sawidis et al. 1995a,b; Saiki et al. 1995; Absil and van Scheppingen 1996; Hu et al. 1996; Mason 1996; Samecka-Cymerman and

Kempers 1996; Wong 1996; Toli et al. 1997) and the majority of the studies have been associated with various industrial and agricultural activities. However, generally speaking, agricultural or industrial activities result in more diffuse contamination of the environment than does the natural occurrence. Nevertheless, in many cases of naturally high occurrence of heavy metals, there is often a close link with human-derived contamination (e.g. mining, smelting).

From the point of view of environmental pollution the heavy metals which have received the most attention, both in terms of source and effect, are those which are considered either as essential or toxic, or both, or those which show a high geochemical abundance, e.g. zinc, iron, copper, molybdenum, lead, mercury and cadmium. However, there are other heavy metals present in significant quantities in the terrestrial environment, e.g. arsenic, silver, nickel, chromium, manganese, thallium, tin and vanadium, important from the point of view of health hazards and environmental pollution. The most important among them (latter category) is arsenic, which has long been employed for its medical virtue in the form of organic arsenicals. Arsenic is also a unique representative of health hazard as a result of the natural occurrence of a metal, without any involvement of human activity; the source of the metal is naturally contaminated drinking water. Another metal of great environmental concern because of its high natural abundance is Al, although it does not fall under the category of heavy metal taking into consideration only its specific gravity. However, it does qualify by virtue of the other criteria and characteristics attached to heavy metals (see Sect. 4.2.4) and hence is given importance on a par with heavy metals as far as its eco-toxicological and environmental pollution aspects are concerned. With regard to other heavy metals, the health hazard is always linked with one or other kinds of human involvement and anthropogenic activity.

4.4 HEAVY METALS, THE TRACE METALS: ESSENTIAL AND NON-ESSENTIAL

The abundance of elements in the lithosphere generally decreases with increasing atomic mass. From Table 4.1 it may be appreciated that the majority of elements in the periodic table have a relative natural occurrence of <0.1%, and contribute together merely a little more than 0.6% to the elemental mass of the earth's crust. The relative abundance of many of them is below ppm level. Because of the inadequacy of analytical techniques in earlier days, although it was possible to detect the presence of such low abundance elements in a given sample, it was not possible to quantify them, and these were simply said to be present at "trace levels" and called trace elements. A reasonable definition of a trace element is one that occurs at a level of a ppm or less. In fact, many of the heavy metals occur in traces in our body, constituting less than 0.01% of the body mass, and some of them are essential for normal functioning of the cellular machinery, and hence called essential trace metals.

From the list of essential metals given in Table 4.4 a distinction can be drawn between the alkali and alkaline earth metals on the one hand, and the heavy metals exceeding the atomic mass of calcium on the other. It may be noted that heavy metals are present in much lower concentrations than are the alkali and alkaline earth metals; the maximum concentration among the heavy metals in the human body is that of Fe (0.005% w/w). As

Table 4.4 Essential metals and their concentration in humans (after Ochiai 1977)

Essential elements	Concentration in body (mg g^{-1} of homogeneous mass of body of a person of 70 kg body wt.)
Sodium	2.6
Potassium	2.2
Calcium	14
Magnesium	0.4
Iron	5×10^{-2}
Zinc	2.5×10^{-2}
Manganese	1×10^{-3}
Copper	4×10^{-3}
Molybdenum	2×10^{-4}
Cobalt	4×10^{-5}
Vanadium	3×10^{-5}

the essential heavy metals are required only in traces, these are also simply termed trace metals. Nevertheless, non-essential heavy metals do also remain present in traces in living systems. It has been estimated that heavy metals or trace metals are not required in concentrations greater than 0.01% of the mass of the organism for a healthy growth and reproduction, and the trace metals present in the body in concentrations above this, 40- to 200-fold, depending on the metals, are toxic to the organism (Venugopal and Luckey 1975).

4.5 DOSE-RESPONSE RELATIONSHIP OF TRACE ELEMENTS

An essential trace metal affects a living organism harmfully in two different ways: (1) as a result of deficient supply, and (2) as a result of over supply. Under the deficient supply of an essential metal an organism shows poor yield and growth. With increase in the supply and uptake of the metal, the growth and yield increase, until they reach a maximum value. Further supply of the metal does not change the growth of the organism up to a certain concentration. However, after this, the metal shows toxic effects with a consequent decrease in growth and yield, ultimately becoming lethal (Fig. 4.3a). The plateau region reflects the concentrations for optimum growth, health and reproduction. The nature of the plateau reflects the toxic potential of a metal: a wide plateau corresponds to a low inherent toxicity of the metal, whereas a narrow plateau reveals a small difference between required and harmful doses (Fig. 4.3a).

In the case of a non-essential element, the deficiency-toxicity symptom is not exhibited, nor does the organism show any growth or vigour at the initial exposure concentration of the metal. Above a certain concentration, the organism exposed starts showing toxic symptoms, which increase with increase in the metal concentration until the organism dies (Fig. 4.3b). The dose of the metal which produces lethality is referred to as the lethal concentration whereas the concentration of the metal up to which no toxic symptoms are produced represent the "no effect" level or the threshold of safety. In between lie the sublethal concentrations where toxic symptoms are observed but no lethality occurs (Fig. 4.3b).

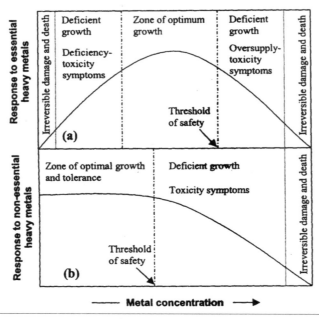

Fig. 4.3 Hypothetical dose-response curves depicting the effect of *a* essential and *b* non-essential heavy metals on plants

4.6   UPTAKE OF HEAVY METALS

4.6.1   Organization of Plants

Plants consist of two main structural components: (1) the root system, which anchors the plant in the soil, and is the site for absorption of materials from it, and (2) the shoot system consisting of the stem and leaves. The stem supports the plant above the ground, and is the part through which minerals and water absorbed from roots are conveyed to leaves and the carbohydrates produced in leaves are transferred to roots.

4.6.2   Routes of Uptake

An element in soil may move to the root surface either through diffusion along with the water in soil, provided there is a concentration gradient, or by ion exchange between the clay particles and the root in contact with it. The concentration gradient for diffusion may be created by continuous absorption of ions by roots. Roots and the associated microorganisms also exude various compounds, which are very effective in releasing the metals from the soil particles making the metal species available to the plant for absorption. The release of the bound metal into solution also results in establishment of a diffusion gradient between the absorbing root and the soil particles from where the metal is released.

The movement of elements into roots occurs either by passive diffusion through the cell membrane, or by the more common process of active transfer against concentration and/or electrochemical potential gradients (Clarkson and Luttge 1989; Fergusson 1990; Stroinski 1999). The latter is mediated by carriers, which could be a complexing agent, such as an organic acid or protein, that binds to the metal species, transports it across the membrane and then dissociates freeing the metal species to move into the cell. The active uptake process is adapted by plants for the uptake of essential trace metals but, simultaneously, the other available elements are also taken up (Silver 1983). A metal can also move inside the cell along a concentration gradient through a cation channel in the cell membrane meant for a different metal (Kochian 1995). The requirement for the purpose is that the ionic radius of the concerned heavy metal should be close in size to the ion for which the channel is meant for; for example, Al^{3+} is closest in size to Fe^{3+} and Mg^{2+} (Martin 1988, 1992). Thus, Al^{3+} might cross the plasma membrane by sluggishly permeating the divalent cation channel normally functioning in Mg^{2+} uptake (Kochian 1995). Furthermore, in grasses, the roots release non-protein amino acids, phytosiderophores (low molecular weight iron-chelating ligands produced by plants), in response to Fe^{3+} deficiency that not only binds Fe^{3+}, but can complex other divalent and polyvalent cations fairly effectively (Kochian 1995). Because of the similarity of ionic radii of Al^{3+} and Fe^{3+}, the phytosiderophores can facilitate the absorption of the former into the cytoplasm through the Fe(III)-phytosiderophore transport system (Kochian 1995).

The foliar uptake is another route of entry of metals into plant cells. The entry may be either through stomata, or leaf cuticle, or both. The entry of metals into plant cells through leaves is of particular significance from the pollution point of view because of aerosol deposit (Bowen 1979; Kabata-Pendias and Pendias 1984; Lodenius 1990; Misra and Mani 1991; Suszcynsky and Shann 1995).

4.6.3 Factors Influencing Heavy Metal Uptake

Many factors influence the uptake of metals, and include the growing environment, such as temperature, soil pH, soil aeration, E_h condition (particularly of aquatic environment) and fertilization, competition between the plant species, the type of plant, its size, the root system, the availability of the elements in the soil or foliar deposits, the type of leaves, soil moisture and the plant energy supply to the roots and leaves (Shaw and Panigrahi 1986; Fergusson 1990; Lodenius 1990; Misra and Mani 1991; Albers and Camardese 1993a,b; Baker et al. 1994; Chuan et al. 1996; Yamamoto 1996). As far as the growing environment is concerned the increase in pH, i.e. the environment becoming more alkaline, and decrease in Eh (redox potential), i.e. the environment becoming more reducing, result in decrease in availability of heavy metals, or metals in general, to plants (Fergusson 1990; Misra and Mani 1991). However, under a given environmental condition, the uptake of a metal by a plant can be estimated from the biological absorption coefficient (BAC):

$$BAC = [Mp]/[Ms]$$

where [Mp] is the concentration of the element in the plant and [Ms] is its concentration in the soil (Brooks 1983; Fergusson 1990). However, in field conditions, the relationship works best only when the concentration of the metal in the soil is not too high (Shaw and Panigrahi 1986).

Because of the influence of environmental factors, and the type of plant itself, the levels of heavy metals in plants (both terrestrial and aquatic) vary widely (Shaw et al. 1985, 1986; Shaw and Panigrahi 1986; Lodenius 1990; Baker et al. 1994; Moral et al. 1994; Yurukova and Kochev 1994; Sawidis et al. 1995a; Samecka-Cymerman and Kempers 1996;

Wong 1996). The range of heavy metals observed in plants is presented in Table 4.5. If the other factors are constant, the uptake of a metal by different plant species may be compared.

Table 4.5 Range of a few environmentally important heavy metals in plants (after Allaway 1968; Bowen 1979; Kabata-Pendias and Pendias 1984; Shaw and Panigrahi 1986; Shaw et al. 1986; Fergusson 1990; Misra and Mani 1991)

Elements	Land plants (μg g^{-1} dry wt.)	Elements	Land plants (μg g^{-1} dry wt.)
As	0.02–7	Cu	4-15
Cd	0.1–2.4	Fe	140
Hg	0.005–0.02	Mn	15–100
Pb	1–13	Mo	1–10
Sb	0.002–0.06	Ni	1
Co	0.05–0.5	Sn	0.30
Cr	0.2–1	Zn	8–100

4.6.4 Indicator and Accumulator Species

With respect to a metal, plants may be classified as either accumulators or excluders. An accumulator plant may accumulate a metal in a concentration-dependent manner, or the accumulation by it may be independent of the external metal concentration; the former is termed monitor and the latter simply indicator. Unlike accumulators, there are plant species which can discriminate between metal ions, and are able to check the uptake of one or more elements. These are termed 'excluders' for the metal ions whose uptake by them is restricted.

Among plant types growing in the same environment, fungi, lichen and mosses accumulate more metals than the others (Table 4.6). Nevertheless, the accumulation depends on the metal species as well as on the type of plant. The data in Table 4.6 reflect that, with respect to lead, lichens and mosses show more accumulation than the others, whereas with regard to Cd the accumulation by lichen, and also by fungi and mosses, is not much different from that by other plants. For a particular species the concentration levels of a metal generally decreases in the order root>stem>leaves>fruit>seed when the source of the metal is only the soil (Shaw and Panigrahi 1986; Fergusson 1990).

Table 4.6 Levels of cadmium and lead in plants from pine and birch forest ecosystems in Poland (after Kabata-Pendia and Pendias 1984; Fergusson 1990)

Plants	Cd (μg g^{-1} dry wt.)	Pb (μg g^{-1} dry wt.)
Grass	0.6	1.2
Clover	0.7	2.8
Plantain	1.9	2.4
Mosses	0.8, 0.7	22.4, 13.0
Lichen	0.4, 0.5	17.0, 28.0
Edible fungi	1.0, 2.7	1.2, <0.1
Inedible fungi	1.6, 1.0	0.4, 1.0

4.7 TOXICITY OF HEAVY METALS IN PLANTS: A GENERAL CONSIDERATION

4.7.1 Toxicity Sequence of Heavy Metals

The toxicity sequence for heavy metals varies with the taxonomical group of plants; for flowering plants (barley) the sequence observed is Hg>Pb>Cu>Cd>Cr>Ni>Zn, for algae (*Chlorella vulgaris*) it is Hg>Cu>Cd>Fe>Cr>Zn>Ni>Co>Mn, and for fungi the observed sequence is Ag>Hg>Cu>Cd>Cr>Ni>Pb>Co>Zn>Fe (see Nieboer and Richardson 1980). It has also been reported that root and shoot also differ in their toxic response to heavy metals: for the shoot of the test plant (*Triticum aestivum* L.) the toxicity of the metals was in the order Mn<Zn<B<Fe=Ga<La<Cu, and for the root it was in the order Mn<Ga<Zn<Fe=La<Cu<B (Wheeler and Power 1995). Reports on the toxicity of heavy metals simultaneously on several plant species are scant. However, Fergusson (1990) after a review of the literature came to the conclusion that the toxicity sequence of heavy metals may take the shape As(III)~Hg>Cd>Tl>Se(IV) >Te (IV)>Pb>Bi~Sb for flowering plants in general. The toxicity sequence framed nevertheless depends greatly on the information available to the reviewing author(s). Moreover, the order of toxicity should also highly depend on the properties of the soil, besides the type of plants.

The relative toxicity of heavy metals, besides depending on the soil property and the plant species, also depends on the age of the plant (Shaw 1995a,b; Shaw and Rout 1998, 2002); *Phaseolus aureus* seeds germinated over Hoagland's solution containing Hg or Cd showed significant inhibition of root elongation, and the inhibition was much more in the seedlings germinated and grown over Hg-containing nutrient solution (EC=3 μM) than those germinated and grown over Cd-containing nutrient solution (EC=13.2 μM; Fig. 4.4). No death of seedlings occurred after observation for 10 days. In contrast to the germination stage treatment described above, when the seeds were allowed to germinate over Hoagland's solution and more than 5-day-old germinated seedlings were exposed to various concentrations of Hg or Cd, all the seedlings died within 24 h of exposure to Cd at a concentration as low as 30 μM, but not even a single death was recorded in the seedlings exposed to Hg at concentrations as high as 200 μM. Nevertheless, the observed lethality of the fully grown seedlings to Cd was found to be only species-specific (Shaw and Rout 2002). Hence, despite a few isolated observations of lesser toxicity of Hg than other heavy metals in plants, Hg is generally recognized as the most toxic among the heavy metals.

The dependence of toxicity of heavy metals on the age of plants (seedlings) has also been observed by other workers. Maksymeic and Baszynski (1996a,b) and Maksymiec et al. (1995) observed that when the seedlings of *Phaseolus coccineus* were exposed to copper at the stationary growth stage of the primary leaves, a significant decrease in Chl *a/b* and Chl/Car ratios occurred, but not when they were exposed at the intensive growth stage of the primary leaves. Similarly, Skorzynska-Polit and Baszynski (1997) found that the sensitivity of the seedlings of *Phaseolus aureus* to Cd increased with increase in their age. However, to date, this type of age (of the plant) dependent toxicity of heavy metals has not been reported in any species belonging to genera other than *Phaseolus*. And, hence, the age-dependent toxic response (of *Phaseolus* sp.) cannot be generalized for the entire class of flowering plants.

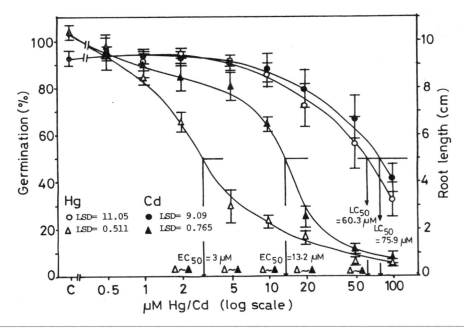

Fig. 4.4 Effects of Hg (*open symbols*) and Cd (*closed symbols*) on the germination of seeds of *P. aureus* (*circles*, observation on the eighth day of germination) and the growth of roots of the seedlings (*triangles*, observation on the fifth day of germination). The symbols in open and closed pairs at the bottom of the graph against concentration indicate significant difference between the effects of the two metals in that concentration. LC_{50} Lethal concentration inhibiting 50% seed germination; EC_{50} effective concentration inhibiting 50% root elongation. (Reproduced from Shaw and Rout 1998)

4.7.2 Visible Symptoms of Toxicity of Heavy Metals

The visible symptoms of toxicity for some of the heavy metals are listed in Table 4.7. As can be seen from the table, stunting and chlorosis are the most common visible symptoms of heavy metal toxicity. Necrosis and yellowing of leaves are also generally observed, except in the case of exposure to Al (not a heavy metal on the basis of specific gravity) and Pb where the leaves turn dark green. The visible symptoms observed are in fact a result of deficiency in the availability of one or more of the essential elements created upon exposure of the plants to the heavy metals. For example, the overall stunting - small, dark green leaves and late maturity, purpling of stems, leaves and leaf veins, and yellowing and death of leaf tips are symptoms of P deficiency. Similarly, curling or rolling of young leaves and collapse of growing points or petioles are the symptoms of Ca deficiency. In addition, the symptom of the much observed chlorosis appears to be due to a direct or indirect interaction of heavy metals with Fe resulting in its depletion in leaves, although Fe has no direct or indirect role in chlorophyll synthesis. The relationship between chlorosis and foliar iron deficiency is strengthened by the fact that painting chlorotic leaves with $FeSO_4$ alleviates the chlorosis (Foy et al. 1978, and references therein).

Table 4.7 Visible symptoms of injuries in flowering plants for some important heavy metals (after Foy et al. 1978; Jastow and Koeppe 1980; Aasami 1984; Kabata-Pendias and Pendias 1984; Adriano 1986; Padmaja et al. 1990; Parekh et al. 1990; Davies et al. 1991; Leita et al. 1991; Jalil et al. 1994; Delhaize and Ryan 1995; Shaw 1995a,b; Arduini et al. 1996; Mocquot et al. 1996; Quariti et al. 1997; Shaw and Rout 1998; Bonnet et al. 2000; Kukkola et al. 2000)

Metals	Toxic symptoms
Arsenic	Red brown necrotic spots on old leaves, yellow browning of roots, growth reduction
Cadmium	Brown margin to leaves, chlorosis, necrosis, curled leaves, brown stunted roots, reddish veins and petioles, reduction in growth, purple colouration
Mercury	Severe stunting of seedlings and roots, chlorosis, browning of leaf points, reduction in growth
Lead	Dark green leaves, stunted foliage, increased amounts of shoots
Selenium	Interveined chlorosis, black spots at high selenium, bleaching and yellowing of young leaves, pink spots on roots
Zinc	Chlorosis, stunting, reduction of root elongation
Copper	Chlorosis, yellow colouration, purple colouration of the lower side of the mid rib, less branched roots, inhibition of root growth
Nickel	Chlorosis, necrosis, stunting, inhibition of root growth, decrease in leaf area
Aluminum	Stunting, dark green leaves, purpling of stems, leaves and leaf vein, yellowing and death of leaf tips, curling of young leaves and collapse of growing points or petioles, thickening of root tips and later roots, inhibition of root elongation
Manganese	Marginal chlorosis and necrosis of leaves, leaf puckering and necrotic spots on leaves, crinkled leaves
Iron	Dark green foliage, stunted top and root growth, thickening of roots, brown spots on leaves, starting from the tip of lower leaves, dark brown and purple leaves, sometimes in same plant

4.8 TOXIC EFFECTS: THE BIOCHEMICAL BASIS

Heavy metals, both in acute and chronic exposure, interact with many different cellular components, thereby interfering with the normal metabolic functions, causing cellular injuries, and, in extreme cases, death of the organism. Toxicological and biochemical investigations have led to the identification of a large number of potential or real target compounds whose normal activity is inhibited, enhanced, or modified by heavy metals. Table 4.8 presents a list of monodentate and pluridentate ligands commonly occurring in cells as components of biomolecules with which heavy metals can form complexes. The strength of the complexes is determined by their stability constants (equilibrium constant) governed by the HASB principle. By way of illustration, the apparent stability constants $\log k$, pH 7[$=\log k$-$\log (1+10^{-7}/k_a)$, where k_a is the acid dissociation constant of the ligand] for Hg (soft acceptor) and Pb (intermediate acceptor) with different monodentate and pluridentate ligands are also given in the table. It may be noted that very strong complexes of Hg and Pb are formed with the thiol group and the pluridentate ligands containing a mercapto group. Thus, for the soft acceptors, and also the intermediate acceptors,

Table 4.8 List of some common monodentate and pluridentate ligands present in cells (after Kagi and Hapke 1984)

Monodentate ligands	Apparent stability constant (log k, pH 7)		Pluridentate ligands	Apparent stability constant (log k, pH 7)	
	Hg	Pb		Hg	Pb
Carboxyl group	5.6	1.9	Glycine	7.6	2.8
Amino group	6.0	−0.5	Histidine	5.4	4.4
Imidazole group	3.7	2.2	Cysteine	9.9	7.1
Thiol group	10.2	4.9	D-penicillaminate	11.8	8.0
Phosphate group		3.1	(β,β-dimethylcysteinate)		
Chloride	7.3	1.6			

k is the equilibrium constant of the reaction $M+L \Leftrightarrow ML$, where ML is the 1:1 complex of a metal ion (M) and a ligand (L); $k=[ML]/[M][L]$.

the most potential targets for action are the enzymes and proteins, which contain several mercapto ligands close enough to form chelate structures with the metals, and thereby losing their functional property.

$$\text{Enzyme} \Big\langle {\text{SH} \atop \text{SH}} + \underset{\text{(a metal ion)}}{M^{2+}} \longrightarrow \text{Enzyme} \Big\langle {S \atop S} \Big\rangle M + 2H^+$$

Among the enzymes whose activity is destroyed by binding of metals with the sulfhydryl group, those producing cellular energy in the citric acid cycle are particularly affected. In fact the first enzyme in this cycle, pyruvate dehydrogenase, is also inhibited. The inhibition is by the oxide of arsenic, ASO_3^{3-}, a complex of a hard acceptor and a hard donor, and is due to the formation of a complex by ASO_3^{3-} with the two neighbouring thiol groups of a-lipoic acid, the co-factor of the enzyme, resulting in a change in its conformation (Manahan 1990).

$$O\text{—}As \Big\langle {O^- \atop O^-} + {\text{HS—CH} \atop \vdots} \longrightarrow O\text{=}As \Big\langle {S\text{—CH} \atop \vdots}$$

Dihydrolipoic acid-protein   Inactivated protein complex with arsenic

In addition to interaction with enzymes, the toxicity of a metal may also result from its interaction with other biomolecules. For example, the highly electropositive Al ion displaces Ca in cross-linking negatively charged phospholipid heads and leads to decreased phospholipid fluidity (Cumming and Taylor 1990). Al^{3+} also binds almost 10^7 times more strongly to ATP than does Mg^{2+}; therefore, less than nanomolar (nM) amounts of Al^{3+} can compete with Mg^{2+} for the P sites (Martin 1988).

Besides the general toxicity mechanisms for metal ions described above, i.e. (1) blocking of the essential biological functional groups of enzymes, and (2) modification of the active conformation of biomolecules, a third important general mechanism also exists, i.e. (3) displacement of the essential metal ions in the biomolecules (Ochiai 1977). A number of enzymes contain metal(s), and displacement or substitution of one of these metal ions by another metal ion, maybe with the same charge and a similar size, results in inhibition in the activity of the enzyme. Zn is a component of many metalloenzymes, like alcohol dehydrogenase, adenosinetriphosphatase, amylase, carbonic anhydrase, etc., and Cd, which is directly below Zn in the periodic table, may substitute for Zn in the enzymes. Despite the chemical similarity between Zn^{2+} and Cd^{2+} ions, the Cd-containing enzymes do not function properly. There are also enzymes containing Fe, Co, Mn, Mg, etc., which could be inhibited by substitution.

The 'displacement' mechanism of toxicity is best explained by classification of the metal ions according to their binding preferences, i.e. oxygen-seeking class A metals, nitrogen- or sulfur-seeking class B metals, and the intermediate metals that show more or less the same preference of bonding to O_2–, S– or N-containing ligands (Nieboer and Richardson 1980). The classification also explains the reason for the similarity in the patterns of toxicity of metal ions even in dissimilar organisms. According to the classification, class B metals are more toxic than the borderline metals, which are more toxic than class A metals. The borderline metal ions frequently act to displace other endogenous borderline ions or class A metal ions from the biomolecules; e.g. the displacement of Zn^{2+} by Ni^{2+} in enzymes such as carbonic anhydrase, alcohol dehydrogenase, etc. Also, Ca^{2+} displacement in the membrane proteins results in functional disorders; the inhibitory effect of Ni on secretory functions in mammals is due to the displacement of Ca^{2+} by the metal (Nieboer et al. 1988).

The most toxic among the heavy metals are the metal ions of class B, which exhibit a broad spectrum of toxicity mechanisms. None of the class B metal ions are components of any enzyme. However, they readily displace the borderline elements from their enzymes. Besides, they are the most effective binders to SH groups (e.g. of cysteine) and nitrogen-containing groups (e.g. of lysine and histidine imidazole) at the catalytically active centres in the enzymes.

Unlike the class B and borderline metals, the metals of class A can result in toxicity only by substitution in its own group. For example, a preferential substitution of Be^{2+} for Mg^{2+} in certain enzymes requiring Mg^{2+} as a cofactor, like Rubisco, and also the displacement of Mg^{2+} by Al^{3+} in biological systems because of similarity in ionic size; the complex of Al^{3+} with ATP^{4-} binds 1000 times more strongly to the enzyme hexokinase than does the Mg^{2+} complex, which is the reason for inhibition of the enzyme by Al^{3+} (Martin 1988). Similarly, Al^{3+} may interact with the Ca^{2+}-binding sites in biomolecules, like that of calmodulin, and in fact the primary target of Al phytotoxicity might be the Ca^{2+}-binding protein, calmodulin (Siegel and Haug 1983; Siegel et al. 1983).

Apparent anomalies in the above classification are, however, expected because of several factors. The formation of complexes of the metals with some ligands (given in Table 4.8) can prevent or greatly reduce their uptake when the complex formation is outside the cell, or reduce the redistribution of the metal ions within an organism when the complex is formed inside the cell after their uptake. Furthermore, the development of tolerance mechanisms for specific ions in various organisms can result in exclusion or innocuous accumulation of a potentially toxic ion (Connell and Miller 1984).

4.9 OXIDATIVE PHYTOTOXICITY

4.9.1 Oxygen and Reactive Oxygen Intermediates

Oxygen, which appeared in the earth's atmosphere mainly as a product of photosynthesis, is a two-edged sword for aerobic organisms: it enables efficient energy production by enzymatic combustion of organic compounds, but at the same time leads to damage to aerobic cells due to the formation of reactive oxygen intermediates (Bartosz 1997). The reactive oxygen intermediates (RIOs) generally encountered are superoxide radical ($O_2^{\cdot-}$), hydrogen peroxide (H_2O_2), hydroxyl radical ($HO^{\cdot}$) and singlet oxygen ($^1O_2$). These are called ROIs because they are more prone to participate in chemical reactions than molecular O_2. The greater reactivity of two of them, $O_2^{\cdot-}$ and $HO^{\cdot}$, is because of their "free radical" nature: a free radical is any species capable of independent existence that contains one or more unpaired electrons.

Going by the definition of free radical, in fact O_2 itself qualifies as a free radical (Fig. 4.5); the ground state O_2 has two unpaired electrons, one each in the $2P\pi^*$ (antibonding) molecular orbitals (Halliwell and Gutteridge 1985). However, its reactivity is restricted because both the unpaired electrons are in same spin, and thus it must receive only one electron at a time, making the molecule react only sluggishly with many non-radicals.

4.9.2 Formation of ROIs and Their Toxic Action

4.9.2.1 Superoxide Radical and Hydrogen Peroxide

Superoxide radical ($O_2^{\cdot-}$) and hydrogen peroxide (H_2O_2) are formed by reduction of the O_2 molecule. Addition of one electron to the ground state O_2 molecule, which finds it place on one of the $2P\pi^*$ (antibonding) orbitals (Fig. 4.5), results in the formation of $O_2^{\cdot-}$. The redox potential E'_o (pH 7) of the reaction is –0.33 V, and hence can proceed only with input of energy (Naqui et al. 1986). With only one unpaired electron, $O_2^{\cdot-}$ is actually less of a radical than O_2, but is somewhat more reactive than the latter because it can easily undergo a one-electron redox reaction. The chemical properties of $O_2^{\cdot-}$ depend greatly on the nature of the solvent. In aqueous solution, $O_2^{\cdot-}$ behaves as a weak reducing agent (a donor of electrons):

$$\text{Cytochrome c } (Fe^{3+}) + O_2^{\cdot-} \rightarrow O_2 + \text{Cyt c } (Fe^{2+})$$
$$\text{Plastcyanin } (Cu^{2+}) + O_2^{\cdot-} \rightarrow O_2 + \text{Plastocyanin } (Cu^+)$$

as well as a weak oxidizing agent (an acceptor of electrons):

$$\text{Ascorbate} + O_2^{\cdot-} \rightarrow H_2O_2 + \text{ascorbate radical (Asc}^{\cdot})$$
$$\text{Enzyme-NADH} + H^+ + O_2^{\cdot-} \rightarrow \text{enzyme-NAD}^{\cdot} + H_2O_2.$$

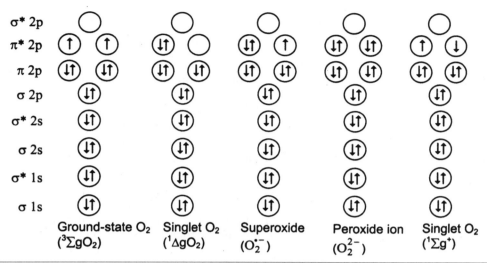

Fig. 4.5 Electronic configuration of the oxygen molecule and its derivatives. In covalent compounds the atomic orbitals interact to form molecular orbitals and the electrons occupy these molecular orbitals. The numbers of molecular orbitals are twice that of the number of atomic orbitals; for example, interaction of three 2p orbitals ($2p_x$, $2p_y$ and $2p_z$) of two oxygen atoms will result in the formation of six 2p orbitals, three bonding, designated as $\sigma 2p_x$, $\pi 2p_y$ and $\pi 2p_z$, and three antibonding, designated as $\sigma^* 2p_x$, $\pi^* 2p_y$ and $\pi^* 2p_z$ (please note one of the 2p orbitals forms a s bond and the other two p bonds). The energy of the antibonding orbitals is higher than the respective bonding orbitals, and that of σ^* is greater than of π^*. The energy of the molecular 1s, 2s and 2p bonding and antibonding orbitals is in the order $\sigma 1s < \sigma^* 1s < \sigma 2s < \sigma^* 2s < \sigma 2p < \pi 2p < \pi^* 2p < \sigma^* 2p$. The orbital with lowest energy level is filled first (Aufbau principle), and all orbitals with equal energy levels receive one electron before any receives two (Hund's rule). Presence of electrons in the antibonding orbitals energetically cancels the bonding of the respective bonding orbital(s). For example, the presence of two electrons, one each in the two antibonding orbitals, cancels out one of the $\pi 2p$ bonding orbitals, and hence two oxygen atoms are effectively joined by a double bond. In fluorine three bonding and two antibonding 2p orbitals are occupied, and hence two fluorine atoms in the fluorine molecule are effectively bound by only a single bond. (Reproduced from Halliwell and Gutteridge 1985 with permission of the publisher)

The oxidation of NADH, however, occurs only when it remains bound to the enzyme, and that too only to lactate dehydrogenase, and not to other dehydrogenases. $O_2^{\cdot-}$ in aqueous solution also dismutates slowly to H_2O_2 and O_2 through the dismutation reaction:

$$O_2^{\cdot-} + O_2^{\cdot-} + 2H^+ \rightarrow H_2O_2 + O_2$$

In fact, the rate constant of this dismutation reaction is virtually zero (<0.3 M^{-1} s^{-1}), and hence it is believed that the dismutation probably proceeds through the formation of the hydroperoxy radical ($HO_2^{\cdot}$). This is possible because $O_2^{\cdot-}$ in aqueous solution, in addition to acting as a weak reducing and oxidizing agent, also behaves as a weak base (an acceptor of H^+), leading to the reaction:

$$O_2^{\cdot-} + H^+ \Leftrightarrow HO_2^{\cdot}$$

Subsequently, $HO_2^{\cdot}$ species react among themselves or with $O_2^{\cdot-}$ to complete the dismutation:

$$HO_2^{\cdot} + O_2^{\cdot-} + H^+ \rightarrow H_2O_2 + O_2 \; (k_2 = 8 \times 10^7 M^{-1}S^{-1})$$
$$HO_2^{\cdot} + HO_2^{\cdot} \rightarrow H_2O_2 + O_2 \quad (k_2 = 8 \times 10^5 M^{-1}S^{-1})$$

The protonated form of $O_2^{\cdot-}$ ($HO_2^{\cdot}$) is a much more powerful reducing and oxidizing agent than is $O_2^{\cdot-}$ itself; unlike $O_2^{\cdot-}$, $HO_2^{\cdot}$ oxidizes NADH directly without needing the involvement of any dehydrogenases, and it reduces cytochrome c at a much faster rate compared to $O_2^{\cdot-}$. Fortunately, however, the pK_a of the protonation reaction leading to the formation of $HO_2^{\cdot}$ is acidic (4.8) while the pH of the cellular fluid is maintained at around neutrality, varying from slightly acidic to slightly basic, indicating that excess formation of $HO_2^{\cdot}$ species is not possible. However, this also means that the dismutation of $O_2^{\cdot-}$ molecules in the cellular milieu will be very slow, and this will accumulate in the cells, particularly when its formation increases, which is also of great concern.

Unlike in aqueous solution, $O_2^{\cdot-}$ in an organic solvent acts as a strong base and reducing agent, and a powerful nucleophile (an agent that is attracted to the centre of the positive charge in a molecule) too. As the interior of the biological membrane is hydrophobic, being similar in chemical nature and viscosity to a "light oil", the accumulation of $O_2^{\cdot-}$ could be extremely damaging for the cells. Besides interfering in the normal redox reactions of the cells, $O_2^{\cdot-}$ can destroy phospholipids by nucleophilic attack upon the carbonyl group of the ester bonds linking the fatty acids to the glycerol 'backbone' of the phospholipids (Halliwell and Gutteridge 1985).

Thus, while the prevailing pH of the cellular environment decreases the risk of damage to the cells by preventing the formation of $HO_2^{\cdot}$, the accumulation of $O_2^{\cdot-}$ is detrimental to the cells.

The addition of one more electron to $O_2^{\cdot-}$ gives rise to the peroxide ion, O_2^{2-}; the redox potential of the electron transfer reaction is positive (+0.94 V), and hence requires no input of energy (Naqui et al. 1986). Here no unpaired electron is left, and thus it is not a radical. However, O_2^{2-} is more reactive than the ground state O_2 molecule since the strength of the oxygen-oxygen bond decreases with the electrons being added to the antibonding molecular orbitals. The two oxygen atoms in O_2^{2-} are bound by a single covalent bond, in contrast to the double bond in the ground state oxygen molecule, and hence the oxygen-oxygen bond in O_2^{2-} is quite weak. This increases the reactivity of O_2^{2-} over the ground state oxygen. In biological systems the two-electron reduction product of oxygen is hydrogen peroxide (H_2O_2).

4.9.2.2 Hydroxyl Radical and Singlet Oxygen

Hydroxyl radical (HO$^\bullet$) and singlet oxygen ($^1O_2$), unlike $O_2^{\bullet-}$ and H_2O_2, are not formed as a result of direct addition of an electron to the ground state oxygen molecule. While HO$^\bullet$ is formed as a result of subsequent reaction of $O_2^{\bullet-}$ and H_2O_2 in biological systems (discussed in Sect. 4.9.5.1.1), $^1O_2$ is generated by input of energy.

$^1O_2$ has two forms, $^3\Sigma gO_2$ and $^1\Delta gO_2$ (Fig. 4.5). $^3\Sigma gO_2$ is highly energetic having 37.5 kcal energy above the ground state. It is similar to the ground state O_2 in electronic configuration, except that the spin of one of the electrons in the $2P\pi^*$ orbital is reversed due to the input of energy, and hence behaves as a highly reactive free radical. However, its lifetime is very short, and hence is not of much importance from the point of view of toxicity. It decays rapidly to a slightly more stable second form, $^1\Delta gO_2$, with the loss of some energy. $^1\Delta gO_2$ has energy 22.4 kcal above the ground state O_2. During the state conversion, the unpaired electron from one of the $2P\pi^*$ (antibonding) molecular orbital moves to the other $2P\pi^*$ molecular orbital. Thus, $^1\Delta gO_2$ is not a free radical, but is highly reactive because in this case also the spin restriction is removed, and so is its oxidizing ability; it has no restriction in receiving a pair of electrons from any atom, unlike the ground state O_2.

Although all ROIs are more or less highly reactive and are toxic to living organisms, the ultimate damaging effect is, however, mainly by $^1O_2$ and HO$^\bullet$. Both the species are extremely reactive, reacting instantly at the site of their generation. Their rapid and non-specific reaction leads to damage of all classes of biomolecules including lipids, proteins, enzymes and DNA (Fridovich 1978; Asada 1992, 1994; Stadtman 1992; Breen and Murphy 1995). Reaction of HO$^\bullet$ and $^1O_2$ with unsaturated fatty acids causes peroxidative degradation of essential lipids in the plasma membrane or the intracellular organelles leading to rapid desiccation and cell death (Halliwell and Gutteridge 1985). Intracellular membrane damage in turn can affect respiratory activity in mitochondria, cause pigment breakdown, and loss of carbon-fixing ability in chloroplasts. Damage to proteins and DNA can often lead to irreparable metabolic dysfunction and cell death (Halliwell and Gutteridge 1985; Bartosz 1997).

HO$^\bullet$ has a lifetime of approximately 1 μs (Hippeli and Elstner 1997), and the reaction involving it has a redox potential of about +2 V (Naqui et al. 1986; Hippeli and Elstner 1997), and hence is thermodynamically favourable. HO$^\bullet$ can react with a biomolecule in three ways: (1) hydrogen abstraction, (2) addition, and (3) electron transfer (Halliwell and Gutteridge 1985). Abstraction of a hydrogen atom from biomolecules, like lipids, alcohols, etc., results in production of a free radical of the biomolecule.

$$CH_3 + HO^\bullet \rightarrow {}^\bullet CH_2OH + H_2O$$

The carbon radicals react further with O_2 or among themselves to form non-radical products. Attack of HO$^\bullet$ on a sugar, such as deoxyribose, found in DNA, produces a huge variety of products, some of which are even mutagenic in bacterial test systems (Halliwell and Gutteridge 1985).

Addition of HO$^\bullet$ to aromatic ring structures, like that of purine and pyrimidine bases of DNA and RNA, occurs at the double-bond carbon atom, resulting in the formation of a free radical of the compound, which later degrades by a series of further reactions.

Thymine

The electron transfer reaction generally occurs with organic or inorganic anions, for example with Cl^-, resulting in the formation of a radical of the anion.

$$Cl^- + HO^{\cdot} \rightarrow Cl^{\cdot} + OH^-$$

$$Cl^{\cdot} + Cl^- \rightarrow Cl_2^{\cdot-}$$

In fact the reactivity of $HO^{\cdot}$ is so great that it reacts immediately with whatever biological molecule remains present in its vicinity, producing secondary radicals of variable reactivity: it is an important principle of radical chemistry that reaction of a free radical with a non-radical produces a different free radical, which may be more, or less, reactive than the original radical (Cadenas 1989).

$^1O_2$ ($^1\Delta gO_2$) is even more reactive than $HO^{\cdot}$, reacting with most biological molecules at near diffusion control rates (Knox and Dodge 1985; Cadenas 1989). Its high reactivity is because of its high energy, and the presence of an empty antibonding molecular orbital (Fig. 4.5) ready to accept a pair of electrons from any atom or molecule. It can interact with other molecules in essentially two ways: it can either combine chemically with them, or else it can transfer its excitation energy to them returning itself to ground state while the reacting molecules enter into an excited state. Sometimes both can happen. The best-studied chemical reactions of singlet oxygen are those involving compounds that contain carbon-carbon double bonds ($>C=C<$). Such bonds are present in many biological molecules, such as carotenes, chlorophylls, and the fatty acid side chains present in the membrane lipids. The compounds containing conjugate double bonds (two double bonds separated by a single bond) often react to give endoperoxide, and those with single double bonds produce hydroperoxide.

endoperoxide

$$
\begin{array}{c}
CH_2 \\
CH_2 \\
| \\
CH_2 \\
\backslash H
\end{array}
\; + \; {}^1O_2 \; \longrightarrow \;
\begin{array}{c}
CH_2 \\
CH \quad\quad O \\
|| \quad\quad\quad | \\
CH_2 \quad\quad O \\
H
\end{array}
$$

Hydroperoxide

These compounds subsequently degrade. Damage to proteins by ${}^1O_2$ is often due to oxidation of essential methionine, tryptophan, histidine or cysteine residues.

${}^1O_2$ may react with biomolecules to form dioxetane if there is an electron-donating atom, such as N or S, adjacent to the double bond.

$$
\begin{array}{c}
| \\
CH \\
|| \\
CH \\
N
\end{array}
\; + \; {}^1O_2 \; \longrightarrow \;
\begin{array}{c}
| \\
HC-O \\
| \quad | \\
HC-O \\
N
\end{array}
\; \longrightarrow \;
HC{=}O \; + \; H{-}C{=}O \\
\begin{array}{c}
| \\
N
\end{array}
$$

Dioxetane   Carbonyl compounds

Dioxetanes are unstable. They decompose to compounds containing carbonyl group, ($\rangle$C=O). Amino acids like tryptophan and histidine react in this way.

4.9.3 Lipid Peroxidation

The process of lipid peroxidation by HO$^{\bullet}$ and ${}^1O_2$ needs special mention, as this is generally considered a reflection of the oxidative stress an organism might be facing. This is initiated by abstraction of a hydrogen atom from the methylene group (—CH$_2$—) of the polyunsaturated fatty acid of the membrane lipid by HO$^{\bullet}$ (Fig. 4.6). The presence of a double bond in the fatty acid weakens the C—H bonds on the carbon atom adjacent to the double bond and so makes H$^{\bullet}$ removal easier. Abstraction of a hydrogen atom leaves behind an unpaired electron on the carbon (—$^{\bullet}$CH$_2$—). The carbon radical tends to become stabilized by molecular rearrangement producing a conjugated diene (two double bonds occurring consecutively), which then reacts with an oxygen molecule to give a peroxy radical, R—OO$^{\bullet}$ (Fig. 4.6). The peroxy radical in turn can abstract a hydrogen atom from another lipid molecule creating chain propagation. The combination of the peroxy radical with the hydrogen atom it abstracts gives a lipid hydroperoxide, R—OOH. An alternative fate for the peroxy radical is to form cyclic peroxides (Halliwell and Gutteridge 1985). However, all of them finally get fragmented to aldehydes, including malondialdehyde (MDA), and yield various polymerization products. The level of MDA in tissue is considered a measure of lipid peroxidation status, or, in other words, the oxidative stress an organism is experiencing. ${}^1O_2$ reacts with unsaturated fatty acids directly to give lipid hydroperoxide, R-OOH. No chain propagation occurs.

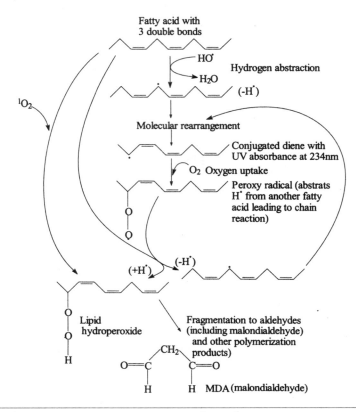

Fig. 4.6 Steps involved in lipid peroxidation by HO$^•$ and $^1O_2$, and propagation of the peroxidation process initiated by HO$^•$. (Reproduced from Halliwell and Gutteridge 1985, with permission of the publisher)

4.9.4 Oxidative Stress in Plants

Significant quantities of ROIs are in fact commonly produced in various compartments or organelles even under normal conditions. To countermine the toxicity of ROIs, living organisms possess highly efficient defense systems, called antioxidative or antioxidant systems, comprising of both non-enzymatic and enzymatic constituents (Fig. 4.7). The non-enzymatic antioxidants are generally small molecules that include the tripetides glutathione, cysteine, hydroxyquinone, ascorbate (vitamin C), the lipophilic antioxidants α-tocopherol, carotenoid pigments, alkaloids, and a variety of other compounds (Larson 1988). The enzymatic antioxidant components include the enzymes capable of removing, neutralizing or scavenging ROIs, such as catalase (Cat), peroxidase (Px), ascorbate peroxidase (APx), superoxide dismutase (SOD), glutathione reductase (GR), monodehydroascorbate reductase (MDHAR) and dehydroascorbate reductase (DHAR). The non-enzymatic and enzymatic components work in close coordination for the effective removal of ROIs. Thus, a sort of balance is maintained between their formation and destruction. A shift in the balance between the prooxidative and antioxidative reactions in favour of the former, or inhibition of the functioning of the antioxidative system, will lead to accumulation of the toxic ROIs, otherwise called oxidative stress.

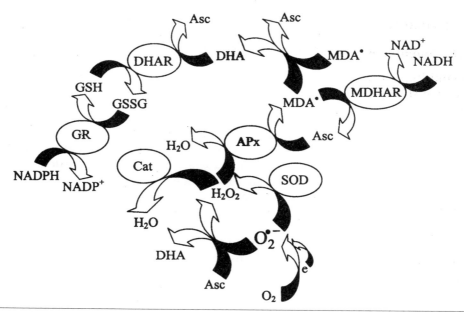

Fig. 4.7 Coordinated functioning of various antioxidative components in plants. *Apx* Ascorbate peroxidase; *Cat* catalase; *SOD* superoxide dismutase; *MDHAR* monodehydroascorbate reductase; *DHAR* dehydroascorbate reductase; *GR* glutathione reductase; *Asc* ascorbic acid; *MDA*• monodehydroascorbate radical; *DHA* dehydroascorbate; *GSH* glutathione; *GSSG* oxidized glutathione

4.9.5 Heavy Metal Induced Oxidative Damage

Oxidative toxicity of heavy metals in plants is a relatively recent concept, although it is well documented in animal systems; studies have shown that heavy metals such as iron, copper, cadmium, chromium, lead, mercury, nickel and vanadium exhibit the ability to produce ROIs, resulting in lipid peroxidation, DNA damage, depletion of sulfhydryls, and altered calcium homeostasis (Stohs and Bagchi 1995). The growing body of evidence, nevertheless, suggests that in plants also the toxicities associated with heavy metals may be due at least in part to oxidative damage (Lidon and Henriques 1993; Shaw 1995a,b; Stohs and Bagchi 1995; Maksymiec 1997; Shaw and Rout 1998). In fact, plants, most of which are green, face greater danger of oxidative damage upon exposure to heavy metals than animals because of the photosynthetic process they use, as stated earlier.

The threat of oxidative damage to living tissues by heavy metals may be due to two reasons: (1) as a result of enhancement in production of ROIs, and (2) as a result of slowing down or inhibition of the removal/scavenging of ROIs. Both lead to enhanced accumulation of ROIs. As far as plants are concerned, heavy metals may enhance generation of ROIs by interfering with the respiratory processes, similar to that in animal systems, and in addition by impairing the photosynthetic processes, specific to plants.

4.9.5.1 Oxidative Damage: Enhanced Production of ROIs

4.9.5.1.1 Impairment of Respiratory Processes

The O_2-requiring phase of the respiratory reactions is carried out in mitochondria where NADH and FADH produced during the Kreb's cycle are oxidized through a chain of electron carriers leading to the production of ATP and H_2O. The end reaction of the electron transport chain is catalyzed by cytochrome oxidase, which uses four electrons to reduce a molecule of O_2 to two molecules of water (Fig. 4.8). Some other components of the electron transport chain are, however, less well designed, and leak out a few electrons to O_2 departing from the original route of transfer. The leakage produces a univalent reduction of O_2 to give $O_2^{\cdot-}$. The main site of leakage seems to be the NADH-coenzyme reductase complex, complex I (Halliwell and Guteridge 1985; Naqui et al. 1986; Moller 2001). This is because the complex is placed at a more negative redox potential than the redox potential of $O_2/O_2^{\cdot-}$ (–0.33 V). Another probable site is the reduced form of coenzyme Q: electrophiles, like quinoid compounds, when reduced to semiquinones can subsequently reduce O_2 via one-electron transfer within a process termed redox cycling (Cadenas 1989).

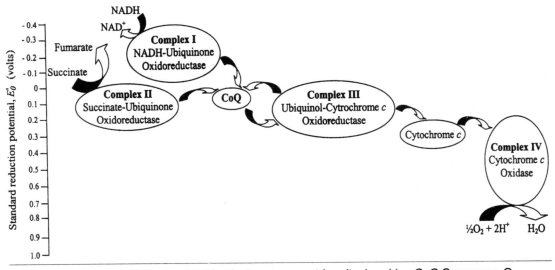

Fig. 4.8 Redox complexes involved in electron transport in mitochondria. *CoQ* Coenzyme Q

It is believed that the presence of heavy metals somehow enhances these leakages of electrons to O_2. However, no direct evidence is available in support of this view, at least in plant systems. However, in animal systems, it has been found that when rat kidney mitochondria are incubated with mercuric ion in vitro, an approximately 4-fold increase in H_2O_2 formation occurs at the ubiquinone-cytochrome *b* region, and a 2-fold increase in the NADH dehydrogenase region (Lund et al. 1991), which gives somewhat direct evidence for enhanced formation of H_2O_2 due to impairment of respiratory chain electron transport. This is accompanied by a 3.5-fold increase in lipid peroxidation at the NADH dehydrogenase region and a small amount at the ubiquinone-cytochrome b region. An enhancement in the state 3 (substrate and ADP present in sufficient amounts) and state 4 (substrate present in sufficient amount but ADP in negligible amount) uptake of O_2 by mitochondria isolated from the roots of Al-treated wheat seedling, when compared with that isolated from the roots of untreated plants, in the presence of NADH as substrate has also been reported (de Lima and Copeland 1994). Although no measurement of H_2O_2 or lipid peroxidation was conducted, the increase in the state 4 O_2 uptake, which occurs in the presence of exhausted ADP, could be due to the leakage of electrons to O_2.

Although lipid peroxidation is linked to the production of $O_2^{\cdot-}$, it is not directly involved in the process. In aqueous solution it decays spontaneously, although very slowly, to H_2O_2 and O_2 through dismutation, as stated earlier.

$$O_2^{\cdot-} + O_2^{\cdot-} + 2H^+ \rightarrow H_2O_2 + O_2$$

The slow reaction is, however, accelerated greatly by the enzyme superoxide dismutase (SOD). The H_2O_2 produced is, in turn, broken down by catalase (Fig. 4.7). At the same time, however, H_2O_2 produced may react with $O_2^{\cdot-}$ to produce the hydroxyl radical (HO$^\cdot$), the most reactive of the ROIs.

$$H_2O_2 + O_2^{\cdot-} \rightarrow O_2 + HO^\cdot + HO^-$$

The reaction is called the Haber-Weiss reaction. In fact it is an iron-catalyzed reaction and proceeds via the following steps:

$$Fe^{3+} \rightarrow O_2^{\cdot-} \rightarrow Fe^{2+} + O_2$$

$$\underline{Fe^{2+} + H_2O_2 \rightarrow Fe^{3+} + HO^\cdot + HO^-}$$

$$O_2^{\cdot-} + H_2O_2 \rightarrow O_2 + HO^\cdot + HO^-$$

$\qquad$ Fe catalyzed

The second step of the reaction is called the Fenton reaction. Another transitional (d-block) element, Cu, has also been reported to catalyze the reaction (Wardman and Cadeias 1996).

Thus, the greater the generation of $O_2^{\cdot-}$, the higher will be the chances of HO$^\cdot$ formation, and, in turn, the greater will be the chance of peroxidative damage of the membrane lipids. Considering this relationship of $O_2^{\cdot-}$ generation and lipid peroxidation, although not a direct one, the elevated level of MDA observed in the roots of *P. aureus* seedlings exposed to Hg and Cd for 48 h (Shaw 1995b), in the roots and shoots of *Pisum sativum* seedlings exposed to Cd for 15 days (Lozano-Rodriguez et al. 1997), in the roots of *Phaseolus vulgaris* seedlings exposed to Cd and Zn for 96 h (Chaoui et al. 1997), in the roots of Cd-sensitive cultivar Totley of *Holcus lanatus* upon exposure to Cd (Hendry et al. 1992) and that in the root tips of soybean exposed to Al for 24 h (Cakmak and Horst 1991) could be a result of enhanced generation of $O_2^{\cdot-}$ due to impairment of mitochondrial electron transport chain by the metals. The highly significant increase in the levels of MDA in root, as high as 200% (de Vos et al. 1993), and in leaf, as high as 100% (Luna et al. 1994), upon exposure to copper (leaves exposed in dark) could also be the result of impairment of the respiratory chain electron transport leading to generation of $O_2^{\cdot-}$. Of course, the presence of high amounts of transitional metals, like copper or iron, would also favour enhanced generation of HO$^\cdot$ from $O_2^{\cdot-}$ through the Fenton reaction. This is reflected by the fact that Al, up to a concentration of 300 μM, and Fe(II), up to a concentration of 120 μM, were not toxic to cultured tobacco cells in nutrient solution, but the presence of 100 μM Fe(II) along with even only 100 μM of Al reduced their viability by as much as 90% with a concomitant highly significant increase in lipid peroxidation (Ono et al. 1995; Yamamoto et al. 1997). While the role of Fe(II), or the transitional elements in general, in this synergistic reaction is very much clear, the nature of the contribution of Al in stimulating lipid peroxidation in the presence of Fe(II) has only been discovered recently; Verstraeten et al. (1997) reported that Al^{3+} and Al^{3+}-related cations (Sc^{3+}, Ga^{3+}, In^{3+}, Y^{3+}, La^{3+}, Bi^{3+}) without redox capacity can stimulate Fe(II)-initiated peroxidation of lipids by increasing the lipid packing and by promoting the formation of rigid clusters, both of which may bring the phospholipid acyl chains closer together, thus favouring the propagation step of lipid peroxidation.

4.9.5.1.2 *Impairment of Photosynthetic Processes*

In addition to the generation of $O_2^{\cdot-}$ by leakage of electrons from the respiratory chain, plants also face the risk of generation of $O_2^{\cdot-}$ because of the photosynthetic processes they conduct, which use another electron transport chain, for capturing light energy from the sun. The electron transport chain of chloroplasts, like that of mitochondria, is leaky and can lead electrons to O_2; therefore, the chances of formation of $O_2^{\cdot-}$ increases greatly because of the fact that the green cells in light remain saturated with O_2, which is produced as a byproduct of photosynthesis. Isolated illuminated chloroplast thylakoids have been observed to take up oxygen in the absence of added electron acceptors, which is referred as the "Mehler reaction" (Halliwell 1981). This is due to the reduction of O_2 to $O_2^{\cdot-}$ by the electron acceptors of photosystem I (PS I). Addition of the stromal protein ferredoxin (Fd) increases the amount of oxygen uptake, as Fd is reduced by PS I much more quickly than is O_2, and the reduced ferredoxin can then reduce the O_2 (Fig. 4.9).

The photosynthetic process is also the cause of generation of singlet oxygen ($^1O_2$) in illuminated plants. $^1O_2$ is generated most often by photosensitization: if a photosensitizer molecule is illuminated with high light of a given wavelength, it absorbs the light energy and the energy raises the molecule into an excited state. The excitation energy can then be transferred onto an adjacent oxygen molecule converting it to singlet state whilst the photosensitized molecule returns to ground state. Chlorophylls *a* and *b* are good examples of photosensitizers. In intact thylakoid, the light energy absorbed by the light-harvesting array of pigments, Chl *a*, Chl *b*, carotenoids, etc., called the pigment bed, is channeled to the reaction centre Chl *a* molecules of PS I and PS II by a series of excitation and deexcitation processes (Fig. 4.10). The chlorophyll molecules in the pigment bed get excited to singlet state upon absorption of a photon of light. Carotenoids also transfer their excitation energy to Chl molecules raising them to singlet state. The lifetime of the singlet state is very short (less than 10^{-11}s), but this state of the chlorophylls is used for the transfer of energy to another molecule of Chl (the former return to the ground state and the latter get excited to singlet state) until the energy of excitation reaches the reaction centre Chl *a* of PS I or PS II. The reaction centre Chl transfers the excited electron to the neighbouring electron acceptor, a process called charge separation (Halliwell 1981; Demmig-Adams and Adams 1992; Foyer et al. 1994). However, the decay of the singlet excited state of chlorophylls, including that of the reaction center chlorophylls, can also be by (1) radiative decay as fluorescence, (2) dissipation of the energy as heat, and

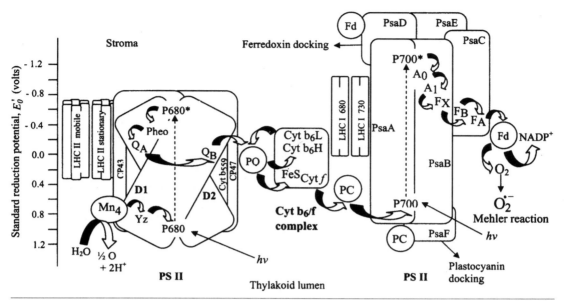

Fig. 4.9 Schematic representation of photosynthetic light reactions showing composition (partial) of photosystem II (*PSII*) and photosystem I (*PSI*) and 'Z-scheme' of photosynthetic electron transport. Both PSII and PSI are composed of numerous gene products, but only those whose functions are understood in some detail are shown. The polypeptides *D1* and *D2* bind all the electron carriers of PSII involved in the transfer of electrons from Yz to plastoquinone. *Yz* Tyrosine at 161 position of the D1 protein; *P680* reaction centre chlorophyll *a*; *P680** excited P680; *Pheo* pheophytin; Q_A and Q_B bound plastoquinone; *PQ* reduced plastoquinone; *Cyt b_{559}* b-type cytochrome; Mn_4 water-oxidizing complex; *CP43* and *CP47* Chl *a* binding proteins of PSII core complex; *LHCII* light-harvesting complex, stationary and mobile, associated with PSII. The gene products *PsaA*, *PsaB* and *PsaC* all bind the electron carriers of PSI involved in the transfer of electrons from P700 to Fd. In addition, PsaA and PsaB also bind Chl *a* and carotenoids, and form the intrinsic antenna of PSI. The proteins PsaD and PsaF are the docking sites for Fd and PC, respectively, while PsaE is believed to be involved in cyclic electron flow (not shown) and Fd reduction. *P700* Reaction centre Chl *a*; *P700** excited P700; A_0 chlorophyll *a*; A_1 phylloalloquinone; F_X, F_A and F_B iron-sulfur (FeS) clusters; *Fd* ferredoxin; *LHCI 680* and *LHCI 730* light-harvesting complexes associated with PSI; *Cyt b_6f* complex of cytochrome b_6 and cyt *f*; *Cyt b_6H* cytochrome b_6 high potential; *Cyt b_6L* cytochrome b_6 low potential; *NADP+* nicotinamide adenine dinucleotide phosphate; *PC* plastocyanin

(3) conversion to triplet state (Fig. 4.10). Triplet Chl (*Chl) has no role in photosynthesis. Its lifetime is also much longer than the singlet Chl state, which facilitates its interaction with the ground state O_2 to produce $^1O_2$ (Foyer et al. 1994).

The production of $^1O_2$ occurs when the transfer of electrons along the photosynthetic electron transport chain does not occur on a par with the excitation of the pigment molecules in the pigment bed (Demmig-Adams and Adams 1992). This occurs also in nature, as plants are exposed to a much greater photon flux density (nearly 2000 μmol m^{-2} s^{-1}) than required to saturate the dark reaction (less than 1000 μmol m^{-2} s^{-1}) of photosynthe-

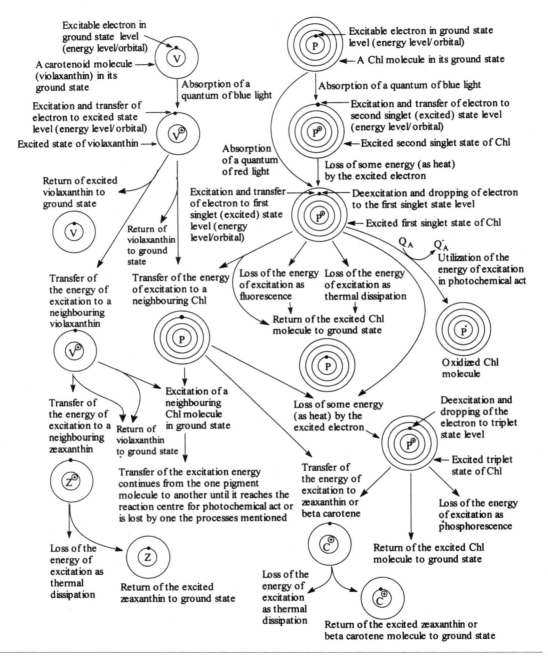

Fig. 4.10 Excitation energy transfer and possible fates of a quantum of light energy absorbed by photosynthetic pigments. *P* A chlorophyll molecule; *V* a violaxanthin molecule; *Z* a zeaxanthin molecule; *C* a β-carotene or zeaxanthin molecule. Superscript (+) against P, V, Z or C denotes their excited state

sis. To take care of this natural stress, which increases the chances of the formation of $^1O_2$, plants have adapted two strategies: the first is the rapid quenching of both triplet chlorophyll states and $^1O_2$ by membrane-bound quenchers, and the second is the regulation of the light-harvesting apparatus to minimize triplet chlorophyll production (Foyer et al. 1994; Logan et al. 1999).

The membrane-bound quenchers or antioxidants include carotenoids and α-tocopherol (vitamin E). α-Tocopherol is a lipophilic antioxidant capable of quenching singlet oxygen. The product of the reaction of α-tocopherol with singlet oxygen is a free radical, α-tocopherylquinone, which can be reduced back by ascorbic acid present in stroma at high concentration (Halliwell and Gutteridge 1985). It also quenches the reducing peroxidized membrane lipids (ROO$^{\cdot}$), thus breaking the lipid peroxidation cascades (Battle et al. 1976; Fryer 1992). The phenols present in plants in high amounts are also involved in quenching singlet oxygen. Amino acids like methionine, cysteine, histidine and tryptophan also react with singlet oxygen, reducing the risk of probable damage by the molecule (Halliwell and Gutteridge 1985).

The membrane-bound carotenoids not only quench $^1O_2$, but also prevent the formation of $^1O_2$ by quenching the triplet excited state of chlorophylls (Fig. 4.10) that otherwise would lead to the formation of $^1O_2$. For example, β-carotene, and also other higher plant carotenoids, react with $^1O_2$, preferably over the chlorophylls and proteins of the photosystems, quenching the energy of excitation. They also quench the energy of triplet chlorophyll, preventing the formation of $^1O_2$ (di Mascio et al. 1989).

$$Car + {}^*Chl \text{ (or } {}^1\Delta gO_2) \rightarrow {}^*Car + Chl \text{ (or } {}^3\Sigma gO_2)$$

Subsequently, the excited carotenoid (*Car) undergoes radiationless decay to ground state (Car).

The second process, the regulation of the light-harvesting apparatus, involves thermal dissipation of the excess excitation energy resulting in quenching of singlet-excited chlorophyll to the ground state. The carotenoids of the xanthophyll cycle, violaxanthin, antheraxanthin and zeaxanthin, play a major role in the process. During exposure to excess light the concentration of H^+ in the thylakoid lumen increases because of a decrease in the ATPase-mediated efflux of H^+ due to the reduced availability of ADP for ATP synthesis and continued electron transport mediated influx of H^+, creating a transthylakoid membrane proton-gradient, ΔpH (Demmig-Adams and Adams 1996a,b; Yamamoto and Bassi 1996a; Gilmore and Govindjee 1999). This ΔpH, and the decrease in the intrathylakoid pH, induce enzymatic de-epoxidation of violaxanthin (a de-epoxide) to antheraxanthin (a mono-epoxide) and further to zeaxanthin, which is epoxide free (Fig. 4.11; Demmig-Adams and Adams 1992, 1996b; Eskling et al. 1997; Gilmore and Govindjee 1999; Logan et al. 1999). The enzyme-catalyzed removal of the epoxide group is a reductive process utilizing ascorbic acid as a co-substrate (Yamamoto 1979; Bratt et al. 1995; Eskling et al. 1997). Strong correlation exists between leaf zeaxanthin (and antheraxanthin) content and the level of energy dissipation, measured as non-photochemical quenching of chlorophyll fluorescence (Gilmore and Yamamoto 1993; Adams et al. 1995; Demmig-Adams and Adams 1996a).

Zeaxanthin (and antheraxanthin) mediated deexcitation of singlet Chl to ground state with the release of energy as heat occurs as a direct or indirect method. Recent work on the photophysiology of carotenoids of the xanthophyll cycle has shown that a direct

Lumen	Thylakoid membrane	Stroma

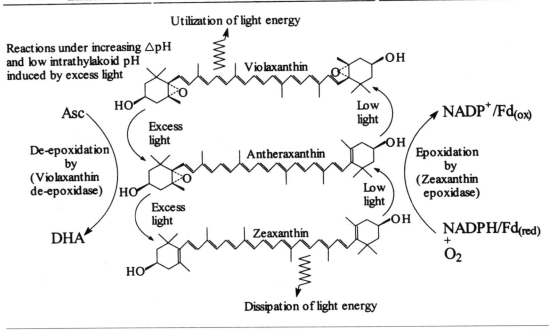

Fig. 4.11 Model of the xanthophyll cycle and its regulation by light. *Asc* Ascorbate; *DHA* dehydroascorbate; *Fd* ferredoxin. (Adapted from Eskling et al. 1997, with permission of the publisher)

transfer of excitation energy from chlorophyll to zeaxanthin (and possibly antheraxanthin) is thermodynamically possible (Frank et al. 1994; Owens 1996). Carotenoids possess a low-lying excited state, the energy level of which is inversely proportional to the number of conjugated carbon-carbon double bond (two covalent double bonds separated by a single bond) within the molecule (9 for violaxanthin, 10 for antheraxanthin and 11 for zeaxanthin). These energy levels are very near the singlet-excited energy level for chlorophyll; violaxanthin's low-lying excited level is apparently slightly above that of singlet-excited chlorophyll, antheraxanthin's is equal to or just below, and zeaxanthin's is below (Frank et al. 1994). After the transfer of excitation energy from singlet Chl to zeaxanthin (and possibly also to antheraxanthin), the latter undergoes a rapid radiationless deexcitation safely releasing the energy as heat (Fig. 4.10).

In the indirect facilitation of deexcitation of singlet Chl, increasing lumen pH causes protonation of the minor Chl *a+b* binding proteins (CPs; Horton et al. 1996; Ruban et al. 1996) leading to a change in their conformation with activation of a binding site on the lumenal side to which zeaxanthin or antheraxanthin binds, and the photosystem unit (PS II) turns into an unit with increased rate constant of heat dissipation (Gilmore and Govindjee 1999). It has also been suggested that protonation of antenna proteins may lead to conformational changes that bring chlorophyll and zeaxanthin molecules into orbital overlap which is necessary for effective energy transfer and its dissipation (Logan et al. 1999).

Looking into the reasons of production of $O_2^{\cdot-}$ and $^1O_2$ in plants, it is very conceivable that the generation of these ROIs can be enhanced greatly by the environmental factors impairing the photosynthetic processes. For example, if the Calvin cycle is inhibited, or the supply of CO_2 is decreased, it will lead to accumulation of NADPH and deficiency of NADP$^+$, which in turn would increase the chances of transfer of electrons from reduced ferredoxin to O_2 molecules leading to the formation of $O_2^{\cdot-}$. Furthermore, as the redox potential of $O_2/O_2^{\cdot-}$ (–0.33 V) is only slightly more negative than that of the Fd-NADPH reductase, and the normal preferred route of transfer of electrons from Fd is NADP$^+$ (Fig. 4.9), reduced ferredoxin will accumulate leading to back pressure on the photosynthetic electron transport chain for the transport of electrons. This will increase the gap between the transfer of electrons along the photosynthetic electron transport chain and the excitation of Chl molecules further, enhancing the chances of formation of triplet Chl, and in turn the formation of $^1O_2$. Similarly, the inhibition of the electron transport chain will also enhance the chances of generation of $^1O_2$, but there should not be any formation of $O_2^{\cdot-}$.

Heavy metals have been reported to impair the photosynthetic process in terrestrial plants at several steps, and hence it is expected that plants under heavy metal stress must face the challenge of oxidative stress as well. For example, Stiborova et al. (1987, 1988) reported significant inhibition of the activity of Rubisco in maize and barley seedlings by Cu, Co and Zn, both in vitro and in vivo. Similarly, Sheoran et al. (1990) observed inhibition of several enzymes of the photosynthetic carbon reduction cycle, such as Rubisco, 3-PGA kinase, NADP and NAD-glyceraldehyde 3-P dehydrogenase, aldolase and FDPase, to various levels (2-61%) by Cd and Ni, accompanied by a reduction in photosynthesis. The authors also observed a significant reduction in the chlorophyll contents of the leaves. Although they did not take any measurements of oxidative stress, the decrease in the chlorophyll content by as much as 25% in response to exposure to both the metals may be taken as a clear indication of damage of Chl by generation of $^1O_2$. A significant decrease in Chl has also been reported in seedlings of *Phaseolus aureus* exposed to Cd (Shaw 1995a) and *Phaseolus vulgaris* exposed to Ni (Krupa et al. 1993a). In both cases, the Chl a/b ratio also decreased significantly with increase in concentration of the metal indicating preferential degradation of Chl a over Chl b; Chl a being present in greater quantity and remaining in close association with the reaction centres, bound to CP43 and CP47 in PS II (Yamamoto and Bassi 1996b) and to PsaA in PSI (Irrgang 1999) (Fig. 4.9), is liable to be destroyed in greater number by the attack of $^1O_2$ when compared with Chl b remaining located in LHC II and LHC I constituting the outer antenna of PS II and PS I, respectively.

Working on isolated oat leaves, Luna et al. (1994) observed a significant decrease in their chlorophyll and carotenoid contents upon exposure to increasing concentrations of Cu under light. In contrast, however, Maksymiec and Baszynski (1996b) reported an increase in the levels of chlorophylls in *Phaseolus coccineus* (runner bean) grown hydroponically on $CuSO_4$ solution. Nevertheless, the increase was accompanied by a decrease in Chl a/b and Chl/Car ratios, suggesting oxidative damage of Chl a; the increase in the levels of chlorophylls could be because of inhibition of the tissue elongation processes by Cu (Maksymiec and Baszynski 1996b). Increase in chlorophyll contents in growing seedlings in response to heavy metals (Hg and Cd) has also been suggested as an adaptive process to keep up the photosynthetic activity equivalent to that of control (Shaw and Rout 1998). In reality, however, the photosynthetic activity normalized to Chl decreases (Shaw and Rout 1998).

The decrease of photosynthesis in plants exposed to heavy metals could be due to inhibition of transport of electrons across the photosynthetic electron transport chain (PS I and PS II activity), as observed in in vitro studies of Hg, Zn, Cd, Co, Cr and Cu toxicity (Murthy et al. 1989; Husaini et al. 1991; Prasad et al. 1991; Rai et al. 1991a,b; Babu et al. 1992; Yruela et al. 1992, 1996; Maksymiec 1997). However, in vivo studies of Ni, Cd and Cu toxicity have indicated that, although these metals inhibit photosynthetic electron transport in the test plants, the inhibition is not the result of a direct effect of the metals on the photosystems (Krupa et al. 1993a,b; Maksymiec et al. 1994; Maksymiec and Baszynski 1996a,b). Rather the inhibition is indirect, as a result of inhibition of Calvin cycle reactions, which leads to accumulation of ATP and NADPH, and in turn down regulation or feedback inhibition of the photosynthetic electron transport (Krupa et al. 1993a, b).

Irrespective of the steps at which photosynthesis is inhibited in the plants challenged by heavy metals, it is implicit, however, that the inhibition would lead to generation of $O_2^{.-}$ and $^1O_2$, and the plant would be facing oxidative stress. Thus, oxidative damage may be considered as inherent with heavy metal toxicity, at least in green plants. In fact, oxidative damage of green tissue upon exposure to heavy metals under light has been reported by many workers. For example, rice plants grown over Cu-containing nutrient solution showed elevated levels of malondialdehyde when compared with control (Lidon and Henriques 1993). Baryla et al. (2000) and Mazhoudi et al. (1997) reported a significant increase in lipid peroxidation in plantlets of *Brassica napus* and in the leaves of tomato seedlings, respectively, exposed to Cu. Again, Luna et al. (1994) working on oat leaves observed a significant increase in MDA content in illuminated leaves exposed to Cu, and that the presence of Tiron (4, 5-dihydroxy-1, 3-benzenedisulfonic acid—a scavenger of $O_2^{.-}$) and sodium benzoate and mannitol (scavengers of $HO^.$) in the treatment solution inhibited the formation of MDA significantly. Furthermore, the protection provided by sodium benzoate and mannitol was greater than that provided by Tiron, suggesting that the oxidative damage was largely through the production of the $HO^.$ radical (Luna et al. 1994). Similarly, the content of thiobarbituric acid reactive metabolites (TBArm) was significantly increased upon exposure of the seedlings to Zn, and this was accompanied by a significant increase in the levels of H_2O_2 in the leaf tissue when compared with the control (Weckx and Clijsters 1997), suggesting oxidative damage of leaf tissue by the metal in light.

4.9.5.2 Oxidative Damage: Inhibition of Removal of ROIs

The oxidative damage induced by heavy metals as a result of inhibition of the removal of ROIs is mediated through the inhibition of the functioning of one or more of the enzymes and/or depletion of one or more of the antioxidant molecules of the antioxidative system by the heavy metals. Several reports are available to substantiate this view. For example, Cakmak and Horst (1991) observed that the increase in the MDA content of the root tips of soybean (*Glycine max*) exposed to Al for 48 h was concomitant with a significant decrease in catalase activity. They also observed a significant increase in SOD activity. They concluded that, while the increase in SOD activity resulted in enhanced formation of H_2O_2, the latter accumulated as a result of a decrease in the catalase activity, and the accumulated H_2O_2 was mostly consumed in oxidative processes leading to enhanced accumulation of MDA. It is well established that H_2O_2 gives rise to the highly reactive

HO˙ radical through the Fenton reaction. Working on detached oat leaves, Luna et al. (1994) also observed a significant increase in MDA level with concomitant decrease in the activity level of catalase and ascorbate peroxidase upon exposure of the leaves to Cu^{2+}; a significant increase in the level of SOD was also observed. The observation further supports the view that the inhibition of the antioxidative enzymes involved in scavenging H_2O_2 could be one of the reasons for oxidative damage to plant tissues exposed to heavy metals. In fact, the accumulation of MDA in plant tissues exposed to heavy metals, as in the case of oat leaves exposed to copper (Luna et al. 1994), roots of tomato seedlings exposed to Cu (Mazhaudi et al. 1997), roots and leaves of *Phaseolus vulgaris* seedlings exposed to Cd and Zn (Chaoui et al. 1997) and roots of soybean exposed to Al for 72 h (Cakmak and Horst 1991), has also been observed to be significantly correlated with an increase in the H_2O_2 level, which in turn was accompanied by a significant decrease in the activity of catalase.

In addition to the inhibition of the H_2O_2-scavenging enzymes catalase and ascorbate peroxidase, Lidon and Henriques (1993) observed inhibition of SOD as well in the leaves of rice seedlings exposed to Cu^{2+}. From the inhibition in the activity of SOD it may appear that the accumulation of H_2O_2 would decrease, and so also the oxidative damage. However, in fact, the inhibition of SOD would also enhance the chances of formation of the HO˙ radical by increasing the availability of $O_2^{·-}$ for reaction with H_2O_2 (Haber-Weiss reaction), as reflected by the increase in MDA content of the leaf tissue despite inhibition of the SOD activity (Lidon and Henriques 1993). Gallego et al. (1999) reported a decrease in the activity of most of the antioxidative enzymes in sunflower cotyledons in response to Cd treatment.

Depletion of the antioxidant molecule GSH has been reported in *Silene cucubalus* and the marine planktonic diatom *Ditylum brightwellii* upon exposure to Cu (de Vos et al. 1992; Rijstenbil et al. 1994), which is accompanied by oxidative stress in the plants. This is probably because GSH is responsible for converting the oxidized ascorbate, oxidized during ROI-scavenging reactions (Fig. 4.7), back into the reduced form, and hence its depletion is bound to lead to accumulation of H_2O_2 and $O_2^{·-}$. The depletion that resulted is due to synthesis of phytochelatin, for which GSH serves as the only substrate, in plants upon exposure to heavy metals, particularly to Cd, Cu, Zn, etc. (Ahner and Morel 1995; Rauser 1995; Chen et al. 1997). Information on the response of ascorbic acid, another important antioxidant component of the antioxidative cycle (Fig. 4.7), to heavy metals is scant.

The concept of inhibition of the removal of ROIs due to inhibition of the antioxidative enzymes as the cause of oxidative damage by heavy metals is, however, not widely accepted, or is considered of much less significance than that due to additional generation of ROIs. This is because there are reports of highly significant increases in the level of MDA despite no significant changes in the activity of the antioxidative enzymes, for example, of catalase and SOD in the leaves of seedlings of *Phaseolus vulgaris* exposed to Zn (Weckx and Clijsters 1997), of catalase in the root tips of soybean exposed to Al for 24 h (Cakmak and Horst 1991), and of catalase in the stems of *Phaseolus vulgaris* seedlings exposed to Cu and Zn (Chaoui et al. 1997). A significant increase in the level of MDA has also been observed despite a significant increase in the activity of the H_2O_2-scavenging enzymes catalase and ascorbate peroxidase in *Phaseolus aureus* exposed to Hg or Cd (Shaw

1995b). Furthermore, in many studies, the activity of the antioxidative enzymes, including that of catalase, has been found to increase in response to heavy metal treatment (Subhadra et al. 1991; Weckx and Clijsters 1996; Gwozdz et al. 1997; Shaw and Rout 1998; Rucinska et al. 1999; Schicker and Caspi 1999; Teisseire and Guy 2000). In fact, the activity of the antioxidative enzymes has been reported to increase significantly in response to environmental stress in general, and the increase in their activity is considered as circumstantial evidence of induction of oxidative stress by an environmental stress (Foyer et al. 1994; Kangasjarvi et al. 1994; Polle et al. 2000; Rout and Shaw 2001).

4.10 CONCLUSIONS

Thus, from the foregoing discussion it is very clear that, although the term heavy metal represents a group of metals, classified based on their specific gravity, their chemical characteristics differ greatly, from hard acceptor to soft acceptor on the 'softness' scale of Pearson, and from S and N seeking to that of O_2 seeking. Accordingly, their interactions with biomolecules also differ greatly. There are, of course, the borderline or intermediate category elements, as the case may be, showing the characteristics of both the contrasting groups, although of less intensity than that shown by the elements of the respective group. In the toxicity series we find that the metals placed at the higher toxic levels are those showing preference for binding with ligands containing S or N, and those placed at the lower levels show preference for binding with ligands containing oxygen. However, irrespective of the toxicity, many of the metals of both the major groups, as well as that of the intermediate or borderline categories, are essential for a plant's life, and in general for the life of living organisms, at low concentration.

It has also been demonstarted that most of the heavy metals, the metals of environmental concern, belong to group B and borderline elements (Nieboer and Richardson 1980), and their toxic effect is mostly as a result of their interaction with enzymes or proteins, resulting in changes in their properties and functions. However, the ultimate toxic effect does not remain confined to the impairment of the functions that the enzymes or the protein molecules might be performing. Rather it is carried beyond. If the metabolic functions affected happen to be linked with energy metabolism (respiration or photosynthetic processes), then accumulation of ROIs may occur, leading to oxidative damage to the cells and tissues. In fact, in plants involved actively in capturing solar energy, oxidative damage to cells and tissues could be the measure cause of heavy metal induced toxic injury. Probably, this is why the phenotypic responses of plants upon exposure to most of the heavy metals are very similar. The concept of oxidative-phytotoxicity of heavy metals, nevertheless, needs further consolidation supported by more data and investigations.

ACKNOWLEDGEMENTS

The authors express their gratitude to the Director, Dr. S.K. Gupta and the former director, Prof. A. P. Dash of the Institute for providing the basic infrastructure necessary to complete the article, and to all the authors and publishers that kindly permitted the reproduction of material from their publications. Financial support for this work was drawn from projects from NRSA, SAC and DBT, Govt. of India.

REFERENCES

Aasami T (1984) Pollution of soil by cadmium. In: Nriagu JO (ed) Changing metal cycles and human health, Dahlem Konferenzen. Springer, Berlin Hiedelberg New York, pp 95-111

Absil MCP, van Scheppingen Y (1996) Concentration of selected heavy metals in benthic diatoms and sediment in the Westerschelde estuary. Bull Environ Contam Toxicol 56: 1008-1015

Adams WW III, Demmig-Adams B, Verhoeven AS, Baker DH (1995) 'Photoinhibition' during winter stress: involvement of sustained xanthophylls in cycle-dependent energy dissipation. Aust J Plant Physiol 22: 261-276

Adriano DC (1986) Trace elements in the terrestrial environment. Springer, Berlin Heidelberg New York

Ahner BA, Morel FMM (1995) Phytochelatin production in marine algae. 2. Induction by various metals. Limnol Oceanogr 40: 658-665

Ahrland S (1968) Thermodynamics of complex formation between hard and soft acceptors and donors. Nature and scope of the classification of acceptors and donors as hard and soft. Struct Bonding 5: 118-123

Albers PH, Camardese MB (1993a) Effects of acidification on metal accumulation by aquatic plants and invertebrates. 1. Constructed wetlands. Environ Toxicol Chem 12: 999-967

Albers PH, Camardese MB (1993b) Effects of acidification on metal accumulation by aquatic plants and invertebrates. 2. Wetlands, ponds and small lakes. Environ Toxicol Chem 12: 969-976

Allaway WH (1968) Agronomic controls over environmental cycling of trace elements. Adv Agron 20: 235-274

Anonymous (1964) Encyclopedia of chemical science. Van Norstrand, Princeton, p 533

Arduini I, Godbold DL, Antonino O (1996) Cadmium and copper uptake and distribution in Mediterranean tree seedlings. Physiol Plant 97: 111-117

Asada K (1992) Production and scavenging of active oxygen in chloroplasts. In: Scandalios JG (ed) Photoinhibition. Cold Spring Harbor Laboratory Press, Cold Spring Harbor, pp 173-192

Asada K (1994) Production and action of active oxygen species in photosynthetic tissues. In: Foyer CH, Mullineaux PM (eds) Causes of photooxidative stress and amelioration of defense systems in plants. CRC Press, Boca Raton, pp 77-104

Babiarz CL, Andren AW (1995) Total concentrations of mercury in Wisconsin (USA) lakes and rivers. Water Air Soil Pollut 83: 173-183

Babu TS, Sabat SC, Mohanty P (1992) Alterations in photosystem II organization by cobalt treatment in the cyanobacterium *Spirulina plantensis*. J Plant Biochem Biotechnol 1: 61-63

Baker AJM, Reeves RD, Hajar ASM (1994) Heavy metal accumulation and tolerance in British population of the metallophyte *Thalaspi caerulescens* J. & C. Presl (Brassicaceae). New Phytol 127: 61-68

Barak NAE, Mason CF (1990) Mercury, cadmium and lead concentrations in five species of freshwater fish from eastern England. Sci Total Environ 92: 257-263

Bartosz G (1997) Oxidative stress in plants. Acta Physiol Plant 19: 47-64

Baryla A, Laborde C, Montillet J-L, Triantaphylides C, Chagvardieff P (2000) Evaluation of lipid peroxidation as a toxicity bioassay for plants exposed to copper. Environ Pollut 109:131-135

Battle RW, Gaunt JK, Laidman DL (1976) The effect of photoperiod on endogenous γ-tocopherol and plastochromanol in leaves of *Xanthium strumarium* L. (Cocklebur). Biochem Soc Trans 4: 484-486

Bonnet M, Camares O, Veisseire P (2000) Effects of zinc and influence of *Acremonium lolii* on growth parameters, chlorophyll *a* fluorescence and antioxidant enzyme activites of ryegrass (*Lolium perenne* L. cv. Apollo). J Exp Bot 51: 945-953

Bowen HJM (1979) Environmental chemistry of the elements. Academic Press, New York

Bratt CE, Arvidsson P-O, Carlsson M, Akerlund H-E (1995) Regulation of violaxanthin de-epoxidase activity by pH and ascorbate concentration. Photosynth Res 45: 169-175

Breen AP, Murphy JA (1995) Reactions of oxyl radicals with DNA. Free Rad Biol Med 18:1033-1077

Brooks RR (1983) Biological methods of prospecting for minerals. Wiley, New York

Cadenas E (1989) Biochemistry of oxygen toxicity. Annu Rev Biochem 58: 79-110

Cakmak I, Horst WJ (1991) Effect of aluminium on lipid peroxidation, superoxide dismutase, catalase, and peroxidase activities in root tips of soybean (*Glycine max*). Physiol Plant 83: 463-468

Chaoui A, Mazhoudi S, Ghorbal MH, Ferjani EE (1997) Cadmium and zinc induction of lipid peroxidation and effects on antioxidant enzyme activities in bean (*Phaeolus vulagris* L.). Plant Sci 127: 139-147

Chen J, Zhou J, Goldbrough PB (1997) Characterization of phytochelatin synthase from tomato. Physiol Plant 101: 165-172

Chuan MC, Shu GY, Liu JC (1996) Solubility of heavy metals in a contaminated soil: effects of redox potential and pH. Water Air Soil Pollut 90: 543-556

Clarkson DT, Luttge V (1989) Divalent cations, transport and compartmentation. Prog Bot 51: 93-112

Connell DW, Miller GJ (1984) Chemistry and ecotoxicology of pollution. Wiley, New York

Cruz AC, Fomssgaard, IS, Lacayo J (1994) Lead, arsenic, cadmium and copper in Lake Asososca, Nicaragua. Sci Total Environ 155: 229-236

Cumming JR, Taylor GJ (1990) Mechanism of metal tolerance in plants: physiological adaptations for exclusion of metal ions from the cytoplasm. In: Alscher RG, Cumming JR (eds) Stress responses in plants: adaptation and acclimation mechanisms. Wiley-Liss, New York, pp 329-356

Davies MS, Francies D, Thomas JD (1991) Rapidity of cellular changes induced by zinc in a zinc tolerant and non-tolerant cultivar of Festuca rubra L. New Phytol 117: 103-108

De Lima ML, Copeland L (1994) The effect of aluminium on respiration of wheat roots. Physiol Plant 90: 51-58

De Vos CHR, Vonk MJ, Vooijs R, Schat H (1992) Glutathione depletion due to copper-induced phytochelatin synthesis causes oxidative stress in *Silene cucubalus*. Plant Physiol 98: 853-858

De Vos CH, Bookum WMT, Vooiji R, Schat H, de Kok LJ (1993) Effect of copper on fatty acid composition and peroxidation of lipids in the roots of copper tolerant and sensitive *Silene cucubalus*. Plant Physiol Biochem 31: 151-158

Delhaize E, Ryan PR (1995) Aluminum toxicity in plants. Plant Physiol 107: 315-321

Demming-Adams B, Adams WW III (1992) Photoprotection and other responses of plants to high light stress. Annu Rev Plant Physiol Plant Mol Biol 43: 599-626

Demmig-Adams B, Adams WW III (1996a) The role of xanthophylls cycle carotenoids in the protection of photosynthesis. Trends Plant Sci 1: 21-26

Demming-Adams B, Adams WW III (1996b) Xanthophyll cycle and light stress in nature: uniform response to excess direct sunlight among higher plant species. Plant 198: 460-470

Demming-Adams B, Gilmore AM, Adams WW III (1996) In vivo functions of carotenoids in higher plants. FASEB J 10: 403-412

Denton GRW, Burdon-Jones C (1986) Trace metals in algae from the Great Barrier Reef. Mar Pollut Bull 17: 98-107

Di Mascio P, Kaiser S, Sies H (1989) Lycopene as the most efficient biological carotenoid singlet oxygen quencher. Arch Biochem Biophys 274: 532-538

Eskling M, Arvidsson P-O, Akerlund H-E (1997) The xanthophylls cycle, its regulation and components. Physiol Plant 100: 806-816

Fabris JG, Richardson BJ, O'Sullivan, Brown FC (1994) Estimation of cadmium, lead, and mercury concentrations in estuarine waters using the mussel *Mytilus edulis planulatus* L. Environ Toxicol Water Qual 9: 183-192

Fergusson JE (1990) The heavy elements: chemistry, environmental impact and health effects. Pergamon Press, Oxford

Forstner U, Wittmann (1979) Metal pollution in the aquatic environment. Springer, Berlin Heidelberg New York

Fowler SW (1990) Critical review of selected heavy metal and chlorinated hydrocarbon concentrations in the marine environment. Mar Envron Res 29: 1-64

Foy CD, Chaney RL, White MC (1978) The physiology of metal toxicity in plants. Annu Rev Plant Physiol 29: 511-566

Foyer CH, Lelandais M, Kunert KJ (1994) Photooxidative stress in plants. Physiol Plant 92:696-717

Francesconi KA, Moore EJ, Edmonds JS (1994) Cadmium uptake from seawater and food by the western rock lobster Panulirus Cygnus. Bull Environ Contam Toxicol 53: 219-223

Frank HA, Cua A, Chynwat V, Young A, Gosztola D, Wasieleweski MR (1994) Photophysics of carotenoids associated with the xanthophylls cycle in photosynthesis. Photosynth Res 41: 389-395

Fridovich I (1978) The biology of oxygen radicals. Science 201: 875-880

Fryer MJ (1992) The antioxidant effects of thylakoid vitamin E (α-tocopherol). Plant Cell Environ 15: 381-392

Gallego SM, Benavides MP, Tomaro ML (1999) Effect of cadmium ions on antioxidant defense system in sunflower cotyledons. Biol Plant 42: 49-55

Gilmore AM, Govindjee (1999) How higher plants respond to excess light: energy dissipation in photosystem II. In: Singhal GS, Renger G, Sopory SK, Irrgang K-D, Govindjee (eds) Concepts in photobiology: photosynthesis and photomorphogenesis. Narosa Publishing House, New Delhi, pp 513-548

Gilmore AM, Yamamoto HY (1993) Linear models relating xanthophylls and lumen acidity to nonphotochemical fluorescence quenching. Evidence that antheraxanthin explains zeaxanthin-independent quenching. Photosynth Res 35: 67-78

Gwozdz EA, Przymusinski R, Rucinska R, Deckert J (1997) Plant cell responses to heavy metals: molecular and physiological aspects. Acta Physiol Plant 19: 459-465

Halliwell B (1981) Chloroplast metabolism: the structure and function of chloroplasts in green cells. Clarendon Press, Oxford

Halliwell B, Gutteridge JMC (1985) Free radicals in biology and medicine. Clarendon Press, Oxford

Hamasaki T, Nagase H, Yoshioka Y, Sato T (1995) Formation, distribution, and ecotoxicology of methylmetals of tin, mercury, and arsenic in the environment. Crit Rev Environ Sci Technol 25: 45-91

Hendry GAF, Baker AJM, Ewart CF (1992) Cadmium tolerance and toxicity, oxygen radical processes and molecular damage in cadmium-tolerant and cadmium-sensitive clones of Holcus lanatus L. Acta Bot Neerl 41: 271-291

Hippeli S, Elstner FE (1997) OH-radical-type reactive oxygen species: a short review on the mechanisms of OH-radical- and peroxynitrite toxicity. Z Naturforsch 52c: 555-563

Horton P, Ruban AV, Walters RG (1996) Regulation of light harvesting in green plants. Annu Rev Plant Physiol Plant Mol Biol 47: 655-684

Hu S, Tang CH, Wu M (1996) Cadmium accumulation by several seaweeds. Sci Total Environ 187: 65-71

Husaini Y, Singh AK, Rai LC (1991) Cadmium toxicity to photosynthesis and associated electron transport system of *Nostoc linckia*. Bull Environ Contam Toxicol 46: 146-150

Irrgang K-D (1999) Architecture of the thylakoid membrane. In: Singhal GS, Renger G, Sopory SK, Irrgang K-D, Govinjee (eds) Concepts in photobiology: photosynthesis and photomorphogenesis. Narosa Publishing House, New Delhi, pp 139-180

Jalil A, Selles F, Clarke JM (1994) Growth and cadmium accumulation in two durum wheat cultivars. Commun Soil Sci Plant Anal 25: 2597-2611

Jastow JD, Koeppe DE (1980) Uptake and effect of cadmium in higher plants. In: Nriagu JO (ed) Cadmium in the environment, part 1. Ecological cycling. Wiley, New York, pp 607-638

Kabata-Pendias A, Pendias H (1984) Trace elements in soils and plants. CRC Press, Boca Raton

Kagi JHR, Hapke H-J (1984) Biochemical interactions of mercury, cadmium, and lead. In: Nriagu JO (ed) Changing metal cycles and human health, Dahlem Konferenzen. Springer, Berlin Heidelberg New York, pp 237-250

Kangasjarvi J, Talvinen J, Utriainen M, Karjalainen R (1994) Plant defence systems induced by ozone. Plant Cell Environ 17: 783-794

Knox JP, Dodge AD (1985) Singlet oxygen and plants. Phytochemistry 24: 889-896

Kochian LV (1995) Cellular mechanism of aluminum toxicity and resistance in plants. Annu Rev Plant Physiol Plant Mol Biol 46: 237-260

Krupa Z, Siedlecka A, Maksymiec W, Baszynski T (1993a) In vivo response of photosynthetic apparatus of *Phaseolus vulgaris* L. to nickel toxicity. J Plant Physiol 142: 664-668

Krupa Z, Quist G, Huner NPA (1993b) The effects of cadmium on photosynthesis of *Phaseolus vulgaris* - a fluorescence analysis. Physiol Plant 88: 626-630

Kukkola E, Rautio P, Huttunen S (2000) Stress indications in copper- and nickel-exposed Scots pine seedlings. Environ Exp Bot 43: 197-210

Lapedes DN (1974) Dictionary of scientific and technical terms. McGraw Hill, New York, p 674

Larson RA (1988) The antioxidants of higher plants. Phytochemistry 27: 969-978

Leita L, Contin M, Maggioni A (1991) Distribution of cadmium and induced Cd-binding proteins in root, stem and leaves of *Phaseolus vulgaris*. Plant Sci 77: 139-147

Lidon FC, Henriques FS (1993) Copper-mediated toxicity in rice chloroplasts. Photosynthetica 29:385-400

Lodenius M (1990) Environmental mobilization of mercury and cadmium. Publ Dept Environ Conserv Univ Helsinki, no 13, Helsinki

Logan BA, Demmig-Adams B, Adams WW III (1999) Acclimation of photosynthesis to the environment. In: Singhal GS, Renger G, Sopory SK, Irrgang K-D, Govindjee (eds) Concepts in photobiology: photosynthesis and photomorphogenesis. Narosa Publishing House, New Delhi, pp 477-512

Lozano-Rodriguez E, Hernandez LE, Bonay P, Carpena-Reiz RO (1997) Distribution of cadmium in shoot and root tissue of maize and pea plants: physiological disturbances. J Exp Bot 48:123-128

Luna CM, Gonzales CA, Trippi VS (1994) Oxidative damage caused by an excess of copper in oat leaves. Plant Cell Physiol 35: 11-15

Lund BO, Miller DM, Woods JS (1991) Mercury-induced H_2O_2 production and lipid peroxidation in vitro in rat kidney mitochondria. Biochem Pharmacol 42: S181-S187

Maksymiec W (1997) Effect of copper on cellular processes in higher plants. Photosynthetica 34: 321-342

Maksymeic W, Baszynski T (1996a) Chlorophyll fluorescence in primary leaves of excess Cu-treated runner bean plants depends on their growth stages and the duration of Cu-action. J Plant Physiol 149: 196-200

Maksymeic W, Baszynski T (1996b) Different susceptibility of runner bean plants to excess copper as a function of growth stages of primary leaves. J Plant Physiol 149: 217-221

Maksymiec W, Russa R, Urbanik-Spyniewska T, Baszynski T (1994) Effect of excess Cu on the photosynthetic apparatus of runner bean leaves treated at two different growth stages. Physiol Plant 91: 715-721

Maksymiec W, Bednara J, Baszynski T (1995) Responses of runner bean plants to excess copper as a function of plant growth stages: effects on morphology and structure of primary leaves and their chloroplast ultrastructure. Photosynthetica 31: 427-435

Manahan SE (1990) Environmental chemistry. Lewis Publishers, Boston

Martin MH, Coughtrey PJ (1982) Biological monitoring of heavy metal pollution. Applied Science Publishers, London

Martin RB (1988) Bioinorganic chemistry of aluminum. In: Siegel H, Siegel A (eds) Metal ions in biological systems: aluminum and its role in biology, vol 24. Dekker, New York, pp 2-57

Martin RB (1992) Aluminum speciation in biology. In: Cjadwik DJ, Shelan J (eds) Aluminum in biology and medicine. Wiley, New York, pp 5-25

Mason CF (1996) Biology of freshwater pollution. Longman, London

Mason G (1958) Principles of geochemistry. Wiley, New York

Mazhoudi S, Chaoui A, Ghorbal MH, Ferjani EE (1997) Response of antioxidant enzymes to excess copper in tomato (*Lycopersicon esculentum*, Mill.). Plant Sci 127: 129-137

Misra SG, Mani D (1991) Soil pollution. Ashish Publishing House, New Delhi

Mocquot B, Vangronsveld J, Clijsters H, Mench M (1996) Copper toxicity in young maize (*Zea mays* L.) plants: effects on growth, mineral and chlorophyll contents, and enzyme activities. Plant Soil 182: 287-300

Moller IM (2001) Plant mitochondria and oxidative stress: electron transport, NADPH turnover, and metabolism of reactive oxygen species. Annu Rev Plant Physiol Plant Mol Biol 52: 561-591

Moral R, Palacious G, Gomez I, Navarro-Pedreno J, Mataix J (1994) Distribution and accumulation of heavy metals (Cd, Ni and Cr) in tomato plants. Fresenius Environ Bull 3: 395-399

Muller HW, Schwaighofer B, Kalman W (1994) Heavy metal contents in river sediments. Water Air Soil Pollut 72: 191-203

Murthy SDS, Sabat SC, Mohanty P (1989) Mercury-induced inhibition of photosystem II activity and changes in the emission of fluorescence from phycobilisomes in intact cells of the cyanobacterium, *Spirulina platensis*. Plant Cell Physiol 30: 1153-1157

Naqui A, Chance B, Cadenas E (1986) Reactive oxygen intermediates in biochemistry. Annu Rev Biochem 55: 137-166

Nieboer E, Richardson DHS (1980) The replacement of the non-descriptive term «heavy metals» by a biologically and chemically significant classification of metal ions. Environ Pollut Ser B 1: 3-26

Nieboer E, Rossetto FE, Menon R (1988) Toxicology of nickel compounds. In: Siegel H, Siegel A (eds) Nickel and its role in biology. Dekker, New York, pp 359-402 (Metal ions in biological systems, vol 23)

Ochiai E-I (1977) Bioinorganic chemistry: an introduction. Allyn and Bacon, Boston

Ono K, Yamamoto Y, Hachiya A, Matsumoto H (1995) Synergistic inhibition of growth by Al and iron of tobacco (*Nicotiana tabacum* L.) cells in suspension culture. Plant Cell Physiol 36: 115-125

Owens TG (1996) Processing of excitation energy by antenna pigments. In: Baker NR (ed) Photosynthesis and the environment. Kluwer, Dordrecht, pp 1-23

Padmaja K, Prasad DDK, Prasad ARK (1990) Inhibition of chlorophyll synthesis in *Phaseolus vulgaris* L. seedlings by cadmium acetate. Photosynthetica 24: 399-405

Parekh D, Puranik RM, Srivastava HA (1990) Inhibition of chlorophyll biosynthesis by cadmium in greening maize leaf segments. Biochem Physiol Pflanz 186: 239-242

Pearson R (1968a) Hard and soft acids and bases, HSAB, part I. Fundamental principles. J Chem Educ 45: 581-587

Pearson R (1968b) Hard and soft acids and bases, HSAB, part II. Underlying theories. J Chem Educ 45: 643-648

Polle A, Matyssek R, Gunthardt-Goerg MS, Maurer S (2000) Defense strategies against ozone in trees: the role of nutrition. In: Agrawal SB, Agrawal M (eds) Environmental pollution and plant responses. Lewis Publishers, Boca Raton

Prasad SM, Singh JB, Rai LC, Kumar HD (1991) Metal-induced inhibition of photosynthetic electron transport chain of the cyanobacterium *Nostoc muscorum*. FEMS Microbiol Lett 82: 95-100

Prudente MS, Ichihashi H, Tatsukawa R (1994) Heavy metal concentrations in sediments from Manila Bay, Philippines, and inflowing rivers. Environ Pollut 86: 83-88

Quariti O, Gouia J, Ghorbal MH (1997) Responses of bean and tomato plants to cadmium: growth, mineral nutrition, and nitrate reduction. Plant Physiol Biochem 35: 347-354

Rai LC, Singh AK, Mallick N (1991a) Studies on photosynthesis, the associated electron transport system and some physiological variables of *Chlorella vulgaris* under heavy metal stress. J Plant Physiol 137: 419-424

Rai LC, Mallick N, Singh JB, Kumar HD (1991b) Physiological and biochemical characteristics of a copper tolerant and a wild type strain of *Anabaena doliolum* under copper stress. J Plant Physiol 68-74

Rauser WE (1995) Phytochelatins and related peptides. Plant Physiol 109: 1141-1149

Rijstenbil JW, Derksen JWM, Gerringa LJA, Poortvliet TCW, Sandee A, van den Berg M, van Drie J, Wijnholds JA (1994) Oxidative stress induced by copper: defense and damage in the marine planktonic diatom Ditylum brightwellii, grown in continuous cultures with high and low zinc levels. Mar Biol 119: 583-590

Rout NP, Shaw BP (2001) Salt tolerance in aquatic macrophytes: possible involvement of the antioxidative enzymes. Plant Sci 160: 415-423

Ruban AV, Young AJ, Horton P (1996) Dynamic properties of the minor chlorophyll a/b binding proteins of photosystem II, an in vitro model for photoprotective energy dissipation in the photosynthetic membrane of green plants. Biochemistry 35: 674-678

Rucinska R, Waplak S, Gwozdz EA (1999) Free radical formation and activity of antioxidant enzymes in lupin roots exposed to lead. Plant Physiol Biochem 37: 187-194

Saiki MK, Castleberry DT, May TW, Martin BA, Bullard FN (1995) Copper, cadmium, and zinc concentrations in aquatic food chain from the upper Sacramento river (California) and selected tributaries. Arch Environ Contam Toxicol 29: 484-491

Samecka-Cymerman A, Kempers AJ (1996) Bioaccumulation of heavy metals by aquatic macrophytes around Wroclaw, Poland. Ecotoxicol Environ Safety 35: 242-247

Sawidis T, Chettri MK, Zachariadis GA, Stratis JA (1995a) Heavy metals in aquatic plants and sediments from water systems in Macedonia, Greece. Ecotoxicol Environ Safety 32: 73-80

Sawidis T, Chetri MK, Zachariadis GA, Strtis JA, Seaward MRD (1995b) Heavy metal bioaccumulation in lichens from Macedonia in northern Greece. Toxicol Environ Chem 50: 157-166

Schicker H, Caspi H (1999) Response of antioxidative enzymes to nickel and cadmium stress in hyperaccumulator plants of the genus *Alyssum*. Physiol Plant 105: 39-44

Shaw BP (1995a) Changes in the levels of photosynthetic pigments in Phaseolus aureus Roxb. exposed to Hg and Cd at two stages of development: a comparative study. Bull Environ Contam Toxicol 55: 574-580

Shaw BP (1995b) Effects of mercury and cadmium on the activities of antioxidative enzymes in the seedlings of *Phaseolus aureus* Roxb. Biol Plant 37: 587-596

Shaw BP, Panigrahi AK (1986) Uptake and tissue distribution of mercury in some plant species collected from a contaminated area in India: its ecological implications. Arch Environ Contam Toxicol 15: 439-446

Shaw BP, Panigrahi AK (1987) Geographical distribution of mercury around a chlor-alkali factory. J Environ Biol 8: 227-281

Shaw BP, Rout NP (1998) Age-dependent responses of *Phaseolus aureus* Roxb. to inorganic salts of mercury and cadmium. Acta Physiol Plant 20: 85-90

Shaw BP, Rout NP (2002) Hg and Cd induced changes in the level of proline and the activity of proline biosynthesizing enzymes in *Phaseolus aureus* Roxb. and *Triticum aestivum* L. Biol Plant 45: 267-271

Shaw BP, Sahu A, Panigrahi AK (1985) Residual mercury concentration in brain, liver and muscle of contaminated fish collected from an estuary near a caustic-chlorine industry. Curr Sci 54: 810-812

Shaw BP, Sahu A, Panigrahi AK (1986) Mercury in plants, soil and water from a caustic-chlorine industry. Bull Environ Contam Toxicol 36: 299-305

Shaw BP, Sahu A, Choudhuri SB, Panigrahi AK (1988) Mercury in the Rushikulya river estuary. Mar Pollut Bull 19: 233-234

Sheoran IS, Singal HR, Singh R (1990) Effect of cadmium and nickel on photosynthesis and the enzymes of the photosynthetic carbon reduction cycle in pigeonpea (*Cajanus cajan* L.). Photosynth Res 23: 345-351

Siegel N, Haug A (1983) Aluminum interaction with calmodulin: evidence for altered structure and function from optical and enzymatic studies. Biochem Biophys 744: 36-45

Siegel N, Coughlin RT, Haug A (1983) A thermodynamic and electron paramagnetic resonance study of structural changes in calmodulin induced by aluminum interaction with calmodulin: evidence for altered structure and function from optical and enzymatic studies. Biochem Biophys Acta 744: 36-45

Silver S (1983) Bacterial interactions with mineral cations and anions: good ions and bad. In: Wesbrock P, de Jong EW (eds) Biomineralization and biological metal ion accumulation. Reidel, Amsterdam, pp 439-457

Skorzynska-Polit E, Baszynski T (1997) Difference in sensitivity of the photosynthetic apparatus in Cd-stressed runner bean plants in relation to their age. Plant Sci 128: 11-21

Stadtman ER (1992) Protein oxidation and aging. Science 257: 1220-1224

Stiborova M, Ditrichova M, Brezinova A (1987) Effect of heavy metal ions on growth and biochemical characteristics of photosynthesis of barley maize seedlings. Biol Plant 29: 453-467

Stiborova M, Dtrichova M, Brezinova A (1988) Mechanism of action of Cu^{2+}, Co^{2+} and Zn^{2+} on ribulose-1,5-bisphosphate carboxylase from barley (*Hordeum vulgare* L.). Photosynthetica 22: 161-167

Stohs SJ, Bagchi D (1995) Oxidative mechanisms in the toxicity of metal ions. Free Rad Biol Med 18: 321-336

Stroinski A (1999) Some physiological and biochemical aspects of plant resistance to cadmium effect. I. Antioxidative system. Acta Physiol Plant 21: 175-188

Subhadra AV, Nanda AK, Behera PK, Panda BB (1991) Acceleration of catalase and peroxidase activities in *Lemna minor* L. and *Allium cepa* L. in response to low levels of aquatic mercury. Environ Pollut 69: 169-179

Suszcynsky EM, Shann JR (1995) Phytotoxicity and accumulation of mercury in tobacco subjected to different exposure routes. Environ Toxicol Chem 14: 61-67

Teisseire H, Guy V (2000) Copper-induced changes in antioxidant enzymes activities in fronds of duckweed (*Lemna minor*). Plant Sci 153: 65-72

Thompson DR, Stewart FM, Furness RW (1990) Using seabirds to monitor mercury in marine environments. Mar Pollut Bull 21: 339-342

Toli K, Misaelides P, Godelitsas A (1997) Distribution of heavy metals in the aquatic environment of the kerkini lake (N. Greece): an exploratory study. Fresenius Environ Bull 6: 605-610

Venugopal B, Luckey TD (1975) Toxicology of non-radioactive heavy metals and their salts. In: Luckey TD, Venugopal B, Hutchenson D (eds) Heavy metal toxicity, safety and hormology. Thieme, Stuttgart, pp 4-73

Verstraeten SV, Nogueira LV, Schreier S, Oteiza PI (1997) Effect of trivalent metal ions on phase separation and membrane lipid packing: role in lipid peroxidation. Arch Biochem Biophys 388: 121-127

Wardman P, Cadeias LP (1996) Fenton chemistry: an introduction. Radiat Res 145: 525-531

Weckx JEJ, Clijsters HMM (1996) Oxidative damage and defense mechanisms in primary leaves of *Phaeolus vulgaris* as a result of root assimilation of toxic amounts of copper. Physiol Plant 96: 506-512

Weckx JEJ, Clijsters HMM (1997) Zn phytotoxicity induces oxidative stress in primary leaves of *Phaseolus vulgaris*. Plant Physiol Biochem 35: 405-410

Wheeler DM, Power IL (1995) Comparison of plant uptake and plant toxicity of various ions in wheat. Plant Soil 172: 167-173

Wong JWC (1996) Heavy metal contents in vegetables and market garden soils in Hong Kong. Environ Technol 17: 407-414

Wood JM (1974) Biological cycles for toxic elements in the environment. Science 183: 1049-1052

Yamamoto HY (1979) The biochemistry of the violaxanthin cycle in higher plants. Pure Appl Chem 51: 639-648

Yamamoto HY, Bassi R (1996a) Carotenoids: localization and function. In: Ort DR, Yocum CF (eds) Oxygenic photosynthesis: the light reactions. Kluwer, Dordrecht, pp 539-563 (Advances in photosynthesis, vol 4)

Yamamoto HY, Bassi R (1996b) Carotenoids: localization and function. In: Donald R, Yocum CF (eds) Photosynthesis: the light reactions. Kluwer, London, pp 539-563

Yamamoto M (1996) Stimulation of elemental mercury oxidation in the presence of chloride ion in aquatic environments. Chemosphere 32: 1217-1224

Yamamoto Y, Hachiya A, Matsumoto H (1997) Oxidative damage to membranes by a combination of aluminum and iron in suspension-cultured tobacco cells. Plant Cell Physiol 38: 1333-1339

Yan CTDC, Schofield CL, Munson R, Holsapple J (1994) The mercury cycle and fish in the Adirondack lakes. Environ Sci Technol 28: 136-143

Yruela I, Motoya G, Picorel R (1992) The inhibitory mechanism of Cu(II) on the photosystem II electron transport from higher plants. Photosynth Res 33: 227-233

Yruela I, Gatzen G, Picorel R, Holzwarth AR (1996) Cu(II)-inhibitory effect on photosystem II from higher plants. A picosecond time-resolved fluorescence study. Biochemistry 35: 9469-9474

Yurukova L, Kochev L (1994) Heavy metal concentrations in freshwater macrophytes from the Aldomirovsko swamp in Sofia district, Bulgaria. Bull Environ Contam Toxicol 52: 627-632

Van Camp W, Willekens H, Bowler C, Van Montagu M, Inze D (1994) Elevated levels of superoxide dismutase protect transgenic plants against O_3 damage. Biotechnology 12: 165-168

Venedictov RS, Krivosheyava AA (1983) The mechanism of fatty acid inhibition of electron transport in chloroplasts. Planta 159: 411-414

Vuletic M, Kohler K (1990) Effect of aluminum on the channels in plant membranes. Stud Biophys 138: 185-188

Weckx JEJ, Clijsters HMM (1996) Oxidative damage and defense mechanisms in primary leaves of *Phaseolus vulgaris* as a result of root assimilation of toxic amounts of copper. Physiol Plant 96: 506-512

Weckx JEJ, Clijsters HMM (1997) Zn phytotoxicity induces oxidative stress in primary leaves of *Phaseolus vulgaris*. Plant Physiol Biochem 35: 405-410

Wise RR, Naylor AW (1987) Chilling induced photoperoxidation. The peroxidative destruction of lipids during chilling injury. Plant Physiol 83: 272-277

Wolter FP, Schmidt R, Heinz E (1992) Chilling sensitivity of *Arabidopsis thaliana* with genetically engineered membrane lipids. EMBO J 11: 4685-4692

Xu S, Patterson G (1985) The biochemical effects of cadmium on sterol biosynthesis by soybean suspension culture. Curr Top Plant Biochem Physiol 4: 245-248

Xu Y, Siegenthaler PA (1996) Effect of non-chilling temperatures and light intensity during growth of squash cotyledons on the composition of thylakoid membrane lipids and fatty acids. Plant Cell Physiol 37: 471-479

Yamamoto Y, Hachiya A, Matsumoto H (1997) Oxidative damage to membranes by a combination of aluminum and iron in suspension cultured tobacco cells. Plant Cell Physiol 38: 1333-1339

Yamamoto Y, Kobayashi Y, Matsumoto H (2001) Lipid peroxidation is an early symptom triggered by aluminum, but not the primary cause of elongation inhibition in pea roots. Plant Physiol 125: 199-208

Yamamoto Y, Kobayashi Y, Devi SR, Rikiishi S, Matsumoto H (2002) Aluminum toxicity is associated with mitochondrial dysfunction and the production of reactive oxygen species in plant cells. Plant Physiol 128: 63-72

Yaneva IA, Vunkova-Radeva RV, Stefanova KL, Tsenov AS, Petrova TP, Petkov GO (1995) Changes in lipid composition of winter wheat leaves under low temperature stress: effect of molybdenum supply. Biol Plant 37: 561-566

Yang X, Baliger VC, Martens DC, Clark RB (1996) Cadmium effects on influx and transport of mineral nutrients in plant species. J Plant Nutr 19: 643-656

Zel J, Svetek J, Crne H, Schara M (1993) Effects of aluminum on membrane fluidity of the mycorrhizal fungus *Amanita muscaria*. Physiol Plant 89: 172-176

Zhang G, Slaski JJ, Archambutt DJ, Taylor GJ (1997) Alteration of plasma membrane lipids in aluminum resistant and aluminum sensitive wheat genotypes in response to aluminum stress. Physiol Plant 99: 302-308

Zhao XJ, Sucoff E, Stadelmann EJ (1987) Al^{2+} and Ca alteration of plasma membrane permeability of *Quercus rubra* root cortical cells. Physiol Plant 83: 159-162

Zwaizek JJ, Blake JJ (1990) Effects of preconditioning on electrolyte leakage and lipid composition in black spruce (*Picea mariana*) stresses with polyethylene glycol. Physiol Plant 79: 71-77

Chapter 6

Photosynthesis in Heavy Metal Stressed Plants

B. Myśliwa -Kurdziel[1], M.N.V. Prasad[2], K. Strzałka[1]
[1]Department of Plant Physiology and Biochemistry, Faculty of Biotechnology,
Jagiellonian University, ul. Gronostajowa 7, 30-387 Kraków, Poland
[2]Department of Plant Sciences, University of Hyderabad, Hyderabad 500046, India

6.1 INTRODUCTION

Use of phytotoxicity of metallic compounds dates back to 1896, when French farmers applied Bordeaux mixture (copper sulfate, lime and water) to control fungal pests (Martin and Woodcock 1983). Currently, global heavy metal (HM) pollution is a serious environmental concern.

Heavy metals (HMs) are available to plants through (Ross 1994; see also Chap. 1):

1. mining activities (smelting, river dredging, mine spoils and tailings, metal industries etc.);
2. industries (plastics, textiles, microelectronics, wood preservatives, refineries, etc.);
3. atmospheric deposition (urban refuse disposal, pyrometallurgical industries, automobile exhausts, fossil fuel combustion, etc.); and
4. excessive use of agrochemicals (fertilizers and pesticides) and waste disposal (sewage sludge, leachate when the former is used as landfill).

When taken up by plants, HMs can result in a wide variety of toxic effects. The photosynthetic apparatus appears to be very sensitive to the toxicity of HMs. Cadmium, copper, mercury, lead, zinc, etc. ions invariably affect photosynthetic functions either directly or indirectly. Different aspects of their influence on photosynthetic activity have already been reviewed by Clijsters and van Assche (1985), van Assche and Clijsters (1990), Krupa and Baszyński (1995), Prasad (1995a, 1997), and Myśliwa-Kurdziel et al. (2002b). The scale and character of changes observed in plants after HM application was shown to be dose-dependent and it can vary for different plant species even for identical metal treatment depending on individual plant tolerance. For example, corn and soybean exhibited different degrees of sensitivity of photosynthesis to excess Pb (Bazzaz et al. 1974a, 1975).

A critical examination of the literature reveals that HMs react with the photosynthetic apparatus at various levels of organization and architecture:

— accumulation of metals in leaves (main photosynthetic organ);
— partitioning in leaf tissues like stomata, mesophyll and bundle sheath;
— metal interaction with cytosolic enzymes and organics;
— alteration of the functions of chloroplast membranes;
— supramolecular level action, particularly on PS II, PS I, membrane acyl lipids, and carrier proteins in vascular tissues;
— molecular level interactions, particularly with photosynthetic carbon reduction cycle enzymes, xanthophyll cycle and adenylates.

Cadmium was shown to be the most effective inhibitor of photosynthetic activity. Cadmium influenced the net photosynthesis in tomato, rice and maize (Baszyński et al. 1980; Misra et al. 1989; Ferretti et al. 1993; Moya et al. 1993; Rascio et al. 1993; Prasad 1995b). Even if only small amounts of Cd^{2+} actually enter the chloroplasts, many direct and indirect effects are observed, which result in a strong inhibition of photosynthesis. This phenomenon is often described as an "effect of multiplication" (for reviews, see Krupa and Baszyński 1995; Prasad 1995a; Krupa 1999). Toxicity of Cd^{2+} is related in part to the Cd-induced Fe deficiency (for reviews, see Krupa and Siedlecka 1995; Siedlecka and Krupa 1999).

Photosynthesis has also been found to be one of the most sensitive processes in plants treated with mercury (for reviews, see Boening 2000; Patra and Sharma 2000) and lead (for review, see Singh et al. 1997).

Copper is an essential biometal for the photosynthetic apparatus, among others as a component of plastocyanin (for reviews, see Droppa and Horváth 1990; Maksymiec 1997). Both excess and deficiency of Cu alter photosynthetic activity in higher plants (Baszyński et al. 1978, 1982; Baszyński and Tukendorf 1984; Lidon and Henriques 1993a; Lidon et al. 1993; Ouzounidou 1995). Wheat plants (*Triticum aestivum* L. cv. Vergina) growing in the field on ore bodies containing copper at about 3050 $\mu g/g$ showed stunted growth, reduced leaf fresh and dry weights, area, protein, chlorophyll (Chl) and Rubisco activity compared with control plants grown in garden soil with a copper content of 140 $\mu g/g$ (Lanaras et al. 1993).

Deficiency of iron, another biometal, also affected photosynthesis (Terry and Abadia 1986; Geider and La Roche 1994). This effect was attributed to the reduced number of photosynthetic units per leaf area (Spiller and Terry 1980) or to the lower efficiency of photosynthetic energy conversion found for algal suspension (Guikema 1985; Greene et al. 1992) or in higher plants (Morales et al. 1990, 1991). It was also shown that in Fe-deficient leaves of sugar beet (*Beta vulgaris*) the plastoquinone pool became reduced in the dark and that the extent of the reduction depended on the chlorophyll concentration in the leaf, time of preillumination, as well as on the duration of dark-adaptation (Belkhodja et al. 1998).

Zinc and manganese are also microelements that can be strongly toxic and can disturb photosynthesis if applied in higher concentrations (reviewed by Krupa and Baszyński 1995). Due to the similarities of ion radii of bivalent cations: Mn, Fe, Cu, Mg and Zn, excess Zn can shift equilibrium of different physiological processes, for instance, the equilibrium between CO_2 and O_2 binding by Rubisco (van Assche and Clijsters 1990).

Zinc decreased the net photosynthetic rate in ryegrass (*Lolium perenne*; Monnet et al. 2001). Increasing Mn concentrations in the nutrient medium were found to decrease the net photosynthesis rate in ricebean (*Vigna umbellata*; Subrahmanyam and Rathore 2000).

In this chapter, the current knowledge about the influence of HMs on the functioning of the photosynthetic apparatus is described. It concerns direct effects on light and dark photosynthetic reactions as well as some indirect changes in photosynthetic efficiency caused by, amongst others, lower content of photosynthetic pigments and alterations in stomatal functions.

6.2 PIGMENT BIOSYNTHESIS AND CHLOROPLAST STRUCTURE

Retardation of plant growth and chlorosis of leaves are often observed symptoms of HM toxicity in metal-polluted environments. Lower content of photosynthetic pigments induces some changes in plastid development, photosynthetic efficiency as well as in general metabolism.

6.2.1 Influence of HMs on Pigment Content

Reduced Chl content due to the toxicity of Cd^{2+}, Cu^{2+}, Pb^{2+}, Ni^{2+}, Hg^{2+}, and Zn^{2+} in different plant species has been well documented (for review, see Myśliwa -Kurdziel and Strzatka 2002a). Cadmium-induced inhibition of Chl biosynthesis was even suggested to be a primary event as compared to the inhibition of photosynthesis (Baszyński et al. 1980). Inhibition of the biosynthesis of photosynthetic pigments results in a lower pigment content as well as in different proportions of various pigments.

It is difficult to compare the scale of a metal-induced reduction in Chl content described in different reports, as it seems impossible to determine a universal dose of a metal producing toxic effect on Chl accumulation. This is due to large differences in the experimental conditions used by different authors (plant species, metal concentration and the way of its introduction into the plant, plant age, composition of growing medium, etc.). However, some comparative studies concerning the influence of Cu^{2+}, Cd^{2+} and Pb^{2+} on Chl accumulation have already been performed. Cu^{2+} appeared to affect the Chl biosynthesis more strongly than Cd^{2+} in *Lemna trisulca* (Prasad et al. 2001) and in *Chlamydomonas reinhardtii* (Prasad et al. 1998). Cd^{2+} was found to affect Chl biosynthesis when applied at lower concentration than Pb^{2+} (50 mg/l and 500 mg/l, respectively) in two wheat varieties (*Triticum aestivum* L. cv. Gerek-79 and Bolal 2973; Öncel et al. 2000).

In general, changes in Chl accumulation induced by Cd^{2+}, Cu^{2+} and Hg^{2+} strongly depend on the growth stage of plant that is treated with the metal. Cadmium affected the biosynthesis of Chl more in mature leaves having properly organized inner membranes than in younger, developing leaves (Barceló et al. 1988; Padmaja et al. 1990; Skórzyńska-Polit and Baszyński 1995). In rye seedlings, Chl accumulation was inhibited mostly in the older parts of leaves (Krupa and Moniak 1998). Long-term exposure of whole plants to Cd^{2+} affected Chl and chloroplast development in young leaves. Cd^{2+} (200 ppm) applied through the rooting medium to 30-day-old wheat plants considerably decreased Chl content (Malik et al. 1992a,b). A decrease in Chl *a* and *b* content was also observed in cultures of *Chlamydomonas reinhardtii* treated with Cd^{2+} and Cu^{2+}, both in wild-type (WT 2137) as well as in cell wall deficient mutant strains (CW15; Prasad et al. 1998). The heavy metals reduced both cell volume and cell multiplication rate. Cook et al. (1997) found a correlation between the copper accumulation and Chl content in bean leaves. In contrast, Stiborova et al. (1986a) observed lower total Chl content for higher internal concentration of metals like Cd^{2+}, Cu^{2+} and Pb^{2+} in barley leaves. Copper treatment of bean plants at the stationary growth phase of primary leaves resulted in a decrease in Chl content while Cu^{2+} applied at the early growth stage caused an increase in the level of Chl calculated per leaf area, concomitant with the strong inhibition of leaf expansion (Maksymiec et al. 1994, 1995; Maksymiec and Baszyński 1996b). In 30-day-old tomato seedlings treated with Hg^{2+} (10 and 50 μM) for 10 consecutive days, Chl content was decreased significantly in the first and the second leaves and was unchanged in the youngest third leaves (Cho and Park 2000). In different studies, mercury resulted in a decrease in Chl level or it showed no effect, except in the case of plants treated with Hg^{2+} during germination or at the very early stage of seedling development, in which it induced higher Chl accumulation (Prasad and Prasad 1987a,b; Schlegel at al. 1987; Shaw and Rout 1998). The increase in Chl content in young leaves observed as a result of Cd^{2+}, Cu^{2+} and Hg^{2+} treatment may be regarded as an adaptive mechanism to the toxic conditions. In contrast, some fluorescence measurements performed on intact leaves and measurements of O_2 evolution showed a decrease in the photosynthetic efficiency in metal-treated plants even if a significant increase in the level of Chl was observed (Krupa et al. 1993b; Maksymiec and Baszyński 1996a; Shaw and Rout 1998).

It was found that Chl accumulation in Cu^{2+}-treated plants depended also on the individual tolerance of the plant toward copper toxicity. Copper excess induced an increase in total Chl content in Cu-tolerant spinach (Baszyński et al. 1982), whereas, in Cu-sensitive spinach, a decrease in Chl content was found (Baszyński et al. 1988). In Cu-tolerant *Silene cucubalus* no changes in Chl level were found whereas, in a Cu-sensitive variety, a decreased Chl level was observed (Lolkema and Vooijs 1986). Cu-tolerant *Silene compacta* and non-tolerant *Thlaspi ochroleucum* differed in total Chl in response to excess Cu. In *S. compacta*, Chl was increased by 34% whereas decreased by 43% in *T. ochroleucum* by 8 μM Cu. However, an elevated concentration of Cu, i.e. 160 μM, showed toxic symptoms in both the species. The Chl content of the non-tolerant species (*T. ochroleucum*) decreased by 65% compared with *S. compacta* (Cu-tolerant) in which the effect was only 17% (Ouzounidou 1993). In the case of Cu, Cu-Fe antagonism was suggested for Cu-induced chlorosis, as excess Cu can induce Fe deficiency and thereby chlorosis (van Assche and Clijsters 1990; Lanaras et al. 1993).

Not only an excess of copper but also a deficiency induced the inhibition of pigment synthesis and affected the development of the photosynthetic apparatus (for review, see Maksymiec 1997; also in: Baszyński et al. 1978; Droppa et al. 1984a,b; Henriques 1989).

It was shown that the toxic effect of Cd^{2+} and Pb^{2+} on Chl accumulation did not depend on the availability of nitrogen in the growing medium (Parekh et al. 1990; Sengar and Padney 1996). For Cd^{2+}, the inhibitory effect also did not depend on the presence of chelating agents (e.g. EDTA) in the nutrient medium (Fodor et al. 1996). On the other hand, the decrease in Chl content caused by Pb was enhanced in the presence of iron and it was higher in plants supplied with Fe-citrate than Fe-EDTA (Fodor et al. 1996). Cd- and Pb-induced decrease of Chl content in wheat seedlings was found to be enhanced by higher temperature probably due to more effective Cd^{2+} intake from the medium (Öncel et al. 2000).

The mechanism(s) of Cd-induced chlorosis is not clear at present. It may act indirectly through the Cd-induced deficiency of Fe and Mg (Greger and Lindberg 1987; Siedlecka and Krupa 1999) or by direct influence of Cd^{2+} on enzyme(s) of Chl biosynthesis (see below). Several factors like manganese (Baszyński et al. 1980), Calcium (Skórzyńska-Polit et al. 1998) as well as kinetin and uniconazole (triazole derivative; Gadallah 1995b; Thomas and Singh 1996), and triacontanol [$CH_3(CH_2)_{28}CH_2OH$; Muthuchelian et al. 2001] can reduce or even partly reverse the inhibitory effect of Cd^{2+} on Chl biosynthesis.

Shalygo et al. (1999) found a strong increase in Chl accumulation during greening of etiolated barley seedlings treated with Ni^{2+} with respect to non-treated plants for low Ni^{2+} concentration (μM range), progressive inhibition of this process for higher metal concentrations, and complete inhibition of Chl synthesis for 100 mM Ni^{2+}. A similar effect on Chl accumulation was observed for Fe^{2+}; however, for the highest Fe^{2+} concentration used (100 mM), the Chl amount in the treated seedlings reached 94% of the Chl amount in control seedlings.

Zinc applied in concentrations below 1 mM was found not to affect or slightly stimulate Chl accumulation in barley (Agarwala et al. 1977; Stiborova et al. 1986a). At higher concentration, it induced a decrease in Chl content in spruce seedlings (Schlegel et al. 1987) and a common grass *Lolium perenne* (Monnet et al. 2001).

Several experiments were performed to reveal the site(s) of metal action in the process of Chl biosynthesis. The current knowledge of the influence of HMs on individual steps in this process is summarized in Table 6.1. Mechanisms of the inhibition are not yet understood thoroughly; however, sulfhydryl group interaction has been proposed as the mechanism for the inhibition of δ-aminolevulinic acid (ALA) synthase, ALA dehydrogenase and protochlorophyllide reductase (POR).

HMs also influence the Chl *a/b* ratio in higher plants. Whether this change results from alteration of the rate of pigment biosynthesis or its degradation remains to be elucidated. Cadmium was found to decrease the Chl *a/b* ratio in tomato, pigeon pea and bean (Baszyński et al. 1980; Sheoran et al. 1990b; Gadallah 1995a); however, Ferretti et al. (1993) observed an increase in this ratio for Cd^{2+} concentrations higher than 200 μM in maize. A slight increase in Chl *a/b* ratio was also found for both excess and deficiency of copper in spinach, oat, sugar beet and pea (Baszyński et al. 1978; Droppa et al. 1984a; Henriques 1989; Angelov et al. 1993). In contrast, in bean plants treated with Cu^{2+} at the stationary growth stage of primary leaves, a decrease in Chl *a/b* ratio was measured (Maksymiec and Baszyński 1996b). Stiborova et al. (1986a) observed an increase in Chl *a/b* ratio in barley seedlings treated with Cu^{2+}, Pb^{2+} or Zn^{2+}. The effect of Ni^{2+} on the Chl *a/b* ratio was concentration-dependent; a slight decrease in the Chl *a/b* ratio was observed for 500 μM Ni^{2+}, whereas no effect was found for 100-200 μM Ni^{2+}. Recently, Monnet et al. (2001) found a significant reduction in the Chl *a/b* ratio in *Lolium perenne* growing in Zn excess.

Besides the influence on Chl biosynthesis, Cd^{2+} also causes a decrease in the carotenoid, mainly β-carotene, content in metal-treated plants (Naguib et al. 1982; Baszyński et al. 1980; El-Shintinawy 1999). This inhibition was found to be partly reversed by addition of manganese to the medium (Baszyński et al. 1980) as well as in the presence of uniconazole (a triazole derivative) or uniconazole with kinetin (Thomas and Singh 1996). In greening radish seedlings, Cd^{2+} enhanced the synthesis of carotenoids and the effect was larger for β-carotene than for xanthophylls (Krupa et al. 1987).

Table 6.1 Inhibition of the activity of the enzymes of Chl biosynthesis pathway by HMs

Enzyme/metals	Concentration	Taxon	Reference
ALA synthase			
Cd	100 μM	*Hordeum vulgare*	Stobart et al. (1985)
Pb[a]	0.1-1 mM	*Cucumis sativus*	Burzyński (1985)
Cd	10-50 μM	*Phaseolus vulgaris*	Padmaja et al. (1990)
Fe, Co, Ni, Mn	0.1 M	*Hordeum vulgare*	Shalygo et al. (1999)
ALA dehydratase			
Pb	1 mM	*Cucumis sativus*[c]	Burzyński (1985)
Hg, Pb	50-250 μM	*Phaseolus vulgaris*	Prasad and Prasad (1987a)
Hg, Pb	50-250 μM	*Penissetum typhoideum*	Prasad and Prasad (1987b)
Se	62 μM	*Phaseolus vulgaris*	Padmaja et al. (1989)
Cd	10-50 μM	*Phaseolus vulgaris*[c]	Padmaja et al. (1990)
Co, Ni, Mn, Fe	0.01-0.1 M	*Hordeum vulgare*[c]	Shalygo et al. (1999)
Cr	1-200 μM	*Nymphaea alba*	Vajpayee et al. (2000)
Porphobilinogenase[b]			
Mn	1.7 μM	*Anacystis nidulans*	Csatorday et al. (1984)
Mn, Fe, Co, Ni	0.01-0.1 M	*Hordeum vulgare*[c]	Shalygo et al. (1999)
Uroporphyrinogen III decarboxylase			
Cs	10 mM	*Hordeum vulgare*[c]	Shalygo et al. (1997, 1998)
Mg-chelatase			
Co	1.7 μM	*Anacystis nidulans*	Csatorday et al. (1984)
NADPH:protochlorophyllide oxidoreductase			
Cd	1 mM	*Hordeum vulgare*[c]	Stobart et al. (1985)
Cd	1-1000 μM	*Triticum aestivum*[c]	Böddi et al. (1995)
Cd, Fe, Cr	10^{-5}-10^{-2}M	*Triticum aestivum*[c]	Berska et al. (2001)

[a]However, for 50-250 mM Pb or Hg, no effects on ALA synthase were observed in mung bean and bajra (Prasad and Prasad 1987a,b).
[b]This is a complex of two enzymes: porphobilinogen deaminase and uroporphyrinogen III syntase.
[c]Experiments performed on greening seedlings.

HMs affect accumulation of carotenoids and Chl in a different way; thus, the Chl/carotenoid ratio is often changed in metal-treated plants. However, the direction as well as the magnitude of these changes seem to depend on several experimental conditions.

It was also found that the substitution of the central Mg in Chl by HMs (Hg, Cd, Cu, Ni, Zn, and Pb) in vivo was an important type of damage in submerged aquatic plants growing in a metal-contaminated environment. This substitution prevents photosynthetic light harvesting in the affected Chl molecules, and results in a breakdown of photosynthesis (Küpper et al. 1996, 1998). The extent of damage varies with light intensity. In low light irradiance, all the central atoms of the Chl are accessible to HMs, with HM-Chl complexes being formed, some of which are much more stable towards irradiance than Mg-Chl; consequently, plants remain green even when they are dead. In high light irradiance, however, almost all Chl decays, showing that under such conditions most of the Chl are inaccessible to HM ions (Küpper et al. 1996).

6.2.2 INFLUENCE OF HMS ON CHLOROPLAST STRUCTURE

HMs were found to cause structural and ultrastructural changes in chloroplasts of plants growing in a metal-polluted environment. Thus, chloroplast membranes, particularly thylakoids, are also investigated as the sites of action of HMs.

Chloroplasts of wheat (*Triticum aestivum*) leaves growing on copper-contaminated soils in Greece (Cu concentration range between 100-1200 mg/kg depending on the location) were compared with control plants (copper in the mineral nutrient ranged between 50-80 mg/kg). The numbers of chloroplasts, starch grains, plastoglobuli per chloroplast, surface area of chloroplasts, volume fraction of internal membrane system and starch grains were significantly different (Eleftheriou and Karataglis 1989). In another study, excess Cu modified the morphology and structure of primary leaves and their chloroplast ultrastructure in runner beans (*Phaseolus coccineus* L.; Maksymiec et al. 1994, 1995). Copper altered leaf ultrastructure and photosynthesis in *Thlaspi ochroleucum* L. The critical toxic Cu level in leaves of almost all plant species is about 20-30 μg/g dry weight (Ouzounidou et al. 1992). It was also shown that the inhibition of the photosynthetic electron transport caused by copper excess was often accompanied by changes in the structure and composition of the thylakoid membranes (Ouzounidou et al. 1992; Lidon et al. 1993). Baszyński et al. (1988) showed that in Cu-sensitive spinach (*Spinacia oleracea* L. cv. Matador) changes observed in chloroplast structure preceeded the loss of photochemical activity of PS I and II. Cu was also shown to cause slower Chl incorporation into PS I and II in greening barley seedlings (Caspi et al. 1999).

Baryla et al. (2001) demonstrated that leaf chlorosis observed in oilseed rape (*Brassica napus* L.) growing from seeds on a reconstituted soil contaminated with Cd^{2+} (100 mg Cd/ kg dry soil) was related to a reduced number of chloroplasts per cell and a change in cell size. In these plants, the chlorosis was rather due to the decrease in chloroplast replication and cell division than to the direct interaction of Cd^{2+} with Chl biosynthesis. Cd-induced decrease of pigment content was larger at the leaf surface than in the leaf interior. Chloroplast ultrastructure, as well as the photosynthesis efficiency, were found to be unchanged in this study. In Cd-treated bush bean plants (*Phaseolus vulgaris* L.), chlorosis was not noticed in the primary leaves, but electron microscopic studies revealed disorder in the position and stacking of thylakoid membranes in chloroplasts of trifoliate leaves being more affected than those of primary leaves (Barceló et al. 1986a,b, 1988). Cd^{2+} altered the ultrastructure of developing chloroplasts of soybean, corn and tomato (Baszyński et al. 1980; Ghoshrony and Nadakavukaren 1990). Cd-treated tomato plants had smaller chloroplasts as observed by electron microscopy. The volume of chloroplasts and their number per unit surface of leaf decreased. The fine structure of chloroplasts in Cd-treated plants was degraded in a similar way to the senescence response, marked by the occurrence of large plastoglobuli and disorganization of the lamellar structure, mainly the grana stacks (Baszyński et al. 1980; McCarthy et al. 2001). A reduction in the intercellular spaces of pea (*Pisum sativum* L.) leaves growing in the presence of 1-50 μM Cd^{2+}, concomitant with a reduction in chloroplast number as well as disruption of chloroplast ultrastructure, e.g. disorganized thylakoid systems and increased size of plastoglobuli and starch grains, was also observed by Sandalio et al. (2001).

In radish cotyledons, Cd^{2+} induced changes in the composition and structure of the light-harvesting Chl a/b protein complex II (Krupa et al. 1987; Krupa 1988). Cd^{2+} reduced the Chl and accessory pigments before photosynthetic function. Ahmed and Tajmir-Riahi (1993) observed interaction of Cd^{2+}, Hg^{2+}, and Pb^{2+} (0.01-0.1 mM) with the light-harvesting proteins (LHC II) in spinach thylakoid membranes using Fourier transform infrared (FTIR) spectroscopy. The highest structural changes of proteins, even for low concentration, were found for Hg^{2+}, lower for Cd^{2+}, and the lowest for Pb^{2+}. Cd^{2+} was also shown to damage the structure of chloroplasts in wheat (*Triticum aestivum* L.) leaves, which was seen as a disturbed shape and dilation of thylakoid membranes and swollen intrathylakoidal spaces (Ouzounidou et al. 1997). The observed changes were probably related to the Cd-induced premature senescence.

In contrast to some results indicating the strong influence of HMs on Chl biosynthesis described in the previous section, Horváth et al. (1996) demonstrated that in the greening leaves of barley, Cd^{2+} did not primarily affect either the biosynthetic pathway from ALA to protochlorophyllide or the protochlorophyllide phototransformation. In their study, cadmium inhibited the incorporation of Chl by interfering with the organization of pigment protein complexes that are essential for optimal function of PS II. In greening bean (*Phaseolus vulgaris*), Cd^{2+} was also found to reduce the LHC II accumulation by reducing strongly the steady-state level of *Lhcb* transcripts (Tziveleka et al. 1999).

Changes in chloroplast ultrastructure in cabbage leaves treated with Ni as organic complexes [Ni(II)-Glu, Ni(II)-citrate, Ni(II)-EDTA] were observed by Molas (2002). Density of chloroplast stroma, number of grana as well as the size and shape of thylakoids changed depending on the type of Ni source in that study. Doncheva et al. (2001) found some alterations in the chloroplast structure, mainly smaller granal thylakoids in pea plants growing in the presence of 0.67-1000 μM Zn^{2+}, and observed that the photochemical activity of PS II was less affected. Effect of Zn^{2+} toxicity was reversed by succinate (200 μM) in this study.

In the case of barley (*Hordeum vulgare*) leaves treated with Pb ranging from 0.04-4.0 mM for 8-24 h, the Pb content in etiolated leaves remained the same regardless of the time of exposure. Mesophyll cells did not accumulate Pb. However, apoplasm, guard cells and cuticle showed deposits of Pb. Lead reduced Chl b content and number of grana. Kinetin treatment recovered Pb toxicity when treated together with Pb (Woźny et al. 1995). Pb inhibited the transformation of etioplasts into chloroplasts in etiolated wheat leaves exposed to light (Wrischer and Meglaj 1980).

6.3 INFLUENCE OF HMs ON PHOTOSYNTHETIC LIGHT REACTIONS

Three different but complementary experimental approaches were mostly used in the studies on the influence of HMs on photosynthetic light reactions. Experiments were performed on organelles isolated from untreated plants (chloroplasts, thylakoids or PSII particles) that were subsequently treated with rather high metal concentration for a short period of time. Alternatively, the whole plants were grown in the presence of HMs. The first approach gives an opportunity to find precisely the site(s) of metal inhibition, whereas the other enables investigation of the real influence of HMs on the photosynthetic apparatus in the presence of all detoxifying mechanisms (accumulation of HMs in root system, inhibition of HMs translocation, etc.). However, in the latter case, the internal metal concentration differs from that which is applied in the nutrient medium. In the third approach, excised leaves from untreated plants floating in a solution of HMs were used. This method allows application of known metal concentration to the photosynthetic apparatus in a whole leaf.

It had already been found that cadmium altered photosynthesis and transpiration of excised silver maple leaves (*Acer saccharum*; Lamoreaux and Chaney 1978). Cadmium changed or inhibited the light reactions of isolated chloroplasts (Bazzaz and Govindjee 1974; Weigel 1985a,b). According to Weigel (1985a,b), Cd^{2+} affects photosynthesis by inhibition of different reaction steps of the Calvin cycle, and not by interaction with photosynthetic reactions in the thylakoid membranes. The membrane-bound photosynthetic reactions in isolated mesophyll protoplasts were not impaired by Cd^{2+} concentrations that drastically inhibited CO_2 fixation (Weigel 1985a,b). At low concentrations, Cd^{2+} also acts as a photophosphorylation inhibitor in spinach chloroplasts (Lucero et al. 1976).

Cadmium and Zn inhibited photosynthetic CO_2 fixation and Hill reaction activity of isolated spinach chloroplasts (Hampp et al. 1976; Barua and Jana 1986). Zinc inhibited photosynthetic electron transport in isolated barley chloroplasts (Tripathy and Mohanthy 1980).

A relationship between the influence of HMs on photosynthesis and the stage of plant growth has not been intensively studied so far. Some reports, however, which have appeared recently underline the differential response of plants to an excess of Cd^{2+}, Cu^{2+} and Ni^{2+} depending on plant age (Sheoran et al. 1990a,b). Copper excess causes the strongest decline in photosynthesis in dicotyledonous plants (bean and cucumber) when they are treated with the metal at advanced leaf growth stage (Maksymiec et al. 1994; Maksymiec and Baszyński 1996a,b; Vinit-Dunand et al. 2002). Higher susceptibility of bean at the final growth stage of the primary leaves than that of younger plants towards cadmium toxicity was also observed (Skórzyńska-Polit and Baszyński 1995). In young leaves, the toxic effect of cadmium was mainly visible as growth inhibition while in older leaves it was more connected to the functioning of the photosynthetic apparatus (Skórzyńska-Polit and Baszyński 1997). In the oldest leaves, disruption of the thylakoid membranes was observed. Krupa and Moniak (1998) also found higher inhibition of photosynthetic light reactions in the older leaf sections of rye seedlings. However, more experiments are needed to reveal if mono- and dicotyledonous plants show the same response to HM toxicity.

Cadmium was shown to be a very effective inhibitor of the whole photosynthetic electron transport chain (see Table 6.2). Its toxicity towards the photosynthetic apparatus was found to be dose- and time-dependent (Chugh and Sawhney 1999). PS II appears to be more sensitive than PS I, thus a decrease in its activity precedes damage to PS I (see below). Toxic Cd^{2+} effect on PS I and PS II efficiency is enhanced under nutrient medium Fe deficiency (Siedlecka and Krupa 1996). In runner bean leaves, it was also shown to be strongly dependent on the Ca^{2+} level in the nutrient medium and on the plant growth stage (Skórzyńska-Polit et al. 1998). Calcium deficiency enhanced the toxic effect of Cd^{2+}, whereas Ca excess reduced this effect. Ca-Cd interaction was earlier shown by Greger and Lindberg (1987) in sugar beet roots.

Table 6.2 Sites of HM action in light reactions of photosynthesis

Metal	Site of action	Reference
Cd, Hg, Pb	Light-harvesting chlorophyll a/b protein complex	Krupa (1988); Ahmed and Tajmir-Riahi (1993); Krupa and Baszyński (1995)
Cd, Co, Cu, Hg, Ni, Pb, Zn	Photosystem II	Miles et al. (1972); Bazzaz and Govindjee (1974); Shioi et al. (1978b); Tripathy and Mohanty (1980); Clijsters and van Assche (1985); Baszyński (1986); Hsu and Lee (1988); Samson et al. (1988); Mohanty et al. (1989); Rashid et al. (1991); Yruela et al. (1991, 1993, 2000); Malik et al. (1992a); Bernier et al. (1993); Maksymiec et al. (1994); Schröder et al. (1994); Drazkiewicz (1994); Jegerschöld et al. (1995); Chugh and Sawhney (1999); Boucher and Carpentier (1999); Szalontai et al. (1999)
Cd, Cr, Cu, Hg, Zn	Oxygen-evolving complex	Van Duijvendijk-Matteoli and Desmeta (1975); Miller and Cox (1983); Bernier et al. (1993); Ouzounidou et al. (1993); Skórzyńska and Baszyński (1993); Maksymiec et al. (1994); Krupa and Baszyński (1995); Seršeň et al (1996, 1998); Prasad et al. (2001); Burda et al. (2002)
Cd, Cr, Cu, Zn	Plastoquinone pool	Tripathy and Mohanthy (1980); Baszyński et al. (1980, 1982)
Cu, Ni	Cytochrome b_6/f	Veeranjaneyulu and Das (1982); Lidon and Henriques (1993a); Rao et al. (2000)
Cu, Hg, Ni	Plastocyanin	Kimimura and Katoh (1972); Veeranjaneyulu and Das (1982); Lidon and Henriques (1993a)
Cu, Pb, Zn	Photosystem I	Miles et al. (1972); Radmer and Kok (1974); Wong and Govindjee (1976); Chugh and Sawhney (1999)
Cd, Cu, Hg, Ni, Zn	Ferredoxin	Honeycutt and Krogmann (1972); Siedlecka and Baszyński (1993); Seršeň et al. (1996)
Cd, Cu, Hg, Zn	Ferredoxin NADP$^+$ oxidoreductase	Shioi et al. (1978b); Siedlecka and Baszyński (1993)
Cd, Cu, Hg, Ni, Pb	Chloroplast coupling factor$_1$	Clijsters and van Assche (1985); Baszyński (1986)

Excess of copper induces several deleterious effects on plants, including the inhibition of photosynthetic electron transport, which was observed in higher plants (Hsu and Lee 1988; Mohanty et al. 1989; Yruela et al. 1991, 1993), green algae (Shioi et al. 1978a; Samson et al. 1988) and cyanobacteria (Lu and Zhang 1999). High Cu concentrations were also found to inhibit O_2 evolution and to enhance heat emission in Cu-treated *Thlaspi ochroleucum* leaves in vivo, while less pronounced changes in the activity of PS I were observed (Ouzounidou 1996). Varying concentrations of Cu also limited the different phases of photosynthesis in rice plants (Lidon and Henriques 1993a-c). As shown already for Cd^{2+}, the effect of Cu excess on higher plants can be modified by different Ca^{2+} levels in the growing medium (Ouzounidou et al. 1995; Maksymiec and Baszyński 1998b, 1999a,b). PS II is regarded as the main target of the copper toxicity (see below). In a study by Rao et al. (2000), Cu^{2+} inhibited the spinach cyt b_6f complex by competing with quinol.

Copper deficiency also inhibits the photosynthetic electron transport, probably due to the decreased level of plastocyanin or due to the lower mobility of plastoquinone, which is related to the copper-induced changes in the membrane fluidity (Droppa et al. 1984a). Copper deficiency resulted in a decrease in PSI and PS II photochemical activity and the inhibition was higher for PSI than PSII (Henriques 1989).

It has been shown that the photosynthetic apparatus of oats (*Avena sativa*) growing on a contaminated site with excess Cu and Pb was not damaged, but its functions were disrupted (Moustakas et al. 1994). Treated plants showed reduced height and biomass as well as reduced Chl ($a+b$) and changes in Rubisco in comparison to control plants.

Mercury was found to be an effective inhibitor of the rate of photosynthetic electron transport (Kimimura and Katoh 1972; Bernier et al. 1993; Šeršeň et al. 1998). On the other hand, chromium and cobalt ions applied to bean seedlings, either individually or mixed together in the nutrient medium, stimulated the activity of the Hill reaction if used in a concentration of 10^{-6}-10^{-4} M; however, higher concentrations of these metal ions inhibited plant growth (Zeid 2001).

Besides interaction with the components of the electron transport chain, HMs (Cd^{2+}, Ni^{2+}) inhibited Mg-ATPase activity and also interfered with membrane integrity (Ros et al. 1990).

6.3.1 Photosystem I

PS I has been extensively studied during the past decade and has been shown to be homologous in all photosynthetic organs of the higher plants. Its core complex was found to be highly conserved through evolution from cyanobacteria to higher plants (Almog et al. 1991). PS I is a membrane-bound protein complex which catalyzes oxidation of plastocyanin and reduction of ferredoxin under light conditions (Strzatka and Ketner 1997). A photon of light is captured by the trap Chl of PS I (P_{700}) which passes an excited electron to the electron acceptor called A_0 and then to ferredoxin by several carriers of increasing redox potential.

A site of Cd^{2+} toxicity was found on the reducing side of PS I. Siedlecka and Baszyński (1993) compared electron transport activities (DCIP→MV and DCIP→NADP$^+$) in Cd^{2+}-treated (10, 20 and 30 μM Cd^{2+}) isolated chloroplasts of 21-day-old maize. They suggested that the site of Cd^{2+} inhibition in PS I is between primary electron acceptor X and NADP$^+$. Exogenous ferredoxin [4 nM ferredoxin (mg Chl)$^{-1}$] restored NADP$^+$ photoreduction (Fig. 6.1). Cd^{2+} treatment caused Fe deficiency, indicating that the light phase of photosynthesis was affected in the treated plants due to Cd-induced Fe deficiency. Replacement of Fe^{3+} by Al^{3+} in FE-S centers was also the reason for pronounced inhibition of PS I activity

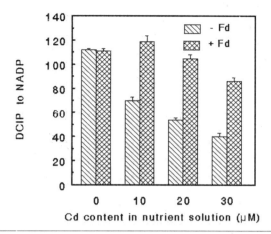

Fig. 6.1 Restoration of NADP$^+$ photoreduction in Cd-treated maize by exogenous ferredoxin [4 nM ferredoxin (mg Chl)$^{-1}$]. (Data from Siedlecka and Baszyński 1993)

in cyanobacterium *Nostoc linckia* at increasing acidity (Husaini and Rai 1992). Additionally, a pH-dependent inhibition of O_2 evolution and CO_2 fixation was observed.

Ferredoxin was found to be the place of Hg^{2+} (Honeycutt and Krogmann 1972) and Cu^{2+} (Shioi et al. 1978a,b) action. It was shown in EPR investigations that Cu^{2+} interrupted the electron transport from ferredoxin to NADP$^+$ in *Chlorella vulgaris* (Šeršeň et al. 1996); however, no interruption in the electron transport was observed in isolated spinach chloroplasts (Králová et al. 1994). Mercury induced the process of P$_{700}$ oxidation in darkness but it did not inhibit the electron transport around PS I (Šeršeň et al. 1998). In other studies, mercury interacted with the donor side of PS I (Singh et al. 1989) and with plastocyanin (Radmer and Kok 1974). Lead inhibited PS I activity in isolated bundle sheath and mesophyll chloroplasts of *Zea mays* (Wong and Govindjee 1976).

Shainberg et al. (2001) observed an inhibition of PS I activity measured as the kinetics of P$_{700}$ photooxidation in far-red light in bean (*Phaseolus vulgaris*) overloaded with Cu. This reduction in photosynthetic electron transport, together with the observed increase in activity of antioxidative mechanisms, may function as a protective mechanism toward an oxidative stress induced by Cu ions.

6.3.2 Photosystem II

PS II is a multi-protein complex located in the grana or appressed lamellae. It consists of at least 25 different subunits (Aro et al. 1993). Many HMs have effectively blocked photosynthetic electron transport at the level of PS II both at oxidizing (donor) and reducing (acceptor) sides (Clijsters and van Assche 1985; Krupa and Baszyński 1995; Myśliwa-Kurdziel et al. 2002; Table 6.2). The majority of reported experiments on the influence of HMs on PS II activity were performed for Cd^{2+} and Cu^{2+}.

In isolated chloroplasts, Cd^{2+}, Pb^{2+} and Cu^{2+} inhibited PS II activity (Cedeno-Maldonado et al. 1972; Miles et al. 1972; Li and Miles 1975; Shioi et al. 1978b; Hsu and Lee 1988). Bazzaz and Govindjee (1974) observed a 50% inhibition of PS II electron flow after a 10-min dark incubation of chloroplasts from *Zea mays* with 0.2 μM CdNO$_3$. More recently,

El-Shintinawy (1999) observed stimulation of Hill reaction activity in isolated soybean chloroplasts by 5 μM $CdCl_2$ and inhibition of this activity by higher Cd^{2+} concentrations (up to 200 μM). The inhibition was reversed in the presence of glutathione (1 mM). Based on the analysis of absorption and fluorescence spectra, it was concluded that Cd^{2+} affected both the PS II reaction center and the light-harvesting complex and caused an inefficient energy transfer from the light-harvesting complex to the reaction center.

Cadmium (200 ppm) applied through the rooting medium to 30-day-old wheat plants decreased net CO_2 exchange and PS II activity (Malik et al. 1992a,b).

The oxygen-evolving complex (OEC), located on the lumenal side of the thylakoids, which donates electrons from water to the PS II reaction center, was reported to be the primary target of HM toxicity (Bernier et al. 1993; Ouzounidou et al. 1993; Skórzyńska and Baszyński 1993; Maksymiec et al. 1994; Krupa and Baszyński 1995). The primary effect is the inhibition of photosynthetic O_2 evolution, reduction of $NADP^+$ and photophosphorylation. Secondary effects include delayed Chl degradation, a decreased Chl a/b ratio, and changes in nitrogen metabolism (Bishnoi et al. 1993a; Gadallah 1995a; Shalaby and Al-Wakeel 1995). Decrease in O_2 evolution was observed in Lemna trisulca treated with increasing Cd^{2+} concentrations up to 10 mM and for Cu^{2+} 2-50 μM (Prasad et al. 2001). Judging by the level of toxic concentration, Cu^{2+} appeared to be more toxic than Cd^{2+} for O_2 evolution activity. In a comparative study on the toxic effects of cadmium and mercury on Phaseolus aureus seedlings, it was found that Cd^{2+} inhibited the photosynthetic O_2 evolution more than Hg^{2+} did (Shaw and Rout 1998). This was seen for seedlings at older stages of their development (6-day-old), whereas no effect on the O_2 evolution was observed for younger seedlings (4-day-old). Decrease in O_2 evolution was also found in Cu-treated algal suspension (Chlorella vulgaris; Šeršeň et al. 1996).

Cadmium inhibited PS II, probably at the level of manganoprotein of the water splitting system (van Duijvendijk-Matteoli and Desmeta 1975; Atal et al. 1991; Greger and Ögren 1991). The mechanism of this destruction is not known; cadmium ions may act directly or indirectly by interaction with ions like Mn^{2+}, Ca^{2+} and Cl^- that are needed for proper functioning of the OEC (Skórzyńska and Baszyński 1993; Maksymiec and Baszyński 1998a). Inhibition of O_2 evolution together with the release of 17, 23 and 33 kDa proteins were observed also in PS II membranes isolated from tobacco (Nicotiana tabacum) after treatment with lanthanide ions (Dy^{3+} and Eu^{3+}; Burda et al. 1995). Cadmium and UV-B are supposed to act at a similar site in the chloroplast. In Vigna unguiculata, Cd^{2+} treatment for 3 days (3, 6, and 9 mM $CdCl_2$) degraded PS II polypeptides of 43, 33, 23 and 17 kDa. Polypeptides of 33, 23 and 17 kDa are needed for long-term stability of PS II and for proper functioning of the OEC. PS I activity in experimental variants did not change significantly compared with PS II (Nedunchezhian and Kulandaivelu 1995).

PS II was found to be highly sensitive also to Cu toxicity (for review, see Baron et al. 1995) which was not the case for PS I (Ouzounidou 1996, 1997). The site(s) of Cu binding and mechanism(s) of its action are still not understood thoroughly. Copper binding site(s) was found either on various parts of the acceptor (Mohanty et al. 1989; Yruela et al. 1991, 1993; Maksymiec et al. 1994) or donor (Shioi et al. 1978a,b; Samson et al. 1988) side of PS II (Fig. 6.2). Copper inhibited both the donor and the acceptor side of PS II in PS II preparations isolated from Spinacia oleracea (spinach) in an EPR spectroscopic study (Jegerschöld et al. 1995). Hsu and Lee (1988) suggested that it might influence directly the PS II core.

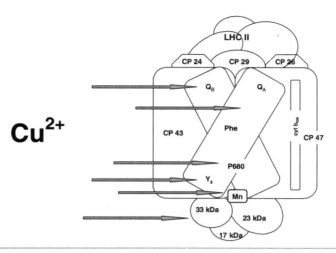

Fig. 6.2 Copper binding site(s) of PSII

On the acceptor (reducing) side of PS II, the Cu^{2+} inhibition site was found at the secondary quinone acceptor Q_B level (Mohanty et al. 1989). Yruela et al. (1991, 1992, 1993) showed precisely the site of Cu binding between pheophytin and Q_A in both higher plants and photosynthetic bacteria. A possible mechanism of copper toxicity was proposed based on measurements of O_2 evolution activity, in which Cu^{2+} was non-competitive with respect to DCBQ (2,6-dichlorobenzoquinone) and DCMU [3-(3,4-dichlorophenyl)-1,1-dimethylurea] and competitive with respect to protons (Yruela et al. 1992). Thus, copper-induced inhibition of O_2 evolution was related to the Cu^{2+} interaction with amino acid group(s), which can be protonated/deprotonated during the inhibition.

On the donor (oxidizing) side of PS II, copper probably acts close to the reaction center or at the level of OEC. The kinetics of $P680^+$ reduction in PS II membrane particles became markedly slower in the presence of 10 μM $CuSO_4$; however, charge separation ($P680+Q_A$) remains unchanged. It was concluded that Cu^{2+} blocks electron transport to $P680^+$ by specific modification of tyrosine Z and/or its microenvironment (Schröder et al. 1994). According to Arellano et al. (1995), the site of action of Cu^{2+} is located at the donor side of PS II, close to the reaction center. Another place of copper action is the manganese cluster in the OEC. The possible mechanism of copper toxicity may be the replacement of Mn^{2+} ions by Cu^{2+}. The release of free Mn^{2+} was observed using the EPR technique in copper-treated *Chlorella vulgaris* (Šeršeň et al. 1996) as well as isolated spinach chloroplasts (Králová et al. 1994). What is more, Šeršeň et al. (1996) observed the disappearance of EPR signals belonging to tyrosine cation radicals from D1 and D2 protein as well as an decrease in Chl *a* fluorescence.

Copper treatment of spinach thylakoids and PS II preparation resulted in a release of 17, 24 and 33 kDa proteins from the OEC for a Cu^{2+} concentration of 300 μM, which corresponded to 1400 Cu^{2+} ions per PS II reaction center unit (Yruela et al. 2000). No effects were observed on antenna complexes as well as on the D1 protein. In this study, the inactivation of the OEC was the primary effect of copper toxicity because a 50% inhibition was already observed for 7 and 27 μM Cu^{2+} for PS II particles and thylakoids, respectively.

Reversibility of Cu-induced inhibition of PS II activity is still under debate. In some experiments it was found that this effect is irreversible (Schröder et al. 1994; Arellano et al. 1995; Jegerschöld et al. 1995). However, in some studies, the inhibition was found to be reversed by the removal of excess copper (Hsu and Lee 1988; Yruela et al. 1993).

It has also been shown that a low concentration of copper might stimulate photosynthetic O_2 evolution. Prasad et al. (2001) observed this effect in the water plant *Lemna trisulca* for 1 μM Cu^{2+}, which correlated with an increase in Chl *a* concentration. Higher Cu^{2+} concentration (2-50 μM) resulted in lower efficiency of O_2 evolution. Similar results were obtained by Burda et al. (2002) for PS II particles isolated from tobacco. They observed stimulation of O_2 evolution for an equimolar Cu^{2+}/PS II proportion, and its suppression for higher Cu^{2+} concentrations.

HMs may also lower the photosynthetic yield due to the decrease in the number of active PS II centers caused by changing the equilibrium among the photoinhibition, D1 synthesis and D1 degradation. HMs interfere with the degradation of the D1 protein by a protease, occurring during normal operation of the PS II reaction center, and the damaged D1 protein cannot be replaced. Lower rate of the PS II repair cycle due to the copper treatment was observed in the green alga *Chlorella pyrenoidosa* (Vavilin et al. 1995) and in higher plants (Maksymiec et al. 1994; Yruela et al. 1996; Pätsikkä et al. 1998). In some studies, light was found to be necessary for excess-copper toxicity (Cedeno-Maldonado et al. 1972; Yruela et al. 1993). Lu and Zhang (1999) confirmed the essential role of light for Cu toxicity in an in vivo investigation of Chl fluorescence from the cyanobacterium *Spirulina platensis*. They found that Cu^{2+} inhibited the PS II photochemistry in the light but produced no significant effect for dark-incubated cells (12 h) and demonstrated that the inhibitory effect of Cu^{2+} increased for higher light intensity. In Cu^{2+}-treated thylakoid membranes isolated from bean (*Phaseolus vulgaris* L.) and pumpkin (*Cucurbita pepo* L.), light caused irreversible damage to PS II, which was observed at its donor site and led to the oxidation of the OEC (Pätsikkä et al. 2001). After prolonged illumination, a gradual collapse of the thylakoid structure was observed.

Cadmium also influences processes of synthesis, assembly and degradation of the D1 protein. Initial stimulation of the D1 synthesis followed by its inhibition was observed in Cd-treated pea (*Pisum sativum*) and broad beans (*Vicia faba*); however, the mechanism of this action has not been described (Geiken et al. 1998; Franco et al. 1999). Lower sensitivity towards photoinhibition was observed in cadmium-tolerant mutants of *Chlamydomonas reinhardtii* (Voigt et al. 1998).

Mercury is reported to inhibit photosynthetic electron transport, with PS II being the most sensitive target. Oxygen evolution was strongly inhibited and variable Chl fluorescence was severely quenched by Hg. Chloride, an inorganic cofactor known to be essential for the optimal function of PS II, significantly reversed the inhibitory effect of Hg^{2+}. However, Ca, another essential cofactor, showed no reversal capacity (Bernier et al. 1993). Mercury acts on the donor side of PS II and exerts its action by perturbing chloride binding. Another possible mechanism of mercury action that was found for Hg-treated spinach chloroplasts was the release of Mn^{2+} from the OEC observed as an EPR signal from free Mn^{2+} ions (Šeršeň et al. 1998).

Zinc inhibited O_2 evolution in isolated barley chloroplasts (Tripathy and Mohanthy 1980) and lettuce thylakoids (Miller and Cox 1983). In the latter, Zn^{2+} action caused the appearance of an EPR signal from free Mn^{2+} and the full inhibition of the O_2 evolution corresponded to the release of two manganese atoms per PS II unit. Rashid et al. (1991) also reported the interaction of Zn^{2+} with the donor side of PS II.

HMs also impair the functions of PS II indirectly via the plastoquinone pool, inhibiting photosynthetic carbon reduction cycle enzymes via changes in ATP level and feedback regulation.

6.4 ENZYMOLOGY

Enzymes of the photosynthetic carbon reduction (PCR) cycle are known to be inhibited by HMs. All three key steps of the Calvin cycle: carboxylation, reduction and regeneration, were found to be affected by HMs, with the first being the most sensitive (reviewed by Krupa and Baszyński 1995; Prasad 1995a, 1997; Table 6.3). Cadmium exerts its toxicity through membrane damage and inactivation of enzymes, possibly through reaction with sulfhydryl groups of proteins (Mathys 1975; Fuhrer 1988). Sulfhydryl group inactivation was suggested to explain the inhibitory effects of Pb^{2+}, Cd^{2+}, Zn^{2+} and Cu^{2+} on the activity of the chloroplast enzyme Rubisco and phosphoribulokinase in vitro (Stiborova et al. 1986b; for review, see van Assche and Clijsters 1990). Full inhibition of these enzyme activities was only observed in the presence of millimolar metal concentrations. In some cases, HM phytotoxicity is reflected by an increase in the activity of enzymes. This has been found for malic enzyme, glucose-6-phosphate dehydrogenase and peroxidase in a leaf (van Assche and Clijsters 1987).

Table 6.3 Inhibition of certain enzymes under heavy metal stress (after Krupa and Baszyński 1995, modified)

Enzyme/Taxa	Metal	Reference
Ribulose-1,5-bisphosphate carboxylase/oxygenase		
Betulla platyphylla	Mn	Kitao et al. (1997)
Cajanus cajan	Cd, Ni, Cu	Sheoran et al. (1990a,b)
Hordeum vulgare	Cd, Cu, Pb	Stiborova et al. (1986b)
Hordeum vulgare	Cd	Stiborova (1988)
Lolium perenne	Zn	Monnet et al. (2001)
Nicotiana tabacum	Mn	Houtz et al. (1988)
Oryza sativa	Cu	Lidon and Henriques (1993a-c)
Phaseolus vulgaris	Cd	Siedlecka et al. (1997, 1999)
Phaseolus vulgaris	Zn	Van Assche and Clijsters (1986)
Pisum sativum	Cu	Angelov et al. (1993)
Triticum aestivum	Cd	Malik et al. (1992b)
Vigna umbellata	Mn	Subrahmanyam and Rathore (2000)
PEP-carboxylase, and 3-phosphoglyceric acid kinase		
C. cajan	Cd, Ni	Sheoran et al. (1990b)
P. vulgaris	Zn	Van Assche and Clijsters (1987, 1990)
Zea mays	Cd, Cu, Pb, Zn	Stiborova et al. (1986b)

(Contd.)

Table 6.3 (*Contd.*)

Enzyme/Taxa	Metal	Reference
NADP-dependent glyceraldehyde 3-phosphate dehydrogenase		
C. cajan	Cd, Ni	Sheoran et al. (1990b)
P. vulgaris	Zn	Van Assche and Clijsters (1987, 1990)
Fructose-1,6-bisphosphatase		
C. cajan	Cd, Ni	Sheoran et al. (1990b)
T. aestivum	Cd, Ni	Malik et al. (1992b)
Aldolase		
C. cajan	Cd, Ni	Sheoran et al. (1990b)
Fructose 6-phosphate, 2 kinase		
T. aestivum	Cd	Malik et al. (1992b)
Adenosine-diphosphate glucose pyrophosphorylase		
T. aestivum	Cd	Malik et al. (1992b)
Peroxisomal enzymes		
P. vulgaris	Zn	Van Assche and Clijsters (1987, 1990)

Bivalent cations play a major role in the activation of Rubisco and in the equilibrium between CO_2 and O_2 binding by this protein (Lorimer 1981; Strzałka and Ketner 1997). Incubation of purified enzyme with millimolar concentrations of Co^{2+}, Mn^{2+}, and Ni^{2+} resulted in a decrease in carboxylation capacity and in a decrease in the carboxylation/oxygenation ratio, whereas incubation with micromolar concentrations of Zn^{2+}, Co^{2+} or Cu^{2+} caused a slight reactivation of carboxylase capacity (for review, see van Assche and Clijsters 1990).

Cadmium has been the most intensively studied inhibitor of the photosynthetic dark reactions (for reviews, see Krupa and Baszyński 1995; Krupa 1999). It was shown in isolated protoplasts treated with Cd^{2+} that the main target of this metal action was the reactions of the Calvin cycle and that activation of Rubisco was not affected (Weigel 1985a,b). In contrast, Sheoran et al. (1990a,b) showed significant reduction of Rubisco activity in vivo in pigeon pea (*Cajanus cajan* L.) plants if they were treated with Cd^{2+} at an early growth stage while in older plants activity of Rubisco was not affected. They concluded that the reduction in photosynthesis was through a decrease in Chl content and effects on stomatal conductance and the electron transport system. In another in vitro study, it was shown that Cd^{2+} treatment caused changes in the structure of Rubisco and resulted in dissociation of its small subunits (Stiborova 1988).

In some in vitro as well as in vivo investigations (Weigel 1985a,b; Krupa et al. 1992, 1993a; Krupa and Moniak 1998), the inhibitory effect of Cd^{2+} toward the process of photosynthesis was ascribed to the PCR. Cd-induced inhibition of photosynthetic dark reactions influences the photosynthetic electron transport by feedback reactions induced by lower consumption of ATP and NADPH (reviewed by Krupa and Siedlecka 1995; Krupa 1999). The inhibitory effect of Cd^{2+} on Rubisco activity as well as the content of Calvin cycle metabolites depend on Cd^{2+} concentration and Fe/Cd interactions (Siedlecka et al. 1997). Recent studies have confirmed that the inhibition of Rubisco carboxylase activity might be considered as the primary plant response to Cd toxicity (Siedlecka et al. 1997, 1999). At low Cd^{2+} concentration, the inhibitory Cd^{2+} effects were overcome by an increase in the Rubisco activation system and by enhancement of carboxyanhydrase activity. Pankovic et al. (2000) showed that in Cd-treated young sunflower leaves, Cd^{2+} mainly affected the regeneration of ribulose-1,5-bisphosphate in the Calvin cycle and that the amount of Rubisco increased, which indicated that a significant amount of the enzyme is not active in photosynthesis and could have another function.

At present, it is known that, depending on the metal concentration in the nutrient medium, plant growth stage and plant susceptibility to Cd^{2+}, this cation may inhibit the activity of Rubisco by damaging its structure, by replacement of Mg^{2+} ions or by shifting its activity towards oxygenation (Krupa 1999).

In pea leaves growing in the presence of 50 μM $CdCl_2$, some oxidized proteins were identified by using antibodies, e.g. Rubisco, glutathione reductase, manganese superoxide dismutase, and catalase, which in turn were more efficiently degraded (20% increase in proteolitic activity; Romero-Puertas et al. 2002). The results showed that Cd^{2+} might induce an oxidative stress, which is similar to that observed during treatment of leaf extracts with increasing H_2O_2 concentrations. Oxidative stress due to the Cd^{2+} treatment was also reported for pea leaves (McCarthy et al. 2001; Sandalio et al. 2001) as well as in green and greening barley seedlings (Hegedüs et al. 2001, and references therein). Cadmium modifies the activity of various antioxidative enzymes depending on plant species, plant growth stage and environmental conditions. The oxidative stress together with proteolytic degradation of Rubisco is related to the Cd-induced senescence (see Sect. 6.7). It has also been demonstrated that, in *Zea mays*, Cd^{2+} altered the photosynthesis and also the activities of the enzymes of photosynthetic sulfate and nitrate assimilation pathways (Ferretti et al. 1993).

Muthuchelian et al. (2001) found that the inhibitory effect of Cd^{2+} in *Erythrina variegata* could be effectively counterbalanced by application of triacontanol. Increases in growth, photosynthetic pigments, Rubisco, nitrate reductase and photosynthetic activity were observed in plants treated with Cd^{2+} and triacontanol in comparison with Cd-treated plants.

High sensitivity of the dark reactions of photosynthesis toward copper toxicity was also reported in Cu-sensitive rice and runner bean (Lidon and Henriques 1991; Maksymiec et al. 1994; Maksymiec and Baszyński 1996a). A progressive decrease of both carboxylase and oxygenase Rubisco activity was observed with increasing Cu level in the nutrient medium which in turn down-regulated the photosynthetic electron transport chain. Recently, Cu-induced inhibition of the dark phase of photosynthesis together with an unaffected maximal photochemical yield of PS II was found in mature Cu-treated cucumber (*Cucumis sativus* L.) leaves (Vinit-Dunand et al. 2002).

Copper is also known to interfere with oxidative enzymes in wheat (*Triticum durum*), oat (*Avena sativa*) and bean (*Phaseolus vulgaris*) leaves (Luna et al. 1994; Navari-Izzo et al. 1998; Shainberg et al. 2001). A similar reaction was observed for mercury in 30-day-old tomato seedling, in which Hg^{2+} induced oxidative stress manifested by increased H_2O_2 formation and higher activity of antioxidative enzymes: superoxide dismutase, catalase and peroxidase (Cho and Park 2000).

In *Phaseolus vulgaris* and *Lolium perenne*, Zn inhibited Rubisco carboxylase activity without affecting its oxygenase activity (van Assche and Clijsters 1986; Monnet et al. 2001). In another study, inhibition of net photosynthetic CO_2 fixation rate and increase of the CO_2 compensation point were observed in intact *Phaseolus vulgaris* after Zn treatment (van Assche et al. 1988). A 100% increase in activity of detoxifying enzymes was found in Zn-treated ryegrass (*Lolium perenne*) for 50 mM Zn^{2+} concentration in the nutrient medium (Bonnet et al. 2000).

Activity of the PCR cycle was found to be the primary effect of Mn toxicity in ricebean (Subrahmanyam and Rathore 2000). Decrease in PS II photochemistry was the secondary effect caused indirectly by accumulation of ATP and NADPH.

6.5 INVESTIGATION OF HM INFLUENCE ON PHOTOSYNTHESIS USING FLUORESCENCE TECHNIQUES

Fluorescence spectroscopy has proved to be very useful for studying the toxic functions of HMs at extremely low concentrations of environmentally realistic levels (Maksymiec and Baszyński 1996a). Photosynthetic reactions in wheat seedlings at low concentrations of Cd^{2+} were investigated by analyzing electron transport and changes in fluorescence yield (Atal et al. 1991). Similarly, whole leaf fluorescence was used as a technique for measuring tolerance of plants to HMs (Homer et al. 1980). Based on parameters obtained from a PAM (pulse-amplitude-modulation) fluorimeter, Juneau et al. (2002) compared the relative sensitivity of *Chlorella vulgaris*, *Selenastrum capricornutum* and *Chlamydomonas reinhardii* to copper toxicity. *C. reinhardtii* was the most sensitive species followed by *S. capricornutum* and *C. vulgaris*. The authors discussed the method as useful and sensitive for aquatic toxicology. The determination of the most sensitive algal species to different pollutants is essential for the development of new bioassay methods.

In *Phaseolus vulgaris*, photosynthetic functions were studied by fluorescence analysis (Krupa et al. 1992, 1993a,b). Chl fluorescence has been used as a reliable and rapid screen for Al tolerance in cereals (Moustakas et al. 1993).

Copper (160 μM) inhibited variable fluorescence $F_v=(F_m - F_0)$, where F_v is variable fluorescence, F_m is maximal fluorescence, and F_0 is initial fluorescence, by 27% in *Silene compacta* (Cu-tolerant) and by 81% in *Thlaspi ochroleucum* (Cu-sensitive), suggesting inhibition on the photooxidizing side of PS II (Fig. 6.3; Ouzounidou 1993).

In another study, in Cu-treated leaves of *Thlaspi ochroleucum*, a decrease in maximal chlorophyll fluorescence level in dark- and light-adapted leaves (F_m and $F_{m'}$, respectively) and changes in variable fluorescence (F_v) were also observed (Ouzounidou 1996). Moreover, in that study, the vitality index (Rfd) dramatically decreased, while a moderate decrease in photosynthetic energy storage and a marginal decrease in the photochemical yield of open PS II reaction centers in light were observed.

Wheat plants (*Triticum aestivum* L. cv. Vergina) growing in the field on ore bodies containing Cu^{2+} (3050 μg/g) and control plants grown in garden soil with 140 μg/g were investigated for Chl fluorescence parameters like F_m, F_v, and $t_{1/2}$ (this last parameter is a function of the rate of the photochemical reactions and the pool size of electron acceptors on the reducing side of PS II, including the plastoquinone pool and it is used as a simple indicator of the pool size). F_v/F_m was lower in the ore-grown plants than control plants, while F_0 was higher. These values suggest that there is a decrease in the pool size of the electron acceptors on the reducing side of PS II (Lanaras et al. 1993).

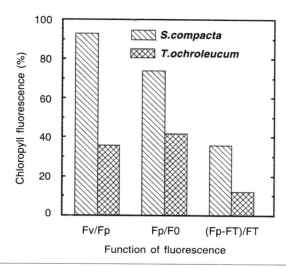

Fig. 6.3 Function of fluorescence in Cu-treated (160 μM) *Silene compacta* (Cu-tolerant) and *Thlaspi ochroleucum* (Cu-sensitive) plant species. F_v variable fluorescence; F_m maximal fluorescence; F_0 initial fluorescence; F_t terminal steady-state fluorescence. The reduction in the vitality index (Rfd)=[$(F_m - F_t)/F_t$] was 64% in *S. compacta* and 88% in *T. ochroleucum*. (Data from Ouzounidou 1993)

In oats (*Avena sativa*) growing on a contaminated site with excess Cu^{2+} and Pb^{2+}, the quantum yield decreased (7%) as given by the ratio of F_v/F_m measured in dark-adapted leaves in field conditions. The half-rise time from the initial (F_0) to maximal (F_m) Chl fluorescence increased, suggesting that the amount of active pigments associated with the photosynthetic apparatus decreased and that the functional Chl antennae size of the photosynthetic apparatus was smaller compared with control plants (Moustakas et al. 1994).

In a comparative study of toxic concentrations of Cu^{2+}, Zn^{2+} and Cd^{2+} on the fast Chl fluorescence induction kinetics of PS II in primary bean leaves in vivo, different effects of these three metals on the shape of the fluorescence induction curve were found (Ciscato et al. 1999). The metals, however, did not have strong effects on the time course of the F_v/F_m ratio, which indicated no changes in PS II efficiency in this experiment. The observed changes in the fluorescence induction kinetics were interpreted in terms of different sites and/or mechanisms of action of these three metals in PS II, and as a different potential of plant adaptation to the metals: complete recovery for Cu^{2+}, partial recovery for Zn^{2+} and no recovery for Cd^{2+} toxicity.

Chl fluorescence analysis was also applied to monitor the effect of Hg^{2+} on photosynthesis in the cyanobacterium *Spirulina platensis* after 2-h treatment with Hg^{2+} up to 20 μM (Lu et al. 2000). A decrease in the maximal photochemical efficiency of PS II (F_v/F_m), the efficiency of excitation energy capture by open PS II reaction centers ($F_{v'}/F_{m'}$), quantum yield of PS II electron transport, photochemical (q_P) and non-photochemical (q_N) quenching was observed. Hg^{2+} treatment resulted in a significant increase in the proportion of the Q_B-non-reducing PS II reaction centers. Changes in fluorescence parameters occurred before any other visible symptoms of Hg toxicity appeared; thus fluorescence analysis can be used for the analysis of early stages of photosynthetic response to HM toxicity.

This method was also used to investigate the influence of Mn on photosynthesis in leaves of tree species differing in successional traits (Kitao et al. 1997, 1998). Mn toxicity was also investigated in ryegrass (*Lolium perenne*) where an increase in non-photochemical quenching (q_N) concomitant with a decrease in photochemical quenching (q_P) was observed (Subrahmanyam and Rathore 2000). Recently, Chl *a* fluorescence was used in the investigation of zinc toxicity in this plant species (Bonnet et al. 2000; Monnet et al. 2001). Monnet et al. (2001) found the decrease in the F_v/F_m ratio to be mainly due to an increase in the F_0 level and ascribed it to a Zn-induced disturbance in PS II reaction centers or in an energy transfer from antenna to the reaction center.

Chl *a* fluorescence measurements were also used to study the effect of Cr on the photosynthetic activity of *Spirodella polyrhiza* in vivo (Appenroth et al. 2001). A decreased yield of primary photochemistry due to a decrease in the number of active reaction centers as well as due to damage to the OEC was observed as a result of Cr treatment.

6.6 STOMATAL AND GAS EXCHANGE FUNCTIONS

Stomatal movements provide the leaf with an opportunity to change the partial pressure of CO_2 and the rate of transpiration. Increased stomatal resistance in Cd-treated plants has been reported (Bazzaz et al. 1974a,b; Poschenrieder et al. 1989). Interference with stomatal function is considered to be a primary mode of action of Cd^{2+} and several other metal ions (Becerril et al. 1989). In contrast, Subrahmanyam and Rathore (2000) suggested that in Mn-treated ricebean (*Vigna umbellata*) the reduction of both dark and light reactions of photosynthesis was not due to the direct effect of Mn on stomatal regulation because a decrease in photosynthesis preceded a decrease in stomatal conductance and transpiration rate.

The inhibition of stomatal opening in plants exposed to Cd^{2+} may depend on both Cd^{2+} concentration and exposure time, and also on the degree of toxicity suffered by the plants (Barceló et al. 1986a,b). A direct effect of Cd^{2+} on the ion and water movement in the guard cells has yet to be proven, but cannot be ignored, particularly in relation to K^+ and Ca^{2+} (Barceló et al. 1986a,b). Cadmium influenced the transpiration of plants more via its effect on the water flow through the roots than via effects on stomatal aperture (Hatch et al. 1988). Cadmium treatment induced symptoms of Fe deficiency. Net CO_2 assimilation, water use efficiency and stomatal conductance were depressed in Cd-treated (10 μM) wheat plants (Abo-Kassem et al. 1995).

Cadmium is an effective inhibitor of plant metabolism, particularly of the photosynthetic process in higher plants. The linear relationship between net photosynthesis and inhibition of transpiration suggests that the response to Cd^{2+} might be due to the closure of stomata (Bazzaz et al. 1974a,b; Huang et al. 1974; Greger and Johansson 1992). Cadmium inhibits net photosynthesis by increasing both stomatal and mesophyll resistance to CO_2 uptake (Lamoreaux and Chaney 1978).

Long-term exposure (120 h) of bush bean to Cd^{2+} increased stomatal resistance (Poschenrieder et al. 1989). Cadmium also inhibited abscisic acid accumulation during drying of excised leaves. In clover and lucerne, Cd^{2+} and Pb^{2+} induced several changes in gas exchange and water relations (Becerril et al. 1989). Plant water relations as affected by HMs have been critically reviewed (Barceló and Poschenrieder 1990). Wheat leaves of different insertion level have different degrees of sensitivity to Cd^{2+} and Ni^{2+} as a function of photosynthesis and water relations (Bishnoi et al. 1993b). Cadmium induced a decrease in water stress resistance in bush bean plants (*Phaseolus vulgaris* cv. Contender) by affecting endogenous abscisic acid, water potential, relative water content, and cell wall elasticity (Barceló et al. 1986a,b; Poschenrieder et al. 1989). Cadmium-induced changes in water economy in beans were attributed to the involvement of ethylene formation (Fuhrer 1988). Net CO_2 exchange decreased in Cd-treated (200 ppm) plants (30-day-old; Malik et al. 1992b).

Plants exposed to excess of HMs had increased stomatal resistance, decreased transpiration rate and altered water relations (Bazzaz et al. 1974a,b). Cadmium generally decreased the water stress tolerance of plants, causing turgor loss at higher relative water content and leaf water potential than in control plants. Cadmium increased the bulk elastic modulus, and therefore decreased cell wall elasticity. Low cell wall elasticity seems to be an important cause of the low water stress tolerance in plants affected by Cd toxicity (Becerril et al. 1989). Xylem obstruction by Cd-induced cell wall degradation products has been suggested as a cause of reduced water transport (Lamoreaux and Chaney 1978; Fuhrer 1988). Wilting in Cd-treated plants is exclusively due to decreased water transport and probably reduced water absorption (principally because of reduced root growth). Vassilev et al. (1998) showed that a slightly reduced photosynthetic rate in barley (*Hordeum vulgare*) treated with Cd^{2+} (45 mg/kg soil) during its ontogenesis was not due to stomatal resistance.

Cadmium and lead adversely affected gas exchange in lucerne (*Medicago sativa*) and clover (*Trifolium pratense*; Becerril et al. 1989). However, the water potentials measured and the decrease in water use efficiency calculated after Pb treatment indicated that, despite strong stomatal closure, the transpiration rate was still remarkable. Low concentrations of Cd^{2+} in certain experiments increased transpiration. High concentrations of Cd^{2+} reduced the transpiration rate, probably as a result of root damage. Transpiration behavior indicates that the effect of water vapor pressure on the degree of stomatal opening is small (Hagemeyer and Waisel 1989a). Under specific conditions, certain HMs reduce plant transpiration by direct interference with stomatal regulation. Light had a strong effect on transpiration rates of the Cd-treated plants (Hagemeyer and Waisel 1989a,b). In *Zea mays*, simultaneous inhibition of photosynthesis and transpiration by Ni suggested its primary effect should be on stomatal function.

6.7 FOLIAR ACCUMULATION OF HEAVY METALS AND SENESCENCE

Plants exhibit variable degrees of HM accumulation in their parts. HMs are primarily taken up by plants through roots. However, foliar absorption and stem uptake also represent potential modes of entry (Greger et al. 1993). In a comparative study, it was demonstrated that cuticular permeability of Cd^{2+} in pea (*Pisum sativum* L.) leaves was higher than in beet (*Beta vulgaris*). Low pH (3.6) in the external medium decreased the net Cd^{2+} uptake compared with higher pH (5.6). Higher pH opened up more stomata on the adaxial surface than the abaxial, favoring the entry of Cd^{2+} (Greger et al. 1993; see also Chap. 1, this Vol.). Leaf cuticles function as weak cation exchangers, due to the negative charge of pectin and cutin polymers. Experiments revealed that stomata are not directly involved in uptake of Cd^{2+}.

The level of HMs detected in leaves was often not more than 10% of the total plant accumulation (Ernst et al. 1992). The level of HMs translocated to chloroplasts is estimated to be only about 1%; however, there is a high variation between species, which needs to be investigated. HM increase in leaves enhanced the production of ethylene, a promoter of senescence (Fuhrer 1988; Pnnazio and Roggero 1992; Fig. 6.4). Both permanent stomatal closure and increased ethylene production may be responsible for senescence induction by Cd^{2+} (Fuhrer 1988). In soybean seedlings growing in the presence of Cd^{2+}, ethylene evolution was increased by 10 times and peroxidase activity by 3 times in comparison to the untreated plants (El-Shintinavy 1999). This effect was counteracted by addition of glutathione.

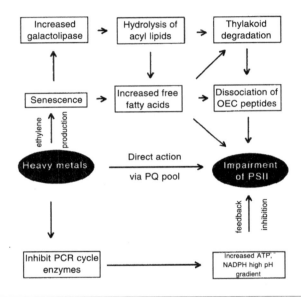

Fig. 6.4 Direct and indirect influence of heavy metals on PS II (after Krupa and Baszyński 1995, modified)

Spinach plants treated with lead acetate for 15 days accumulated 45% of Pb in roots and 55% in leaves. More than 90% of HMs are generally adsorbed to cell walls. The oldest leaves accumulated the highest concentration of HMs. HMs have differing degrees of mobility in plant tissues and are probably regulated by carrier proteins of the vascular tissues which serve as ligands for transportation (Prasad 1995a). Lead in senescent leaves remains constant throughout the year. However, bioaccumulation of Zn gradually increases as senescence progresses (Ernst et al. 1992). Rice plants subjected to Cu treatment (0.002-6.25 g m^{-3} in a nutrient solution), grown over a period of 30 days, showed a decreased Fe content, i.e. Cu inhibited Fe uptake (Lidon and Henriques 1993a,b).

In barley, the leaf epidermis is reported to play a significant role in Zn compartmentation. When the seedlings were raised at normal (1 μM) and inhibitory (200 μM) $ZnSO_4$ concentrations in low strength Hoagland solution, accumulation of Zn in the latter case was 10-fold in the epidermis, 11-fold in the vacuole, eight times in the apoplasm and almost the same in the cytosol (Dietz and Hartung 1996). Variable allocation of HMs in the phytomass of crop plants has been reported (Wagner 1993; Siedlecka 1995; Prasad 1996; Vassilev et al. 1998).

Pigment degradation, which is observed during plant senescence, was also found to be influenced by HMs. Hg, Cu and Zn ions changed the hydrolytic activity of chlorophyllase in rice leaves, with mercury increasing chlorophyllase activity most out of the three metals tested. Zinc increased chlorophyllase activity moderately, and Cu affected it least, the order being Hg>Zn>Cu (Drążkiewicz 1994). Chlorophyllase activity was also enhanced by heavy metals like Cd^{2+}, Pb^{2+}, Mn^{2+}, Co^{2+}, and Ni^{2+}, and their mixture in algal suspensions of *Chlorella fusca* or *Kirchneriella lunaris* (Abdel-Basset et al. 1995). The effect was different for the two algal species and particular heavy metal as well, as it was changed by the presence of Ca^{2+}.

6.8 CONCLUSIONS

Photosynthesis is one of the physiological processes most susceptible to HM toxicity. Due to its own complexity, it is not possible to show a unique mechanism of HM action. HMs were found to inhibit both light and dark reactions of photosynthesis. It is known that in many cases PS II is mainly inhibited in HM-treated plants. It can be affected by different direct and indirect mechanisms (Fig. 6.4).

Concerning the dark photosynthetic reactions, several steps of the Calvin cycle were found to be affected by HMs; however, the carboxylation step appeared to be the most sensitive.

It should also be pointed out that the majority of experiments were performed to investigate the toxic effect of copper and cadmium. Thus it seems that at present the mechanism of metal action can be discussed in more detail only in the case of these two metals. However, common conclusions cannot be drawn due to the fact that copper is a microelement that is required for physiological plant growth, whereas cadmium is a nonessential element.

Differentiation between the primary and secondary effects of Cd on photosynthesis has been under discussion for more than 20 years (reviewed by Krupa and Baszyński 1995; Krupa 1999). At present, it is known that activity of the PCR cycle may be regarded as a primary target of cadmium toxicity. In many cases, inhibition of the photosynthetic electron transport chain is a secondary event caused in part by lower consumption of ATP and NADPH. Nevertheless, direct inhibition of photosynthetic electron transport was also shown in many experiments (see Table 6.1). The results depended on whether the experiments were performed on a whole plant or on isolated organelles. What is more, for plants growing in the presence of Cd^{2+}, the toxic effect depends on the duration of metal treatment, metal concentration, uptake and individual plant sensitivity, and many other factors (Krupa and Moniak 1998; Krupa et al. 1992, 1993a; Krupa 1999). It is also probable in the case of other HMs. Due to the fact that in vitro/in vivo results of HM toxicity do not exactly match in all investigations, more studies on intact leaves are required. Fluorescence spectroscopy seems to be one of the most useful approaches in such investigations.

Chlorosis of leaves which is often observed after HM treatment may be a consequence of an inhibition of chlorophyll synthesis as well as of an increase in chlorophyll degradation. In both cases, however, lower Chl content results in lower efficiency of photosynthesis. The environmental relevance of HM-substituted Chl, particularly in aquatic plants, is also of considerable significance (Küpper et al. 1996, 1998). The role of single/multiple metals (essential and non-essential) in etiolated seedlings and their effects on pigment biosynthesis need critical study. Organization and functional properties of the photosynthetic apparatus during the greening process (in seedlings) under HM stress is also a promising area to investigate HM interaction with the photosynthetic apparatus, which has not been explored yet.

Cadmium at low concentrations, for example 2-9 ppm, reduced dry matter production up to 50% in some field crops. However, other non-essential HMs caused no significant reduction in crop yield (Baszyński 1986). Hence, sewage sludge subjected to partial leaching containing permissible concentrations of HMs may be acceptable for supplementing deficient micronutrients (Morsing 1994), e.g. amelioration of soil Al toxicity by Si (Hodson and Evans 1995).

Interaction between toxic HMs and plant mineral nutrients is also an important aspect that needs detailed investigation (Burzyński and Buczek 1989, 1994; Rubio et al. 1994; Krupa and Siedlecka 1995; Siedlecka 1995). Toxic HMs like Cd, Ni, Co, Cr, Zn and Pb are reported to cause Fe deficiency either by decreased uptake, or by causing immobilization in roots. Interaction of Cu^{2+} and Cd^{2+} ions with Ca^{2+} has also been recently reported (Maksymiec and Baszyński 1998b, 1999a,b; Skórzyńska-Polit et al. 1998). HMs induced lipid/waxy coating on the leaf surface, and thus influenced photosynthesis. The leaf epidermis and its ecophysiological functions might deepen our insight into the multifaceted subject of metal interactions with photosynthetic machinery in higher plants (Dietz and Hartung 1996).

Permanent stomatal closure, structural damage to chloroplasts, increased synthesis of ethylene (promoter of senescence) and imbalance in water relations might be acting synergistically on HM-exposed plants, resulting in impairment of photosynthetic functions.

ACKNOWLEDGEMENTS

MNVP is thankful to DST Indo-Polish (DST-KBN) cooperation (No. INT/POL/P-4/2001 dt. 26.7.2001). KS gratefully acknowledges the financial support of KBN grant No 6P04A 028 19 to undertake this revision.

REFERENCES

Abdel-Basset R, Issa AA, Adam MS (1995) Chlorophyllase activity: effects of heavy metal and calcium. Photosynthetica 31: 421-425

Abo-Kassem E, Sharaf-el-din A, Rozema J, Foda EA (1995) Synergistic effects of cadmium, NaCl on the growth, photosynthesis and ion content in wheat plants. Biol Plant 37: 241-249

Agarwala SC, Bisht SS, Sharma CP (1977) Relative effectiveness of certain heavy metals in producing toxicity and symptoms of iron deficiency in barley. Can J Bot 55: 1299-1307

Ahmed A, Tajmir-Riahi HA (1993) Interaction of toxic metal ions Cd^{2+}, Hg^{2+} and Pb^{2+} with light-harvesting proteins of chloroplast thylakoid membranes. An FTIR spectroscopic study. J Inorg Biochem 50: 235-243

Almog O, Lotan O, Shoham G, Nechushtai R (1991) The composition and organization of photosystem I. J Basic Clin Physiol Pharmacol 2: 123-140

Angelov M, Tsonev T, Uzunova A, Gaidardjieva K (1993) Cu^{2+} effect upon photosynthesis, chloroplast structure, RNA and protein synthesis of pea plants. Photosynthetica 28: 341-350

Appenroth KJ, Stöckel J, Srivastava A, Strasser RJ (2001) Multiple effects of chromate on the photosynthetic apparatus of *Spirodella polyrhiza* as probed by OJIP chlorophyll *a* fluorescence measurements. Environ Pollut 115: 49-64

Arellano JB, Lazaro JJ, Lopez-Gorge J, Baron M (1995) The donor side of photosystem II as the copper-inhibitory binding site. Fluorescence and polarographic studies. Photosynth Res 45: 127-134

Aro EM, Virgin I, Andersson B (1993) Photoinhibition of photosystem II. Inactivation, protein damage and turn-over. Biochim Biophys Acta 1143: 113-134

Atal N, Saradhi PP, Mohanty P (1991) Inhibition of the chloroplast photochemical reactions by treatment of wheat seedlings with low concentrations of cadmium. Analysis of electron transport activities and changes in fluorescence yield. Plant Cell Physiol 32: 943-951

Barceló J, Poschenrieder C (1990) Plant water relations as affected by heavy metals: a review. J Plant Nutr 13: 1-37

Barceló J, Cabot C, Poschenrieder C (1986a) Cadmium-induced decrease of water stress resistance in bush bean plants (*Phaseolus vulgaris* L. cv. Contender). II. Effects of Cd on endogenous abscisic acid levels. J Plant Physiol 125: 27-34

Barceló J, Poschenrieder C, Andreu I, Gunse B (1986b) Cadmium-induced decrease of water stress resistance in bush bean plants (*Phaseolus vulgaris* L. cv. Contender). I. Effects of Cd on water potential, relative water content and cell wall elasticity. J Plant Physiol 125: 17-25

Barceló J, Vazquez, MD, Poschenrieder C (1988) Structural and ultrastructural disorders in cadmium-treated bush bean plants (*Phaseolus vulgaris* L.). New Phytol 108: 37-49

Baron M, Arellano JB, Gorge JL (1995) Copper and photosystem II: a controversial relationship. Physiol Plant 94: 174-180

Barua B, Jana S (1986) Effect of heavy metals on dark induced changes in Hill action activity, chlorophyll and protein contents, dry matter and tissue permeability in detached *Spinacia oleracea* L. leaves. Photosynthetica 20: 74-76

Baryla A, Carrier P, Franck F, Coulomb C, Sahut C, Havaux M (2001) Leaf chlorosis in oilseed rape plants (*Brasicca napus*) grown on cadmium polluted soil: causes and consequences for photosynthesis and growth. Planta 212: 696-709

Baszyński T (1986) Interference of Cd^{2+} in functioning of the photosynthetic apparatus of higher plants. Acta Soc Bot Pol 55: 291-304

Baszyński T, Tukendorf A (1984) Copper in the nutrient medium of higher plants and their photosynthetic apparatus activity. Folia Soc Sci Lubliensis 26: 31-39

Baszyński T, Ruszkowska M, Król M, Tukendorf A, Wolińska D (1978) The effect of copper deficiency on the photosynthetic apparatus of higher plants. Z Pflanzenphysiol 89: 207-216

Baszyński T, Wajda L, Król M, Wolińska D, Krupa Z, Tukendorf A (1980) Photosynthetic activities of cadmium treated tomato plants. Physiol Plant 48: 365-370

Baszyński T, Król M, Krupa Z, Ruszkowska M, Wojcieska U, Wolińska D (1982) Photosynthetic apparatus of spinach exposed to excess copper. Z Pflanzenphysiol 108: 385-395

Baszyński T, Tukendorf A, Ruszkowska M, Skórzyńska E, Maksymiec W (1988) Characteristics of the photosynthetic apparatus of copper non-tolerant spinach exposed to excess copper. J Plant Physiol 132: 708-713

Bazzaz FA, Rolfe GL, Carlson RW (1974a) Effect of cadmium on photosynthesis and transpiration of excised leaves of corn and sunflower. Plant Physiol 32: 373-376

Bazzaz FA, Rolfe GL, Carlson RW (1974b) Differing sensitivity of corn and soybean photosynthesis and transpiration to lead contamination. J Environ Qual 3: 156-158

Bazzaz FA, Carlson RW, Rolfe GL (1975) Inhibition of corn and sunflower photosynthesis by lead. Physiol Plant 34: 326-329

Bazzaz MB, Govindjee (1974) Effects of cadmium nitrate on spectral characteristics and light reactions of chloroplasts. Environ Sci Lett 6: 1-12

Becerril JM, Gonzalez-Murua C, Munoz-Rueda R, de Felipe MR (1989) The changes induced by cadmium and lead in gas exchange and water relations in clover and lucerne. Plant Physiol Biochem 27: 913-918

Belkhodja R, Morales F, Quilez R, López-Millán AF, Abadia A, Abadia J (1998) Iron deficiency causes changes in chlorophyll fluorescence due to the reduction in the dark of the photosystem II acceptor side. Photosynth Res 56: 265-276

Bernier M, Popovic R, Carpentier R (1993) Mercury inhibition at the donor side of photosystem II is reversed by chloride. FEBS Lett 321: 19-23

Berska J, Myśliwa-Kurdziel B, Strzałka K (2001) Transformation of protochlorophyllide to chlorophyllide in wheat under heavy metal stress. In: PS2001 Proceedings of the 12th international congress on photosynthesis, CSIRO 2001, S2-015

Bishnoi NR, Chug LK, Sawhney SK (1993a) The effect of chromium on photosynthesis, respiration and nitrogen fixation in pea (Pisum sativum L.) seedlings. J Plant Physiol 142: 25-30

Bishnoi NR, Sheoran IS, Singh R (1993b) Influence of cadmium and nickel on photosynthesis and water relations in wheat leaves of differential insertion level. Photosynthetica 28: 473-479

Boening DW (2000) Ecological effects, transport, and fate of mercury: a general review. Chemosphere 40: 1335-1351

Bonnet M, Camares O, Veisseire P (2000) Effect of zinc and influence of Acremonium lolii on growth parameters, chlorophyll a fluorescence and antioxidant enzyme activities of ryegrass (Lolium perenne L. cv Apollo). J Exp Bot 51: 945-953

Boucher N, Carpentier R (1999) Hg^{2+}, Cu^{2+}, Pb^{2+} induced changes in photosystem II photochemical yield and energy storage in isolated thylakoid membranes: a study using simultaneous fluorescence and photoacoustic measurements. Photosynth Res 59: 167-174

Böddi B, Oravecz AR, Lehoczki E (1995) Effect of cadmium on organization and photoreduction of protochlorophyllide in dark-grown leaves and etioplast inner membrane preparations of wheat. Photosynthetica 31: 411-420

Burda K, Strzałka K, Schmid GH (1995) Europium and dysprosium ions as probes for the study of calcium binding sites in photosystem II. Z Naturforsch 50: 220-230

Burda K, Kruk J, Radunz A, Schmid GH, Strzałka K (2002) Stimulation and inhibition of oxygen evolution in photosystem II by copper(II) ions. Z Naturforsch 57c: 853-857

Burzyński M (1985) Influence of lead on chlorophyll content and on initial steps of its synthesis in greening cucumber seedlings. Acta Soc Bot Pol 54: 95-107

Burzyński M, Buczek J (1989) Interaction between cadmium and molybdenum affecting the chlorophyll content and accumulation of some heavy metals in the second leaf of Cucumis sativus L. Acta Physiol Plant 11: 137-145

Burzyński M, Buczek J (1994) The influence of Cd, Pb, Cu and Ni on NO_3 uptake by cucumber seedlings. II. In vitro and in vivo effects of Cd, Pb, Cu and Ni on the plasmalemma ATPase and oxidoreductase from cucumber seedling roots. Acta Physiol Plant 16: 297-302

Caspi V, Droppa M, Horváth G, Malkin S, Marder JB, Raskin VI (1999) The effect of copper on chlorophyll organization during greening of barley leaves. Photosynth Res 62: 165-174

Cedeno-Maldonado A, Swader JA, Heath RL (1972) The cupric ion as an inhibitor of photosynthetic electron transport in isolated chloroplasts. Plant Physiol 50: 698-701

Cho UH, Park JO (2000) Mercury-induced oxidative stress in tomato seedlings. Plant Sci 156: 1-9

Chugh LK, Sawhney SK (1999) Photosynthetic activities of Pisum sativum seedlings grown in presence of cadmium. Plant Physiol Biochem 37: 297-303

Ciscato M, Vangronsveld J, Valcke R (1999) Effect of heavy metals on the fast chlorophyll fluorescence induction kinetics of photosystem II: a comparative study. Z Naturforsch 54c: 735-739

Clijsters H, van Assche F (1985) Inhibition of photosynthesis by heavy metals. Photosynth Res 7: 41-40

Cook CM, Kostidou A, Vardaka E, Lanaras T (1997) Effects of copper on the growth, photosynthesis and nutrient concentrations on phaseolus plants. Photosynthetica 34: 179-193

Csatorday K, Gombos Z, Szalontai B (1984) Mn^{2+} and Co^{2+} toxicity in chlorophyll biosynthesis. Proc Natl Acad Sci USA 81: 476-478

Dietz KJ, Hartung W (1996) Leaf epidermis: its ecophysiological significance. Prog Bot 57: 32-53

Doncheva S, Stoyanova Z, Velikova V (2001) Influence of succinate on zinc toxicity of pea plants. J Plant Nutr 24: 789-804

Drażkiewicz M (1994) Chlorophyllase: occurrence, functions, mechanism of action, effects of external and internal factors. Photosynthetica 30: 321-331

Droppa M, Horváth G (1990) The role of copper in photosynthesis. Crit Rev Plant Sci 9: 111-123

Droppa M, Terry N, Horvath G (1984a) Variation in photosynthetic pigments and plastoquinone contents in sugar beet chloroplasts with changes in leaf copper content. Plant Physiol 74: 717-720

Droppa M, Terry N, Horváth G (1984b) Effects of Cu deficiency on photosynthetic electron transport. Proc Natl Acad Sci USA 81: 2369-2373

El-Shintinawy F (1999) Glutathione counteracts the inhibitory effect induced by cadmium on photosynthetic process in soybean. Photosynthetica 36: 171-179

Eleftheriou EP, Karataglis S (1989) Ultrastructural and morphological characteristics of cultivated wheat growing on copper-polluted fields. Bot Acta 102: 134-140

Ernst WHO, Verkleij JAC, Schat H (1992) Metal tolerance in plants. Acta Bot Neerl 41: 229-248

Ferretti M, Ghisi R, Merlo L, Vecchia FD, Passera C (1993) Effect of cadmium on photosynthesis and enzymes of photosynthetic sulphate and nitrate assimilation pathways in maize (*Zea mays* L.). Photosynthetica 29: 49-54

Fodor F, Sarvari E, Lang F, Szigeti Z, Cseh E (1996) Effects of Pb and Cd on cucumber depending on the Fe-complex in the culture solution. J Plant Physiol 148: 434-439

Franco E, Alessandrelli S, Masojidek J, Margonelli A, Giardi MT (1999) Modulation of D1 protein turnover under cadmium and heat stress monitored by [S-35]methionine incorporation. Plant Sci 144: 53-61

Fuhrer J (1988) Ethylene biosynthesis and cadmium toxicity in leaf tissue of beans *Phaseolus vulgaris* L. Plant Physiol 70: 162-167

Gadallah MAA (1995a) Interactive effect of heavy metals and temperature on the growth and chlorophyll, saccharides and soluble nitrogen in *Phaseolus* plants. Biol Plant 36: 373-382

Gadallah MAA (1995b) Effects of cadmium and kinetin on chlorophyll content, saccharides and dry matter accumulation in sunflower plants. Biol Plant 37: 233-240

Geider RJ, La Roche J (1994) The role of iron in phytoplancton photosynthesis, and the potential for iron-limitation of primary productivity in the sea. Photosynth Res 39: 275-301

Geiken B, Masojidek J, Rizutto M, Pompili ML, Giardi MT (1998) Incorporation of [S-35]methionine in higher plants reveals that stimulation of the D1 reaction centre II protein turnover accompanies tolerance to heavy metal stress. Plant Cell Environ 21: 1265-1273

Ghoshrony S, Nadakavukaren KJ (1990) Influence of cadmium on the ultrastructure of developing chloroplasts of soybean and corn. Environ Exp Bot 30: 187-192

Greene RM, Geider RJ, Kolber Z, Falkowski PG (1992) Iron-induced changes in light harvesting and photochemical energy conversion processes in eukaryotic marine algae. Plant Physiol 100: 565-575

Greger M, Johansson M (1992) Cadmium effects on leaf transpiration of sugar beet (*Beta vulgaris*). Physiol Plant 86: 465-473

Greger M, Lindberg S (1987) Effects of Cd^{2+} and EDTA on young sugar beet (*Beta vulgaris*). II. Net uptake and distribution of Mg^{2+}, Ca^{2+} and Fe^{2+}/Fe^{3+}. Physiol Plant 69: 81-86

Greger M, Ögren E (1991) Direct and indirect effects of Cd^{2+} on photosynthesis in sugar beet (*Beta vulgaris*). Physiol Plant 83: 129-135

Greger M, Johansson M, Stihl A, Hamza K (1993) Foliar uptake of Cd by pea (*Pisum sativum*) and sugar beet (*Beta vulgaris*). Physiol Plant 88: 563-570

Guikema JA (1985) Fluorescence induction characteristics of *Anacystis nidulans* during recovery from iron deficiency. J Plant Nutr 8: 891-908

Hagemeyer J, Waisel Y (1989a) Excretion of ions (Cd, Li, Na and Cl) by *Tamarix aphylla*. Physiol Plant 75: 280-284

Hagemeyer J, Waisel Y (1989b) Influence of NaCl, $Cd(NO_3)_2$ and air humidity on transpiration of *Tamarix aphylla*. Physiol Plant 77: 247-253

Hampp R, Beulich K, Zeigler H (1976) Effects of zinc and cadmium on photosynthetic CO_2 fixation and Hill activity of isolated spinach chloroplasts. Z Pflanzenphysiol 77: 336-344

Hatch DJ, Jones LHP, Burau RG (1988) The effect of pH on the uptake of cadmium by four plant species grown in flowing solution culture. Plant Soil 105: 212-126

Hegedüs A, Erdei S, Horváth G (2001) Comparative studies of H_2O_2 detoxifying enzymes in green and greening barley under cadmium stress. Plant Sci 160: 1085-1093

Henriques FS (1989) Effect of copper deficiency on the photosynthetic apparatus of sugar beet (*Beta vulgaris* L.). J Plant Physiol 135: 453-458

Hodson MJ, Evans DE (1995) Aluminium/silicon interactions in higher plants. J Exp Bot 46: 161-171

Homer JR, Cotton R, Evans EH (1980) Whole leaf fluorescence as a technique for measurement of tolerance of plants to heavy metals. Oecologia 45: 88-89

Honeycutt RC, Krogmann DW (1972) Inhibition of chloroplasts reactions with phenylmercuric acetate. Plant Physiol 49: 376-381

Horváth G, Droppa M, Oravecz A, Raskin VI, Marder JB (1996) Formation of the photosynthetic apparatus during greening of cadmium-poisoned barley leaves. Planta 199: 238-243

Houtz RL, Nable RO, Cheniae GM (1988) Evidence for the effects on the in vivo activity of ribulose-bisphosphate carboxylase/oxygenase during development of Mn toxicity in tobacco. Plant Physiol 86: 1143-1149

Hsu B, Lee J (1988) Toxic effects of copper on photosystem II of spinach chloroplasts. Plant Physiol 87: 116-119

Huang CY, Bazzaz FA, Vanderhoeff LN (1974) The inhibition of soybean metabolism by cadmium and lead. Plant Physiol 54: 122-124

Husaini Y, Rai LC (1992) pH dependent aluminium toxicity to *Nostoc linckia*: studies on phosphate uptake, alkaline and acid phosphatase activity, ATP content, photosynthesis and carbon fixation. J Plant Physiol 139: 703-707

Jegerschöld C, Arellano JB, Schroder WP, Van-Kan PJ, Baron M, Styring S (1995) Copper(II) inhibition of electron transfer through photosystem II studied by EPR spectroscopy. Biochemistry 34: 12747-12754

Juneau P, El Berdey A, Popovic P (2002) PAM fluorometry in the determination of the sensitivity of *Chlorella vulgaris*, *Selenastrum capricornutum* and *Chlamydomonas reinhardtii* to copper. Arch Environ Contam Toxicol 42: 155-164

Kimimura M, Katoh S (1972) Studies on electron transport associated with photosystem I. I. Functional site of plastocyanin: inhibitory effects of $HgCl_2$ on electron transport and plastocyanin in chloroplasts. Biochim Biophys Acta 283: 279-292

Kitao M, Lei TT, Koike T (1997) Effect of manganese toxicity on photosynthesis of white birch (*Betula platyphylla* var. *Japonica*) seedlings. Physiol Plant 101: 249-256

Kitao M, Lei TT, Koike T (1998) Application of chlorophyll fluorescence to evaluate Mn tolerance of deciduous broad-leaved tree seedlings native to northern Japan. Tree Physiol 18: 135-140

Králová K, Šerše ň F, Blahová M (1994) Effects of Cu(II) complexes on photosynthesis in spinach chloroplasts. Aqua(aryloxyacetato)copper(II) complexes. Gen Physiol Biophys 13: 483-491

Krupa Z (1988) Cadmium-induced changes in the composition and structure of the light harvesting chlorophyll a/b protein complex II in radish cotyledons. Physiol Plant 73: 518-524

Krupa Z (1999) Cadmium against higher plant photosynthesis - a variety of effects and where do they possibly come from? Z Naturforsch 54c: 723-729

Krupa Z, Baszy ński T (1995) Some aspects of heavy metal toxicity towards photosynthetic apparatus - direct and indirect effects on light and dark reactions. Acta Physiol Plant 17: 177-190

Krupa Z, Moniak M (1998) The stage of leaf maturity implicates the response of the photosynthetic apparatus to cadmium toxicity. Plant Sci 138: 149-156

Krupa Z, Siedlecka A (1995) Cd/Fe interactions and its effects on photosynthetic capacity of primary bean leaves. In: Mathis P (ed) Photosynthesis: from light to biosphere, vol 4, Kluwer, Dordrecht, pp 621-624

Krupa Z, Skórzy ńska E, Maksymiec W, Baszy ński T (1987) Effect of cadmium treatment on the photosynthetic apparatus and its photochemical activities in greening radish seedlings. Photosynthetica 21: 156-164

Krupa Z, Öquist G, Huner NPA (1992) The influence of cadmium on primary PS II photochemistry in bean as revealed by chlorophyll fluorescence - a preliminary study. Acta Physiol Plant 14: 71-76

Krupa Z, Öquist G, Huner NPA (1993a) The effects of cadmium on photosynthesis of *Phaseolus vulgaris* L. A fluorescence analysis. Physiol Plant 88: 626-630

Krupa Z, Siedlecka A, Maksymiec W, Baszy ński T (1993b) In vivo response of photosynthetic apparatus of *Phaseolus vulgaris* to nickel toxicity. J Plant Physiol 142: 664-668

Küpper H, Küpper F, Spiller M (1996) Environmental relevance of heavy metal substituted chlorophylls using the example of water plants. J Exp Bot 47: 259-266

Küpper H, Küpper F, Spiller M (1998) *In situ* detection of heavy metal substituted chlorophylls in water plants. Photosynth Res 58: 123-133

Lamoreaux RJ, Chaney WR (1978) The effect of cadmium on net photosynthesis, transpiration and dark respiration of excised silver maple leaves. Physiol Plant 43: 231-236

Lanaras T, Moustakas M, Symenoides L, Diamantoglou S, Karataglis S (1993) Plant metal content, growth responses and some photosynthetic measurements on field-cultivated growing on ore bodies enriched in Cu. Physiol Plant 88: 307-314

Li EH, Miles CD (1975) Effects of cadmium on photoreaction II of chloroplasts. Plant Sci Lett 5: 33-40

Lidon FC, Henriques FS (1991) Limiting step on photosynthesis of rice plants treated with varying copper levels. J Plant Physiol 138: 115-118

Lidon FC, Henriques FS (1993a) Changes in the contents of the photosynthetic electron carriers, RNAse activity and membrane permeability triggered by excess copper in rice. Photosynthetica 28: 99-108

Lidon FC, Henriques FS (1993b) Changes in the thylakoid membrane polypeptide patterns triggered by excess Cu in rice. Photosynthetica 28: 109-117

Lidon FC, Henriques FS (1993c) Copper-mediated oxygen toxicity in rice chloroplasts. Photosynthetica 29: 385-400

Lidon FC, Ramalho JC, Henriques FS (1993) Copper inhibition of rice photosynthesis. J Plant Physiol 142: 12-17

Lolkema PC, Vooijs R (1986) Copper tolerance in *Silene cucubalus*. Planta 167: 30-36

Lorimer GH (1981) The carboxylation and oxygenation of ribulose 1,5-bisphosphate, the primary events in photosynthesis and photorespiration. Annu Rev Plant Physiol 32: 349-388

Lu CM, Zhang JH (1999) Copper-induced inhibition of PSII photochemistry in cyanobacterium *Spirulina platensis* is stimulated by light. J Plant Physiol 154: 173-178

Lu CM, Chau CW, Zhang JH (2000) Acute toxicity of excess mercury on the photosynthetic performance of cyanobacterium *S. platensis* - assessment by chlorophyll fluorescence analysis. Chemosphere 41: 191-196

Lucero HA, Andreo CS, Vallejos RH (1976) Sulphydryl groups in photosynthetic energy conservation. II. Inhibition of photophosphorylation in spinach chloroplasts by $CdCl_2$. Plant Sci Lett 6: 309-313

Luna CM, Gonzalez CA, Trippi VS (1994) Oxidative damage caused by an excess of copper in oat leaves. Plant Cell Physiol 35: 11-15

Maksymiec W (1997) Effect of copper on cellular processes in higher plants. Photosynthetica 34:321-342

Maksymiec W, Baszyński T (1996a) Chlorophyll fluorescence in primary leaves of excess Cu-treated runner bean plants depends on their growth stages and the duration of Cu-action. J Plant Physiol 149: 196-200

Maksymiec W, Baszyński T (1996b) Different susceptibility of runner bean plants to excess copper as a function of the growth stages of primary leaves. J Plant Physiol 149: 217-221

Maksymiec W, Baszyński T (1998a) The effect of Cd^{2+} on the release of proteins from thylakoid membranes of tomato leaves. Acta Soc Bot Pol 57: 465-474

Maksymiec W, Baszyński T (1998b) The role of Ca ions in changes induced by excess Cu^{2+} in bean plants. Growth parameters. Acta Physiol Plant 20: 411-417

Maksymiec W, Baszyński T (1999a) The role of Ca^{2+} ions in modulating changes induced in bean plants by an excess of Cu^{2+} ions. Chlorophyll fluorescence measurements. Physiol Plant 105: 562-568

Maksymiec W, Baszyński T (1999b) Are calcium ions and calcium channels involved in the mechanisms of Cu^{2+} toxicity in bean plants? The influence of leaf age. Photosynthetica 36: 267-278

Maksymiec W, Russa R, Urbanik-Sypniewska T, Baszyński T (1994) Effect of excess Cu on the photosynthetic apparatus of runner bean leaves treated at two different growth stages. Physiol Plant 91: 715-721

Maksymiec W, Bednara J, Baszyński T (1995) Responses of runner bean plants to excess copper as a function of plant growth stages: effects on morphology and structure of primary leaves and their chloroplast ultrastructure. Photosynthetica 31: 427-435

Malik D, Sheoran IS, Singh R (1992a) Lipid composition of thylakoid membranes of cadmium treated wheat seedlings. Indian J Biochem Biophys 29: 350-354

Malik D, Sheoran IS, Singh R (1992b) Carbon metabolism in leaves of cadmium treated wheat seedlings Plant Physiol Biochem 30: 223-229

Martin H, Woodcock D (1983) The scientific principles of crop protection, 7th edn. Arnold, London

Mathys W (1975) Enzymes of heavy-metal-resistant and non-resistant populations of *Silene cucubalus* and their interactions with some heavy metals in vitro and in vivo. Physiol Plant 33: 161-165

McCarthy I, Romero-Puertas MC, Palma JM, Sandalio LM, Corpas FJ, Gómez M, del Rio LA (2001) Cadmium induces senescence symptoms in leaf peroxisomes of pea plants. Plant Cell Environ 24: 1065-1073

Miles CD, Brandle JR, Daniel DJ, Chu-Der O, Schnore PD, Uhlik DJ (1972) Inhibition of photosystem II in isolated chloroplasts by lead. Plant Physiol 49: 820-825

Miller M, Cox RP (1983) Effect of Zn^{2+} on photosynthetic oxygen evolution and chloroplast manganese. FEBS Lett 155: 331-333

Misra AK, Sarkunan V, Miradas SK, Nayak SK, Nayar PK (1989) Influence of N, P on cadmium toxicity on photosynthesis in rice. Curr Sci 58: 1398-1400

Mohanty N, Vass I, Demeter S (1989) Copper toxicity affects photosystem II electron transport at the secondary quinone acceptor Q_B. Plant Physiol 90: 175-179

Molas J (2002) Changes of chloroplast ultrastructure and total chlorophyll concentration in cabbage leaves caused by excess of organic Ni(II) complexes. Environ Exp Bot 47: 115-126

Monnet F, Vaillant N, Vernay P, Coudret A, Sallanon H, Hitmi A (2001) Relationship between PS II activity, CO_2 fixation, and Zn, Mn and Mg contents of *Lolium perenne* under zinc stress. J Plant Physiol 158: 1137-1144

Morales F, Abadia A, Abadia J (1990) Characterization of the xanthophyll cycle and other photosynthetic pigment changes induced by iron deficiency in sugar beet (*Beta vulgaris* L.). Plant Physiol 94: 607-613

Morales F, Abadia A, Abadia J (1991) Chlorophyll fluorescence and photon yield of oxygen evolution in iron-deficient sugar beet (*Beta vulgaris* L.) leaves. Plant Physiol 97: 886-893

Morsing M (1994) The use of sludge in forestry and agriculture. A comparison of the legislation in different countries. Forskningsserien, no 5. Danish Forest and Landscape Research Institute, Lyngby, Denmark, p 66

Moustakas M, Ouzounidou G, Lannoye G (1993) Rapid screening for aluminium tolerance in cereals by use of chlorophyll fluorescence test. Plant Breed 111: 343-346

Moustakas M, Lanaras T, Symeonidis L, Karataglis S (1994) Growth and some photosynthesis characteristics of field grown *Avena sativa* under copper and lead stress. Photosynthetica 30: 389-396

Moya JL, Ros R, Picazo I (1993) Influence of cadmium and nickel on growth, net photosynthesis and carbohydrate distribution in rice plants. Photosynth Res 6: 75-80

Muthuchelian K, Bertamini M, Nedunchezhian N (2001) Triacontanol can protect *Erythrina variegata* from cadmium toxicity. J Plant Physiol 158: 1487-1490

Myśliwa-Kurdziel B, Strzatka K (2002a) Influences of metals on biosynthesis of photosynthetic pigments. In: Prasad MNV, Strzatka K (eds) Physiology and biochemistry of metal toxicity and tolerance in plants. Kluwer, Dortrecht, pp 201-227

Myśliwa-Kurdziel B, Prasad MNV, *Strzatka K* (2002b) Consequences of heavy metal exposure to the processes related to the light phase of photosynthesis. In: Prasad MNV, Strzatka K (eds) Physiology and biochemistry of metal toxicity and tolerance in plants. Kluwer, Dortrecht, pp 229-255

Naguib MI, Hamed AA, Al-Wakeel SA (1982) Effect of cadmium on growth criteria of some crop plants. Egypt J Bot 25: 1-12

Navari-Izzo F, Quartacci MF, Pinzino C, Vecchia FD, Sgherri CLM (1998) Thylakoid-bound and stromal antioxidative enzymes in wheat treated with excess copper. Physiol Plant 104: 630-638

Nedunchezhian N, Kulandaivelu G (1995) Effect of Cd and UV-B radiation on polypeptide composition and photosystem activities of *Vigna unguiculata* chloroplasts. Biol Plant 37: 437-441

Öncel I, Keleş Y, Üstün AS (2000) Interactive effects of temperature and heavy metal stress on the growth and some biochemical compounds in wheat seedlings. Environ Pollut 107: 315-320

Ouzounidou G (1993) Changes in chlorophyll fluorescence as a result of Cu-treatment: dose-response relations in *Silene* and *Thlaspi*. Photosynthetica 29: 455-462

Ouzounidou G (1995) Cu-ions mediated changes in growth, chlorophyll and other ion contents in a Cu-tolerant *Koeleria splendens*. Biol Plant 37: 71-79

Ouzounidou G (1996) The use of photoacoustic spectroscopy in assessing leaf photosynthesis under copper stress: correlation of energy storage to photosystem II fluorescence parameters and redox change of P-700. Plant Sci 113: 229-237

Ouzounidou G (1997) Sites of copper in the photosynthetic apparatus of maize leaves: kinetic analysis of chlorophyll fluorescence, oxygen evolution, absorption changes and thermal dissipation as monitored by photoacoustic signals. Aust J Plant Physiol 24: 81-90

Ouzounidou G, Eleftheriou EP, Karataglis S (1992) Ecophysiological and ultrastructural effects of copper in *Thlapsi ochroleucum* (Cruciferae). Can J Bot 70: 947-957

Ouzounidou G, Lannoye R, Karataglis S (1993) Photoacoustic measurements of photosynthetic activities in intact leaves under copper stress. Plant Sci 89: 221-226

Ouzounidou G, Moustakas M, Lannoye R (1995) Chlorophyll fluorescence and photoacoustic characteristics in relationship to changes in chlorophyll and Ca^{2+} content of a Cu-tolerant *Silene compacta* ecotype under Cu treatment. Physiol Plant 93: 551-557

Ouzounidou G, Moustakas M, Eleftheriou EP (1997) Physiological and ultrastructural effects of cadmium on wheat (*Triticum aestivum* L.) leaves. Arch Environ Contam Toxicol 32: 154-160

Padmaja K, Prasad DDK, Prasad ARK (1989) Effect of selenium on chlorophyll biosynthesis in mung bean seedlings. Phytochemistry 28: 3321-3324

Padmaja K, Prasad DDK, Prasad ARK (1990) Inhibition of chlorophyll synthesis in *Phaseolus vulgaris* L. seedlings by cadmium acetate. Photosynthetica 24: 399-405

Pankovic D, Plesnicar M, Arsenijevic-Maksimovic I, Petrovic N, Sakac Z, Kastori R (2000) Effects of nitrogen nutrition on photosynthesis in Cd-treated sunflower plants. Ann Bot 86: 841-847

Parekh D, Puranik RM, Srivastava HS (1990) Inhibition of chlorophyll biosynthesis by cadmium in greening maize leaf segments. Biochem Physiol Pflanzen 186: 239-242

Patra M, Sharma A (2000) Mercury toxicity in plants. Bot Rev 66: 379-422

Pätsikkä E, Aro EM, Tyystjärvi E (1998) Increase in the quantum yield of photoinhibition contributes to copper toxicity in vivo. Plant Physiol 117: 619-627

Pätsikkä E, Aro EM, Tyystjärvi E (2001) Mechanism of copper-enhanced photoinhibition in thylakoid membranes. Physiol Plant 113: 142-150

Pnnazio S, Roggero P (1992) Effect of cadmium and nickel on ethylene biosynthesis in soybean. Biol Plant 34: 345-349

Poschenrieder C, Gunse B, Barceló J (1989) Influence of cadmium on water relations, stomatal resistance, and abscisic acid content in expanding bean leaves. Plant Physiol 90: 1365-1371

Prasad DDK, Prasad ARK (1987a) Effect of lead and mercury on chlorophyll synthesis in mung bean seedlings. Phytochemistry 26: 881-883

Prasad DDK, Prasad ARK (1987b) Altered delta-aminolevulinc acid metabolism by lead and mercury in germinating seedlings of bajra (*Pennisetum typhoideum*). J Plant Physiol 127: 241-249

Prasad MNV (1995a) Cadmium toxicity and tolerance in vascular plants. Environ Exp Bot 35: 525-540

Prasad MNV (1995b) The inhibition of maize leaf chlorophylls, carotenoids and gas exchange functions by cadmium. Photosynthetica 31: 635-640

Prasad MNV (1996) Variable allocation of heavy metals in phytomass of crop plants—human health implications. Plze Lék Sborn Suppl 71: 19-22

Prasad MNV (1997) Trace metals. In: Prasad MNV (ed) Plant ecophysiology. Wiley, New York, pp 207-249

Prasad MNV, Drej K, Skawiñska A, Skawiñska K (1998) Toxicity of cadmium and copper in *Chlamydomonas reinhardtii* wild-type (WT2137) and cell wall deficient mutant strain (CW15). Bull Environ Contam Toxicol 60: 306-311

Prasad MNV, Malec P, Waloszek A, Bojko M, Strza l ka K (2001) Physiological responses of *Lemna trisulca* L. (Duckweed) to cadmium and copper bioaccumulation. Plant Sci 161: 881-889

Radmer R, Kok B (1974) Kinetic observation of the photosystem II electron acceptor pool isolated by mercuric ion. Biochim Biophys Acta 357: 177-180

Rao BK, Tyryshkin AM, Roberts AG, Bowman MK, Kramer DM (2000) Inhibitory copper binding site on the spinach cytochrome b_6f complex: implications for Q_o site catalysis. Biochemistry 39: 3285-3296

Rascio N, Dallavecchia F, Ferretti M, Merlo I, Ghisi R (1993) Some effects of cadmium on maize plants. Arch Environ Contam Toxicol 25: 244-249

Rashid A, Bernier M, Pazdernick L, Carpentier R (1991) Interaction of Zn^{2+} with the donor side of photosystem II. Photosynth Res 30: 123-130

Romero-Puertas MC, Palma JM, Gomez M, Del Rio LA, Sandalio LM (2002) Cadmium causes the oxidative modification of proteins in pea plants. Plant Cell Environ 25: 677-686

Ros R, Cooke DT, Burden RS, James CS (1990) Effect of the herbicide MCPA, and the heavy metals cadmium and nickel, on the lipid composition, Mg-ATPase activity and fluidity of plasma membranes from rice, *Oryza sativa* cv. Bahia shoots. J Exp Bot 41: 457-462

Ross S (1994) (ed) Toxic metals in soil plant systems. Wiley, Chichester, p 469

Rubio MI, Escrig I, Martinezcortina C, Lopezbenet FJ, Sanz A (1994) Cadmium and nickel accumulation in rice plants - effects on mineral nutrition and possible interactions of abscisic and gibberellic acids. Plant Growth Regul 14: 151-157

Samson G, Morisette JC, Popovic R (1988) Copper quenching of the variable fluorescence in *Dunaliella tertiolecta*. New evidence for a copper inhibition effect on PS II photochemistry. Photochem Photobiol 48: 329-332

Sandalio LM, Dalurzo HC, Gómez M, Romero-Puertas MC, del Rio LA (2001) Cadmium-induced changes in the growth and oxidative metabolism of pea plants. J Exp Bot 52: 2115-2126

Schlegel H, Godbold DL, Huttermann A (1987) Whole plant aspects of heavy metal induced changes in CO_2 uptake and water relations of spruce (*Picea abies*) seedlings. Physiol Plant 69: 265-270

Schröder WP, Arellano JB, Bittner T, Baron M, Eckert HJ, Renger G (1994) Flash-induced absorption spectroscopy studies of copper interaction with photosystem II in higher plants. J Biol Chem 269: 32865-32870

Sengar RS, Padney M (1996) Inhibition of chlorophyll biosynthesis by lead in greening *Pisum sativum* leaf segments. Biol Plant 38: 459-462

Šerše ň F, Králová K, Blahová M (1996) Photosynthesis of *Chlorella vulgaris* as affected by diaqua(4-chloro-2-methylphenoxyacetato)copper(II) complex. Biol Plant 38: 71-75

Šerše ň F, Králová K, Bumbálová A (1998) Action of mercury on the photosynthetic apparatus of spinach chloroplasts. Photosynthetica 35: 551-559

Shainberg O, Rubin B, Rabinowitch HD, Tel-Or E (2001) Loading beans with sublethal levels of copper enhances conditioning to oxidative stress. J Plant Physiol 158: 1415-1421

Shalaby AM, Al-Wakeel SAM (1995) Changes in nitrogen metabolism enzyme activities of *Vicia faba* in response to aluminum and cadmium. Biol Plant 37: 101-106

Shalygo NV, Averina NG, Grimm B, Mock HP (1997) Influence of cesium on tetrapyrrole biosynthesis in etiolated and greening barley leaves. Physiol Plant 99: 160-168

Shalygo NV, Mock HP, Averina NG, Grimm B (1998) Photodynamic action of uroporphyrin and protochlorophyllide in greening barley leaves treated with cesium chloride. J Photochem Photobiol B Biol 42: 151-158

Shalygo NV, Kolesnikova NV, Voronetskaya VV, Averina NG (1999) Effects of Mn^{2+}, Fe^{2+}, Co^{2+} and Ni^{2+} on chlorophyll accumulation and early stages of chlorophyll formation in greening barley seedlings. Russ J Plant Physiol 46: 496-501

Shaw BP, Rout NP (1998) Age-dependent responses of *Phaseolus aureus* Roxb. to inorganic salts of mercury and cadmium. Acta Physiol Plant 20: 85-90

Sheoran IS, Aggarwala N, Singh R (1990a) Effect of cadmium and nickel on in vivo carbon dioxide exchange rate of pigeon pea (*Cajanus cajan* L.). Plant Soil 129: 243-249

Sheoran IS, Singal HR, Singh R (1990b) Effect of cadmium and nickel on photosynthesis and the enzymes of the photosynthetic carbon reduction cycle in pigeon pea (*Cajanus cajan*). Photosynth Res 23: 345-351

Shioi Y, Tamai H, Sasa T (1978a) Inhibition of photosystem II in the green alga *Ankistrodesmus falcatus* by copper. Physiol Plant 44: 434-438

Shioi Y, Tamai H, Sasa T (1978b) Effects of copper on the photosynthetic electron transport systems in spinach chloroplasts. Plant Cell Physiol 19: 203-209

Siedlecka A (1995) Some aspects of interactions between heavy metals and plant minerals. Acta Soc Bot Pol 64: 265-272

Siedlecka A, Baszy ński T (1993) Inhibition of electron flow around photosystem I in chloroplasts of Cd-treated maize plants is due to Cd-induced iron deficiency. Physiol Plant 87: 199-202

Siedlecka A, Krupa Z (1996) Interaction between cadmium and iron and its effects on photosynthetic capacity of primary leaves of *Phaseolus vulgaris*. Plant Physiol Biochem 34: 833-842

Siedlecka A, Krupa Z (1999) Cd/Fe interaction in higher plants - its consequences for the photosynthetic apparatus. Photosynthetica 36: 321-331

Siedlecka A, Krupa Z, Samuelsson G, Oquist G, Gardestrom P (1997) Primary carbon metabolism in *Phaseolus vulgaris* plants under Cd/Fe interaction. Plant Physiol Biochem 35: 951-957

Siedlecka A, Gardestrom P, Samuelsson G, Kleczkowski LA, Krupa Z (1999) A relationship between carbonic anhydrase and rubisco in response to moderate cadmium stress during light activation of photosynthesis. Z Naturforsch 54c: 759-763

Singh DP, Khare P, Bisen PS (1989) Effect of Ni^{2+}, Hg^{2+} and Cu^{2+} on growth, oxygen evolution and photosynthetic electron transport in *Cylindospermum* IU 942. J Plant Physiol 134: 406-412

Singh RP, Tripathi RD, Sinha SK, Maheshwari R, Srivastava HS (1997) Response of higher plants to lead contaminated environment. Chemosphere 34: 2467-2493

Skórzyńska E, Baszyński T (1993) The changes in PS II complex peptides under cadmium treatment: are they of direct or indirect nature? Acta Physiol Plant 15: 263-269

Skórzyńska-Polit E, Baszyński T (1995) Photochemical activity of primary leaves in cadmium stressed *Phaseolus coccineus* depends on their growth stage. Acta Soc Bot Pol 64: 273-279

Skórzyńska-Polit E, Baszyński T (1997) Differences in sensitivity of the photosynthetic apparatus in Cd-stressed runner bean plants in relation to their age. Plant Sci 128: 11-21

Skórzyńska-Polit E, Tukendorf A, Selstam E, Baszyński T (1998) Calcium modifies Cd effect on runner bean plants. Environ Exp Bot 40: 275-286

Spiller S, Terry N (1980) Limiting factors in photosynthesis. II. Iron stress diminishes photochemical capacity by reducing the number of photosynthetic units. Plant Physiol 65: 121-125

Stiborova M (1988) Cd^{2+} ions affect quaternary structure of ribulose-1,5-biphosphate carboxylase from barley leaves. Biochem Physiol Pflanzen 183: 371-378

Stiborova M, Doubravova M, Brezinova A, Friedrich A (1986a) Effect of heavy metal ions on growth and biochemical characteristics of photosynthesis of barley. Photosynthetica 20: 418-425

Stiborova M, Doubravova M, Leblova, S (1986b) A comparative study of the effect of heavy metal ions on ribulose-1,5-bisphosphate carboxylase and phosphoenol pyruvate carboxylase. Biochem Physiol Pflanz 181: 373-379

Stobart AK, Griffiths WT, Ameen-Bukhari I, Sherwood RP (1985) The effect of Cd^{+2} on the biosynthesis of chlorophyll in leaves of barley. Physiol Plant 63: 293-298

Strzałka K, Ketner P (1997) Carbon dioxide. In: Prasad MNV (ed) Plant ecophysiology. Wiley, New York, pp 393-456

Subrahmanyam D, Rathore VS (2000) Influence of manganese toxicity on photosynthesis in ricebean (*Vigna umbellata*) seedlings. Photosynthetica 38: 449-453

Szalontai B, Horváth L, Debreczeny M, Droppa M, Horváth G (1999) Molecular rearrangements of thylakoids after heavy metal poisoning, as seen by Fourier transform infrared (FTIR) and electron spin resonance (ESR) spectroscopy. Photosynth Res 61: 241-252

Terry N, Abadia J (1986) Function of iron in chloroplasts. J Plant Nutr 9: 609-646

Thomas RM, Singh VP (1996) Reduction of cadmium-induced inhibition of chlorophyll and carotenoid accumulation in *Cucumis sativus* L. by uniconazole (S. 3307). Photosynthetica 32: 145-148

Tripathy BC, Mohanthy P (1980) Zinc inhibited electron transport of photosynthesis in isolated barley chloroplasts. Plant Physiol 66: 1174-1178

Tziveleka L, Kaldis A, Hegedüs A, Kissimon J, Prombona A, Horváth G, Argyroudi-Akoyunoglou J (1999) The effect of Cd on chlorophyll and light-harvesting complex II biosynthesis in greening plants. Z Naturforsch 54c: 740-745

Vajpayee P, Tripathi RD, Rai UN, Ali MB, Singh SN (2000) Chromium(VI) accumulation reduces chlorophyll biosynthesis, nitrate reductase activity and protein content in *Nymphaea alba* L. Chemosphere 41: 1075-1082

Van Assche F, Clijsters H (1986) Inhibition of photosynthesis in *Phaseolus vulgaris* by treatment with toxic concentration of zinc, effect on ribulose-1,5-bisphosphate carboxylase/oxygenase. J Plant Physiol 125: 355-360

Van Assche F, Clijsters H (1987) Enzyme analysis in plants as a tool for assessing phytotoxicity of heavy metal polluted soils. Med Fac Landbouw Rijksuniv Gent 52: 1819-1824

Van Assche F, Clijsters H (1990) Effects of metals on enzyme activity in plants. Plant Cell Environ 13: 195-206

Van Assche F, Cardinaels C, Clijsters H (1988) Induction of enzyme capacity in plants as a result of heavy metal toxicity, dose-response relations in *Phaseolus vulgaris* L. treated with cadmium. Environ Pollut 52: 103-115

Van Duijvendijk-Matteoli MA, Desmeta GM (1975) On the inhibitory side of PS II in isolated chloroplasts. Biochim Biophys Acta 408: 164-169

Vassilev A, Tsonev T, Yordanov I (1998) Physiological response of barley plants (*Hordeum vulgare*) to cadmium contamination in soil during ontogenesis. Environ Pollut 103: 287-293

Vavilin DV, Polynov VA, Matorin DN, Venediktov PS (1995) Sublethal concentrations of copper stimulate photosystem II photoinhibition in *Chlorella pyrenoidosa*. J Plant Physiol 146: 609-614

Veeranjaneyulu K, Das VSR (1982) In vitro chloroplast localization of ^{65}Zn and ^{63}Ni in a Zn-tolerant plant *Ocimum basilicum* Benth. J Exp Bot 33: 1161-1165

Vinit-Dunand F, Epron D, Alaoui-Sossé B, Badot PM (2002) Effect of copper on growth and on photosynthesis of mature and expanding leaves in cucumber plants. Plant Sci 163: 53-58

Voigt J, Nagel K, Wrann D (1998) A cadmium-tolerant *Chlamydomonas* mutant strain impaired in photosystem II activity. J Plant Physiol 153: 566-573

Wagner GJ (1993) Accumulation of cadmium in crop plants and its consequences to human health. Adv Agron 51: 172-212

Weigel HJ (1985a) Inhibition of photosynthetic reactions of isolated intact chloroplast by cadmium. J Plant Physiol 119: 179-189

Weigel HJ (1985b) The effect of cadmium on photosynthetic reaction of mesophyll protoplast. Physiol Plant 63: 192-200

Wong D, Govindjee (1976) Effects of lead ions on photosystem I in isolated chloroplasts: studies on the reaction centre P_{700}. Photosynthetica 10: 241-254

Woźny A, Schneider J, Gwóźdź EA (1995) The effects of lead and kinetin on greening barley leaves. Biol Plant 35: 541-552

Wrischer M, Meglaj D (1980) The effects of lead on the structure and function of wheat plastids. Acta Soc Bot Pol 39: 33-40

Yruela I, Montoya G, Alonso PJ, Picorel R (1991) Identification of the pheophytin-QA-Fe domain of the reducing side of the photosystem II at the Cu(II)-inhibitory binding side. J Biol Chem 266: 22847-22850

Yruela I, Montoya G, Picorel R (1992) The inhibitory mechanism of Cu(II) on the photosystem II electron transport from higher plants. Photosynth Res 33: 227-233

Yruela I, Alfonso M, Ortiz de Zarate I, Montoya G, Picorel R (1993) Precise location of Cu(II)-inhibitory binding site in higher plant and bacterial photosynthetic reaction centres as probed by light-induced absorption changes. J Biol Chem 268: 1684-1689

Yruela I, Pueyo J, Alonso PJ, Picorel R (1996) Photoinhibition of photosystem II from higher plants. J Biol Chem 271: 27408-27415

Yruela I, Alfonso M, Barón M, Picorel R (2000) Copper effect on the protein composition of photosystem II. Physiol Plant 110: 551-557

Zeid IM (2001) Responses of *Phaseolus vulgaris* to chromium and cobalt treatments. Biol Plant 44: 111-115ń

Chapter 7

Plant Mitochondrial Respiration Under the Influence of Heavy Metals

R. Lösch

Abt. Geobotanik, H. Heine-Universität, Universitätsstr. 1, 40225 Düsseldorf, Germany

7.1 INTRODUCTION

Cell respiration like other metabolic processes is catalysed by many enzymes that require metals as cofactors, whereas higher concentrations of these metals and other non-essential metals inhibit enzyme activity. Enzymes of dissimilation processes are less affected by higher doses of heavy metals than the enzymes of the energy fixation and carbon assimilation pathways (van Assche and Clijsters 1990). This may be one reason why the majority of research on heavy metal physiology is more concentrated on processes of metal exclusion from the symplast and metal sequestration in the vacuole (e.g. Ernst 1969, 1976; Denny and Wilkins 1987; Hall 2002), on metal toxicity effects on photosynthesis (e.g. Vallee and Ulmer 1972; Clijsters and van Assche 1985), and in particular on metal interactions with gene expression (e.g. Tomsett and Thurman 1988; Cobbett 2000; Clemens 2001). Despite progress in knowledge in all these fields, lack of detailed insight into heavy metal interactions with metabolic processes still hinders a complete causal understanding of toxicity and tolerance mechanisms (e.g. Foy et al. 1978a; Jackson et al. 1990; Mukhopadhyay and Sharma 1991; Barceló and Poschenrieder 1992), and this statement is particularly true for plant respiration.

Heavy metal effects on mitochondrial respiration—photorespiration is not considered in this chapter—become obvious by changed O_2 consumption rates and altered amounts of CO_2 release from metal-treated plants. At the biochemical level heavy metal disturbances of the catabolic pathways result in lowered ATP production and changed amounts of intermediate substances, e.g. of the citrate cycle.

Respiration rates remain unchanged or even increase if sensitive organelles in the symplast remain protected from the ionic stressor due to metal exclusion or sequestration in the vacuole. The increase in respiration rates can be a consequence of an increased energy demand for ion uptake or export at the cytoplasmic membranes (Meharg 1993) or for the production of specific metal-binding polypeptides (Grill et al. 1985; Rauser 1995, 1999) and the removal of the resulting metal-phytochelatin or -thionein complexes from the cytoplasm and its organelles (Vögeli-Lange and Wagner 1990). The energetic costs for the metabolic control of the symplastic ion concentrations have been assumed to be the reason for the low productivity of metal-tolerant species or ecotypes, which show 20-50% lower production than related sensitive plants from non-metalliferous soils (Ernst 1976; Wu and Antonovics 1978; Fernandes and Henriques 1991; Ernst et al. 1992). However, a direct link between metabolic costs and metal tolerance has not been shown up to now (Harper et al. 1997).

The sites in which respiratory processes take place are separated from potentially toxic metals in the environment by barriers, in the first place the cation-exchange capacity of the cell wall (e.g. Peterson 1969; Poulter et al. 1985) and the plasma membrane. Hydrolytic and glycolytic enzymes of the cytoplasm can be protected from toxic metals through metal chelation or precipitation by SH-rich molecules, organic anions or phosphate.

Starch-degrading enzymes providing substrates for catabolism were found to be sensitive to increased heavy metal concentrations in vitro and in vivo (Cr: Dua and Sawhney 1991; Cd: Chugh and Sawhney 1996; Bansal et al. 2001; Pb: Bansal et al. 2001): α-amylase activity from germinating pea seeds was approximately halved under the influence of Cr(VI) ions (1-5 mM) and Cd^{2+} ions (0.5 and 1 mM) 1 week after imbibition (Dua and Sawhney 1991; Cd: Chugh and Sawhney 1996). β-Amylase was depressed transiently by the influence of chromium but continuously by the cadmium treatment. As a result, after 1 week treatment, total amylolytic activity of seeds germinated in 0.25, 0.5 and 1 mM cadmium was 56, 42 and 33% of that of control seeds. Under the influence of chromium, amylolytic activity was depressed by 60% between days 2 and 6 after imbibition, but recovered to the control level a few days later. Enhancement of α-amylase activity most probably was based on de novo synthesis (Mayer and Poljakoff-Mayber 1989). Starch phosphorylase activity was hardly affected by chromium. Cr(VI) treatment first increased maltase activity but decreased the activity considerably compared with the control values in later stages. The invertase activity was decreased by Cr, whereas the time course of enzyme activity during germination was unaltered compared to the control. A reduced release of sugars from the storage macromolecules under the influence of heavy metal ions will reduce the available amount of substrates for the mitochondrial respiration processes. The same must be the case when heavy metals inhibit photosynthesis thus reducing the amount of energy-rich substrates for cell catabolism.

Heavy metal ions will affect the catabolic processes in the mitochondria directly, when metals are transfered through the outer and the inner mitochondrial membrane. Only then will the ions have immediate access to the functional elements of the citrate cycle, electron transport chain and ATPsynthase. The mitochondrial membranes are thus the next barriers protecting sensitive macromolecules from detrimental contact with too high metal concentrations. In situ determinations of mitochondrial heavy metal contents, e.g. by EDAX techniques, are not available. The resolution of such approaches is simply not high enough.

In microorganisms, synthesis or activation of specific, less metal sensitive isoenzymes has been found under metal stress (for general overview of heavy metal effects on bacteria, see Silver 1996). For example, in *Mycobacterium tuberculosis avium* (Horio et al. 1955) and in *Saccharomyces cerevisiae* (Murayama 1961), several enzymes of the Krebs cycle are more copper-resistant in tolerant strains than in sensitive strains. In higher plants, however, no example is known for constitutive or induced expression of enzymes with higher metal tolerance in metal-tolerant genotypes in comparison to sensitive genotypes. A comparative study of Wu et al. (1975) about copper effects on malate dehydrogenase activity in roots of tolerant and intolerant *Agrostis stolonifera* clones could not demonstrate such a differential tolerance of respiratory enzymes. The metal concentrations that resulted in 50% inhibition of the activities of isocitrate dehydrogenase (ICDH: 0.16-0.47 mM Zn, 0.4-0.42 mM Cu) and malate dehydrogenase (MDH: 1.4-1.46 mM Zn, 0.38-0.39 mM Cu) were independent of the origin of the enzymes from metal-tolerant or intolerant populations of *Silene cucubalus* (Ernst 1976). However, the detectable number of MDH isoenzymes was different in tolerant and intolerant *Silene* populations grown on media with and without zinc (Ernst et al. 1975).

7.2 HEAVY METAL INFLUENCE ON RESPIRATORY GAS EXCHANGE

The effect of increased metal concentrations on respiration rates depends on the degree of metal stress. Mild metal stress increases dark respiration rates in comparison to control conditions, whereas O_2 consumption or CO_2 release decreases under severe metal stress indicating metabolic damage. For example, 1.5 mM Zn in seawater stimulated *Ulva lactuca* respiration, while Zn concentrations above 5 mM inhibited respiration (Webster and Gadd 1996). In Zn-tolerant species, the respiration may be higher at normal micronutrient concentrations of Zn than at elevated Zn concentrations that do not result in growth inhibition. In this case, increased respiration may result from micronutrient deficiency effects in consequence of a lower metal efficiency in Zn-tolerant populations (e.g. various *Armeria* ecotypes; Köhl 1995; cf. Table 7.1). Some data on O_2 consumption or CO_2 release from plants treated with heavy metals, mostly reported only incidentally, are compiled in Table 7.1. To facilitate comparison, values are expressed as percent of control.

Excess Cu or Mn concentrations decrease respiration rates (Wu et al. 1975; Foy et al. 1978b; Sirkar and Amin 1979). Similarly, Zn^{2+}, Cd^{2+} and Hg^{2+} treatments result in lower rates of oxygen consumption in all cases in the unicellular alga *Euglena* (de Filippis et al. 1981) and in other aquatic plants (Jana and Choudhuri 1982). In this case, as well as in studies with suspension-cultured cells (Reese and Roberts 1985; Poulter et al. 1985), the heavy metals probably affect directly the cytoplasm and damage the mitochondrial structures. In terrestrial plants, where Cd or Zn treatments result in increased respiration, exclusion or sequestration of the metals may efficiently protect sensitive enzymes from access of metal ions at relatively high external concentrations. In this situation, an increased energy demand for active exclusion or sequestration of the heavy metals is met by increased net respiration rates. Additionally, a reduced photophosphorylation in the heavy metal sensitive chloroplasts may enhance the demands on mitochondria-based energy supply (Lamoreaux and Chaney 1978). Higher demands for ATP may result in a decreased electron flow through the alternative pathway of the respiratory chain (Marschner 1995). Additionally, oxygen consumption and NADH oxidation may result

Table 7.1 Heavy metal effects on respiratory gas exchange rates of plants. *manom.* manometry; *gas exch.* gas exchange; *tol.* tolerant; *intol.* intolerant; *ecot.* ecotype; *susp.* suspension; *treatm.* treatment

Heavy metal	Plant species	Heavy metal conc. of substrate or medium of cultivation	Gas exchange (Respiration) [control = 100%]	Method	Reference
Cd	*Pisum sativum* (seeds)	0.25 mM 0.50 mM 1 mM	76% 52% 29%	O_2 electrode	Chugh and Sawhney (1996)
Cd	*Vicia faba*	0.45 μM 0.90 μM 1.35 μM	112% 163% 153%	O_2 manom.	Lee et al. (1976)
Cd	*Acer saccharinum* (detached leaves, 16/40 h after treatm.)	45 μM 90 μM 180 μM	95/129% 86/136% 99/142%	CO_2 gas exch.	Lamoreaux and Chaney (1978)
Cd	*Picea abies*	1 μM 5 μM	76% 82%	CO_2 gas exch.	Schlegel et al. (1987)
Cd	*Oryza sativa* (roots)	0.1 mM 1 mM	88% 74%	O_2 manom.	Llamas et al. (2000)
Cd	*Potamogeton pectinatus*	0.1 mM 1 mM	87% 26%	O_2 electrode	Jana and Choudhuri (1982)
Cd	*Vallisneria spiralis*	10 μM 0.1 mM	93% 47%	O_2 electrode	Jana and Choudhuri (1982)
Cd	*Hydrilla verticillata*	10 μM 0.1 mM	95% 50%	O_2 electrode	Jana and Choudhuri (1982)
Cd	*Nicotiana tabacum* (cell susp. cultures)	44.5 μM 89 μM 178 μM	93% 86% 64%	O_2 manom.	Reese and Roberts (1985)
Cd	*Euglena gracilis*	0.10 μM	53%	O_2 manom.	De Filippis et al. (1981)
Cd	*Picea abies*	30 μM 60 μM	88% 104%	CO_2 gas exch.	Schlegel et al. (1987)
Zn	*Viola arvensis* (leaves)	1.53 mM 3.06 mM	135% 110%	CO_2 gas exch.	Evenschor and Lösch, unpubl.

(Contd.)

Table 7.1 (*Contd.*)

Heavy metal	Plant species	Heavy metal conc. of substrate or medium of cultivation	Gas exchange (Respiration) [control = 100%]	Method	Reference
Zn	*Viola calaminaria* Zn tolerant	1.53 mM 3.06 mM	145% 151%	CO_2 gas exch.	Evenschor and Lösch, unpubl.
Zn	*Armeria elongata* (leaf/root) mesophyte	0 mM 0.14 mM 0.71 mM	100/100% 109/101% 102/100%	O_2 manom.	Köhl (1995)
Zn	*A. halleri* (leaf/root) Zn tolerant	0 mM 0.14 mM 0.71 mM	100/100% 102/88% 83/60%	O_2 manom.	Köhl (1995)
Zn	*Armeria maritima* (leaf/root) halophyte	0 mM 0.14 mM 0.71 mM	100/100% 100/104% 129/115%	O_2 manom.	Köhl (1995)
Zn	*Anthoxanthum od.* (cell susp.) tol./intol. ecot.	0.61 mM	100/47%	O_2 electrode	Poulter et al. (1985)
Zn	*Euglena gracilis*	50 μM	19%	O_2 manom.	De Filippis et al. (1981)
Cu	*Agrostis stolonifera* root; tol. ecot.	10 μM 25 μM	93% 29%	O_2 manom.	Wu et al. (1975)
Cu	*A. stolonifera* root; intol. ecot.	10 μM 25 μM	28% 12%	O_2 manom.	Wu et al. (1975)
Cu	*Potamogeton pectinatus*	10 μM 100 μM	100% 85%	O_2 electrode	Jana and Choudhuri (1982)
Cu	*Vallisneria spiralis*	1 μM 10 μM	94% 49%	O_2 electrode	Jana and Choudhuri (1982)
Cu	*Hydrilla verticillata*	1 mM 10 mM	96% 47%	O_2 electrode	Jana and Choudhuri (1982)
Cu	*Anthoxanthum od.* (cell susp.) tol./intol.	0.15 mM	70/31%	O_2 electrode	Poulter et al. (1985)

(*Contd.*)

Table 7.1 (*Contd.*)

Heavy metal	Plant species	Heavy metal conc. of substrate or medium of cultivation	Gas exchange (Respiration) [control = 100%]	Method	Reference
Cu	*Euglena gracilis* (2 h treatm.)	200 μM	180%	O_2 electrode	Edjlali and Calvayrac (1991)
Pb	*Potamogeton pectinatus*	0.1 mM 1 mM	94% 75%	O_2 electrode	Jana and Choudhuri (1982)
Pb	*Vallisneria spiralis*	1 mM 10 mM	95% 50%	O_2 electrode	Jana and Choudhuri (1982)
Pb	*Hydrilla verticillata*	1 mM 10 mM	90% 48%	O_2 electrode	Jana and Choudhuri (1982)
Pb	*Anthoxanthum od.* (cell susp.) tol./intol.	0.24 mM	93/45%	O_2 electrode	Poulter et al. (1985)
Mn	*Gossypium hirsutum* (leaf disks)	0.16 mM 0.49 mM 1.48 mM	111% 84% 67%	O_2 manom.	Sirkar and Amin (1974)
Hg	*Picea abies*	0.1 μM	105%	CO_2 gas exch.	Schlegel et al. (1987)
Hg	*Potamogeton pectinatus*	0.1 mM 1 mM	97% 75%	O_2 electrode	Jana and Choudhuri (1982)
Hg	*Vallisneria spiralis*	0.1 mM 1 mM	100% 97%	O_2 electrode	Jana and Choudhuri (1982)
Hg	*Hydrilla verticillata*	0.1 mM 1 mM	100% 92%	O_2 electrode	Jana and Choudhuri (1982)
Hg	*Euglena gracilis*	0.01 μM	27%	O_2 manom.	De Filippis et al. (1981)

from peroxidase activities (van der Werf et al. 1991). As a result, there is no stoichiometric relationship between net O_2 or CO_2 exchange and tissue energy consumption, and the metabolic costs of heavy metal stress cannot be directly concluded from its effect on gas exchange.

7.3 RESPIRATORY ENERGY SUPPLY FOR HEAVY METAL EXCLUSION FROM THE SYMPLAST

Controlled ion transport through membranes is mediated by transmembrane proteins that act as ion carriers or ion channels (Clarkson and Lüttge 1989) and transfer ions actively or passively depending on the gradient of the electrochemical potential. Most divalent cations (except Mn and Mg), including most of the heavy metals reviewed here, are supposed to share a common carrier system that is energised by a H^+-ATPase (Cataldo et al. 1983; Pfeffer et al. 1987; Clarkson and Lüttge 1989). However, electrophysiological studies and, more recently, heterologous gene expression in oocytes and yeast complementation systems have identified specific transporters for many divalent metals, including Zn, Cu and Fe (e.g. Kampfenkel et al. 1995; Eide et al. 1996; Grotz et al. 1998; Arazi et al. 1999; Saier 2000; Clemens 2001).

At high metal supply, heavy metal will be transported passively along the electrochemical potential gradient into the cytoplasm. Pulse-chase experiments with tracer-loaded cells and tissues strongly indicate a highly efficient efflux across the plasmalemma (Co^{2+}: Macklon and Sim 1976, 1987; Zn^{2+}: Santa María and Coggliatti 1988) into the apoplast. In contrast, the rate of reflux of trace element cations from the vacuole into the cytoplasm seems to be negligible (Clarkson and Lüttge 1989). Whereas some years ago rather unspecified transporter proteins were assumed for membrane passages of divalent cations, more recently element-specific transporters have been found (Clemens 2001). The extrusion of divalent cations out of the cell seems to depend on ATPase activities. At high external metal concentration, the energy requirement for active metal extrusion increases the ATP demand by 10-30% of the normal ATP turnover. Under sublethal toxic ion stress, this increased ATP demand is generally met by an increased supply from enhanced catabolic processes. Insufficient ATP supply does not seem causal for the onset of toxicity symptoms (Al^{3+} effects on wheat roots: Kinraide 1988).

The high efficiency of such an efflux-based ion-homeostasis system becomes obvious when the ion exclusion mechanism is inactivated by metabolic inhibitors. Cumming and Taylor (1990) report several studies about aluminium accumulation in root cells caused by passive influx when Al^{3+} extrusion at the plasmalemma is inhibited by metabolic inhibitors (e.g. wheat: Zhang and Taylor 1989; cabbage, lettuce, kikuyu grass: Huett and Menary 1979; radish, buckwheat, cucumber and rice: Wagatsuma 1983). In most studies, ATP production is inhibited by the uncoupling protonophore 2,4-dinitrophenol (DNP), which, however, has an effect on membrane potential and permeability as well (Cutler and Rains 1974; Huett and Menary 1979). Therefore, it is still uncertain whether the increased net uptake of heavy metals into the symplast after application of this inhibitor results from membrane structural changes or, as it is suggested in this context, from a reduced energy supply to the membrane-bound ion-extrusion mechanisms. The latter process probably brought about increased *Ulva lactuca* respiration rates when 2,4-DNP together with moderately elevated Zn levels was administered. Similarly, cyanide effects upon *Ulva* O_2 consumption were counteracted by cadmium (Webster and Gadd 1996).

Under high metal supply, toxic metals are not completely excluded from the symplast. Harmful metal interaction with sensitive molecules and structures may be prevented by specific chelation of metals in the cytoplasm and their transport into the vacuole.

Compounds that form stable metal-chelates are phytochelatins and metallothioneins (Zenk 1996), amino acids (Krämer et al. 1996), organic acids (Ma 2000) and phenolic compounds (Barceló and Poschenrieder 2002).

The efficiency of a specific metal chelator to protect essential metabolic elements from heavy metal depends on the chemical nature of the metal ions, which affects their affinities to macromolecules and protective chelators (Woolhouse 1983). Thus, the protective system can be expected to be element-specific. With respect to metabolic processes in mitochondria the role of organic acids (citrate, isocitrate, malate, aconitate, oxaloacetate) as detoxifying substances outside these organelles may be of particular interest. It has been suggested that these acids are involved in metal tolerance (e.g. malate-Zn, citrate-Zn complexes: Mathys 1977; Thurman and Rankin 1982; Godbold et al. 1984; Harrington et al. 1996; citrate-Ni complexes: Lee et al. 1977, 1978; Al-removal from sensitive membrane structures by organic acids: Suhayda and Haug 1986, apoplastic Al-transport and Al-sequesteration: Ma 2000, Cu and Zn binding by organic acids: Parker et al. 2001). Zinc resistance was thought to depend on malate-mediated Zn transport into the vacuole (Mathys 1977; Ernst 1982; specific transporter: Verkleij et al. 1998). More recent studies on Zn complexation by organic acids have been carried out, e.g., by Tolrà et al. (1976) and Chardonnes et al. (1999). There were no strong stoichiometric relationships between specific metal tolerances and availability of organic acids, but obviously the latter at least contributed to the cell-internal detoxification of heavy metals. Moreover, an increased energy demand for vacuolar zinc sequestration was evidenced by the study of Chardonnes et al. (1999). While a higher cellular ATP demand for heavy metal detoxification by sequestration is beyond any doubt, there seems to be no obvious need to sluice out of the mitochondria higher amounts of citric acid cycle intermediates in order to meet cytoplasmic and vacuolar demands for heavy metal complexing anions. Studies on Cd and Zn accumulation and sequestration in suspension-cultured tobacco cells (Krotz et al. 1989) led to the conclusion that the concentration of organic acids (mostly malate and citrate) in the cells is sufficiently high to complex potentially toxic metal amounts even at high heavy metal supplies. The withdrawal of citrate cycle intermediates by metal-chelation processes is, therefore, unlikely to have a negative effect on the citrate cycle.

7.4 MITOCHONDRIAL STRUCTURE AND PERFORMANCE UNDER THE INFLUENCE OF HEAVY METALS

Most studies on ion effects on mitochondria have been performed on animal mitochondria, mainly from liver and heart (Brierley 1977). Some studies have also been carried out on plant mitochondria (e.g. Bittell et al. 1974; Kesseler and Brand 1995), which differ in response from animal mitochondria (Prebble 1981). Experimental results mainly concentrate on the effect of heavy metals on structure and function of the respiratory organelles in vitro, which is not necessarily identical to the situation in vivo.

The capacity of animal and plant mitochondria to accumulate ions and thus control the mitochondrial ionic milieu is important for their regular function (Millard et al. 1964; Hanson et al. 1965; Zaitseva 1978). The outer membrane of mitochondria has an overall high permeability for substances of a size smaller than approx. 10-13 kDa (Brierley 1977; Earnshaw and Cooke 1984). The inner mitochondrial membrane, in contrast, has a very low permeability in particular for charged substances, but is permeated easily by water molecules. Increased ion uptake, therefore, results in swelling, which occurs easily and without special metabolic control in the presence of high external concentrations of the univalent cations Na^+ and Li^+ (together with suitable anions), but not in the presence of high K^+ concentrations. Na and Li transport across the membrane seems to be based on a cation/proton exchange carrier (Tzagoloff 1982), whereas K and Mg accumulation in the mitochondrial matrix depends on a transport system that is energised by oxidative phosphorylation. The hitherto known mitochondrial transport systems for other divalent cations (Ca^{2+}, Sr^{2+}, Mn^{2+}) seem to be energy-dependent electrogenic carriers (Tzagoloff 1982). Divalent metal cations belonging to transition elements have been assumed to pass into the mitochondria through the membrane channels for essential main group elements. Lead, for example, seems to be transported into human and animal mitochondria through the endogenous Ca^{2+} transporters (see studies reviewed by Brierley 1977). A competitive inhibition of the transport of essential macronutrients by heavy metals may disrupt ion homeostasis in mitochondria, as Mg (and probably also Ca) is essential for sequestration effects and transport characteristics of the mitochondrial membrane (Earnshaw and Cooke 1984). However, the hypothesis that heavy metals pass membranes through Ca or Mg transporters will possibly have to be revised in the future, as specific plasma membrane-transporters have been recently identified for Fe, Cu and Zn (see above).

Binding of divalent cations to membranes from animal mitochondria has been reported, e.g., by Chappell et al. (1963) and Scarpa and Azzi (1968). Bittell et al. (1974) found an approximately equal binding capacity of corn mitochondria for Cd^{2+}, Zn^{2+}, Co^{2+}, Ni^{2+} and Mn^{2+} of maximally 58 nmol metal ions per milligram of mitochondrial protein. The tenfold higher capacity for Pb^{2+} sorption in comparison to the other metals is probably due to precipitation of lead phosphate. Both phospholipids (Scarpa and Azzone 1968) and proteins have been suggested as specific sites for cation binding. Cd-membrane interactions seem to involve the binding of Cd^{2+} to sulfhydryl groups of the mitochondrial membrane in *Zea mays* as suggested by the effects of the sulfhydryl protecting agent dithiothreitol (Miller et al. 1973). Cd^{2+} and Zn^{2+} have a higher affinity to mitochondrial structures than Co^{2+}, Ni^{2+} and Mn^{2+} and, therefore, seem to have a stronger effect on mitochondrial metabolism, viz. electron transport and phosphorylation (Bittell et al. 1974).

The mitochondrial electron transport chain, the proton-ATPase and transporters for respiration substrates, adenine nucleotides and phosphate, are in contact both with the mitochondrial intermembrane space and the matrix. These structures are more likely to be affected by heavy metals than enzymes of the citrate cycle, which are matrix-located and, therefore, only affected by those toxic substances that pass the mitochondrial membranes. Table 7.2, based on Brierley (1977), lists the heavy metal species that are known to influence specifically these different functional units of mitochondria.

Higher amounts of trace metal ions at the mitochondrial membrane may increase its permeability by interaction with cation transporters or by influencing the permeability of lipophilic domains of the membrane, as suggested for Cu (Fernandes and Henriques 1991). Models for the effect of heavy metal transport on the plasma membrane (Cumming and Taylor 1990) may also apply to mitochondrial membranes; however, experimental data from in vivo systems are lacking. In vitro, isolated mitochondria from sensitive plant genotypes swell and change their submicroscopic structure when exposed to heavy

Table 7.2 Mitochondria-located metabolic systems of animals, microorganisms and plants and heavy metal species known to influence their performance (after Brierley 1977)

Mitochondrial structure or process	Reactive heavy metals
Substrate transporters	Pb^{2+}, Hg^{2+}
Electron transport systems	Zn^{2+}, Pb^{2+}, Hg^{2+}
Adenine nucleotide and phosphate transporters	Hg^{2+}, Cu^{2+}
Permeability to K^+, H^+, Cl^-	Cu^{2+}, Pb^{2+}, Zn^{2+}, Cd^{2+}, Hg^{2+}
Phosphorylation	Hg^{2+}
Citrate cycle	Cd^{2+}, Pb^{2+}, Hg^{2+}, Cu^{2+}, Ni^{2+}, Zn^{2+}

metals in bathing solution. Presumably, this is due to net influx of the metal ions and, maybe, other solutes, into the mitochondria, which is followed by water influx. In the extreme, this can bring about mitochondrial degeneration (Reese et al. 1986). Some heavy metal species, especially Cu, seem to interact with the K^+ channel proteins and change their transport characteristics (Brierley 1977; Murphy and Taiz 1997). Osmotic disorder under the influence of heavy metals may be the reason for membrane destruction and collapse of cristae, which are seen very rapidly in sensitive *Festuca rubra* root meristematic cells treated with a higher zinc concentration (Davies et al. 1995). An energy-dependent osmotic swelling of isolated heart mitochondria under the influence of metal ions (Zn^{2+}: Brierley and Knight 1967; Pb^{2+}: Scott et al. 1971) results probably from an increased cation permeability of the membrane, whereas pH and anion gradients remain unchanged (Brierley 1976). Mitochondrial swelling is induced by zinc (Davies et al. 1995), lead (Bittell et al. 1974), copper (Brierley 1977), and manganese (Mukhopadhyay and Sharma 1991). In contrast, Cd^{2+} (Kesseler and Brand 1995), Ni^{2+} and Co^{2+} (Bittell et al. 1974) have no effect on mitochondrial volume as measured by the light transmission properties of a mitochondrial suspension (Lorimer and Miller 1969).

Metal ions are essential components of macromolecules of the mitochondrial electron transport chain and the enzymes that catalyse their turnover. Iron is a constitutive part of cytochrome oxidase, cytochromes and flavoproteins and copper is a constituent of cytochrome oxidase (Gupta 1979; Clarkson and Hanson 1980; Fernandes and Henriques 1991). Both metals change their redox state when electrons flow through the electron transport chain. An excess of other, similar metal ions may partly replace these constitutive parts of the respiration chain and thus disturb its function, e.g. Mn^{2+}/Fe^{2+} competition: Twyman (1951); Weinstein and Robbins (1955). However, Mn is also a necessary trace element in mitochondria contributing there the cofactor for superoxide dismutase. In contrast to cytosol-located CuZn-SOD and chloroplast-located Fe-SOD and CuZn-SOD, Mn-SOD is in plants mostly specific for mitochondria (Cakmak 2000). Because reduced iron easily reacts with oxygen-forming radicals, prevention of formation and accumulation in mitochondria of the latter by Mn-SOD is very important to avoid damage to the organelle structures and to guarantee the availability of reduced Fe for redox cycling in the electron transport chain. This detail underlines the delicate balance between the small amounts of different divalent cations needed for normal functioning of the steady-state metabolism in organelles and cytoplasma. Heavy metal homeostasis therefore occurs not only at the level of the whole cell (Krämer et al. 2000: high similarities between cytoplasmic Ni contents of hyperaccumulator and non-accumulator leaf tissues), but also in subcellular compartments. A coordinated supply only of the demanded small amounts seems to be based on particular chaperones. At least for copper such specific proteins are already identified ("copper chaperone", CCH, and "response to antagonist1", RAN1; Himmelblau and Amasino 2000) and have been shown to deliver copper to yeast mitochondria (Glerum et al. 1996). When the supply/demand relation exceeds the homeostatic capacity of the system toxic concentrations of the metal ions will result and alter structural parts of the components of the respiration chain and divert or block the electron flow (Bansal and Sharma 2000). Zn and Cd are suspected to stimulate respiratory O_2 consumption by interaction with the alternative oxidative pathway (Webster and Gadd 1996).

Pb^{2+} stimulates the oxidation of exogenous NADH (at least up to a certain, relatively high concentration: Bittell et al. 1974) and inhibits succinate oxidation (Koeppe and Miller 1970). Relevant data are also reported in studies on human and animal mitochondria (Brierley 1977). Cd^{2+} and Zn^{2+} have a similar, but less strong, effect on succinate oxidation and reduce NADH oxidation; the effects of Co^{2+} and Ni^{2+} are small (Bittell et al. 1974). Literature data on the direct influences of excess heavy metals on the complex III are scarce (some information about zinc effects on intermediary mitochondria electron transport is given by Kleiner 1974). More data exist about the effect of metals on cytochrome oxidase, mainly from studies on animal and microbial mitochondria. Cytochrome oxidase activity is reduced by most metals, namely Hg^{2+}, Cd^{2+} (Vallee and Ulmer 1972) and Mn^{2+} (Weinstein and Robbins 1955; Sirkar and Amin 1974), whereas other metals enhance the activity of cytochrome oxidase (Cd^{2+}, Pb^{2+}: Vallee and Ulmer 1972; Ni^{2+}: Mattioni et al. 1997). The enhancement of cytochrome oxidase by cadmium is particularly remarkable, as negative effects of this metal are emphasised from other studies. Interaction between metals makes the comparison of results especially from in vivo experiments but also from in vitro experiments difficult, as the effects strongly depend on the particular experimental conditions.

The effect of heavy metals on the coupling of electron transport and ATP synthesis is quantified by determination of the ADP/O ratios (=number of ATP molecules synthesised per electron pair transferred to oxygen) in the presence and absence of high metal concentration in the assay solution. Metal concentrations that resulted in uncoupling of phosphorylation from the electron transport were found to be between 0.5 and 1.1 mM for Cd and Zn and 10 mM for Ni or Co (Bittell et al. 1974). In studies on vertebrate renal mitochondria, Hg decreased the ADP/O ratio (Weinberg et al. 1982), and Cu treatment resulted in a decrease in or total breakdown of oxidative phosphorylation in human and animal mitochondria (Brierley 1977). Kesseler and Brand (1995) explained the decrease in phosphorylation efficiency under elevated Cd^{2+} levels by an increase in the permeability of the mitochondrial inner membrane to protons in the presence of cadmium.

Insufficient availability of phosphate for ATP synthesis under heavy metal influence may be a consequence of partial inhibition of the membrane phosphate transporter, which is evidently (Tyler 1969; Fonyo et al. 1974) very Hg-sensitive and is probably affected also by other heavy metal ions. ATP amounts decrease considerably if heavy metal concentrations exceed critical thresholds. For example, ATP concentrations in cotton leaves decrease from 0.076 mg g^{-1} fresh weight to 0.025 mg g^{-1} fresh weight if the manganese concentration of the nutrient solution increases from 0.018 to 2.73 mM (Sirkar and Amin

1974). Exposure of corn root tissue to toxic Al concentrations results in a decrease in ATP concentration of 45-65% in comparison with the control within 20 h (^{31}P NMR measurements by Pfeffer et al. 1986).

The localisation of the citrate cycle enzymes in the mitochondrial matrix reduces their accessibility to heavy metals. In vivo, heavy metal effects on these enzymes are therefore likely to appear later or at higher metal concentrations than metal effects on enzymes in the cytoplasm or the mitochondrial membranes. Bansal et al. (2002) found that Cd^{2+} or Pb^{2+} reduces the activity of extramitochondrial ICDH more than the activity of the mitochondrial enzyme. Information about potential effects of heavy metals on the citrate cycle has been derived from studies on isolated citrate cycle enzymes in vitro. These investigations may reveal particularly sensitive enzymes that become limiting steps under metal stress. Figure 7.1 summarises qualitatively the known heavy metal effects on individual enzymes of the citrate cycle. Isocitrate-dehydrogenase (ICDH) and malate dehydrogenase (MDH) have been studied most intensively, and both increase and decrease in enzyme activity have been found under the influence of metals. Under comparable in vitro conditions, Cd^{2+}, Pb^{2+}, Zn^{2+}, Cu^{2+} and Ni^{2+} reduce the activity of these enzymes, whereas Mn^{2+} and Co^{2+} show no or even positive effects (Mathys 1975; MDH: Wu et al. 1975; ICDH: Anderson and Evans 1956; Mocquot et al. 1996; Bansal et al. 2002). However, even the qualitative effect (enhancement/inhibition) of metals depends on the origin of the enzyme and the specific experimental conditions, which makes generalisations, e.g. on the effects of Cd (Vallee and Ulmer 1972), almost impossible.

In a model by Brooks et al. (1981), enzyme inhibition by nickel is part of a regulation system in hyperaccumulating *Alyssum* species. Mitochondria contain 2% of the whole cell Ni^{2+} content in the hyperaccumulator *Alyssum serpyllifolium* ssp. *malacintanum*. The resulting Ni concentration inhibits the activity of citrate cycle enzymes in vitro and is thought to result in an accumulation of malate and isocitrate, which could act as nickel chelators. A translocation of the Ni complexes to the vacuole may then result in a regeneration of the suppressed metabolism in the mitochondria. The inhibition of the enzyme activity could thus be part of a regulatory pathway involved in metal detoxification.

Van Assche and Clijsters (1990) emphasise that enzyme induction by high heavy metal concentrations is a typical in vivo response of the plant coping with this environmental stress. Stress proteins expressed as a response to the stressor must remain functional when in contact with the toxic ions. Chaperones may play a very important role in these intracellular shifts and sequestration processes (Clemens 2001). The sensitive enzymes of the intermediary metabolism could be protected in this way from contact with the toxic substance. Additionally, the enzyme turnover may be increased, thus responding to the stress. The information cascades, which are needed to shift the cell-internal steady-state equilibria in such a situation, and the dynamics of subcellular damage prevention and repair have been poorly studied up to now. A more detailed knowledge about the mechanisms of interactions between toxic metal ions, cell macromolecules and the structures of the organelles will allow the formulation of a structured model of all these feedback loops and to explain causally the limits of resistance of the cell, tissue, and plant to heavy metal stress.

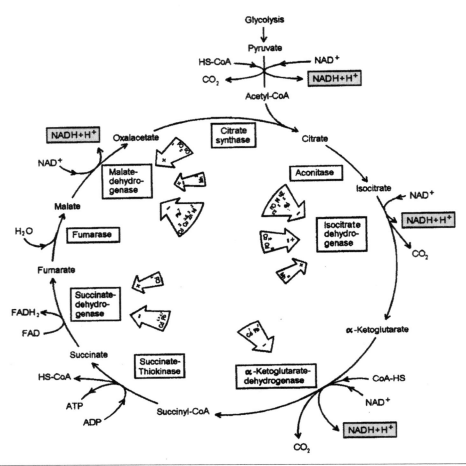

Fig. 7.1   Qualitative effects of heavy metals on citrate cycle enzymes (– = decreased activity, + = increased activity, ± = without obvious effects) according to information given by Me[1] = Vallee and Ulmer (1972), Me[2] = Mathys (1975), Me[3] = Mocquot et al. (1996), Me[4] = Koeppe and Miller (1970), Me[5] = Wu et al. (1975), Me[6] = Anderson and Evans (1956) and Me[7] = Bansal and Sharma 2000; Me = the respective metal ion species

ACKNOWLEDGEMENT

Useful suggestions by Dr. K.I. Köhl, Golm, are gratefully acknowledged.

REFERENCES

Anderson I, Evans HJ (1956) Effect of manganese and certain other metal cations on isocitric dehydrogenase and malic enzyme activities in *Phaseolus vulgaris*. Plant Physiol 31: 22-28

Arazi T, Sunkar R, Kaplan B, Fromm H (1999) A tobacco plasma membrane calmodulin-binding transporter confers Ni^{2+} tolerance and Pb^{2+} hypersensitivity in transgenic plants. Plant J 20: 171-182

Bansal P, Sharma P (2000) Effect of Pb^{2+} and Cd^{2+} on respiration and mitochondrial electron transport chain in germinating pea seeds (*Pisum sativum*). Indian J Environ Ecoplan 3:249-254

Bansal P, Sharma P, Dhindsa K (2001) Impact of Pb^{2+} and Cd^{2+} on activities of hydrolytic enzymes of respiration in germinating pea seeds. Ann Agri-Bio Res 6: 113-122

Bansal P, Sharma P, Goyal V (2002) Impact of lead and cadmium on enzymes of citric acid cycle in germinating pea seeds. Biol Plant 45: 125-127

Barceló J, Poschenrieder C (1992) Respuestas de las plantas a la contaminación por metales pesados. Suelo Planta 2: 345-361

Barceló J, Poschenrieder C (2002) Fast root growth responses, root exsudates, and internal detoxification as clues to the mechanisms of aluminium toxicity and resistance: a review. Environ Exp Bot 48: 75-92

Bittell JE, Koeppe DE, Miller RJ (1974) Sorption of heavy metal cations by corn mitochondria and the effects on electron and energy transfer reactions. Physiol Plant 30: 226-230

Brierley GP (1976) The uptake and extrusion of monovalent cations by isolated heart mitochondria. Mol Cell Biochem 10: 41-62

Brierley GP (1977) Effects of heavy metals on isolated mitochondria. In: Lee SD (ed) Biochemical effects of environmental pollutants. Ann Arbor Sci, Ann Arbor, pp 397-411

Brierley GP, Knight VA (1967) Ion transport by heart mitochondria. X. The uptake and release of Zn^{+2} and its relation to the energy-linked accumulation of Mg^{+2}. Biochemistry 6: 3892-3902

Brooks RR, Shaw S, Marfil AA (1981) The chemical form and physiological function of nickel in some Iberian *Alyssum* species. Physiol Plant 51: 167-170

Cakmak I (2000) Possible roles of zinc in protecting plant cells from damage by reactive oxygen species. New Phytol 146: 185-205

Chardonnes AN, Koevoets PLM, van Zanten A, Schat H, Verkleij JAS (1999) Properties of enhanced tonoplast zinc transport in naturally selected zinc-tolerant *Silene vulgaris*. Plant Physiol 120: 779-785

Cataldo DA, Garland TR, Wildung RE (1983) Cadmium uptake kinetics in intact soybean plants. Plant Physiol 73: 844-848

Chappell JB, Cohn M, Greville GD (1963) The accumulation of divalent ions by isolated mitochondria. In: Chance B (ed) Energy-linked functions of mitochondria. Academic, New York, pp 219-231

Chugh LK, Sawhney SK (1996) Effect of cadmium on germination, amylases and rate of respiration of germinating pea seeds. Environ Pollut 92: 1-5

Clarkson DT, Hanson JB (1980) The mineral nutrition of higher plants. Annu Rev Plant Physiol 31: 239-298

Clarkson DT, Lüttge U (1989) Mineral nutrition: divalent cations, transport and compartmentation. Prog Bot 51: 93-112

Clemens S (2001) Molecular mechanisms of plant metal tolerance and homeostasis. Planta 212: 475-486

Clijsters H, van Assche F (1985) Inhibition of photosynthesis by heavy metals. Photosynth Res 7: 31-40

Cobbett CS (2000) Phytochelatin biosynthesis and function in heavy-metal detoxification. Curr Opin Plant Biol 3: 211-216

Cumming JR, Taylor GJ (1990) Mechanisms of metal tolerance in plants: physiological adaptation for exclusion of metal ions from the cytoplasm. In: Alscher RG, Cumming JR (eds) Stress responses in plants: adaptation and acclimation mechanisms. Wiley-Liss, New York, pp 329-356

Cutler JM, Rains DW (1974) Characterization of cadmium uptake by plant tissue. Plant Physiol 54: 67-71

Davies KL, Davies MS, Francis D (1995) The effects of zinc on cell viability and on mitochondrial structure in contrasting cultivars of *Festuca rubra* L. A rapid test for zinc tolerance. Environ Pollut 88: 109-113

De Filippis LF, Hampp R, Ziegler H (1981) The effects of sublethal concentrations of zinc, cadmium and mercury on *Euglena*. II. Respiration, photosynthesis and photochemical activities. Arch Microbiol 128: 407-411

Denny HJ, Wilkins DA (1987) Zinc tolerance in *Betula* spp. II. Microanalytical studies of zinc uptake into root tissues. New Phytol 106: 525-534

Dua A, Sawhney SK (1991) Effect of chromium on activities of hydrolytic enzymes in germinating pea seeds. Environ Exp Bot 31: 133-139

Earnshaw MJ, Cooke A (1984) The role of cations in the regulation of electron transport. In: Palmer JM (ed) The physiology and biochemistry of plant respiration. Cambridge University Press, Cambridge, pp 177-182

Edjlali M, Calvayrac R (1991) Effects des ions métalliques sur l'intensité respiratoire et sur les capacités catalatiques chez *Euglena gracilis* Z. C R Acad Sci Paris 312(III): 177-182

Eide E, Broderius M, Fett J, Guerinot ML (1996) A novel iron-regulated metal transporter from plants identified by functional expression in yeast. Proc Natl Acad Sci USA 93:5624-5628

Ernst WHO (1969) Zur Physiologie der Schwermetallpflanzen - subzelluläre Speicherungsorte des Zinks. Ber Dtsch Bot Ges 82: 161-164

Ernst WHO (1976) Physiological and biochemical aspects of metal tolerance. In: Mansfield TA (ed) Effects of air pollution on plants. Cambridge University Press, Cambridge, pp 115-133

Ernst WHO (1982) Schwermetallpflanzen. In: Kinzel H (ed) Pflanzenökologie und Mineralstoffwechsel. Ulmer, Stuttgart, pp 472-506

Ernst W, Mathys W, Janiesch P (1975) Physiologische Grundlagen der Schwermetallresistenz - Enzymaktivitäten und organische Säuren. Forschungsberichte des Landes Nordrhein-Westfalen 2496. Westdeutscher Verlag, Opladen, pp 1-50

Ernst WHO, Verkleij JAC, Schat H (1992) Metal tolerance in plants. Acta Bot Neerl 41: 229-248

Fernandes JC, Henriques FS (1991) Biochemical, physiological, and structural effects of excess copper in plants. Bot Rev 57: 246-273

Fonyo A, Palmieri F, Ritvay J, Quagliariello E (1974) Kinetics and inhibitor sensitivity of the mitochondrial phosphate carrier. In: Azzone GF (ed) Membrane proteins in transport and phosphorylation. North Holland Publ, Amsterdam, pp 283-286

Foy CD, Chaney RL, White MC (1978a) The physiology of metal toxicity in plants. Annu Rev Plant Physiol 29: 511-566

Foy CD, Chaney RL, White MC (1978b) The physiology of plant tolerance to excess available aluminum and manganese in acid soils. In: Jung GA (ed) Crop tolerance to suboptimal land conditions. Publ 32. Am Soc Agron, Madison, pp 301-328

Glerum DM, Shtanko A, Tzagoloff A (1996) Characterization of *COX17*, a yeast gene involved in copper metabolism and assembly of cytochrome oxidase. J Biol Chem 271: 14504-14509

Godbold DL, Horst WJ, Collins JC, Thurman DA, Marschner H (1984) Accumulation of zinc and organic acids in roots of zinc tolerant and non-tolerant ecotypes of *Deschampsia caespitosa*. J Plant Physiol 116: 59-69

Grill E, Winnacker E-L, Zenk MH (1985) Phytochelatins: the principal heavy-metal complexing peptides of higher plants. Science 230:674-676

Grotz N, Fox T, Connolly E, Park W, Guerinot ML, Eide D (1998) Identification of a family of zinc transporter genes from *Arabidopsis* that respond to zinc deficiency. Proc Natl Acad Sci USA 95: 7220-7224

Gupta UC (1979) Copper in agricultural crops. In: Nriagu JO (ed) Copper in the environment, part I. Ecological cycling. Wiley, New York, pp 255-288

Hall JL (2002) Cellular mechanisms for heavy metal detoxification and tolerance. J Exp Bot 53:1-11

Hanson JB, Malhotra SS, Stoner CD (1965) Action of calcium on corn mitochondria. Plant Physiol 40: 1033-1040

Harper FA, Smith SE, Macnair MR (1997) Where is the cost in copper tolerance in *Mimulus guttatus*? Testing the trade-off hypothesis. Funct Ecol 11: 764-774

Harrington CF, Roberts DJ, Nickless G (1996) The effect of cadmium, zinc, and copper on the growth, tolerance index, metal uptake, and production of malic acid in two strains of the grass *Festuca rubra*. Can J Bot 74: 1742-1752

Himmelblau E, Amasino RM (2000) Delivering copper within plant cells. Curr Opin Plant Biol 3: 205-210

Horio T, Higashi T, Okunuki K (1955) Copper resistance of Mycobacterium tuberculosis avium. II. The influence of copper ion on the respiration of the parent cells and copper-resistant cells. J Biochem (Tokyo) 42: 491-498

Howden R, Goldsbrough PB, Andersen CR, Cobbett CS (1995) Cadmium-sensitive, cad1 mutants of *Arabidopsis thaliana* are phytochelatin deficient. Plant Physiol 107: 1059-1066

Huett DO, Menary RC (1979) Aluminium uptake by excised roots of cabbage, lettuce and Kikuyu grass. Aust J Plant Physiol 6: 643-653

Jackson PJ, Unkefer PJ, Delhaize E, Robinson NJ (1990) Mechanisms of trace metal tolerance in plants. In: Katterman F (ed) Environmental injury to plants. Academic, San Diego, pp 231-255

Jana S, Choudhuri MA (1982) Senescence in submerged aquatic angiosperms: effects of heavy metals. New Phytol 90: 477-484

Kampfenkel K, Kushnir S, Babiychuk E, Inze D, Vanmontagu M (1995) Molecular characterization of a putative *Arabidopsis thaliana* copper transporter and its yeast homologue. J Biol Chem 270: 28479-28486

Kesseler A, Brand MD (1995) The mechanism of the stimulation of state 4 respiration by cadmium in potato tuber (*Solanum tuberosum*) mitochondria. Plant Physiol Biochem 33: 519-528

Kinraide TB (1988) Proton extrusion by wheat roots exhibiting severe aluminum toxicity symptoms. Plant Physiol 88: 418-423

Kleiner D (1974) The effect of Zn^{+2} on mitochondrial electron transport. Arch Biochem Biophys 165: 121-125

Köhl KI (1995) Ökophysiologische Grundlagen der Sippendifferenzierung bei *Armeria maritima* (Mill.)Willd.: Evolution von Dürre-, Kochsalz- und Schwermetallresistenz. Dissertation, Heinrich-Heine-Universität, Düsseldorf, Germany

Koeppe DE, Miller RJ (1970) Lead effects on corn mitochondrial respiration. Science 167: 1376-1377

Krämer U, Cotter-Howells JD, Charnock JM, Baker AJM, Smith JAC (1996) Free histidine as a metal chelator in plants that accumulate nickel. Nature 379: 635-638

Krämer U, Pickering IJ, Prince RC, Raskin I, Salt DE (2000) Subcellular localization and speciation of nickel in hyperaccumulator and non-accumulator *Thlaspi* species. Plant Physiol 122: 1343-1353

Krotz RM, Evangelou BP, Wagner GJ (1989) Relationships between cadmium, zinc Cd-peptide and organic acid in tobacco suspension cells. Plant Physiol 91: 780-787

Llamas A, Ullrich CI, Sanz A (2000) Cd^{2+} effects on transmembrane electrical potential difference, respiration and membrane permeability of rice (*Oryza sativa* L) roots. Plant Soil 219: 21-28

Lamoreaux RJ, Chaney WR (1978) The effect of cadmium on net photosynthesis, transpiration, and dark respiration of excised silver maple leaves. Physiol Plant 43: 231-236

Lee KC, Cunningham BA, Paulson GM, Liang GH, Moore RB (1976) Effects of cadmium on respiration rate and activities of several enzymes in soybean seedlings. Physiol Plant 36: 4-6

Lee J, Reeves RD, Brooks RR, Jaffré T (1977) Isolation and identification of a citrato-complex of nickel from nickel-accumulating plants. Phytochemistry 16: 1503-1505

Lee J, Reeves RD, Brooks RR, Jaffré T (1978) The relation between nickel and citric acid in some nickel-accumulating plants. Phytochemistry 17: 1033-1035

Lorimer GH, Miller RJ (1969) The osmotic behavior of corn mitochondria. Plant Physiol 44:839-844

Ma JF (2000) Role of organic acids in detoxification of aluminum in higher plants. Plant Cell Physiol 41: 383-390

Macklon AES, Sim A (1976) Cortical cell fluxes and transport to the stele in excised root segments of *Allium cepa* L. III. Magnesium. Planta 128: 5-9

Macklon AES, Sim A (1987) Cellular cobalt fluxes in roots and transport to the shoots of wheat seedlings. J Exp Bot 38: 1663-1677

Marschner H (1995) Mineral nutrition of higher plants, 2nd edn. Academic Press, London

Mathys W (1975) Enzymes of heavy-metal-resistant and non-resistant populations of *Silene cucubalus* and their interaction with some heavy metals in vitro and in vivo. Physiol Plant 33: 161-165

Mathys W (1977) The role of malate, oxalate, and mustard oil glucosides in the evolution of zinc-resistance in herbage plants. Physiol Plant 40: 130-136

Mattioni C, Gabbrielli R, Vangronsfeld J, Clijsters H (1997) Nickel and cadmium toxicity and enzymatic activity in Ni-tolerant and non-tolerant populations of *Silene italica* Pers. J Plant Physiol 150:173-177

Mayer AM, Poljakoff-Mayber A (1989) The germination of seeds, 4th edn. Pergamon Press, Oxford

Meharg AA (1993) The role of the plasmalemma in metal tolerance in angiosperms. Physiol Plant 88: 191-198

Millard DL, Wiskirch JT, Robertson RN (1964) Ion uptake by plant mitochondria. Proc Natl Acad Sci USA 52: 996-1004

Miller RJ, Bittell JE, Koeppe DE (1973) The effect of cadmium on electron and energy transfer reactions in corn mitochondria. Physiol Plant 28: 166-171

Mocquot B, Vangronsveld J, Clijsters H, Mench M (1996) Copper toxicity in young maize (*Zea mays* L.) plants: effects on growth, mineral and chlorophyll contents, and enzyme activities. Plant Soil 182: 287-300

Mukhopadhyay MJ, Sharma A (1991) Manganese in cell metabolism of higher plants. Bot Rev 57: 117-149

Murayama T (1961) Studies on the metabolic pattern of yeast with reference to its copper resistance. Memoirs Ehime Univ Sect II B4: 43-66

Murphy A, Taiz L (1997) Correlation between potassium efflux and coppersensitivity in 10 Arabidopsis ecotypes. New Phytol 136: 211-222

Ortiz DF, Kreppel L, Speiser DM, Scheel G, McDonald G, Ow DW (1992) Heavy metal tolerance in the fission yeast requires an ATP-binding casette-type vacuolar membrane transporter. EMBO J 11: 3491-3499

Parker DR, Fedler JF, Ahnstrom ZAS, Resketo M (2001) Reevaluating the free-ion activity model of trace metal toxicity towards higher plants: experimental evidence with copper and zinc. Environ Toxicol Chem 20: 899-906

Peterson PJ (1969) The distribution of zinc-65 in *Agrostis tenuis*, Sibth. and *A. stolonifera*, L. tissues. J Exp Bot 20: 863-875

Pfeffer PE, Tu SI, Gerasimowicz WV, Canvanaugh JR (1986) *In vivo* ^{31}P NMR studies of corn root tissue and its uptake of toxic metals. Plant Physiol 80: 77-84

Poulter A, Collin HA, Thurman DA, Hardwick K (1985) The role of the cell wall in the mechanism of lead and zinc tolerance in *Anthoxanthum odoratum* L. Plant Sci 42: 61-66

Prebble JN (1981) Mitochondria, chloroplasts, and bacterial membranes. Longman, London

Rauser WE (1995) Phytochelatins and related peptides. Structure, biosynthesis, and function. Plant Physiol 109: 1141-1149

Rauser WE (1999) Structure and function of metal chelators produced by plants. Cell Biochem Biophys 31: 19-48

Reese RN, Roberts LW (1985) Effects of cadmium on whole cell and mitochondrial respiration in tobacco cell suspension cultures (*Nicotiana tabacum* L. var. *xanthi*). J Plant Physiol 120: 123-130

Reese RN, McCall RD, Roberts LW (1986) Cadmium induced ultrastructural changes in suspension-cultured tobacco cells (*Nicotiana tabacum* L. var. *xanthi*). Exp Environ Bot 26: 169-173

Saier MH (2000) A functional-phylogenetic classification system for transmembrane solute transporters. Microbiol Mol Biol Rev 64: 354-411

Salt DE, Rauser WE (1995) MgATP-dependent transport of phytochelatins across the tonoplast of oat roots. Plant Physiol 107: 1293-1301

Santa María GE, Cogliatti DH (1988) Bidirectional Zn-fluxes and compartmentation in wheat seedling roots. J Plant Physiol 132: 312-315

Scarpa A, Azzi A (1968) Cation binding to submitochondrial particles. Biochim Biophys Acta 150: 473-481

Scarpa A, Azzone GF (1968) Ion transport in liver mitochondria. J Biol Chem 243: 5132-5138

Schat H, Kalff MMA (1992) Are phytochelatins involved in differential metal tolerance or do they merely reflect metal-imposed strain? Plant Physiol 99: 1475-1480

Schlegel H, Godbold DL, Hüttermann A (1987) Whole plant aspects of heavy metal induced changes in CO_2 uptake and water relations of spruce (*Picea abies*) seedlings. Physiol Plant 69: 265-270

Scott KM, Hwang KM, Jurkowitz M, Brierley GP (1971) Ion transport by heart mitochondria. XXIII. The effect of lead on mitochondrial reactions. Arch Biochem Biophys 147: 557-567

Silver S (1996) Bacterial resistances to toxic metal ions - a review. Gene 179: 9-19

Sirkar S, Amin JV (1974) The manganese toxicity of cotton. Plant Physiol 54: 539-543

Sirkar S, Amin JV (1979) Influence of auxins on respiration of manganese toxic cotton plants. Indian J Exp Biol 17: 618-619

Suhayda CG, Haug A (1986) Organic acids reduce aluminum toxicity in maize root membranes. Physiol Plant 68: 189-185

Thurman DA, Rankin JL (1982) The role of organic acids in zinc tolerance in *Deschampsia caespitosa*. New Phytol 91: 629-635

Tolrà RP, Poschenrieder C, Barceló J (1996) Zinc hyperaccumulation in *Thlaspi caerulescens*. II. Influence on organic acids. J Plant Nutr 19: 1541-1550

Tomsett AB, Thurman DA (1988) Molecular biology of metal tolerances of plants. Plant Cell Environ 11: 383-394

Twyman ES (1951) The iron and manganese requirements of plants. New Phytol 50:210-226

Tyler DD (1969) Evidence of a phosphate-transporter system in the inner membrane of isolated mitochondria. Biochem J 111: 665-678

Tzagoloff A (1982) Mitochondria. Plenum Press, New York

Vallee BL, Ulmer DD (1972) Biochemical effects of mercury, cadmium, and lead. Annu Rev Biochem 41: 91-128

Van Assche F, Clijsters H (1990) Effects of metals on enzyme activity in plants. Plant Cell Environ 13: 195-206

Van der Werf A, Raaimakers D, Poot P, Lambers H (1991) Evidence for a significant contribution by peroxidase-mediated O_2 uptake to root respiration of *Brachypodium pinnatum*. Planta 183: 347-352

Verkleij JAC, Koevoets PLM, Blakekalff MMA, Chardonnens AN (1998) Evidence for an important role of the tonoplast in the mechanism of naturally selected zinc tolerance in *Silene vulgaris*. J Plant Physiol 153: 188-191

Vögeli-Lange R, Wagner GJ (1990) Subcellular localization of cadmium and cadmium-binding peptides in tobacco leaves. Plant Physiol 92: 1066-1093

Wagatsuma T (1983) Effect of non-metabolic conditions on the uptake of aluminum by plant roots. Soil Sci Plant Nutr 29: 323-333

Wang J, Evangelou BP, Nielsen MT, Wagner GJ (1992) Computer simulated evaluation of possible mechanisms for sequestring metal ion activity in plant vacuoles. II. Zinc. Plant Physiol 99: 621-626

Webster EA, Gadd GM (1996) Stimulation of respiration in *Ulva lactuca* by high concentrations of cadmium and zinc—evidence for an alternative respiratory pathway. Environ Toxicol Water Qual 11: 7-12

Weigel HJ, Jäger HJ (1980) Der Einfluß von Schwermetallen auf Wachstum und Stoffwechsel von Buschbohnen. Angew Bot 54: 195-205

Weinberg JM, Harding PG, Humes HD (1982) Mitochondrial bioenergetics during the initiation of mercuric chloride induced renal injury. J Biol Chem 257: 60-74

Weinstein LH, Robbins WR (1955) The effect of different iron and manganese nutrient levels on the catalase and cytochrome oxidase activities of green and albino sunflower leaf tissues. Plant Physiol 30: 27-32

Woolhouse HW (1983) Toxicity and tolerance in the response of plants to metals. In: Lange OL, Nobel PS, Osmond CB, Ziegler H (eds) Physiological plant ecology II. Responses to the chemical and biological environment. Encyclopedia of plant physiology. New series 12 C. Springer, Berlin Heidelberg New York, pp 245-300

Wu L, Antonovics J (1978) Zinc and copper tolerance of *Agrostis stolonifera* L. in tissue culture. Am J Bot 65: 268-271

Wu L, Thurman DA, Bradshaw AD (1975) The uptake of copper and its effect upon respiratory processes of roots of copper-tolerant and non-tolerant clones of *Agrostis stolonifera*. New Phytol 75: 225-229

Zaitseva MG (1978) The role of cation transport in regulation of the activity of plant mitochondria. In: Ducet G, Lance C (eds) Plant mitochondria. Elsevier/North-Holland Biomedical Press, Amsterdam, pp 183-189

Zenk MH (1996) Heavy metal detoxification in higher plants - a review. Gene 179: 21-30

Zhang G, Taylor GJ (1989) Kinetics of aluminum uptake by excised roots of aluminum-tolerant and aluminum-sensitive cultivars of *Triticum aestivum* L. Plant Physiol 91: 1094-1099

Chapter 8

Ecophysiology of Plant Growth Under Heavy Metal Stress

Jürgen Hagemeyer
Im Bergsiek 43, 33739 Bielefeld, Germany

8.1 INTRODUCTION

The growth of whole plants or of plant parts is frequently used as an easily measurable parameter to monitor the effects of various stressors. Changes in growth are often the first and most obvious reactions of plants under stress. In particular, those organs that have the first direct contact with noxious substances, normally the roots in contaminated soils, show rapid and sensitive changes in their growth characteristics (Baker and Walker 1989).

This chapter describes effects of potentially toxic trace metals on the physiological and ecological aspects of the growth of higher land plants. The large and continually increasing stock of published data on the growth of plants under trace metal stress makes it impossible to give a full account of the available knowledge. Thus, only selected articles are reviewed here as examples from an extremely vast research field. Some studies are focused on growth processes, whereas in many others growth was merely measured as an additional parameter to describe plant reactions to stress. Most plants frequently studied with respect to trace metal stress belong to one of the following groups: (1) crop plants of commercial interest either under conditions of nutrient deficiency or on polluted substrates, (2) plants with an evolved metal resistance (metallophytes), and (3) trees (see also Chap. 12).

8.2 WHAT IS GROWTH?

Growth is a process characteristic of all living organisms. It is a change in size, mass, form or number (Chiariello et al. 1989). On the cellular level, basic growth processes are cell division, differentiation and the production of cellular substances (Fellenberg 1981). An increase in biomass can lead to larger spatial dimensions of the organism, or additional biomass can increase the mass of the organism while its size remains unchanged. Therefore, when we consider effects of trace elements on plant growth we can distinguish between direct effects on biomass production and effects on the biomass allocation, i.e. effects on the size or shape. The development of organisms depends on growth and on differentiation. Growth is related to the quantitative aspects of development, whereas differentiation is concerned with the qualitative changes occurring during the formation of cells, tissues and organs (Causton and Venus 1981).

Growth is usually considered as an irreversible increase in size of an organism or any of its parts (Causton and Venus 1981). However, not all living plants show continually increasing dimensions; they can also reduce biomass during their life cycle. This may be seen in a way as negative growth. The shedding of plant parts, including leaves, shoots and branches, bark, roots, seeds, fruits and pollen, has been discussed in detail by Kozlowski (1973). The loss of plant parts can be an adaptive trait. For example, the shedding of leaves of rosette plants after excessive mineral uptake could be a kind of disposal mechanism for the toxic minerals.

The control of growth processes in plants depends largely on phytohormones (Roberts and Hooley 1988). Many studies of plant growth under trace metal stress are, thus, concerned with the role of plant hormones (e.g. Rubio et al. 1994; Lidon et al. 1995; Moya et al. 1995; Sekimoto et al. 1997).

8.3 HOW IS GROWTH MEASURED?

Changes in biomass are quantified by weighing. Measurements of dry weight changes over periods of time describe the overall growth of a plant. The biomass allocation is determined by measurements of spatial dimensions (e.g. root length) or numbers (e.g. numbers of leaves, lateral roots, etc.). The leaf area is a measure of the plants productive investment, since leaves are the most important photosynthetic producers (Causton and Venus 1981). Examples of other parameters are plant height, leaf numbers, cell numbers or C content. The determination of some parameters is destructive (e.g. shoot dry weight), whereas others can be measured non-destructively and can be monitored continuously. Since the fresh weights of plants or of plant parts change with water status, they are problematic as measures of productivity.

One of the most often used derived quantities in plant growth analysis is the relative growth rate (RGR, unit, e.g. day^{-1}; Smolders et al. 1991). It is calculated with a differential equation: $RGR = (dW/dt).(1/W)$, where W is dry biomass and t is time (Chiariello et al. 1989). The RGR reflects plant productivity and gives the rate of biomass increase relative to the productive mass of the plant. There are two approaches to growth studies (Smolders et al. 1991): (1) In the classical approach (Blackman 1919) growth parameters are directly derived from differences in biometric parameters from different plants of succeeding harvests. (2) In the functional approach (Vernon and Allison 1963) a presumed growth function is fitted to biometric data.

In a typical growth analysis experiment a large number of test species are planted and sampled (harvested) at time intervals. At each harvest dry weights are determined. Each plant can be sampled only once and therefore the variance of the dry weight data is large (Causton 1991). In contrast, fresh weights can be determined non-destructively and repeatedly in the same plants. Smolders et al. (1991) developed a method for continuous shoot growth monitoring in hydroculture. A hydroponic container with plants is placed on a balance and the fresh weight of the plant parts above the nutrient solution is continuously recorded. The basic methods of plant growth research are presented in textbooks, for instance by Evans (1972), Causton and Venus (1981) and Hunt (1990).

8.4 GROWTH RESPONSES TO VARIOUS METALS

Trace metals can be grouped according to their effects on plants in essential micronutrients and non-essential, toxic elements (Derome and Lindroos 1998a,b). Differences in the effects of elements of these groups on plant growth can be seen in dose-response curves (Fig. 8.1).

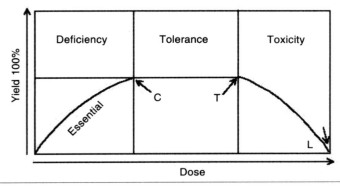

Fig. 8.1 Dose-response curves for non-essential and essential elements; *C* Critical deficiency; *T* toxicity threshold; *L* lethal toxicity. (After Berry and Wallace 1981)

The absence of an essential element causes abnormal growth or failure of the life cycle and essential elements cannot be substituted by others in their biochemical role (Aller et al. 1990; Egli et al. 1999; Mälkönen et al. 1999; Vesk et al. 1999). The dose-response curves of essential elements have three phases: deficiency, tolerance and toxicity. For non-essential elements there is no deficiency phase and the tolerance plateau extends to zero dose (Berry and Wallace 1981).

8.4.1 Essential Metals

8.4.1.1 *Copper*

Concentrations of Cu in normal soils range between 10–80 ppm (Mortvedt et al. 1972). It is strongly bound to organic matter and Cu organic complexes play an important role in the regulation of Cu mobility in the soil (Mengel and Kirby 1987). Uptake in plants is usually low and plant Cu concentrations are commonly level, between 2–20 ppm (Wallnöfer and Engelhardt 1984). Its biochemical function is mainly to be a cofactor in enzymes, e.g. plastocyanin, superoxide dismutase and amine oxidases. Symptoms of damage induced by excess Cu are chlorosis and alterations in root growth like formation of many stunted brownish lateral roots (Wallnöfer and Engelhardt 1984; Doncheva 1998; Gupta et al. 1999). Effects of excess Cu on growth of crop plants, Cu-resistant plant species and on trees have been investigated repeatedly (Kukkola et al. 2000). For instance, Adalsteinsson (1994) studied the effects of Cu on root growth of winter wheat (*Triticum aestivum*, Poaceae). In order to examine the effect of spatial differences in soil Cu concentrations on root architecture, a split-root technique was used. Root systems of hydroculture-grown plants were divided in two equal parts that were both treated with a separate nutrient solution containing Cu concentrations between 0.5 (control) and 10 μM CuCl$_2$. Root growth in low Cu (0.5 μM) compensated for the impeded growth of Cu-treated roots, mainly by an increase in the number of lateral roots. Roots grown in low Cu supported undisturbed shoot growth. The author concluded that photosynthates from the shoot were mainly directed towards newly growing lateral roots. The involvement of a hormonal control mechanism was suggested.

On sites with elevated soil trace metal concentrations, like sites with geogenic contamination, mine dumps or near metal smelters, trace metal resistant plant species (metallophytes) are found. The reaction of such plants to trace metals has been investigated in order to find the mechanisms of resistance. For example, Ouzounidou et al. (1994) studied the response of two populations of *Minuartia hirsuta* (Caryophyllaceae) to increased concentrations of Cu in nutrient solutions. Plants of the Cu-resistant population showed increased root growth above the control level up to concentrations of 80 μM Cu, whereas plants of the Cu-sensitive population had a root growth maximum at the control level of 0.5 μM. The sensitive plants showed chlorosis and necrosis of young leaves in response to Cu stress. Resistant plants had a higher demand for Cu and grew more vigorously at higher Cu concentrations. The authors suggested that the resistance is based on efficient sequestration of Cu in root cells.

Plants of a Zn- and Pb-resistant population of *Alyssum montanum* (Brassicaceae) from a mining site in northern Greece also showed a certain co-resistance to Cu (Ouzounidou 1994a). The plants were grown in nutrient solutions with various Cu concentrations. At levels up to 16 µM Cu in the nutrient solution, root growth remained at the control level, but it decreased sharply at higher Cu treatments. Copper accumulated more in roots than in shoots. High Cu accumulations in the plants had negative effects on uptake and transport of nutrients like Ca, Mg, Fe and K, and on the photosynthetic capacity.

The effect of Cu on a Cu-resistant ecotype of *Silene compacta* (Caryophyllaceae) from a heavy metal contaminated site in northern Greece was studied by Ouzounidou (1994b). At Cu concentrations of 8 μM in the nutrient solution, root elongation, chlorophyll content and nutrient uptake were increased relative to the control. However, at high Cu treatments, growth was significantly reduced. The author suggested the development of an adaptive mechanism of Cu resistance in the studied ecotype of *Silene compacta* which enabled the plants to cope with mild Cu stress.

Some investigators assume that the expression of trace metal resistance involves a cost for the plants in terms of energy or other resources (Sibly and Calow 1989). This would imply that resistant individuals have a lowered fitness in the absence of trace metal stress (Ernst et al. 1990; Harper et al. 1997). The cost of the resistance was seen as a reason for the low frequency of metal-resistant individuals in uncontaminated habitats. However, this concept is challenged by other authors, who claim that the evidence for such a cost is not decisive and the nature of the cost remains unknown (Harper et al. 1997). According to the 'trade-off hypothesis' plants under trace element stress invest energy or other resources in their resistance mechanisms whereas other processes, like growth or reproduction, are depressed. This would reduce the competitive fitness of such plants under conditions without the metal stress. Indeed, it is known that metal-resistant plants have lower growth rates in comparison with non-resistant plants (Wilson 1988). However, the problem is to prove that the growth reductions result from the activity of the resistance genes and not from that of other genes, which enable the plants to cope with other adverse factors, like low nutrient supply or drought. In order to test the 'trade-off hypothesis', Harper et al. (1997) studied the response of the yellow monkey flower (*Mimulus guttatus*, Scrophulariaceae) to Cu in order to search for a cost. In earlier studies it was shown that Cu resistance in this species is controlled by a single major gene (Macnair 1983), which is supported by other genes that modify the level of Cu resistance (Macnair et al. 1993). Under laboratory conditions of Cu stress, *Mimulus* plants were selected, which even exceeded the Cu resistance of field-grown plants on highly contaminated sites. This result may indicate a selection against maximum resistance in the field. Harper et al. (1997) conducted culture experiments with *M. guttatus* plants selected for Cu resistance and measured the response of a large variety of traits of the plants to Cu stress. From a statistical analysis, the authors concluded that no clear evidence for a true cost of Cu resistance could be found. Thus, the 'trade-off hypothesis' of trace metal resistance is still under discussion.

Besides herbaceous plants, long-lived woody plants were also investigated. Utriainen et al. (1997) compared the Cu and Zn resistance of ten different clones of birch trees (*Betula pendula, B. pubescens,* Betulaceae). The clones were obtained from micropropagated birch trees from three sites differing in metal contamination. At an age of 6–8 weeks the plants were transferred to nutrient solutions. The young trees were treated with a variety of concentrations of Cu and Zn for periods of 7 days. The root growth response was determined using two methods: (1) by calculation of the metal tolerance index for different metal concentrations describing the effect of the metal on the increase in length of the longest root on the seventh day of exposure (TI [%]$_c$=(Root growth in metal solution/Root growth in control solution)100; c=metal concentration), and (2) by determining the EC$_{50}$ concentration, that is the concentration which inhibits root growth by 50%. Plants of different clones showed significant differences in metal resistance. Interestingly, a Zn-resistant clone from a site with very low soil Cu contamination also showed resistance to Cu. However, another clone derived from a seed of the same parent birch did not show resistance to Cu or Zn. The authors suggested that resistance to Cu and Zn in birch trees may share a common mechanism. Metal-resistant birch trees may be used for afforestation of industrially contaminated areas (Kozlov et al. 2000).

8.4.1.2 Iron

Iron is a very abundant element comprising about 5 wt% of the earth's crust (Mengel and Kirby 1987). Concentrations in soils range between 0.5-5% (Schachtschabel et al. 1984). However, in aerated soil systems in the physiological pH range, concentrations of ionic Fe (Fe^{2+}, Fe^{3+}) are extremely low reaching 10-10 M (?) or lower (Marschner 1986). Normally, Fe^{2+} is more readily absorbed by plants. Due to its low availability, many plants have developed mechanisms to mobilize Fe in the soil, like siderophores (Römheld and Marschner 1990; Ganmore-Neumann et al. 1992; Cakmak et al. 1994; Marschner and Römheld 1996). Requirements for optimal plant growth are 10^{-9} to 10^{-4} M and in a typical plant leaf concentrations range between 50-100 ppm (Guerinot and Yi 1994). The bio-chemical functions of Fe are numerous and depend on chelate formation and on valency changes. Iron occurs in enzyme systems with a heme group, like catalase, peroxidase or cytochrome oxidase (Mengel and Kirby 1987).

Iron deficiency symptoms are interveinal chlorosis in young leaves caused by an inhi-bition of chloroplast development. A well-known Fe deficiency syndrome is called 'Fe chlorosis' or 'lime induced chlorosis'. It occurs in plants on calcareous soils and, appar-ently, results from a physiological disorder caused by an excess of HCO^{3-} ions. Bicarbon-ate ions appear to immobilize Fe in the plants and Fe translocation to growing plant tissue is inhibited (Mengel and Kirby 1987).

Iron toxicity occurs in rice fields where flooding leads to a rapid increase in soluble Fe concentrations from 0.1 to 50-100 ppm. The excess Fe can cause browning of the leaves, known as 'bronzing' (Mengel and Kirby 1987). The symptoms are diverse among plant species and Fe toxicity is difficult to identify from the outer appearance of the plants (Foy et al. 1978).

Many experimental studies deal with effects of Fe on plant growth. Iron deficiency limits crop production in many regions. Ares et al. (1996) studied the response of an ancient crop, *Colocasia esculenta* (Araceae), to variations in Fe supply. Plants were grown for 7 weeks in nutrient solutions containing 0-20 μM Fe in complexed form (FeEDDHA). Without Fe addition plants developed leaf chlorosis and showed reduced lateral root formation. Such plants showed significant reductions in dry weights of leaves, roots and petioles, total biomass and in the leaf area relative to Fe-supplied plants. The critical Fe concentration in leaves to produce 90% of the maximum leaf dry weight was in the range 55-70 ppm. Such values can be used as a reference for diagnosis of Fe deficiency in *Colocasia* plantations.

On calcareous soils the acquisition of Fe can be a problem for plants. Gries and Runge (1995) studied the uptake strategies of six different grass species. Some of the grasses are calcicoles, common on calcareous soils (*Brachypodium sylvaticum, Hordelymus europaeus, Bromus ramosus*), in which Fe deficiency problems are most likely to occur. The other species are calcifuges not found on $CaCO_3$-rich soils (*Nardus stricta, Avenella flexuosa, Danthonia decumbens*). Plants were treated with concentrations of Fe-EDTA from 0-10 μM in nutrient solutions for 2 months. The calcicole grasses showed higher relative yields at low Fe supply than the calcifuges. The specific root surface and Fe uptake requirements were lower in calcicole species than in calcifuges. Exudation rates of Fe-mobilizing phytosiderophores from the roots of calcicoles were much higher compared with calci-fuges. The authors concluded that the capability of calcicole species to live on calcareous soils was partly due to their higher production of Fe-mobilizing compounds and to their lower tissue Fe requirements.

Toxicity problems from high Fe concentrations have also been described. For instance, in wetland habitats high concentrations of reduced Fe can affect the growth of plants. Therefore, Snowden and Wheeler (1993) studied the toxicity of ferrous Fe on seedlings of 44 fen plant species. The plants were exposed for 2 weeks to Fe sulfate at concentrations between 3.8 (control) and 100 ppm in nutrient solutions. Various growth parameters were recorded, including shoot and root length and dry weight, as well as size and numbers of leaves. Relative growth rates were determined: RGR per day=(final dry weight-mean initial dry weight)/(mean initial dry weight ? 14). The most appropriate resistance index was Σ % RGR, i.e. the summation of standardized relative growth rates over a range of iron concentrations. For standardization, mean RGR for each treatment were expressed as percentage of that of the control. Based on these resistance indices, a species ranking of Fe resistance was derived. The species with the highest resistance were almost exclusively monocots (e.g. *Eriophorum angustifolium*, *Carex echinata*), while the most sensitive species were all dicots (e.g. *Rumex hydrolapathum*, *Epilobium hirsutum*). The higher Fe resistance of monocots may result from two traits: a more efficient detoxification system which reduces direct Fe toxicity and lower growth rates which minimize indirect toxicity effects on mineral nutrition. Iron resistance and relative growth rates were negatively correlated.

Effects of Fe on tree growth were also studied. Göransson (1993) investigated the influence of Fe supply on plant nutrition and on the partitioning of biomass when Fe supply was the growth-controlling factor. Saplings of *Betula pendula* (Betulaceae) were grown in an aeroponic culture system in which the roots of the trees were continuously sprayed with nutrient solution (for technical details, see Waisel 1996). The relative addition rate of Fe was adjusted as the growth-limiting factor, whereas all other nutrients were added in free access. Deficiency symptoms like interveinal chlorosis in younger leaves occurred at low Fe supply. The relation between relative growth rates and internal Fe concentrations was nearly linear at sub-maximum rates of Fe supply. When the plants had free access to Fe, internal concentrations were about three times higher than required for optimal growth. The dry matter allocation to leaves and the specific leaf area were largely unaffected by the Fe supply. However, the net assimilation rates were reduced at limited Fe supply. The author concluded that the acclimation of growth rates of birch trees to limited Fe availability did not involve changes in dry matter partitioning to leaves and leaf expansion, but was associated with reduced net assimilation rates.

8.4.1.3 *Manganese*

Total Mn concentrations in soils vary between 20-3000 ppm (Mortvedt et al. 1972). Concentrations in plant material are between 10-100 ppm (El-Bassam 1978). For plant nutrition, Mn^{2+} and easily reducible Mn species are most relevant. The most important role of Mn in green plants is in the Hill reaction, the H_2O-splitting and O_2-releasing process. A Mn-protein complex appears to catalyze O_2 evolution (Mengel and Kirby 1987). Only a few Mn-containing enzymes, like superoxide dismutase, have been found in plants (Mengel and Kirby 1987). A deficiency symptom is interveinal chlorosis, occurring first in young leaves. Manganese-deficient dicots often show small yellow spots on the leaves and interveinal chlorosis (Maynard 1979). Symptoms of Mn toxicity are diverse among plant species and appear as marginal leaf chlorosis and necrosis, puckering of leaves or necrotic spots. Some specific disorders related to excess Mn are 'crinkle leaf' of cotton, 'stem streak necrosis' of potato and 'internal bark necrosis' of apple trees (Foy et al. 1978).

Manganese toxicity is a growth-limiting factor on acidic soils. Samantaray et al. (1997) studied the effects of different concentrations of Mn on the tropical grass *Echinochloa colona* (Poaceae). Plants were treated with 1.8-910 μM $MnCl_2$ in hydroculture for 3 weeks. Root growth was inversely correlated with Mn concentrations, and in the highest Mn treatment root elongation was completely inhibited. Chlorophyll contents decreased with increasing Mn concentrations in the nutrient solution. With Mn concentrations of 45 μM, plants developed toxicity symptoms, like localized necrotic lesions of the oldest leaves.

Effects of Mn toxicity on CO_2 assimilation of three genotypes of bean (*Phaseolus vulgaris*, Fabaceae) differing in Mn resistance were studied by Gonzalez and Lynch (1997). Plants grown in hydroculture were treated with $MnSO_4$ in concentrations of up to 300 μM. Manganese toxicity decreased the chlorophyll content in immature leaves and the CO_2 assimilation of such leaves was reduced. Mature leaves with toxicity symptoms like brown speckles did not show reduced CO_2 assimilation capacity. The CO_2 assimilation expressed per unit of chlorophyll did not differ between Mn-treated and control plants. Thus, the authors concluded that the observed negative effect of Mn on CO_2 assimilation of young leaves resulted from a reduction in chlorophyll content.

The effect of Mn on trees was also investigated. Langheinrich et al. (1992) described the response of young spruce trees (*Picea abies*, Pinaceae) to $MnSO_4$ concentrations between 3-800 μM in hydroponic solutions. In order to check the influence of the N source, spruce trees were either supplied with NO_3^- or NH_4^+. After 11 weeks of treatment with 800 μM Mn, 50% of the trees had ceased shoot growth and developed terminal buds, irrespective of the N source. No visible symptoms of damage were observed. Root elongation rates were reduced in Mn concentrations of 200 μM with NO_3^- supply but not with NH_4^+. This difference was assumed to result from higher Mn concentrations in root tips of NO_3^--supplied trees. In Mn-treated trees, concentrations of Mg and Ca were reduced, which might explain the growth reductions of such trees. Increased concentrations of starch, phenolic compounds and nitrate in the shoots of such trees pointed to disturbances in carbohydrate and N metabolism.

8.4.1.4 Zinc

Zinc concentrations in most soils are between 10-300 ppm (Mortvedt et al. 1972). In contaminated places near metal smelters soil concentrations of up to 5% Zn are possible (Wallnöfer and Engelhardt 1984). Sites with geogenically elevated soil concentrations of Zn are found on all continents (Antonovics et al. 1971). Such places are inhabited by an adapted flora (metallophytes). About 60% of the soluble Zn in soils occurs in organic complexes (Hodgson et al. 1966). These are formed with amino, organic and fulvic acids (Stevenson and Ardakani 1972).

Zinc concentrations in plants are between 15-100 ppm (El-Bassam 1978). Several enzymes contain Zn, like carbonic anhydrase, alcohol dehydrogenase, superoxide dismutase and RNA polymerase (Mengel and Kirby 1987). The element has an important role in the N metabolism of plants. In Zn-deficient plants protein synthesis is reduced. Deficiency symptoms are interveinal chlorosis of leaves and short internodes (Mengel and Kirby 1987). On a global scale, Zn deficiency is more widespread than that of any other micronutrient and Zn fertilizers are often used in agriculture (Adriano 1986). For instance, in regions of Pakistan with alkaline-calcareous soils, a field survey of *Sorghum* (Poaceae) plants and associated soils revealed Zn deficiency in up to 64% of the fields (Rashid et al. 1997).

Zinc deficiency causing reductions in crop production is an important economic factor and has been studied intensively. Different application methods of Zn fertilizers were compared by Yilmaz et al. (1997). The grain yield of wheat cultivars (*Triticum aestivum, T. durum,* Poaceae) on severely Zn-deficient calcareous soils was determined with different methods of Zn fertilization. Zinc sulfate was sprayed onto the soil or onto different plant parts. Compared with the unfertilized control, increases in grain yield were about 260% with application to the soil, soil+leaf or to seed+leaf, 204% with seed application and 124% with leaf application. The largest biomass increase of above-ground plant parts was obtained with soil application (109%) and the smallest increase with leaf application (44%). The authors concluded that soil fertilization was the most economical method, with long-term effects on grain yield on Zn-deficient soils.

Effects of Zn deficiency on root growth of three wheat genotypes (*Triticum aestivum, T. turgidum*, Poaceae) were examined by Dong et al. (1995). The three cultivars differed in their growth response to low Zn availability. Plants were grown in sandy soils with Zn additions of between 0-0.4 ppm. Four weeks after sowing, root and shoot dry matter were significantly increased at 0.1 ppm compared with the control without Zn addition. The Zn-efficient genotype developed longer and thinner roots compared with the less Zn-efficient plants. The authors suggested that the two traits associated with Zn efficiency were to grow longer and thinner roots and to have a larger proportion of thinner roots in the root biomass early in the growth period. This may enhance Zn uptake at low soil concentrations.

General symptoms of Zn toxicity are turgor loss, necroses on older leaves and reduced growth (Wallnöfer and Engelhardt 1984). Some plant species are adapted to high Zn levels. Brune et al. (1994) studied the mechanism of Zn resistance in barley (*Hordeum* sp., Poaceae). Plants were grown in nutrient solutions with Zn concentrations of between 2-1600 μM for 10 days. Root elongation was strongly inhibited and plants in the highest Zn treatment had only 35% of the root length of control plants. The growth of primary leaves was little affected, although Zn concentrations in leaves of Zn-treated plants increased strongly. The distribution of Zn in the leaf tissue was studied in a compartmental analysis. At a low Zn supply of 2 μM, highest Zn contents were found in the cytoplasm of mesophyll protoplasts. At higher Zn concentrations, which inhibited root growth, Zn contents increased strongly in the apoplastic space. Increased Zn levels also occurred in epidermis cells and in mesophyll vacuoles. The cytoplasmic content of mesophyll protoplasts was not changed. The authors took this as an indication of a perfect homeostasis of Zn in the leaf. They concluded that compartmentation and transport are important mechanisms of Zn homeostasis in the leaves under conditions of high Zn supply.

Some metallophytes have particularly efficient adaptations to elevated soil Zn concentrations. Tolra et al. (1996) studied the response of the metallophyte *Thlaspi caerulescens* (Brassicaceae) to Zn in the substrate. The plant is known as a hyperaccumulator of Zn that can accumulate up to 4% Zn in the leaf dry matter. This trait makes the species a possible candidate for phytoremediation of Zn contaminated soil (see Chap. 15). The growth of the plant is stimulated by Zn concentrations that are toxic to most other plant species, but the mechanism is still not known. Experimental plants were grown in nutrient solutions containing from 1.5-1500 μM Zn (Tolra et al. 1996). Highest dry weights of shoots and of roots were produced with 500 μM Zn. Such plants had Zn concentrations of 0.36% in their shoots. Therefore, the Zn requirement of the plants appeared to be very high. The authors suggested that only a small fraction of the absorbed Zn was involved in metabolic processes and that most of the Zn was either sequestered in vacuoles or detoxified in chelated form. Significant inhibition of root elongation was observed with Zn concentrations >1000 μM. The growth stimulation at medium Zn supply was accompanied by an increased Fe translocation from roots to shoots. The Zn hyperaccumulator *T. caerulescens* was able to maintain adaquate concentrations of essential nutrients even under a high Zn supply.

Another metallophyte often found on metal-rich soils is *Armeria maritima* (Plumbaginaceae). The ecotype growing on metalliferous soils appears to have evolved from a hybrid group between a subspecies from salt marshes and a subspecies growing on sandy soils (Köhl 1997). This author devised an artificial soil culture method using mixtures of ion-exchange resin and an inert sand matrix. Metal ions are buffered by the ion-exchange system like in a natural soil. The growth responses of different genotypes of *Armeria maritima* to Zn stress were studied using this system. In short-term tests, plants of non-metalliferous soil populations were more sensitive to Zn concentrations of 1 mM than plants from metalliferous soil. In the first weeks after sowing, plants from non-metalliferous soils had a lower total biomass and root growth was severely inhibited compared with the control. In contrast, plants from metalliferous soil showed no significant reductions in root and shoot growth at 1 mM Zn. However, over periods of several months, adult plants of the salt marsh and of the sandy soil populations proved as resistant to elevated soil Zn concentrations as those of populations from metalliferous soils. The Zn resistance of all populations was sufficient for survival on Zn mine soils. The author suggested that genes for metal resistance are present in all genotypes. In plants from metalliferous soils these genes are constitutively active, while in plants from non-metalliferous soils the genes for metal resistance are only activated by trace metal imposed stress. Thus, in short-term experiments, the resistance of such plants might be underestimated.

8.4.2 Non-essential Metals

8.4.2.1 *Cadmium*

Cadmium concentrations of uncontaminated soils are usually below 0.5 ppm but can reach up to 3 ppm depending on the geological background (Schachtschabel et al. 1984). Places with higher Cd concentrations from geogenic sources and human mining activities are known (Antonovics et al. 1971). Furthermore, Cd is a pollutant that has been emitted into the environment for decades (Nriagu 1980). Major anthropogenic sources are Cd-containing phosphate fertilizers, sewage sludge and industrial emissions (Adriano 1986).

Plant Cd concentrations are between 0.05-0.2 ppm (El-Bassam 1978), but can be much higher on contaminated sites (Adriano 1986). Cadmium is a toxic element without any known physiological function in plants. Toxicity symptoms are similar to Fe chlorosis and also necroses, wilting, red-orange leaf coloration and growth reductions have been described (Haghiri 1973; Bingham and Page 1975).

Many studies deal with effects of Cd on the growth of crop plants. Inouhe et al. (1994) compared the growth responses of various monocot and dicot species to Cd. Roots of 7-day-old hydroponically grown seedlings of the tested plants were exposed to various Cd concentrations. Fresh weights of roots were recorded. Roots of cereal plants [maize (*Zea mays*), oat (*Avena sativa*), barley (*Hordeum vulgare*), rice (*Oryza sativa*)] grew even with 60 μM Cd, although growth was reduced in some species. In contrast, root growth of most dicot species was more severely inhibited at Cd concentrations of 10-30 μM [azuki bean (*Vigna angularis*), tomato (*Lycopersicon esculentum*), cucumber (*Cucumis sativus*), radish

(*Raphanus sativus*), pea (*Pisum sativum*), lettuce (*Lactuca sativa*), sesame (*Sesamum indicum*)]. Under Cd stress a Cd-binding complex containing phytochelatin was detected in the cytoplasm of cereal roots, but not in roots of dicots. The authors suggested that the synthesis of this complex played a role in the resistance of monocot roots to Cd.

The growth of strawberry plants (*Fragaria* x *ananassa*, Rosaceae) under Cd stress was studied by Cieslinski et al. (1996). Cadmium was added to sandy loam culture soils in concentrations up to 60 ppm. Roots accumulated the highest Cd concentrations of all tested plant parts. However, under Cd stress, leaf weights were more reduced than root weights. More than 90% of the total absorbed Cd was accumulated in roots. Root concentrations were between 2.6 ppm (control) and 506 ppm for plants in the highest treatment. The authors concluded that leaf dry weights were the best indicators of Cd toxicity.

Growth and ultrastructural changes of Cd-stressed wheat plants (*Triticum aestivum*, Poaceae) were described by Ouzounidou et al. (1997). Young plants were grown in nutrient solutions with Cd concentrations up to 1 mM. At the highest Cd concentration elongation of roots and of above-ground parts was reduced to 28 and 40% of the control, respectively. Increased Cd concentrations in leaves of plants under Cd stress were accompanied by declining concentrations of Fe, Mg, Ca and K. The growth inhibition, reduced chlorophyll content and inhibition of photosynthesis in upper plant parts may have resulted from Cd effects on the plant content of essential nutrients. In Cd-treated plants the structure of chloroplasts was changed. The authors suggested that Cd induced premature senescence.

The responses of bean (*Phaseolus vulgaris*, Fabaceae) and of tomato (*Lycopersicon esculentum*, Solanaceae) to Cd stress were compared by Ouariti et al. (1997). Young plants were treated for 7 days with Cd concentrations of 0-50 μM in nutrient solutions. The inhibitory effect of Cd on shoot and root dry weight production was greater in bean than in tomato plants. Cadmium concentrations of the nutrient solution that caused 50% root growth reduction were 2 μM for bean and 50 μM for tomato. In shoots, 50% growth reduction was induced by 5 μM Cd in bean and by 10 μM Cd in tomato plants. Both species had much higher Cd concentrations in roots than in shoots. The resistance of the plants depended mostly on the Cd resistance of their roots, which must retain the capability to absorb balanced amounts of essential nutrients.

Much experimental evidence indicates the involvement of metal-binding low molecular weight compounds (phytochelatins) in the resistance of plants to toxic trace metals (Prasad 1995; Zenk 1996; Chap. 2). Tukendorf et al. (1997) showed that the accumulation of homophytochelatins in bean plants (*Phaseolus coccineus*, Fabaceae) under Cd stress depended on their growth stage. Plants were exposed to 25 μM Cd in nutrient solutions at different developmental stages of the primary leaves. High levels of homophytochelatins were found in primary leaves and roots of plants treated with Cd at an early growth stage. At later stages the capacity for homophytochelatin accumulation was much smaller. The rate of synthesis and the accumulation of the Cd-binding compound depended on the developmental stage of the plant in which the Cd stress occurred. These factors should be considered in studies of plant responses to trace metals.

Effects of Cd on growth of trees were also studied. For instance, Gussarsson (1994) described the response of birch trees (*Betula pendula*, Betulaceae) to Cd. Seedlings were treated with Cd concentrations of up to 2 μM in nutrient solutions. After 8 days, root concentrations of K, Ca, Mg and Mn were reduced, whereas concentrations of Cu and Mo were increased. Cadmium was accumulated particularly in fine roots. Total plant growth was stimulated by 0.05 μM Cd, but was reduced at external Cd concentrations of 0.5 μM. Root growth was stimulated by Cd treatments and growth inhibitions were found only in shoots. The author suggested that an accumulation of Cd, Cu, Ca and S in fine roots combined with a preference for root growth could be part of a mechanism for Cd resistance.

8.4.2.2 Nickel

Nickel concentrations in uncontaminated soils are 5-50 ppm (Schachtschabel et al. 1984) and concentrations in plants growing on such soils range between 0.4-3 ppm (El-Bassam 1978).

The essentiality of Ni for plants is still under dispute (Welch 1981). Nickel is a component of the enzyme urease and is essential for its functioning (Klucas et al. 1983). Urease catalyzes the decomposition of urea to ammonia. In tissue culture experiments with soybean it was shown that cell growth with urea as the N source in the absence of Ni was poor, but was enhanced by Ni addition (Polacco 1977). When the cells were supplied with other sources of N, Ni had no effect on growth. Effects of Ni on growth, N metabolism and leaf urease activity of six crop species [rye (*Secale cereale*), wheat (*Triticum aestivum*), soybean (*Glycine max*), rape (*Brassica napus*), zucchini (*Cucurbita pepo*), sunflower (*Helianthus annuus*)] were studied by Gerendas and Sattelmacher (1997). Plants were grown in urea-based nutrient solutions, either Ni-free or with the addition of 0.04-0.09 μM Ni. All tested species showed shoot growth reductions in Ni-free nutrient solutions compared to Ni-supplied solutions. Activity of urease was hardly detectable in plants grown without Ni addition. Consequently, such plants accumulated urea, showed reduced dry matter production and reduced total N concentrations. The plants appeared chlorotic as a result of metabolic N deficiency. Such results demonstrate the need of Ni for urease activation and for the growth of plants in urea-based media. Apparently, Ni is essential for some plant species that use ureides in their metabolism, but not for other species. The element can be classified as 'beneficial', although its possible essentiality requires further investigation (Marschner 1986; Gries and Wagner 1998).

In most plant species excess Ni concentrations cause chlorosis (Zornoza et al. 1999). Cereals show yellow stripes along the leaves, which may turn white. Necrotic leaf margins can occur. Dicots develop interveinal chlorosis, similar to symptoms of Mn deficiency (Mengel and Kirby 1987).

Effects of Ni were investigated in various crop species. A toxicity test using the root growth of onion (*Allium cepa*, Liliaceae) under Ni stress was presented by Liu et al. (1994). Onion bulbs were placed on test solutions of tap water with Ni additions from 0 (control) to 100 mM for up to 4 days. At Ni concentrations of 100 μM root growth was inhibited after 48 h. A histological examination of root cells of Ni-treated onions showed expanded nucleoli at Ni concentrations of 1-10 μM after 48 h. At higher concentrations of 10 mM Ni, irregularly shaped nucleoli were found in most of the root tip cells. This test method can be used for rapid screening of Ni contamination in environmental studies.

Seed germination of radish (*Raphanus sativus*, Brassicaceae) under Ni stress was investigated by Espen et al. (1997). Seeds were incubated in water containing Ni additions up to 400 μM. At different times growth was measured as fresh weight increase. An external Ni concentration of 200 μM inhibited the fresh weight increase by 53% after 24 h compared with the control. Seeds in 400 μM Ni showed an almost complete inhibition of fresh weight increase. Absorption of Ni by the seeds was correlated to external Ni concentrations. After 10 h of incubation, Ni levels of seeds showed maximum values and then remained constant, reaching an equilibrium with external Ni. Additionally, a variety of metabolic parameters were measured in the seeds, like O_2 uptake, levels of ATP and of nucleic acid and protein synthesis. The authors proposed that in seeds under Ni stress an enhanced activity of transport mechanisms at the plasma membrane (H^+-pumps) occurs which depletes metabolic energy and, thus, inhibits all energy-requiring cellular processes and slows down germination.

Of particular interest are sites with naturally elevated soil concentrations of Ni, the serpentine areas, which occur on all continents (Roberts and Proctor 1992). Serpentine soils derived from ultrabasic igneous rocks can contain very high Ni concentrations of up to 8000 ppm (Schachtschabel et al. 1984). Serpentine soils can also have increased concentrations of Cr or Co as well as unfavorably low Ca/Mg ratios (Menezes de Sequeira and Pinto da Silva 1992; Rodenkirchen and Roberts 1993), which cause severe problems for plant growth. A specialized flora with many endemic species is adapted to serpentine areas (Arianoutsou et al. 1993). Some species have an outstanding potential for trace element accumulation; they are hyperaccumulators (Baker and Brooks 1989; see also Chap. 1). Such plants can have Ni concentrations reaching 3% of the dry weight (Bollard 1983), without visible symptoms of damage. A number of hyperaccumulating species were found in the genus *Alyssum* (Brassicaceae, Morrison et al. 1980). Some accumulating species of this genus showed root growth in solutions with Ni concentrations up to 1 mM, whereas root growth of non-accumulating species was inhibited by traces of Ni. Two serpentine species with differing resistance strategies were studied by Gabbrielli et al. (1990). *Silene italica* (Caryophyllaceae) limits its Ni uptake. Root growth was inhibited by a suppression of mitotic activity in root tips at 7.5 μM Ni in the culture solution. The same concentration did not affect root growth in *Alyssum bertolonii*, which is a Ni accumulator. A Ca supply of 25 mM reversed the effects of Ni on root growth in *Silene*, but in *Alyssum* the addition of Ca reduced root growth. This demonstrates that *Alyssum bertolonii* is also adapted to low Ca concentrations in the substrate, which is typical for serpentine soils.

Growth and elemental composition of an endemic hyperaccumulating serpentine species from north east Portugal, *Alyssum pintodasilvae*, were studied by de Varennes et al. (1996). Plants were grown in clay soils to which various amounts of Ni, Cr, Pb, Cd, Cu or Zn were added. After 2 months, the shoot dry matter production was not affected by the metals except by Cd. Addition of the metals to the soil increased uptake and translocation by the plants, but only Ni was hyperaccumulated. The shoots of plants grown on Ni-enriched soil were cut after 1 month and again after another 2 months. Shoot concentrations of Ni were higher in the second cut. The possibility of continued growth and repeated harvests of the plants suggested, according to the authors, the use of this species for extraction of Ni from polluted soils.

The limiting factors for plant growth on an alpine serpentine soil in Scotland were investigated by Nagy and Proctor (1997). The soil of the study site had toxic concentrations of Mg and Ni. The plant cover was sparse and an important species was *Cochlearia pyrenaica* (Brassicaceae). The aim of the study was to test the hypothesis that low soil nutrient concentrations caused open vegetation cover. Therefore, fertilization experiments with N, P, K and Ca were carried out on marked study plots. Plant cover increased after fertilization. *Cochlearia* plants showed an increase in rosette size, flowering and seed production with fertilizer application. This supported the assumption that a major nutrient deficiency, paticularly of P, rather than metal toxicity, limited plant growth at the serpentine site studied.

The results demonstrate that plant growth reductions on serpentine soils are not induced by trace metal toxicity alone. A complex of adverse factors acting in concert is the major constraint for plant growth.

8.4.3 Interactions of Metals

While in the previous sections effects of single elements on plants were outlined, under field conditions plants are often affected by more than one element, as described for serpentine soils (Ni, Co, Cr, Ca/Mg ratio). Various interactions can occur when plants are exposed to unfavorable concentrations of more than one trace element. Such combination effects were categorized by Berry and Wallace (1981) as independent, additive, synergistic or antagonistic. An independent action of stressors means that the yield of plants exposed to more than one stressor is equal to the yield response with the stressor having the greater effect. A sequentially additive action occurs when the stressors affect different steps of the yield production process. The expected yield is then the arithmetic product of the yields produced with given doses of the single stressors. A similarly additive action occurs when all stressors act on the same step of the yield production process and the effect is given by the added dose of the stressors. An antagonistic action means that the combined effect is smaller than the effect of the most active stressor alone, whereas, in the case of a synergistic action, the total effect is greater than that of the most active stressor.

Many publications report effects of trace metal combinations on plant growth. For instance, Wallace and Romney (1977) measured the growth of bean plants, *Phaseolus vulgaris* (Fabaceae), in soil culture with additions of Cd, Li, Cu and Ni. Leaf dry weights were reduced by 28% with Cu and by 3% with Ni. Combinations of Cu+Ni decreased the yield by 68%, which indicated a synergistic effect of the two metals. When Cd and Li were combined, the effects were less than additive, but in combination treatments of Cu+Ni+Cd the effects were additive. Combination treatments with all four elements had slightly less than additive effects on leaf dry matter production. The same plant species was studied in solution cultures by Chaoui et al. (1997). Combination treatments with Cd+Zn resulted in synergistic and additive effects on the growth of both roots and shoots.

Interactions of Zn and Cu on the growth of three species of *Brassica* (Brassicaceae) were described by Ebbs and Kochian (1997). Plants were grown for 2 weeks in nutrient solutions containing additions of single metals or combinations. In all species shoot dry weights were more strongly reduced by the combined application of Zn+Cu than by single metal treatments. This indicates a synergistic effect of both metals on shoot growth.

Effects of combinations of Al and Mn on growth of wheat (*Triticum aestivum*, Poaceae) were studied by Blair and Taylor (1997). A mathematical analysis of root growth in hydroponic cultures showed that the combined effects of Al+Mn could be described by a multiplicative model. The relative growth of the combination treatment could be calculated as the product of relative yields produced with the various metals in separate treatments. In experiments with the same plant species in hydroculture, Cocker et al. (1998) found that Si induced a significant amelioration of toxic effects of Al on root extension growth. Apparently this was not caused by an Al-Si interaction in the external medium, but was due to an internal component.

An interesting phenomenon observed in many studies is the moderation of trace element toxicity by Ca (Garland and Wilkins 1981; Fuhrer 1983; Hagemeyer 1990; Rengel 1992; Hagemeyer and Breckle 1996). For instance, Wallace et al. (1980) described experiments with soybean plants (*Glycine max*, Fabaceae) in hydroculture. Yields of leaves and roots were little affected by low Ca concentrations of the medium without trace metal additions. However, yields were reduced in treatments with Ni, Cd, Al, Mn or Cu, and such reductions were greater at low Ca concentrations than at a high Ca level.

The described interactions among trace metals or of trace metals and major nutrients obviously complicate studies of trace metal effects on plant growth and other processes. However, in order to reach a realistic understanding of plant responses to stressors under natural conditions, such factors should be considered in experimental studies and in theories of plant reactions to stress.

8.5 EFFECTS OF MYCORRHIZA

Another complicating factor in the relation of plant growth and trace elements is the influence of mycorrhiza, a symbiotic association of fungal hyphae and plant roots (Wilcox 1996; van Tichelen et al. 1999; Brunner and Frey 2000). This widespread phenomenon is often neglected in field studies or in laboratory experiments due to a lack of experience in recognition and handling of mycorrhizal associations.

Nevertheless, effects of mycorrhiza on plant growth under trace metal stress have been observed (Hartley et al. 1997). For example, the Zn resistance of birch seedlings (*Betula* spp., Betulaceae), assessed from dry matter production, was increased by ecto-mycorrhizal associations with the fungi *Amanita muscaria* and *Paxillus involutus* (Brown and Wilkins 1985). The growth stimulation may have resulted from retention of Zn in the mycorrhiza. A similar result was reported by Jones and Hutchinson (1986) for *Betula papyrifera*. In sand cultures with added Ni, ecto-mycorrhizal fungi, particularly *Scleroderma flavidum*, stimulated the growth of birch seedlings compared with non-mycorrhizal controls. The weights of shoots and of roots of mycorrhizal plants were significantly higher than those of non-infected control plants. The beneficial effect of the symbiosis may have resulted from Ni retention in the mycorrhiza. In contrast, mycorrhizal birch plants treated with Cu showed growth reductions compared with non-infected controls. Apparently, the response of mycorrhizal plants to trace metals cannot be generalized.

Mycorrhizae in crop species have also been studied. Ricken and Höfner (1996) described effects of arbuscular mycorrhizal infection of alfalfa (*Medicago sativa*, Fabaceae) and oat (*Avena sativa*, Poaceae) with the fungus *Glomus* sp. Growth of the plants was determined in pot experiments with sewage-sludge-amended soil containing elevated concentrations of various trace metals. Mycorrhizal alfalfa showed only a small increase in biomass compared with the non-infected control. However, mycorrhizal infection of oat increased growth of roots and of shoots by up to 70 and 55%, respectively. The symbiosis protected the plants against adverse effects of soil trace metal pollution.

Under conditions of low nutrient supply, mycorrhiza can stimulate metal uptake, and thus improve plant growth. For instance, in soils with low Zn concentrations, mycorrhizal corn plants (*Zea mays*, Poaceae) grew stronger and absorbed more Zn than non-mycorrhizal plants (Faber et al. 1990).

In metal-contaminated soils the growth of fungi can be inhibited. In this case one must consider the metal resistance not only of the mycorrizal association, but also of the separate partners, since they have to survive at least part of their life cycle alone, before a mycorrhizal symbiosis develops. Fungi must be able to grow hyphae in an adverse environment in order to reach plant roots to form mycorrhiza. To study the development of metal resistance in vesicular-arbuscular mycorrhizal fungi, Griffioen et al. (1994) compared the infection of roots of the grass *Agrostis capillaris* from soils differing in metal contamination. Plants of a Cu-enriched soil had a few mycorrhiza, whereas plants on uncontaminated soil showed infection of about 60%. A negative correlation was found between infection rates of roots and soil Cu contents. In contrast, on a site contaminated with Cd and Zn, roots had infection rates of 40%. The authors concluded that the fungi had evolved resistance against Cd and Zn and that they may play a role in the metal resistance of the grass. Detailed information about responses of various mycorrhizal fungi to toxic metals is presented in a review by Hartley et al. (1997).

The survival of long-lived trees on metalliferous soils may depend on mycorrhiza (Wilkinson and Dickinson 1995). The fungal sheath of mycorrhizal roots can function as a barrier for toxic metals. Furthermore, the shorter life cycle of a fungus may accelerate genetic change within mycorrhizal root systems of long-lived plants with long reproduction cycles. In this way, mycorrhiza can help higher plants to adapt to and to survive in contaminated habitats (Wilkinson and Dickinson 1995).

8.6 FURTHER RESEARCH TOPICS

Numerous publications dealing with effects of potentially toxic trace metals on plants provide information about growth parameters under metal stress in one way or another (Rigina and Kozlov 2000; Schaberg et al. 2000; Winterhalder 2000). Nevertheless, some particular questions are still open.

Why do small amounts of toxic elements stimulate plant growth? For instance, Gussarsson (1994) found growth stimulation of birch trees with low Cd concentrations, although Cd has no known metabolic function.

What is the interrelation between photosynthetic capacity and growth under trace element stress? Trace metals are reported to interference with photosynthesis and carbon fixation (see Chap. 6). However, the link between such effects and observed growth reductions remains to be elucidated. Apparently it is difficult to predict growth from measurements of photosynthesis. Chiariello et al. (1989) proposed that studies of carbon investment (allocation and partitioning) in plants can provide links between the physiological process of photosynthetic carbon fixation and the observed growth.

Are 'growth reductions' necessarily detrimental to plants? For individual plants it seems most important to complete their life cycle and to have large numbers of offspring. Strong growth in terms of biomass production is not necessarily an advantage. Perhaps a temporary or permanent slowing down of growth can be an adaptive trait that helps plants to thrive in an adverse environment. This is particularly true when not the whole plant, but the growth of certain plant parts, e.g. leaves or roots, is considered. It has been proposed that the stress resistance of plants may depend on their ability to divert resources from growth processes to maintenance processes in order to meet the additional energetic cost of resistance (Taylor 1989; see also Sect. 8.4.1.1). Therefore, low growth rates are presumably not always disadvantageous, but may instead increase fitness under certain conditions (Snowden and Wheeler 1993; Sect. 8.4.1.2). Our, often negative, interpretation of growth reductions is perhaps influenced by the prospect of reduced agricultural crop production - but it is only one side of the coin.

REFERENCES

Adalsteinsson S (1994) Compensatory root growth in winter wheat: effects of copper exposure on root geometry and nutrient distribution. J Plant Nutr 17: 1501-1512

Adriano DC (1986) Trace elements in the terrestrial environment. Springer, Berlin Heidelberg New York

Aller AJ, Bernal JL, del Nozal MJ, Deban L (1990) Effects of selected trace elements on plant growth. J Sci Food Agric 51: 447-479

Antonovics J, Bradshaw AD, Turner RG (1971) Heavy metal tolerance in plants. Adv Ecol Res 7: 1-85

Arduini I, Godbold DL, Onnis A (1995) Influence of copper on root growth and morphology of *Pinus pinea* L. and *Pinus pinaster* Ait. seedlings. Tree Physiol 15: 411-415

Ares A, Hwang SG, Miyasaka SC (1996) Taro response to different iron levels in hydroponic solution. J Plant Nutr 19: 281-292

Arianoutsou M, Rundel PW, Berry WL (1993) Serpentine endemics as biological indicators of soil element concentrations. In: Markert B (ed) Plants as biomonitors. VCH, Weinheim, pp 179-189

Baker AJM, Brooks RR (1989) Terrestrial higher plants which hyperaccumulate metallic elements - a review of their distribution, ecology and phytochemistry. Biorecovery 1: 81-126

Baker AJM, Walker PL (1989) Physiological responses of plants to heavy metals and the quantification of tolerance and toxicity. Chem Speciation Bioavail 1: 7-17

Berry WL, Wallace A (1981) Toxicity: the concept and relationship to the dose response curve. J Plant Nutr 3: 13-19

Bingham FT, Page AL (1975) Cadmium accumulation by economic crops. Proceedings of the international conference on heavy metals in the environment, Toronto, pp 433-441

Blackman VH (1919) The compound interest law and plant growth. Ann Bot 33: 353-360

Blair LM, Taylor GJ (1997) The nature of interaction between aluminum and manganese on growth and metal accumulation in *Triticum aestivum*. Environ Exp Bot 37: 25-37

Bollard EG (1983) Involvement of unusual elements in plant growth and nutrition. In: Läuchli A, Bieleski RL (eds) Encyclopedia of plant physiology, new series, vol 15B. Springer, Berlin Heidelberg New York, pp 695-755

Brown MT, Wilkins DA (1985) Zinc tolerance of mycorrhizal *Betula*. New Phytol 99: 101-106

Brune A, Urbach W, Dietz KJ (1994) Compartmentation and transport of zinc in barley primary leaves as basic mechanisms involved in zinc tolerance. Plant Cell Environ 17: 153-162

Brunner I, Frey B (2000) Detection and localization of aluminium and heavy metals in ectomycorrhizal Norway spruce seedlings. Environ Pollut 108: 121-128

Cakmak S, Gülüt KY, Marschner H, Graham RD (1994) Effect of zinc and iron deficiency on phytosiderophore release in wheat genotypes differing in zinc efficiency. J Plant Nutr 17: 1-17

Causton DR (1991) Plant growth analysis: the variability of relative growth rate within a sample. Ann Bot 67: 137-144

Causton DR, Venus JC (1981) The biometry of plant growth. Arnold, London

Chaoui A, Ghorbal MH, Ferjani EE (1997) Effects of cadmium-zinc interactions on hydroponically grown bean (*Phaseolus vulgaris* L.). Plant Sci 126: 21-28

Chiariello NR, Mooney HA, Williams K (1989) Growth, carbon allocation and cost of plant tissues. In: Pearcy RW, Ehleringer J, Mooney HA, Rundel PW (eds) Plant physiological ecology. Chapman and Hall, London, pp 327-365

Cieslinski G, Neilsen GH, Hogue EJ (1996) Effect of soil cadmium application and pH on growth and cadmium accumulation in roots, leaves and fruit of strawberry plants (*Fragaria x ananassa* Duch.). Plant Soil 180: 267-276

Cocker KM, Evans DE, Hodson MJ (1998) The amelioration of aluminium toxicity by silicon in wheat (*Triticum aestivum* L.): malate exudation as evidence for an in planta mechanism. Planta 204: 318-323

De Varennes A, Torres MO, Coutinho JF, Rocha MMGS, Neto MMPM (1996) Effects of heavy metals on the growth and mineral composition of a nickel hyperaccumulator. J Plant Nutr 19: 669-676

Derome J, Lindroos A-J (1998a) Effects of heavy metal contamination on macronutrient availability and acidification parameters in forest soil in the vicinity of the Harjavalta Cu-Ni smelter, SW Finland. Environ Pollut 99: 141-148

Derome J, Lindroos A-J (1998b) Copper and nickel mobility in podzolic forest soils subjected to heavy metal and sulphur deposition in western Finland. Chemosphere 36: 1131-1136

Doncheva S (1998) Copper-induced alterations in structure and proliferation of maize root meristem cells. J Plant Physiol 153: 482-487

Dong B, Rengel Z, Graham RD (1995) Root morphology of wheat genotypes differing in zinc efficiency. J Plant Nutr 18: 2761-2773

Ebbs SD, Kochian LV (1997) Toxicity of zinc and copper to *Brassica* species: implications for phytoremediation. J Environ Qual 26: 776-781

Egli M, Fitze P, Oswald M (1999) Changes in heavy metal contents in and acidic forest soil affected by depletion of soil organic matter within the time span 1969-93. Environ Pollut 105: 367-379

El-Bassam N (1978) Spurenelemente: Nährstoffe und Gift zugleich. Kali-Briefe (Büntehof) 14: 255-272

Ernst WHO, Schat H, Verkleij JAC (1990) Evolutionary biology of metal resistance in *Silene vulgaris*. Evol Trends Plants 4: 45-50

Espen L, Pirovano L, Cocucci SM (1997) Effects of Ni^{2+} during the early phases of radish (*Raphanus sativus*) seed germination. Environ Exp Bot 38: 187-197

Evans GC (1972) The quantitative analysis of plant growth. Blackwell, Oxford

Faber BA, Zasoski RJ, Burau RG, Uriu K (1990) Zinc uptake by corn as affected by vesicular-arbuscular mycorrhizae. Plant Soil 129: 121-130

Fellenberg G (1981) Pflanzenwachstum. Fische, Stuttgart

Foy CD, Chaney RL, White MC (1978) The physiology of metal toxicity in plants. Annu Rev Plant Physiol 29: 511-566

Fuhrer J (1983) Phytotoxic effects of cadmium in leaf segments of *Avena sativa* L., and the protective role of calcium. Experientia 39: 525-526

Gabbrielli R, Pandolfini T, Vergnano O, Palandri MR (1990) Comparison of two serpentine species with different nickel tolerance strategies. Plant Soil 122: 271-277

Ganmore-Neumann R, Bar-Yosef B, Shanzer A, Libman J (1992) Enhanced iron (Fe) uptake by synthetic siderophores in corn roots. J Plant Nutr 15: 1027-1037

Garland CJ, Wilkins DA (1981) Effect of calcium on the uptake and toxicity of lead in *Hordeum vulgare* L. and *Festuca ovina* L. New Phytol 87: 581-593

Gerendas J, Sattelmacher B (1997) Significance of Ni supply for growth, urease activity and the concentrations of urea, amino acids and mineral nutrients of urea-grown plants. Plant Soil 190: 153-162

Gonzalez A, Lynch JP (1997) Effects of manganese toxicity on leaf CO_2 assimilation of contrasting common bean genotypes. Physiol Plant 101: 872-880

Göransson A (1993) Growth and nutrition of small *Betula pendula* plants at different relative addition rates of iron. Trees 8: 31-38

Gries D, Runge M (1995) Responses of calcicole and calcifuge Poaceae species to iron-limiting conditions. Bot Acta 108: 482-489

Gries EG, Wagner GJ (1998) Association of nickel versus transport of cadmium and calcium in tonoplast vesicles of oat roots. Planta 204: 390-396

Griffioen WAJ, Ietswaart JH, Ernst WHO (1994) Mycorrhizal infection of an *Agrostis capillaris* population on a copper contaminated soil. Plant Soil 158: 83-89

Guerinot ML, Yi Y (1994) Iron: nutritious, noxious, and not readily available. Plant Physiol 104: 815-820

Gupta M, Cuypers A, Vangronsveld J, Clijsters H (1999) Copper affects the enzymes of the ascorbate-glutathione cycle and its related metabolites in the roots of Phaseolus vulgaris. Physiol Plant 106: 262-267

Gussarsson M (1994) Cadmium-induced alterations in nutrient composition and growth of *Betula pendula* seedlings: the significance of fine roots as a primary target for cadmium toxicity. J Plant Nutr 17: 2151-2163

Hagemeyer J (1990) Ökophysiologische Untersuchungen zur Salz- und Cadmiumresistenz von *Tamarix aphylla* (L.) Karst. (Tamaricaceae). Diss Bot 155: 1-194

Hagemeyer J, Breckle SW (1996) Growth under trace element stress. In: Waisel Y, Eshel A, Kafkafi U (eds) Plant roots: the hidden half, 2nd edn. Dekker, New York, pp 415-433

Haghiri F (1973) Cadmium uptake by plants. J Environ Qual 2: 93-96

Harper FA, Smith SE, Macnair MR (1997) Where is the cost in copper tolerance in *Mimulus guttatus*? Testing the trade-off hypothesis. Funct Ecol 11: 764-774

Hartley J, Cairney JWG, Meharg AA (1997) Do ectomycorrhizal fungi exhibit adaptive tolerance to potentially toxic metals in the environment? Plant Soil 189: 303-319

Hodgson JF, Lindsay WL, Trierweiler JF (1966) Micronutrient cation complexing in soil solution. Soil Sci Soc Am Proc 30: 723-726

Hunt R (1990) Basic growth analysis. Hyman, London

Inouhe M, Ninomiya S, Tohoyama H, Joho M, Murayama T (1994) Different characteristics of roots in the cadmium-tolerance and Cd-binding complex formation between mono- and dicotyledonous plants. J Plant Res 107: 201-207

Jones MD, Hutchinson TC (1986) The effect of mycorrhizal infection on the response of *Betula papyrifera* to nickel and copper. New Phytol 102: 429-442

Klap JM, Oude Voshaar JH, de Vries W, Erisman JW (2000) Effects of environmental stress on forest crown condition in Europe. IV. Statistical analysis of relationships. Water Air Soil Pollut 119: 387-420

Klucas RV, Hanus FJ, Russell SA (1983) Nickel. A micronutrient element for hydrogen-dependent growth of *Rhizobium japonicum* and for expression of urease activity in soybean leaves. Proc Natl Acad Sci USA 90: 2253-2257

Köhl KI (1997) Do *Armeria maritima* (Mill.) Willd. ecotypes from metalliferous soils and non-metalliferous soils differ in growth response under Zn stress? A comparison by a new artificial soil method. J Exp Bot 48: 1959-1967

Kozlov MV, Haukioja E, Bakhtiarov AV, Stroganov DN, Zimina SN (2000) Root versus canopy uptake of heavy metals by birch in an industrial polluted area: contrasting behaviour of nickel and copper. Environ Pollut 107: 413-420

Kozlowski TT (1973) Shedding of plant parts. Academic Press, New York

Kukkola E, Rautio P, Huttunen S (2000) Stress indications in copper- and nickel-exposed Scots pine seedlings. Environ Exp Bot 43: 197-210

Langheinrich U, Tischner R, Godbold DL (1992) Influence of a high Mn supply on Norway spruce (*Picea abies* (L.) Karst.) seedlings in relation to the nitrogen source. Tree Physiol 10: 259-271

Lidon FC, da Graca Barreiro M, Santos Henriques F (1995) Interactions between biomass production and ethylene biosynthesis in copper-treated rice. J Plant Nutr 18:1301-1314

Liu D, Jiang W, Guo L, Hao Y, Lu C, Zhao F (1994) Effects of nickel sulfate on root growth and nucleoli in root tip cells of *Allium cepa*. Isr J Plant Sci 42: 143-148

Macnair MR (1983) The genetic control of copper tolerance in the yellow monkey flower, *Mimulus guttatus*. Heredity 50: 283-293

Macnair MR, Smith SE, Cumbes QJ (1993) Heritability and distribution of variation in degree of copper tolerance in *Mimulus guttatus* at Copperopolis, California. Heredity 71: 445-455

Mälkönen E, Derome J, Fritze H, Helmisaari H-S, Kukkola M, Kytö M, Saarsalmi A, Salemaa M (1999) Compensatory fertilization of Scots pine stands polluted by heavy metals. Nutr Cycling Agroecosyst 55: 239-268

Marschner H (1986) Mineral nutrition of higher plants. Academic Press, London

Marschner H, Römheld V (1996) Root-induced changes in the availability of micronutrients in the rhizosphere. In: Waisel Y, Eshel A, Kafkafi U (eds) Plant roots: the hidden half, 2nd edn. Dekker, New York, pp 557-579

Maynard DN (1979) Nutritional disorders of vegetable crops. A review. J Plant Nutr 1: 1-23

Menezes de Sequeira E, Pinto da Silva AR (1992) Ecology of serpentinized areas of north-east Portugal. In: Roberts A, Proctor J (eds) The ecology of areas with serpentinized rocks. A world view. Kluwer, Dordrecht, pp 169-197

Mengel K, Kirby EA (1987) Principles of plant nutrition. International Potash Institute, Berne

Morrison RS, Brooks RR, Reeves RD (1980) Nickel uptake by *Alyssum* species. Plant Sci Lett 17: 451-457

Mortvedt JJ, Giordano PM, Lindsay NL (eds) (1972) Micronutrients in agriculture. Soil Science Society of America, Madison, WI

Moya JL, Ros R, Picazo I (1995) Heavy metal-hormone interactions in rice plants: effects on growth, net photosynthesis, and carbohydrate distribution. J Plant Growth Regul 14: 61-67

Nagy L, Proctor J (1997) Plant growth and reproduction on a toxic alpine ultramafic soil: adaptation to nutrient limitation. New Phytol 137: 267-274

Nriagu JO (1980) Production, uses and properties of cadmium. In: Nriagu JO (ed) Cadmium in the environment, part I. Wiley, New York, pp 35-70

Ouariti O, Gouia H, Ghorbal MH (1997) Responses of bean and tomato plants to cadmium: growth, mineral nutrition, and nitrate reduction. Plant Physiol Biochem 35: 347-354

Ouzounidou G (1994a) Copper-induced changes on growth, metal content and photosynthetic function of *Alyssum montanum* L. plants. Environ Exp Bot 34: 165-172

Ouzounidou G (1994b) Root growth and pigment composition in relationship to element uptake in *Silene compacta* plants treated with copper. J Plant Nutr 17: 933-943

Ouzounidou G, Symeonidis L, Babalonas D, Karataglis S (1994) Comparative responses of a copper-tolerant and a copper-sensitive population of *Minuartia hirsuta* to copper toxicity. J Plant Physiol 144: 109-115

Ouzounidou G, Moustakas M, Eleftheriou EP (1997) Physiological and ultrastructural effects of cadmium on wheat (*Triticum aestivum* L.) leaves. Arch Environ Contam Toxicol 32: 154-160

Polacco JC (1977) Nitrogen metabolism in soybean tissue culture. II. Urea utilization and urease synthesis require Ni^{2+}. Plant Physiol 59: 827-830

Prasad MNV (1995) Cadmium toxicity and tolerance in vascular plants. Environ Exp Bot 35:525-545

Rashid A, Rafique E, Bughio N, Yasin M (1997) Micronutrient deficiencies in rain fed calcareous soils of Pakistan. IV. Zinc nutrition of *Sorghum*. Commun Soil Sci Plant Anal 28: 455-467

Rengel Z (1992) Role of calcium in aluminium toxicity. New Phytol 121: 499-513

Ricken B, Höfner W (1996) Bedeutung der arbuskulären Mykorrhiza (AM) für die Schwermetall-toleranz von Luzerne (*Medicago sativa* L.) und Hafer (*Avena sativa* L.) auf einem klärschlammgedüngten Boden. Z Pflanzenernaehr Bodenkd 159: 189-194

Rigina O, Kozlov MV (2000) The impacts of air pollution on the northern taiga forests of the Kola peninsula, Russian federation. In: Innes JL, Oleksyn J (eds) Forest dynamics in heavily polluted regions. Report no 1 of the IUFRO task force on environmental change. CAB International, Wallingford, UK, pp 37-65

Roberts BA, Proctor J (eds) (1992) The ecology of areas with serpentinized rocks. A world view. Kluwer, Dordrecht

Roberts JA, Hooley R (1988) Plant growth regulators. Blackie, Glasgow

Rodenkirchen H, Roberts BA (1993) Soils and plant nutrition on a serpentinized ridge in south Germany. Z Pflanzenernaehr Bodenkd 156: 407-410

Römheld V, Marschner H (1990) Genotypical differences among graminaceous species in release of phytosiderophores and uptake of iron phytosiderophores. Plant Soil 123: 147-153

Rubio MI, Escrig I, Martinez-Cortina C, Lopez-Benet FJ, Sanz A (1994) Cadmium and nickel accumulation in rice plants. Effects of mineral nutrition and possible interactions of abscisic and gibberellic acids. Plant Growth Regul 14: 151-157

Samantaray S, Rout GR, Das P (1997) Manganese toxicity in *Echinochloa colona*: effects of divalent manganese on growth and development. Isr J Plant Sci 45: 9-12

Schaberg PG, Dehayes DH, Hawley GJ, Strimbeck GR, Cumming JR, Murakami PF, Borer CH (2000) Acid mist and soil Ca and Al alter the mineral nutrition and physiology of red spruce. Tree Physiol 20: 73-85

Schachtschabel P, Blume HP, Hartge KH, Schwertmann U (1984) Lehrbuch der Bodenkunde. Enke, Stuttgart

Sekimoto H, Hoshi M, Nomura T, Yokota T (1997) Zinc deficiency affects the levels of endogenous gibberellins in *Zea mays* L. Plant Cell Physiol 38: 1087-1090

Sibly RM, Calow P (1989) A life-cycle theory or responses to stress. Biol J Linn Soc 37: 101-116

Smolders E, Merckx R, Schoovaerts F, Vlassak K (1991) Continuous shoot growth monitoring in hydroponics. Physiol Plant 83: 83-92

Snowden RED, Wheeler BD (1993) Iron toxicity to fen plant species. J Ecol 81: 35-46

Stevenson FJ, Ardakani MS (1972) Organic matter reactions involving micronutrients in soils. In: Mortvedt JJ, Giordano PM, Lindsay NL (eds) Micronutrients in agriculture. Soil Science Society of America, Madison, WI, pp 79-114

Taylor GJ (1989) Maximum potential growth rate and allocation of respiratory energy as related to stress tolerance in plants. Plant Physiol Biochem 27: 605-611

Tolra RP, Poschenrieder C, Barcelo J (1996) Zinc hyperaccumulation in *Thlaspi caerulescens*. I. Influence on growth and mineral nutrition. J Plant Nutr 19:1531-1540

Tukendorf A, Skorzynska-Polit E, Baszynski T (1997) Homophytochelatin accumulation in Cd-treated runner bean plants is related to their growth stage. Plant Sci 129: 21-28

Utriainen MA, Kärenlampi LV, Kärenlampi SO, Schat H (1997) Differential tolerance to copper and zinc of micropropagated birches tested in hydroponics. New Phytol 137: 543-549

Van Tichelen KK, Vanstraelen T, Colpaert JV (1999) Nutrient uptake by intact mycorrhizal Pinys sylvestris seedlings: a diagnostic tool to detect copper toxicity. Tree Phys 19: 189-196

Vernon AJ, Allison JCS (1963) A method of calculating net assimilation rate. Nature 200:814

Vesk PA, Nockolds CE, Allaway WG (1999) Metal localization in water hyacinth roots from an urban wetland. Plant Cell Environ 22: 149-158

Waisel Y (1996) Aeroponics: a tool for root research. In: Waisel Y, Eshel A, Kafkafi U (eds) Plant roots: the hidden half, 2nd☐edn. Dekker, New York, pp 239-245

Wallace A, Romney EM (1977) Synergistic trace metal effects in plants. Commun Soil Sci Plant Anal 8: 699-707

Wallace A, Romney EM, Mueller RT, Alexander GV (1980) Calcium-trace metal interactions in soybean plants. J Plant Nutr 2: 79-86

Wallnöfer PR, Engelhardt G (1984) Schadstoffe, die aus dem Boden aufgenommen werden. In: Hock B, Elstner EF (eds) Pflanzentoxikologie. BI-Wissenschaftsverlag, Mannheim, pp 95-117

Watmough SA, Hutchinson TC, Evans RD (1999) The distribution of 67Zn and 207Pb applied to white spruce foliage at ambient concentrations under different pH regimes. Environ Exp Bot 41: 83-92

Welch RM (1981) The biological significance of nickel. J Plant Nutr 3: 345-356

Wilcox HE (1996) Mycorrhizae. In: Waisel Y, Eshel A, Kafkafi U (eds) Plant roots: the hidden half, 2nd edn. Dekker, New York, pp 689-721

Wilkinson DM, Dickinson NM (1995) Metal resistance in trees: the role of mycorrhizae. Oikos 72: 298-300

Wilson JB (1988) The cost of heavy-metal tolerance: an example. Evolution 42: 408-413

Winterhalder K (2000) Landscape degradation by smelter emissions near Sudbury, Canada, and subsequent amelioration and restoration. In: Innes JL, Oleksyn J (eds) Forest dynamics in heavily polluted regions. Report no.1 of the IUFRO task force on environmental change. CAB International, Wallingford, UK, pp 87-119

Yilmaz A, Ekiz H, Torun B, Gültekin I, Karanlik S, Bagci SA, Cakmak I (1997) Effect of different zinc application methods on grain yield and zinc concentration in wheat cultivars grown on zinc-deficient calcareous soils. J Plant Nutr 20: 461-471

Zenk MH (1996) Heavy metal detoxification in higher plants—a review. Gene 179: 21-30

Zornoza P, Robles S, Martin N (1999) Alleviation of nickel toxicity by ammonium supply to sunflower plants. Plant and Soil 208: 221-226

Chapter 9

Structural and Ultrastructural Changes in Heavy Metal Exposed Plants

J. Barceló, Ch. Poschenrieder

Laboratorio de Fisiología Vegetal, Facultad de Ciencias, Universidad Autónoma de Barcelona, 08193 Bellaterra, Spain

9.1 INTRODUCTION

Visible symptoms of metal toxicity stress in plants are an expression of preceding metal-induced alterations at the structural and ultrastructural level. These changes at the cell, tissue, and organ level, in turn, are either the result of a direct interaction of the toxic metals with structural components at these sites or a more indirect consequence of changes in signal transduction and/or metabolism. Investigations at the structural and ultrastructural level help to identify the sites of the primary toxicity effects and their consequences for whole plant performance.

Electron microscopy (EM), especially when combined with analytical techniques such as energy dispersive X-ray microanalysis (EDXA), laser microprobe mass analysis (LAMMA), electron energy loss spectroscopy (EELS; Lichtenberger and Neumann 1997), secondary ion mass spectrometry (SIMS), or cytochemical methods, are powerful tools yielding valuable information for further experiments on primary mechanisms of metal toxicity and tolerance at the molecular level. The recent development of atomic force microscopy allowing imaging of living cells and, especially, the fast introduction of laser scanning optical microscopy using confocal microscopy or multi-photon excitation in combination with the development of more powerful and user-friendly image analysis programs are also providing new possibilities for investigations into heavy metal effects at the cell and tissue level. It is not the aim of this chapter to exhaustively describe heavy metal effects at the structural and ultrastructural level, but to consider mainly those cases where the structural and ultrastructural approach is contributing to a better understanding of metal toxicity and tolerance mechanisms in plants. In this context, Al will also be included. Aluminum, although not being a "heavy" metal, is one of the major causes for inhibition of plant growth on acid soils in the tropics (Foy 1984). Research on Al provides a clear example for the usefulness of structural and ultrastructural investigations for the recognition of primary sites of injury, and the establishment of mechanisms of metal ion toxicity and tolerance. Although this chapter mainly concerns terrestrial higher plants, aquatic micro- and macrophytes deserve special mention, because of both the importance of these organisms in water pollution monitoring and control, and the numberous investigations at the ultrastructural level performed in these species.

9.2 STRUCTURAL AND ULTRASTRUCTURAL EFFECTS IN ALGAE

The structural simplicity of algae, in comparison with higher vascular plants, is very attractive for investigations of the ultrastructural effects of heavy metal toxicity. In consequence, a considerable number of ultrastructural effects for different metals, alone or in combination, have been described in many different algal species. However, the usefulness for the recognition of primary toxicity mechanisms and tolerance strategies of many investigations in this field is limited due to the poor characterization of metal ion speciation in the experimental solutions, in addition to the failure of relating ultrastructural disturbances to internal effect concentrations and growth under environmentally relevant experimental conditions. Moreover, a clear distinction between acute and chronic effects has only occasionally been addressed (e.g. Visviki and Rachlin 1992, 1994). Early investigations mainly described the severe, general ultrastructural damages in algal cells exposed to unrealistic high metal concentrations that caused extensive membrane damage and loss of cell compartmentation. The introduction of stereological microscopic morphometry, in combination with adequate statistical treatments of the data, however, allowed a much better evaluation of the relatively small structural and ultrastructural differences between control plants and those exposed to environmentally significant metal concentrations (Sicko-Goad and Stoermer 1979). The possibility to quantify cellular responses not only permitted more realistic experimental designs, but also led to a better identification of the metal-induced key lesions that are indicative of primary toxicity mechanisms (Rachlin et al. 1982, 1985). Table 9.1 summarizes a selection of some morphological and ultrastructural effects of heavy metals on algae that will be considered in more detail below.

9.2.1 Effects on Cell Size, Cell Shape, Cell Division, and Cell Walls

Changes in cell size or morphology are common effects of heavy metal toxicity in Cyanophyceae, and in unicellular and pluricellular eucaryotic algae. However, the kind and intensity of the effects depend on the algal species, the metal, and its concentration. In a comparative study with *Plectonema boryanum* (Cyanophyceae) on the effects of eight different metals, all at 100 ppm and 4 h exposure time, Pb and Cu increased the cell size, while the opposite was observed for Co and Ni; Mn, Zn, Hg, and Cd had no effect on cell volume (Rachlin et al. 1982). A significant reduction in cell size has been reported for *Anabaena variabilis* exposed to 20.6 μM Zn. In contrast, increased cell volume was observed in *Dunaliella tertiolecta* exposed to Hg (Davies 1976), and in *D. salina* and *D. minuta* exposed to Cd (Visviki and Rachlin 1992, 1994). Microscopy investigations suggest that metal toxicity induced increase of cell size is rather due to inhibition of cell division and alterations at the cell wall level than to osmotic changes (Rachlin et al. 1982). Triethyl lead

Table 9.1 Summary of selected effects of toxic heavy metals on morphology and ultrastructure of algae

Species	Metal	Concentration/ exposure time	Effects	Reference
Nostoc	Cd	4.89 μM/6 h	Loss of cohesiveness of the polysaccharide layer of the heterocyst envelope	Mateo et al. (1994)
Skeletonema costatum	Cd	0.45 μM/3 days	Vesiculate cytoplasm, myelin whorls broken tonoplast, dilated double membranes, broken plasma membrane	Smith (1983)
Dunaliella minuta	Cd	0.34 μM/96 h	$\downarrow$ r.v. chloroplasts (23%), $\uparrow$ r.v. starch granules (75%)	Visviki and Rachlin (1992)
D. salina	Cd	$4.5 \times 10^{-6} \times \mu$M/ 8 months	$\uparrow$ lipid no. $\uparrow$ lipid r.v.	Visviki and Rachlin (1994)
Crypthe- codinium cohnii	Cd	44.6 μM/3 days	Vacuolation, $\uparrow$ starch and lipid, lysosomes	Prevot and Soyer- Gobillard (1986)
Chara vulgaris	Cd	0.1 μM/14 days	Protuberances of corticating cells, disordered wall microfibrils, vesiculation	Heumann (1987)
Cystoseira barbata	Cd	0.89 μM/3 days	Swelling of chloroplast stroma, electron-dense material in nucleus and cell walls	Pellegrini et al. (1991)
D. minuta	Cu	7.57 μM/96 h	$\uparrow$ cell volume, $\uparrow$ lipid r.v., $\downarrow$ pyrenoid r.v.	Visviki and Rachlin (1992)
Laminaria saccharina	Cu	7.9 μM/7 days	Swelling thylakoids, vacuolation, cell wall alterations	Brinkhuis and Chung (1986)
Chlamydo- monas reinhardtii	Pb$_{(inorg)}$	5 μM/48 h	Vacuolization, swollen thylakoids, $\uparrow$ lipids, $\downarrow$ starch, disinte- gration of mitochondria and pyrenoids	Irmer et al. (1986)
Plectonema boryanum	Pb$_{(inorg)}$	0.1 μM/3 h	$\uparrow$ r.v. polyphosphate bodies; under P starvation degradation of polyphosphate and enhanced Pb toxicity, $\downarrow$ r.v. intrathylakoidal space	Sicko-Goad and Lazinski (1986)

(Contd.)

(*Contd.*)

Species	Metal	Concentration/ exposure time	Effects	Reference
Poterio-ochromonas malahamensis	$Pb_{(org)}$	5-10 μM/3-24 h	Disturbance of cytokinesis, interference with micro-tubules, inhibition of lorica formation	Röderer (1986)
Anabaena flos-aquae	Zn	20.6-115 μM/96 h	↑ thylakoid surface area, ↑ r.v. polyphosphate bodies, ↓ cyanophycin granules, crystalline inclusions	Rachlin et al. (1985)
Coccomyxa minor	Cr(VI)	19.2 μM/24 h	↑ cell volume, ↓ cell number; vacuolation chloroplast damage; electron-dense deposits	Bassi et al. (1990)
Cyclotella meneghiniana	Cr(VI)	0.2 μM/7 days	↓ chloroplast number, ↑ r.v. autophagic vacuole, ↑ r.v. polyphosphate, ↑ lipid	Lazinsky and Sicko-Goad (1990)

↓, decrease; ↑, increase; r.v., relative volume

(TriEL, 10 μM, 24 h exposure) has been found to inhibit both mitosis and cytokinesis in *Poterioochromonas malhamensis* (Röderer 1986). The metaphase was the most sensitive mitotic phase. However, cytokinesis was more effectively disturbed by TriEL than mitosis as shown by the presence of giant multinucleated algal cells. The ultrastructural effects of TriEL were similar to those observed for colchicine, and it was concluded that TriEL selectively interferes with cytoplasmic and mitotic microtubules causing inhibition of lorica formation, mitosis, and cytokinesis (Röderer 1986). Disturbance of microtubule function may also be a target for metal ion toxicity in higher plants and will be further discussed in Section 9.3.2.

Metal-induced alteration of cell wall ultrastructure has been observed in many investigations. Both Cd and $Pb_{(inorg)}$ induced disorders of cell wall microfibrils in *Chara vulgaris* (Heumann 1987). Copper caused a loose arrangement of microfibrils in the inner cell wall region of *Laminaria saccharina* (Brinkhuis and Chung 1986). Chronic exposure to low levels of Cd or Cu decreased the relative cell wall volume in *Chlamydomonas bullosa* (Visviki and Rachlin 1994), and a degradation of the peptidoglycan layer of the wall has been observed in *Anabaena flos aquae* exposed to high Cd levels (Rachlin et al. 1984; Rai et al. 1990). Metal binding to algal cell walls has frequently been described. It is doubtful that alterations to the wall ultrastructure are mainly caused by a direct toxic effect of metal ions in the walls (Heumann 1987). Certainly, an excess of metal cations may disturb the ion selectivity of polysaccharides present in large proportions in algal cell walls, intercellular spaces, or external layers. The displacement of Ca^{2+} ions by toxic metal ions would alter the mechanical properties of algal cells (Percival and McDowell 1981). Cadmium has also been found to cause a severe loss of cohesiveness of the outer polysaccharide layer of the heterocyst envelope of *Nostoc*, and this effect was substantially ameliorated by Ca (Mateo et al. 1994). Nevertheless, metal toxicity induced decrease of cell wall volume and the disorganization of microfibril orientation are probably a consequence of toxicity effects on processes involved in synthesis and/or deposition of cell wall material (Heumann 1987). This hypothesis is also supported by the finding that microtubuli and the secretory activity of the golgi apparatus are rapidly altered by toxic metal ions (Smith 1983; Heumann 1987).

The binding of metals by extracellular polysaccharides and cell walls has frequently been implied in metal resistance of algae (Gaur and Rai 2001). Also, structural alterations in cell walls have been related to metal tolerance. An increase in cell wall thickness was observed in metal-tolerant *Chaetimorpha brachygona*, a filamentous green algae that was isolated from an iron-ore tailing area (Chan and Wong 1987). This increase in cell wall diameter was not caused by swelling due to the displacement of Ca or other stabilizing cations. Enhanced synthesis of new cell wall components, in response to increased heavy metal concentrations, leads to lower metal uptake into the cells. Binding of Zn in cell walls of the Zn-tolerant brown alga *Padina gymnospora* has been claimed to prevent the accumulation of toxic concentrations of the metal in the protoplasm (Filho et al. 1996). A role of metal binding to cell walls in metal tolerance is also supported by the much lower toxicity of Cu and Co in walled strains of *Chlamydomonas reinhardtii* in comparison to wall-less strains, while no differences in effect concentrations between the strains were observed for Ni (Macfie et al. 1994). Nevertheless, the special composition of *Chlamydomonas* cell walls, which are rich in hydroxyproline and do not contain cellulose, makes an extrapolation of these results to other plant species problematic.

9.2.2 Effects on Chloroplasts, Mitochondria and Other Cell Components

Damage to chloroplasts is the most frequently observed ultrastructural effect of toxic heavy metals in algae (Table 9.1). The most common symptoms are swelling of the organelle, distortion of thylakoids leading to loss of the parallel arrangement of the thylakoid membranes, reduction or increase of the thylakoid surface area, and swelling or decrease of the interthylakoidal space (e.g. Rachlin et al. 1982, 1985; Sicko-Goad et al. 1986; Pellegrini et al. 1991; Visviki and Rachlin 1992). These effects have been interpreted as a consequence of toxic metal effects on membranes and osmotic disturbances. It is unclear to what extent these ultrastructural effects are due to direct toxicity of the metal ions in the chloroplast, metal-induced enhancement of active oxygen species, or consequences of metal-induced deficiency of essential nutrients. In *C. reinhardtii* suffering a 50% reduction of photosynthesis by exposure to 5 μM Pb, the severe chloroplast damage was accompanied by the accumulation of Pb in thylakoid membranes (Irmer et al. 1986). Lead has also been detected by LAMMA in chloroplasts of *Phymatodocis nordstedtiana* (Lorch and Schäfer 1981). In plants exposed to lower metal concentrations, chloroplast damage has also been observed, while the accumulation of the metal in the chloroplast remained below the detection limit of EDXA techniques (Sicko-Goad and Lazinsky 1986; Heumann 1987). This fact, however, does not exclude a direct interaction of the metal with chloroplast membranes and metabolism.

It has been suggested that metal-induced damage of the chloroplast fine structure is responsible for decreased algal growth, as a consequence of photosynthesis inhibition. A reduction in the relative volume and structural integrity of pyrenoids, the site of production of starch synthetase, has also frequently been observed. However, at environmentally significant concentrations, Cu had a more intense effect on the growth rate than on chloroplast structure of *Dunaliella minuta*. Cadmium reduced the size of chloroplasts and caused severe damage to the chloroplast ultrastructure at concentrations as low as 0.34 μM in *D. minuta* (Visviki and Rachlin 1992). Nevertheless, the relative volume of the starch grains increased. The authors suggest that the initial effect of Cd is an uncoupling of growth and photosynthesis which results in accumulation of photosynthetic products, followed by a decrease in chloroplast size leading to photosynthesis reduction. According to this, the algal growth seems to be more sensitive to Cu or Cd than chloroplast ultrastructure and photosynthesis.

The effect of toxic metal ions on mitochondrial ultrastructure largely depends on the algal species, the kind of metal ion and its concentration (Table 9.1). As a rule, mitochondria seem less affected than chloroplasts. Mercury at only 0.5 μM caused swelling of mitochondrial tubuli in *Chara vulgaris*, but at this concentration chloroplasts were also severely damaged. Cadmium at 0.2 μM or 5 μM Pb had no effect on mitochondria in this species (Heumann 1987). In contrast, exposure of *C. reinhardtii* to 5 μM Pb caused structural disintegration of mitochondria (Irmer et al. 1986). Cadmium and Cr have also been found to be much more toxic to chloroplasts than to mitochondria in the brown alga *Cystoseira barbata* (Pellegrini et al. 1991) and in *Cyclotella meneghiana*, respectively. The more intense damage in chloroplasts may be related to metal-induced enhancement of the formation of free radicals.

Increased vacuolization, the formation of autophagic vacuoles and vesicles seem to be common features of metal-stressed algae (e.g. Smith 1983; Prevot and Soyer-Gobillard 1986; Sicko-Goad et al. 1986; Heumann 1987; Pellegrini et al. 1991; Visviki and Rachlin 1994). At present, it is not clear if increased vacuolation and vesiculation is a toxicity symptom or, in contrast, a means for metal detoxification. The latter has been suggested by several authors in the view of the occurrence of electron-dense deposits in these vacuoles and vesicles. Unfortunately, only on a few occasions has the presence of metals actually been confirmed by analytical EM (Irmer et al. 1986; Lorch and Schäfer 1981; Nassiri et al. 1997a,b). Investigations using EELS in Cd-treated *Tetraselmis suecica* demonstrate Cd storage in the osmiophilic deposits of vesicles. A significant relationship between Cd, N and S concentrations suggests that Cd was probably bound to organic molecules via S-Cd bonds (Nassiri et al. 1997b). Metal binding in polyphosphate bodies (PPBs) of algae has also been shown by several authors (Donnan et al. 1979; Jensen et al. 1982). Cadmium, Co, Hg, Mn, Ni, Pb and Zn, but not Sr, have been detected in PPBs of different algae exposed to the metals (Jensen et al. 1984). However, the role of PPBs in metal tolerance of algae remains unclear. A Zn-induced increase of PPBs has been observed in *Anabaena variabilis*, but not in *A. flos aquae* (Rachlin et al. 1985), while Cu and Cd prevent the formation of PPBs in *Diatoma tenue* (Sicko-Goad and Stoermer 1979) and *A. flos aquae* (Rachlin et al. 1984; Rai et al. 1990), respectively. Further investigations comparing strains or varieties differing in metal tolerance will help to ascertain the importance of metal binding to PPBs in algae.

9.3 TERRESTRIAL HIGHER PLANTS

The structural complexity of vascular plants makes the investigation into primary mechanisms of metal toxicity and tolerance extremely difficult. During recent decades, many authors have investigated metal-induced alterations at the whole plant, organ, tissue, cell and subcellular level. Most of the experiments have been performed at only one time sample after relatively large exposure times. Such studies have been useful for obtaining a wide description of structural and ultrastructural alterations that are either cause or consequence of metabolic and physiological disfunctions in metal-stressed plants. If performed under environmentally relevant conditions, and especially if varieties differing in metal tolerance are compared, such an approach may be useful for the assessment of a plant's problems for competition and survival in a contaminated habitat. However, long-term investigations provide little information on the primary toxicity and tolerance mechanisms. To approach these mechanisms, exposure-time-dependent structural and ultrastructural alterations must be studied in relation to physiological alterations and metal compartmentation. We will only briefly summarize the description of effects at the structural and ultrastructural level and center our attention on some cases where an attempt has been made to timely relate these effects with physiological misfunction and metal localization.

9.3.1 Effects on Root Morphology and Structure

In contrast to submerged aquatic plants, which may absorb potentially toxic metal ions all over the organism's surface, the main entrance site for metal ions in higher land plants is the root system.

The responses of roots to heavy metals have been reviewed in depth in both herbaceous plant species and trees (Kahle 1993; Punz and Sieghardt 1993; Hagemeyer and Breckle 1996, 2002). The main morphological and structural effects caused by metal toxicity can be summarized as follows (1) decrease in root elongation, (2) root tip damage, (3) collapsing of root hairs or decrease in their number, (4) decrease in root biomass, (5) increase or decrease in lateral root formation, (6) enhancement of suberification and lignification, (7) decrease in vessel diameter, and (8) structural alterations of hypodermis and endodermis (Barceló and Poschenrieder 1990; Punz and Sieghardt 1993; Hagemeyer and Breckle 1996, 2002).

As a rule, inhibition of root elongation is the first visible effect of toxic metal concentrations. In plants exposed to environmentally relevant metal concentrations such an inhibition is measurable by conventional methods only after several hours or days of metal treatment. Monitoring of elongation of tap roots by a highly sensitive linear transducer technique or video imaging has shown that Al-induced inhibition of root elongation occurrs as soon as 30 min after start of exposure in Al-sensitive maize (Llugany et al. 1995; Kidd et al. 2001). Also root hairs respond rapidly to Al toxicity; in *Limnobium stoloniferum*, an aquatic pondweed, decreased root hair growth was detectable within 30 min (Jones et al. 1995). In soybean and peanut, root hair formation was even more sensitive to Al than root elongation (Hecht-Buchholz et al. 1990; Brady et al. 1994). Other effects listed in (4) to (8) usually are only detectable after several hours or days of metal exposure.

9.3.2 Effects on Root Ultrastructure in Relation to Inhibition of Root Elongation

Table 9.2 shows a summary of metal-induced structural and ultrastructural effects in roots. After severe metal stress, especially when caused by metal ions that induce oxidative damage by free radical production, such as Cu and Cr(VI), surface damage and collapse of root hairs and epidermal cells can be observed (Vázquez et al. 1987; Ouzounidou et al. 1992; Corradi et al. 1993). Surface damage (Vogelei and Rothe 1988) and plasmolysis of epidermal cells (Wagatsuma et al. 1987) have also been found after severe Al stress. Increased vacuolation, the occurrence of autophagic vacuoles, accumulation of electron-dense deposits in vacuoles, altered GA activity, and decreased meristem length seem to be common features of metal-stressed root tips (see Table 9.2 for references).

Table 9.2 Selected structural and ultrastructural effects of metal toxicity in roots

Species	Metal	Concentration/time	Effects	Reference
Festuca rubra (Zn sensitive)	Zn	120 μM/12 h	↓ root elongation, ↓ meristem length, ↓ area mitotic cells, ↓ no. of lateral roots	Davies et al. (1991)
Phaseolus vulgaris	Cd	0.5 μM/48 h	↓ root elongation, ↑ volume cortex cells, transfer cell-like wall ingrowth in hypodermis cells, phytoferritin-like deposits in plastids	Vázquez et al. (1992a)
Zea mays	Al[a]	840 μM/2 h	root cap: ↑ vacuolation, alteration of dictyosomes change GA vesicle content, no effect on nucleus within 20 h	Bennet et al. (1985)
Zea mays	Al[a]	2.5 μM/4 h	2 mm root tips: cell wall swelling, Al in electron-dense deposits in vacuoles	Vázquez et al. (1999)
Glycine max	Al[a]	5-12 μM/3 days	↓ root elongation, ↓ meristem length, ↑ volume cortex cells, ↑ vacuolization, electron-dense deposits in vacuoles	Hecht-Buchholz et al. (1990)
Danthonia linkii	Al[a]	17.2 μM/24 h	root cap: ↑ cell size, ↓ dictyosomes, ↓ amyloplasts	Crawford and Wilkens (1997)

(Contd.)

Table 9.2 (*Contd.*)

Species	Metal	Concentration/time	Effects	Reference
Pinus strobus	Al	0.5-3.7 mM pH 3.6/12 weeks	↑ vacuolation of meristem cells; electron-dense deposits in vacuoles of NM roots, not observed in ECT roots; disruption of ECT fungal cytoplasm	Schier and McQuattie (1995)
Tynopyrum bessarabium	Al[b]	1 mM/12-48 h	↓ root cap and meristem area, ↓ amyloplasts change microtubule disposition, distortion of cell walls	Eleftheriou et al. (1993)
Zea mays	Cu	80 μM/14 days	↓ root elongation, damage to epidermis, ↑ vacuolation, ↓ area nucleoli, ↑ lipid bodies	Ouzounidou et al. (1995)
Thlaspi ochroleucum	Cu	80 μM/15 days	↓ root cap length, plasmolisis, damaged nuclear structure	Ouzounidou et al. (1992)
Triticum aestivum	Cu	3.9-31 μM/14 days	↓ root elongation, ↓ mitotic index	Eleftheriou and Karataglis (1989)
Phaseolus vulgaris	Cr(VI)	96 μM/21 days	↓ root elongation, surface damage, plasmolysis of epidermal cells, ameboid plastids, electron-dense deposits in vacuoles	Vázquez et al. (1987)
Salvia sclarea	CrVI	17-34 μM/48 h	↓ root elongation, damage in root cap, inhibition of lateral root development, collapsed root hairs	Corradi et al. (1993)
Allium cepa	Pb(inorg)	72 μM/24 h	Pb deposits in autolytic vacuoles, in dictyosomal vesicles, and in cell walls, damaged mitochondria	Wierzbicka (1987)
Zea mays	Pb(inorg)	5 mM/1-2 h	Pb in coated pits, on GA vesicles, and in dictyosomes of cap cells; endocytosis	Hübner et al. (1985)
Phaseolus vulgaris	[99]Tc	1 μM/6 days	↓ root elongation, swelling of root tips, autophagic vacuoles, ↑ protein bodies in vacuoles	Vázquez et al. (1990)

↓, decrease; ↑, increase; *ECT*, ectomycorrhizal; *NM* non-mycorrhizal; *GA* Golgi apparatus
[a]Al^{3+} activity.
[b]Al in the form of aluminate $[Al(OH)_4^-]$.

Unfortunately, most of the EM observations have been performed after several days of metal exposure, so that the initial ultrastructural alterations that may specifically under-lay the rapidly occurring inhibition of root elongation and of root hair formation were masked by severe cell damages. Root elongation can be reduced by either or both inhibition of root cell division and decreased cell expansion in the elongation zone. Cell division in root apical meristems is highly sensitive to toxic metal ions, a fact that is used in the *Allium cepa* test for environmental risk assessment (Fiskesjö 1997). There are many literature reports on metal-induced alterations of nucleus and nucleoli morphology, anomalous mitoses, decreased mitotic index, and alterations of the cell cycle (e.g. Clarkson 1969; Eleftheriou and Karataglis 1989; Fiskesjö 1990; Corradi et al. 1991; Gabara et al. 1995; Liu et al. 1995; Doncheva et al. 1996; Doncheva 1997). However, as pointed out by Ernst (1998), metal-induced changes in the morphology of the nucleus may not be of primary importance in ecotoxicological studies, yet the fixation techniques used to investigate these morphological parameters may interfere with the metal exposure. Moreover, the relationship between the metal-induced inhibition of root cell division and root elongation after short exposure times has seldom been investigated. A highly significant correlation between root elongation and the mitotic index in onion exposed to different Pb concentrations for 48 h has been observed (Liu et al. 1994). Viability staining of meristem cells of root tips of *Silene armeria* ecotypes differing in Cu tolerance was in good agreement with Cu effects on root elongation (Llugany et al. 2003). However, to date, a clear relationship in time between the onset of the inhibition of root elongation and that of root cell division has not been established.

In Al-stressed plants, there has been an ardent discussion about what comes first, inhibition of root cell elongation or inhibition of root cell division. In Al-stressed *Zea mays*, Al-induced alterations in root cap cells were observed before root elongation was inhibited (Johnson and Bennet 1991). The authors argued that the primary effect of Al would occur in the cap cells and that inhibition of elongation might be due to a cap-derived hormonal signal. Several investigations had shown that Al enters root tip cells only slowly (Wagatsuma et al. 1987), and that Al mainly accumulates in cell walls (Marienfeld and Stelzer 1993; Marienfeld et al. 1995). This, in addition to the observations of rapidly occuring alterations in cell walls, was the main argument for supporting the hypothesis that apoplastic Al can cause toxicity and root growth inhibition. However, studies with SIMS (Lazof et al. 1994, 1996), EDXA (Barceló et al. 1996; Vázquez et al. 1999), or Al-uptake experiments (refs. in Rengel and Reid 1997) have shown that Al may rapidly enter the symplasm (Lazof and Goldsmith 1997). Lead also penetrates meristem cells of onion roots within 3 h or less (Wierzbicka 1987). A rapid entry of Al or other metals into the symplasm, however, neither precludes primary toxicity effects of metals in the apoplast, nor necessarily implies that the inhibition of root growth is caused by a direct interaction of the toxic metal species with the nucleus of meristem cells. There are many other potential targets for primary toxicity effects of metals other than nuclei.

An attractive hypothesis is that many of the rapid metal effects on cell walls and cell shape can be caused by a direct or indirect interaction of the metals with the cytoskeleton. Indirect support for this hypothesis derives from observations in algae (see Sect. 9.2.1) and pollen tubes. Investigations with tobacco pollen tubes, where orientated tube growth is dependent on the presence of microtubuli, report $Pb_{(org)}$-induced alteration of microtubuli

(Kandasamy and Kristen 1989). In pollen tubes of lily, short-term exposure (60-180 min) to different metal ions caused a significant decrease in growth, and severe alterations to cell walls. Except for Cd, no visible ultrastructural damage in cell organells was observed, but there was a disorganization of the polar zonation of cell organelles (Sawidis and Reiss 1995). Low Cd or Al concentrations decrease the length, but increase the volume of root cortical cells (Vázquez et al. 1992a; Gunsé et al. 1997). Such alterations of cell shape seem to be caused by a change in microfibril orientation, a process that is governed by cortical microtubules (Barlow and Parker 1996). Aluminum-induced swelling of root tip hairs is, apparently, identical to the swelling observed in root hairs exposed to chemical agents that disrupt microtubule structure, suggesting that Al-mediated microtubule depolymerization may be involved in growth inhibition (Jones et al. 1995). Investigations in *Zea mays* provide direct evidence for Al-induced alterations of the cytoskeleton in root tip cells after a few minutes of Al exposure (Sivaguru et al. 1999).

Certainly, more short-term investigations that timely relate ultrastructural effects of toxic metals to metal compartmentation and growth must be performed before the mechanisms of metal-induced inhibition of root elongation can be clearly established. Furthermore, recent developments in our knowledge on the basic mechanisms of cell division in relation to cell elongation (Jacobs 1997) are opening new perspectives for the investigation of primary metal toxicity mechanisms in roots.

9.3.3 Effects of Metals on the Vascular System

There are relatively few investigations at the structural and ultrastructural level that describe the influence of metal toxicity on the vascular (xylem and phloem) transport system. Toxic concentrations of Cd, Zn, Al and Cr have been found to decrease the diameter of xylem vessels (Barceló and Poschenrieder 1990). In contrast, a Pb-induced increase in the number of protoxylem vessels was observed in young maize roots (Hof 1984). Under severe Cd toxicity, the vessel diameter may be further decreased by the accumulation of amorphous depositions in the conduits, which have been tentatively identified as lignin-like insoluble phenolics (Fuhrer 1982; Barceló et al. 1988). Recently, Cd-induced production of H_2O_2 in vascular tissue of pea leaves as visualized by means of histochemical techniques has been related to enhanced lignification (Romero-Puertas 2002). Cadmium-induced enhancement of lignification has also been observed in Cd-treated roots of scots pine. This increased lignification was confined to newly formed protoxylem vessels near the root tips and has been interpreted as a consequence of a Cd-induced inbalance in the root redox system causing accumulation of H_2O_2 which triggers xylogenesis (Schützendübel et al. 2001).

Cadmium caused a substantial increase in crystals, probably of calcium oxalate, in paratracheary cells of beans. This, in addition to the Cd-induced decrease in the Ca concentration in the xylem sap, has been interpreted as a Cd-induced inhibition of Ca transport in the xylem vessels (Barceló et al. 1988). Exposure for 48 h to extremely high Zn concentrations causes senescence in xylem and phloem cells of primary leaves of beans. Plasmodesmata of both xylem and phloem parenchyma cells are plugged by electron-opaque material (Robb 1981). Autoradiographic investigations with bean leaf disks have shown that decreased phloem transport due to excess of Co, Ni or Zn may be caused by a metal-induced inhibition of phloem loading, leading to starch accumulation in leaves

(Rauser and Samarakoon 1980). That Zn-induced inhibition of phloem transport is caused at the site of phloem loading, or by decreased phloem mass flow, has recently been confirmed in intact wheat plants exposed to much lower Zn concentrations (Herren and Feller 1997). In contrast, in plants exposed to Cr(VI) toxicity, starch accumulation occurred in pith parenchyma cells of the upper part of the root and the lower part of the stems suggesting that Cr inhibits rather the use of sucrose than phloem loading (Vázquez et al. 1987). Recent investigations have shown that low Cd concentrations that do not affect plant growth inhibit the systemic spread of the turnip vein clearing virus in tobacco plants. This has been interpreted as a Cd-induced inhibition of plasmodesmatal transport of viral particles from vascular to non-vascular tissues (Ghoshroy et al. 1998).

9.3.4 Effects on Photosynthetic Tissues

Inhibition of leaf expansion growth, deformation of leaf tissues, increased number of glandular and/or nonglandular hairs, and chlorosis are common symptoms of metal-stressed plants. These visible foliar effects have induced many investigations into the ultrastructural effects of metal toxicity on chloroplasts. Chlorotic leaves of plants exposed to high metal concentrations for several days have mainly been analyzed. The usual symptoms of chloroplasts in such leaves can be summarized as follows: decreased number and swelling of chloroplasts, reduced area of thylakoid membranes, distortion and/or swelling of thylakoid membranes, reduced number of grana stacks, increase or decrease of starch, and increase of plastoglobuli (Rauser 1978; Rufner and Baker 1984; Vázquez et al. 1987, 1990; Barceló et al. 1988; Eleftheriou and Karataglis 1989; Ghoshroy and Nadakavuraken 1990; Bennàssar et al. 1991; Ouzounidou et al. 1992, 1997; Rascio et al. 1993; Moustakas et al. 1996, 1997; Ciscato et al. 1997; Doncheva et al. 2001; Panou-Filotheou et al. 2001; Sanità di Toppi et al. 2002). Swelling and distortion of thylakoid membranes, the accumulation of plastoglobuli and the progressive disintegration of chloroplasts indicate a metal-induced enhancement of chloroplast senescence. The poor development of thylakoid membranes and grana, and the observation that chloroplast ultrastructure is usually more affected in young than in mature leaves, however, suggests that metal toxicity also has a severe inhibitory effect on chloroplast development.

In Cd- or Cu-stressed plants, chloroplasts exhibited small, regularly spotted bodies that have been described as "pseudo-crystalline bodies" (Ciscato et al. 1997) or "microtubule-like" structures (Ouzounidou et al. 1997). These structures were quite similar to the prothylakoid bodies that were observed in immature plastids of unstressed plants (Wellburn 1984). The light-stimulated development of etioplasts to chloroplasts is severly inhibited by high Cd concentrations (Wrischer and Kunst 1981; Ghoshroy and Nadakavukaren 1990). The occurrence of ameboid plastids in Cr(VI)-treated roots (Vázquez et al. 1987) and in leaves of ^{99}Tc-treated plants (Vázquez et al. 1990) suggests that these metals also may inhibit normal plastid development.

Direct interaction of toxic metals with chloroplast components cannot be excluded. The toxic effects of Ni on chloroplast ultrastructure in cabbage, for example, were in good agreement with the Ni accumulation inside chloroplasts (Molas 2002). However, in intact higher plants, the ultrastructural alterations in chloroplasts frequently seem to be caused by more indirect effects. This is supported by the observation that poorly mobile heavy metals such as Al and Cr can induce chlorosis without a substantial increase in the metal concentration in leaves (Foy 1984; Barceló and Poschenrieder 1997). When exposed to 40 µM Cu, leaves from a Cu-sensitive *Silene cucubalus* showed lower chlorophyll concentrations than those of a tolerant population. However, plants from both populations did not show differences in the Cu concentration of the chloroplasts (Lolkema and Vooijs 1986). Indirect injury mechanisms can be based, among other possibilities, on metal-induced deficiency of Fe, Mn, or other essential mineral nutrients, increased production of free radicals, and metal-induced alteration of the hormone balance leading to enhanced senescence. Biochemical and ultrastructural investigations in Cd-exposed pea leaves, for example, have shown that Cd induces senescence symptoms in peroxisomes and probably a metabolic transition of leaf peroxisomes into glyoxisomes (McCarthy et al. 2001).

9.4   OXIDATIVE BURST AND CELL DEATH

Induction of free radicals (FRs) and reactive oxygen species (ROS) by heavy metals is well documented (Dietz et al. 1999). Metal-induced formation of FRs and ROS can account for many of the deleterious effects of metals on membrane integrity and organelle ultrastructure. In particular, effects in chloroplasts, as described above, and peroxisome proliferation (del Río et al. 2002) inside cells and cell wall reactions such as cell wall thickenings, inlays of polyphenolics and wart-like protrusions in the apoplast (Günthardt-Goerg and Vollenweider 2002) are indicative of metal-induced oxidative stress. This metal-induced oxidative burst may exhibit similarities to the hypersensitive reaction (HR) as a defence against phytopathogens. However, recent investigations on the subcellular localization of Cd-induced H_2O_2, $O_2{}^{\cdot-}$, and $NO^{\cdot}$ in combination with biochemical studies revealed differences between metal-induced ROS production and the biotic stress response. While in the HR H_2O_2 mainly accumulates in cell walls or at the external site of the plasma membrane, in Cd-treated pea leaves H_2O_2 is induced at the internal site of the plasmamembrane. Moreover, in the HR, ROS and $NO^{\cdot}$ are implied in programmed cell death (apoptosis), while Cd did not induce apoptosis in pea leaves (Romero-Puertas 2002).

In contrast, histochemical studies in roots of Al-tolerant wheat found fast, localized cell death after only 8 h exposure while overproduction of H_2O_2 was not induced until 24 h exposure. Such early cell death was not observed in the Al-sensitive variety and the authors suggest that accelerated epidermal cell turnover by means of controlled cell death may represent a detoxification mechanism helping to protect deeper cell layers of the meristematic and elongation zone (Delisle et al. 2001). In a similar way, enhanced border cell production has been implied in Al resistance in bean (Miyasaka and Hawes 2001), and in Cu resistance in *Silene armeria* (Llugany et al. 2003).

9.5   METAL LOCALIZATION, METAL SPECIATION, AND METAL TOLERANCE

The ability to bind heavy metals in a non-toxic form and to sequester them in organs or subcellular compartments with few or no sensitive metabolic activity is considered a key factor in metal tolerance (Ernst 1992; Wang and Evangelou 1995). Conventional cytochemical methods, autoradiography, fluorescence microscopy, EDXA, LAMMA, EELS, and SIMS, can provide highly useful information about the location and speciation of metals. Analytical microscopy techniques that allow the detection of metals at the structural and ultrastructural level are now being widely used to investigate metal toxicity and tolerance mechanisms. It is beyond the scope of this chapter to exhaustively discuss the advantages and limitations of these methods. However, it must be emphasized that sample preparation is critical for ensuring reliable results (Mullins et al. 1985; van Steveninck and van Steveninck 1991; Lazof and Goldsmith 1997).

9.5.1 Wetland Plants

Wetland plants have centered recent research interest mainly because of their usefulness in wastewater cleaning by rhizofiltration technologies. Structural analysis of roots of wetland plants reveals the ubiquitous presence of iron plaque on the roots growing in hypoxic soils. Sequestration of different heavy metals in the iron plaque has been visualized by fluorescence microtomography and two-dimensional microprobe imaging of *Phalaris arundinaceae* roots (Figs 9.1 and 9.2; Hansel et al. 2001). The formation of an iron plaque under hipoxic conditions retains metals on the root surface and seems to contribute to the constitutively high metal tolerance of wetland plants. In contrast, in a free floating aquatic plant, the water hyacinth *Eichhornia crassipes*, only Fe, but not Cu, Zn or Pb, was retained on the root surface (Vesk et al. 1999).

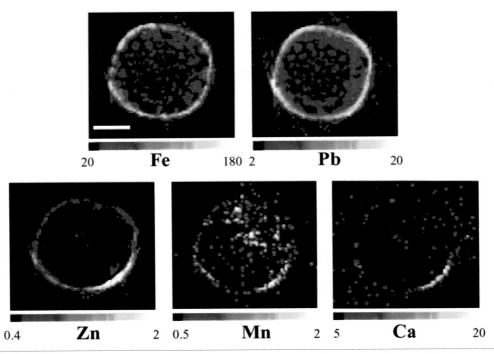

Fig. 9.1 Metal distribution on and within roots of *Phalaris arundinacea* as depicted by fluorescence microtomography. The spatial distribution of Fe, Pb, Zn, Mn and Ca within a cross-section of the root is shown. Bar units are femtograms/μm^3. (Reproduced with permission from Hansel et al. 2001).

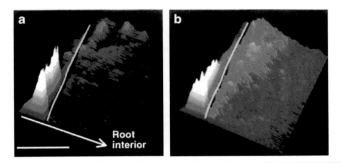

Fig. 9.2 Two-dimensional X-ray microprobe image of **a** Mn and **b** Zn distribution along a transect from the root center to the exterior of the iron plaque of *Phalaris arundinacea*. The *line* represents the root epidermis. Relative element concentrations increase with brightness. (Reproduced with permission from Hansel et al. 2001)

9.5.2 Mycorrhiza

The possible role of mycorrhizal fungi in the protection of plants against heavy metal toxicity has stimulated several microanalytical studies on the distribution of heavy metals in mycorrhizal roots. In *Picea abies* colonized with the fungus *Hebeloma crustuliniforme*, Cd was mainly detected in the Hartig net, Ni in the Hartig net and the cortex cell walls, and Zn in the Hartig net, cortex cell walls and the fungal mantle (Brunner and Frey 2000). Mycorrhizal fungi are essential for orchid growth. In orchids growing on metal-contaminated soil, mycorrhiza seem to play an important role in biofiltering Pb and other heavy metals. Elemental maps from proton-induced X-ray emission revealed that Pb and Zn contents in fungal coils of mycorrhizal roots of an orchid growing on mine tailings was up to five times that of epidermal cells. Fungal cell walls seemed to be the main storage compartment (Jurkiewicz et al. 2001).

9.5.3 Non-metallophytes

9.5.3.1 Lead

The localization of Pb in plant roots has been extensively investigated. Lead can be detected cytochemically by rhodizonate staining. Moreover, the fact that Pb forms electron-dense deposits facilitates localization by X-ray microanalysis. Early investigations with algae (Silverberg 1975), moss (Gullvag et al. 1974), tissue cultures (Ksiazek et al. 1984), and roots from higher land plants (Malone et al. 1974; Lane and Martin 1980, 1982) have shown that Pb mainly accumulates in cell walls. Time-dependent investigations using ^{210}Pb and autoradiographic detection showed that Pb rapidly accumulates on the root surface and in the radial and tangential walls of the root cap of onions. The largest amount of Pb accumulated in ground meristematic and cortex tissues. As soon as 3 h after exposure, Pb was localized in vacuoles of the ground meristematic tissue. In contrast, the central zone of the root tip, comprising the protomeristem, the initial center and the central part of the root cap, remained almost free of Pb (Wierzbicka 1987). The author suggests that there are two major barriers for apoplastic Pb transport in roots: the layers of protoderm and hypodermic meristematic cells in the root meristem zone and the layer of endodermis in the mature root zone. The central zone seems to be a barrier for both apoplastic and symplastic Pb transport (Wierzbicka 1987). Recent investigations with Pb-exposed corn roots have shown that Pb depositions in the protoplast of meristematic cells were significantly less than those in cell walls. The authors suggest that Pb may only enter the protoplast in the primary stage of cell wall development. In cells with mature cell walls, Pb would only enter the protoplast after severe damage to the plasma membrane (Tung and Temple 1996). Several localization studies have demonstrated the presence of Pb in vesicles inside plant cells, and it has been speculated that the Pb that enters the protoplasts may be exported to cell walls by exocytosis (Malone et al. 1974; Wozny et al. 1982). Investigations using EDXA localization of Pb in *Anthoxanthum odoratum* clones that differ in Pb tolerance strongly support a role for cell walls in Pb tolerance (Qureshi et al. 1986). At present, however, it is not clear if the more effective exclusion of Pb from the protoplast in tolerant plants is due to differences in the proper cell wall or to either or both more effective Pb extrusion and Pb chelation.

Only a few authors have addressed Pb accumulation in upper plant parts. Recently, based on EXAFS spectra, cerussite ($PbCO_3$) was proposed as the main Pb form in bean leaves (Sarret et al. 2001).

9.5.3.2 Aluminum

Immobilization of Al in cell walls has also been claimed to play an important role in Al tolerance. Several EDXA studies on plants exposed to high Al concentrations found that Al mainly accumulated in the apoplast, while inside the cells Al was either not detectable or only found after large exposure times (Delhaize et al. 1993; Marienfeld et al. 1995). However, this preferential accumulation of Al in cell walls, at least in part, may be due to the formation of insoluble Al species in the nutrient solutions. The high Al signal in walls may hamper the detection of the Al that has entered the symplasm quite rapidly (Lazof and Goldsmith 1997; see Sect. 9.3.2). Insoluble Al-P deposits were detected by EDXA in root tip cell walls of an Al-tolerant maize variety after only 4 h exposure to Al, when root elongation was inhibited, but not after 24 h when the root elongation rate had recovered (Vázquez et al. 1999). This observation does not support a role for cell walls in varietal differences in Al tolerance of maize, but is in line with the hypothesis that root tip exuda-tion of Al chelators such as organic acids or flavonoid phenolics by root tips may play a major role in Al detoxification in this species (Barceló and Poschenrieder 2002). Com-puter-assisted imaging of elemental maps has been used to visualize the distribution patterns of Al in relation to other elements at the tissue and cell level. As a rule, a good coincidence between the distribution of Al and P, and Al and Si, was observed (Hodson and Wilkins 1991). While the formation of Al-P complexes in root cell walls may be considered rather a sign of Al toxicity than one of tolerance (Ownby 1993), the formation of Al-Si complexes either in the apoplast or inside plant vacuoles may clearly contribute to amelioration of Al toxicity (Corrales et al. 1997).

9.5.3.3 Cadmium

Conflicting results on the subcellular compartmentation of Cd have been reported. Inves-tigations on plants with high Cd supply found that Cd mainly accumulates in cell walls

(Lindsey and Lineberger 1981; Khan et al. 1984) while plants treated with low Cd concentrations seem to accumulate Cd mainly in vacuoles and nuclei (Rauser and Ackerley 1987; Vázquez et al. 1992a). In *Lemna minor* fronds Cd has been found in sheet-like deposits associated with sulfur (van Steveninck et al. 1993). Investigations using EELS in suspension-cultured cells of tomato (Lichtenberger and Neumann 1997) or X-ray absorption spectroscopy in *Brassica juncea* seedlings (Salt et al. 1997) have shown that Cd in these systems, at least in part, is bound to phytochelatins.

9.5.4 Hyperaccumulating and Non-accumulating Metallophytes

Metallophytes, especially those that hyperaccumulate large metal concentrations in leaves, are now rapidly becoming a preferential object for analytical microscopy investigations on metal localization and speciation, in relation to tolerance. The data available, at present, show that metal compartmentation mechanisms may differ according to the metal, the plant species, the nutrient availability and the infection with mycorrhiza. In species that are able to accumulate and tolerate high Al concentrations in leaves, histochemical localization by the aluminon method revealed the presence of Al in cell walls, inside epidermal and stomatal guard cells, as well as in the phloem and xylem (Haridasan et al. 1986).

The first EDXA studies in *Thlaspi caerulescens* roots indicate that Cd mainly accumulates in cell walls and, to a lesser extent, in vacuoles, while the opposite was found for Zn (Vázquez et al. 1992b). In leaves, the highest Zn concentrations were found in vacuolar crystals of epidermal and subepidermal cells (Vázquez et al. 1994). Both the high Zn/P element ratios found in the crystals and the absence of Mg indicate that phytate is not the main storage form of Zn in *T. caerulescens*. In contrast, vacuolar phytate globules have been proposed as the storage form for Zn in non-hyperaccumulator species (van Steveninck et al. 1987). More recently, quantitative EDXMA microanalysis in ultrathin cryosections of soil-grown *T. caerulescens* revealed that Zn in the epidermal cells was not associated with crystals but evenly distributed in the vacuoles and that lower Zn concentrations accumulated in the cells of the stomatal complex. Cell walls also accumulated considerable Zn concentrations (Frey et al. 2000). Data from X-ray absorption spectroscopy indicate that citrate is implied in the storage of Zn in shoots of *T. caerulescens*, while binding to histidine and oxalate is less important (Salt et al. 1999). In heavy metal tolerant *Viola calaminaria* and *T. caerulescens* grown on Zn-contaminated soils, recent investigations have detected electron transluscent deposits in intercellular spaces, and between the cell wall and the plasmalemma; vacuolar crystals and deposits on the tonoplast were also often detected. X-ray microanalysis and EELS showed that the main components of the deposits were silicon and tin. Zn and Cl were also present in these aggregates. The authors suggest a connection between the high Si uptake ability and heavy metal tolerance in these species (Neumann et al. 1997a). In Zn-tolerant *Minuartia verna*, Zn, probably in the form of Zn silicate, has been found tightly bound to cells walls (Neumann et al. 1997b), while, in *Cardaminopsis halleri*, Zn silicate is transiently accumulated in the cytoplasm, being slowly degraded to SiO_2. Zinc is translocated into the vacuole where it is stored in an as yet unidentified form. Zinc and silicate can also be directly transported to the vacuoles by pinocytotic vesicles (Neumann and zur Nieden 2001).

Copper in tolerant *Armeria maritima* spp. *halleri* accumulated in idioblasts, where it seems to be chelated by hydroxyl groups of phenolic compounds (Lichtenberger and

Neumann 1997). Nickel in the hyperaccumulator *T. montanum* has been found to preferentially accumulate in subsidiary cells that surround stomatal guard cells (Heath et al. 1997). Preferential accumulation of Ni in the epidermal cells, most likely in the vacuoles, was also detected in Ni hyperaccumulators *Alyssum lesbiacum*, *A. bertolonii* and *Thlaspi goesingense* (Küpper et al. 2001). While preferential accumulation of Ni in trichomes of *A. lesbiacum* has been reported (Krämer et al. 1997), this was not the case in field samples of *A. bertolonii* (Marmiroli et al. 2002). In seeds of the hyperaccumulator *Thlaspi pindicum*, preferential Ni accumulation in the micropylar area has been observed. As this area is the preferred point of entry of frugivores, the authors suggest that Ni may act as a chemical defense in this area of lowest mechanical strength (Psaras and Manetas 2001).

These examples of the rapid advances in metal localization and speciation by analytical electron microscopy forecast interesting developments in our understanding of metal tolerance mechanisms in the near future.

9.6 CONCLUSIONS

In recent years, a very positive change in the focus of investigations into the structural and ultrastructural effects of heavy metals in plants has been observed. After a period—that was unavoidable, and, in fact, necessary for further advances—of experiments that mainly described the degeneration effects in severely intoxicated plants, short-term, time-dependent studies combining ultrastructural observations with functional approaches have been shown to be extremely useful for the identification of primary toxicity mechanisms. The development and availability of analytical microscopy techniques and their application to plant species with differential metal tolerance are now rapidly increasing our knowledge on the role of metal compartmentation and metal speciation in tolerance. In this context, future developments in sample preparation methods are required for achieving good metal immobilization in combination with better visual resolution of ultrastructural features. Moreover, technical improvements are necessary for a more reliable quantification of low concentrations of soluble metal species. At present, analytical microscopy is very useful for identifying sites and storage forms of metals at the ultrastructural level. However, the metal deposits, especially those in vacuoles, must be considered the consequence of tolerance mechanisms that probably are based on specific differences in the quantity and distribution of metal transport proteins. The analysis of preferential storage sites of metals indicates the most probable sites of the expression of genes encoding such proteins, and provides useful information on further investigations on the molecular mechanisms of metal tolerance. Only when the products of the tolerance genes are characterized, will the production of specific antibodies open the possibility for the application of immuno-techniques to further EM investigations of metal tolerance.

ACKNOWLEDGEMENTS

The work in the author's laboratory is supported by the European Community (ICA4-CT-2000-30017) and the Spanish Government (DGICYTBFI2001-2475-CO2-01).

REFERENCES

Barceló J, Poschenrieder CH (1990) Plant water relations as affected by heavy metal stress: a review. J Plant Nutr 13: 1-37

Barceló J, Vázquez MD, Poschenrieder CH (1988) Cadmium-induced structural and ultrastructural changes in the vascular system of bush bean stems. Bot Acta 101: 254-261

Barceló J, Poschenrieder CH (1997) Chromium in plants. In: Canali S, Tittarelli F, Sequi P (eds) Chromium environmental issues. Franco Angeli Publ, Milano, pp 101-129

Barceló J, Poschenrieder C (2002) Fast root growth responses, root exudates, and internal detoxifications as clues to the mechanisms of aluminium resistance and tolerance: a review. Environ Exp Bot 48: 75-92

Barceló J, Poschenrieder C, Vázquez MD, Gunsé J (1996) Aluminium phytotoxicity. A challenge for plant scientists. Fert Res 43: 217-223

Barlow PW, Parker JS (1996) Microtubular cytoskeleton and root morphogenesis. Plant Soil 187: 23-36

Bassi M, Corradi MG, Favali MA (1990) Effects of chromium in freshwater algae and macrophytes. In: Wang W, Gorsuch JW, Lower WR (eds) Plants for toxicity assessment, ASTM STP 1091. American Society for Testing and Materials, Philadelphia, pp 204-224

Bennàssar A, Vázquez MD, Cabot C, Poschenrieder C, Barceló J (1991) Technetium-99 toxicity in *Phaseolus vulgaris:* ultrastructural evidence for metabolic disorders. Water Air Soil Pollut 57/58: 681-689

Bennet RJ, Breen CM, Fey MV (1985) The primary site of aluminium injury in the root of *Zea mays* L. S Afr J Plant Soil 2: 8-17

Brady DJ, Edwards DG, Asher CJ (1994) Effects of aluminium on the peanut (*Arachis hypogaea* L.)/*Bradyrhizobium* symbiosis. Plant Soil 159: 265-276

Brinkhuis BH, Chung IK (1986) The effects of copper on the fine structure of the kelp *Laminaria saccharina* (L.) Lamour. Mar Environ Res 19: 205-223

Brunner I, Frey B (2000) Detection and localization of aluminium and heavy metals in ectomycorrhizal Norway spruce seedlings. Environ Pollut 108: 121-128

Chan K, Wong SLL (1987) Ultrastructural changes of *Chaetomorpha brachygona* growing in metal environment. Cytologia 52: 97-105

Ciscato M, Valcke R, van Loven K, Clijsters H, Navari-Izzo F (1997) Effects of in vivo copper treatment on the photosynthetic apparatus of two *Triticum durum* cultivars with different stress sensitivity. Physiol Plant 100: 901-908

Clarkson DT (1969) Metabolic aspects of aluminium toxicity and some possible mechanisms for resistance. In: Rorison IH (ed) Ecological aspects of mineral nutrition of plants. Blackwell, Oxford, pp 381-397

Corradi MG, Levi M, Musetti R, Favali MA (1991) The effect of Cr(VI) on different inbred lines of *Zea mays*. I. Nuclei and cell cycle in the root tip tissue. Protoplasma 162: 12-19

Corradi MG, Bianchi A, Albasini A (1993) Chromium toxicity in *Salvia sclarea*. I. Effects of hexavalent chromium on seed germination and seedling development. Environ Exp Bot 33: 405-413

Corrales I, Poschenrieder CH, Barceló J (1997) Influence of silicon pretreatment on aluminium toxicity in maize roots. Plant Soil 190: 203-209

Crawford SA, Wilkens S (1997) Ultrastructural changes in root cap cells of two Australian native species following exposure to aluminium. Aust J Plant Physiol 24: 165-174

Davies AG (1976) An assessment of the basis of mercury tolerance in *Dunaliella tertiolecta*. J Mar Biol Assoc UK 56: 39-57

Davies MS, Francis D, Thomas JD (1991) Rapidity of cellular changes induced by zinc in a zinc tolerant and non-tolerant cultivar of *Festuca rubra* L. New Phytol 117: 103-108

Del Río LA, Corpas FJ, Sandalio LM, Palma JM, Gómez M, Barroso JB (2002) Reactive oxygen species, antioxidant systems and nitric oxide in peroxisomes. J Exp Bot 53: 1255-1272

Delhaize E, Craig S, Beaton CD, Bennet RJ, Jagadish VC, Randall PJ (1993) Aluminum tolerance in wheat (*Triticum aestivum* L). I. Uptake and distribution of aluminum in root apices. Plant Physiol 103: 685-693

Delisle G, Champoux M, Houde M (2001) Characterization of oxalate oxidase and cell death in Al-sensitive and -tolerant wheat roots. Plant Cell Physiol 42: 324-333

Dietz K-J, Baier M, Krämer U (1999) Free radicals and reactive oxygen species as mediators of heavy metal toxicity in plants. In: Prasad MNV, Hagemeyer J (eds) Heavy metal stress in plants. From molecules to ecosystems. Springer, Berlin Heidelberg New York, pp 73-97

Doncheva S (1997) Ultrastructural localization of Ag-NOR proteins in root meristem cells after copper treatment. J Plant Physiol 151: 242-245

Doncheva S, Nikolov B, Ogneva V (1996) Effect of copper excess on the morphology of the nucleus in maize root meristem cells. Physiol Plant 96: 118-122

Doncheva S, Stoynova Z, Velikova V (2001) Influence of succinate on zinc toxicity of pea plants. J Plant Nutr 24: 789-804

Donnan BB, Crang RE, Jensen TE, Baxter M (1979) In situ X-ray energy dispersive microanalysis of polyphosphate bodies in *Aureobasidium pullulans*. J Ultrastruct Res 69: 232-238

Eleftheriou EP, Karataglis S (1989) Ultrastructural and morphological characteristics of cultivated wheat growing on copper-polluted field. Bot Acta 102: 134-140

Eleftheriou EP, Moustakas M, Fragiscos N (1993) Aluminate-induced changes in morphology and ultrastructure of *Tynopyrum roots*. J Exp Bot 44: 427-436

Ernst WHO (1992) Metal tolerance in plants. Acta Bot Neerl 41: 229-248

Ernst WHO (1998) Effects of heavy metals in plants at the cellular and organismic level. In: Schüürmann G, Markert B (eds) Ecotoxicology. Wiley/Spektrum Akademischer Verlag, New York/Heidelberg, pp 587-620

Filho GMA, Karez CS, Pfeiffer WC, Yoneshigue-Valentin Y, Farina M (1996) Accumulation, effects on growth, and localization of zinc in *Padina gymnospora* (Dictyiotales, Phaeophyceae). Hydrobiologia 326/327: 451-456

Fiskesjö G (1990) Occurrence and degeneration of 'Al-structures' in root cap cells of *Allium cepa* L. after Al-treatment. Hereditas 112: 193-202

Fiskesjö G (1997) Allium test for screening chemicals; evaluation of cytological parameters. In: Wang W, Gorsuch JW, Hughes JS (eds) Plants for environmental studies. Lewis Publ, Boca Raton, pp 307-333

Foy CD (1984) Physiological effects of hydrogen, aluminium, and manganese toxicities in acid soil. In: Adams F (ed) Soil acidity and liming, 2nd edn. American Society of Agronomy, Madison, WI, pp 57-97

Frey B, Keller C, Zierold K, Schulin R (2000) Distribution of Zn in functionally different leaf epidermal cells of the hyperaccumulator *Thlaspi caerulescens*. Plant Cell Environ 23: 675-687

Fuhrer J (1982) Early effects of excess cadmium uptake in *Phaseolus vulgaris*. Plant Cell Environ 5: 263-270

Gabara B, Krajewska M, Stecka E (1995) Calcium effect on number, dimension and activity of nucleoli in cortex cells of pea (*Pisum sativum* L.) roots after treatment with heavy metals. Plant Sci 111: 153-161

Gaur JP, Rai LC (2001) Heavy metal tolerance in algae. Physiological, biochemical and molecular mechanisms. In: Rai LC, Gaur JP (eds) Algal adaptation to environmental stresses. Springer, Berlin Heidelberg New York, pp 363-388

Ghoshroy S, Nadakavuraken MJ (1990) Influence of cadmium on the ultrastructure of developing chloroplasts in soybean and corn. Environ Exp Bot 30: 187-192

Ghoshroy S, Freedman K, Lartey R, Citovsky V (1998) Inhibition of plant viral systemic infection by non-toxic concentrations of cadmium. Plant J 13: 591-602

Gullvag BM, Skaar H, Ophus EM (1974) An ultrastructural study of lead accumulation within leaves of *Rhytidiadelphus squarrosum* (Hedw) Warnst. J Bryol 8: 117-122

Gunsé B, Poschenrieder CH, Barceló J (1997) Water transport properties of roots and root cortical cells in proton- and Al-stressed maize varieties. Plant Physiol 113: 595-602

Günthardt-Goerg MS, Vollenweiden P (2002) Cellular injury, heavy metal uptake and growth of poplar, willow and spruce influenced by heavy metals and soil acidity. In: Proceedings cost action 837 WG2 workshop on risk assessment and sustainable land management using plants in trace element contaminated soil. Bordeaux, 2002, pp 103-104

Hagemeyer J, Breckle SW (1996) Growth under trace element stress. In: Waisel Y, Eshel A, Kafkafi U (eds) Plant roots: the hidden half, 2nd edn. Dekker, New York, pp 415-433

Hagemeyer J, Breckle SW (2002) Trace element stress in roots. In: Waisel Y, Eshel A, Kafkafi U (eds) Plant roots: the hidden half, 3rd edn. Dekker, New York, pp 763-785

Hansel CM, Fendorf S, Sutton S, Newville M (2001) Characterization of Fe plaque and associated metals on the roots of mine-waste impacted aquatic plants. Environ Sci Technol 35: 3863-3868

Haridasan M, Paviani TI, Schiavini I (1986) Localization of aluminium in the leaves of some aluminium-accumulating species. Plant Soil 94: 435-437

Heath MS, Southwoth D, D'Allura JA (1997) Localization of nickel in epidermal subsidiary cells of leaves of *Thlaspi montanum* var. Siskiyouense (Brassicaceae) using energy-dispersive X-ray microanalysis. Int J Plant Sci 158: 184-188

Hecht-Buchholz CH, Brady DJ, Asher CJ, Edwards DG (1990) Effects of low activities of aluminium on soybean (*Glycine max*). II. Root cell structure and root hair development. In: van Beusichem ML (ed) Plant nutrition—physiology and applications. Kluwer, Dordrecht, pp 335-343

Herren T, Feller U (1997) Influence of increased zinc levels on phloem transport in wheat shoots. J Plant Physiol 150: 228-231

Heumann HG (1987) Effects of heavy metals on growth and ultrastructure of *Chara vulgaris*. Protoplasma 136: 37-48

Hodson MJ, Wilkins DA (1991) Localization of aluminium in the roots of Norway spruce [Picea abies (L.) Karst.] inoculated with *Paxillus involutus* Fr. New Phytol 118: 273-278

Hof I (1984) Lokalisation und Wirksamkeit des Schwermetalls Blei in juvenilen Wurzelsystemen von *Zea mays* L. Untersuchungen zur Macro- und Mikromorphologie von Primärwurzeln. PhD thesis, University of Vienna, Austria

Hübner R, Depta H, Robinson DG (1985) Endocytosis in maize root cap cells. Evidence obtained using heavy metal salt solutions. Protoplasma 129: 214-222

Irmer U, Wachholz I, Schäfer H, Lorch DW (1986) Influence of lead on *Chlamydomonas reinhardtii* Danegard (Volvocales, Chlorophyta): accumulation, toxicity and ultrastructural changes. Environ Exp Bot 26: 97-105

Jacobs T (1997) Why do plant cells divide? Plant Cell 9: 1021-1029

Jensen TE, Baxter M, Rachlin JW, Jani V (1982) Uptake of heavy metals by *Plectonema boryanum* (Cyanophyceae) into cellular components, especially polyphosphate bodies: an X-ray energy dispersive study. Environ Pollut Ser A 27: 119-127

Jensen TE, Rachlin JW, Baxter M, Warkentine B, Jani V (1984) Heavy metal compartmentalization by algal cells. In: Bailey GW (ed) Proc 42nd Ann Meeting Electron Microscopy Soc Amer. San Francisco Press, San Francisco, pp 294-295

Johnson PA, Bennet RJ (1991) Aluminium tolerance of root cap cells. J Plant Physiol 137: 760-762

Jones DL, Shaff JE, Kochian LV (1995) Role of calcium and other ions in directing root hair tip growth in *Limnobium stoloniferum*. I. Inhibition of tip growth by aluminium. Planta 197: 672-680

Jurkiewicz A, Turnau K, Mesjasz-Przybylowicz J, Przybylowicz W, Godzik B (2001) Heavy metal localization in mycorrhizas of *Epipactis atrorubens* (Hoffm.) Besser (Orchidaceae) from zinc mine tailings. Protoplasma 218: 117-124

Kahle H (1993) Response of roots of trees to heavy metals. Environ Exp Bot 33: 99-119

Kandasamy MK, Kristen U (1989) Influence of triethyllead on growth and ultrastructure of tobacco pollen tubes. Environ Exp Bot 29: 283-292

Khan DH, Duckett JG, Frankland B, Kirkham JB (1984) An X-ray microanalytical study of the distribution of cadmium in roots of *Zea mays* L. J Plant Physiol 115: 19-28

Kidd PS, Llugany M, Poschenrieder C, Gunsé B, Barceló J (2001) The role of root exudates in aluminium resistance and silicon-induced amelioration of aluminium toxicity in three varieties of maize (*Zea mays* L.). J Exp Bot 52: 1339-1352

Krämer U, Grime GW, Smith JAC, Haves CR, Baker AJM (1997) Micro-PIXE as a technique for studying nickel localization in leaves of the hyperaccumulator plant Alyssum lesbiacum. Nucl Instrum Methods Phys Res B 130: 3346-3350

Ksiazek M, Wozny A, Mlodzianowski F (1984) Effect of $Pb(NO_3)_2$ on poplar tissue culture and the ultrastructural localization of lead in culture cells. For Ecol Manage 8: 95-105

Küpper H, Lombi E, Zhao F-J, Wiehammer G, McGrath SP (2001) Cellular compartmentation of nickel in the hyperaccumulators *Alyssum lesbiacum*, *Alyssum bertolonii* and *Thlaspi goesingense*. J Exp Bot 52: 2291-2300

Lane SD, Martin ES (1980) Further observations on the distribution of lead in juvenile roots of *Raphanus sativus*. Z Pflanzenphysiol 97: 145-152

Lane SD, Martin ES (1982) An ultrastructural examination of lead localization in germinating seeds of *Raphanus sativus*. Z Pflanzenphysiol 107: 33-40

Lazinsky D, Sicko-Goad L (1990) Morphometric analysis of phosphate and chromium interactions in *Cyclotella meneghiniana*. Aquatic Toxicol 16: 127-140

Lazof DB, Goldsmith JG (1997) The in situ analysis of intracellular aluminium in plants. Prog Bot 58: 112-149

Lazof DB, Goldsmith JG, Rufty TW, Linton RW (1994) Rapid uptake of aluminum into cells of intact soybean root tips. A microanalytical study using secondary ion mass spectrometry. Plant Physiol 106: 1107-1114

Lazof DB, Goldsmith JG, Rufty TW, Linton RW (1996) The early entry of Al into cells of intact soybean roots. A comparison of three developmental root regions using secondary ion mass spectroscopy imaging. Plant Physiol 112: 1289-1300

Lichtenberger O, Neumann D (1997) Analytical electron microscopy as a powerful tool in plant cell biology: examples using electron energy loss spectroscopy and X-ray microanalysis. Eur J Cell Biol 73: 378-386

Lindsey PA, Lineberger RD (1981) Toxicity, cadmium accumulation and ultrastructural alterations induced by exposure of *Phaseolus* seedlings to cadmium. HortScience 16: 434

Liu D, Jiang W, Wang W, Zhao F, Lu C (1994) Effects of lead on growth, cell division, and nucleolus of *Allium cepa*. Environ Pollut 86: 1-4

Liu D, Zhai L, Jiang W, Wang W (1995) Effects of Mg^{2+}, Co^{2+}, and Hg^{2+} on the nucleus and nucleolus in root tip cells of *Allium cepa*. Environ Contam Toxicol 55: 779-787

Llugany M, Poschenrieder CH, Barceló J (1995) Monitoring of aluminium-induced inhibition of root elongation in four maize cultivars differing in tolerance to Al and proton toxicity. Physiol Plant 93: 265-271

Llugany M, Lombini A, Poschenrieder C, Dinelli E, Barcelo J (2003) Different mechanisms account for enhanced copper resistance in *Silene armeria* ecotypes from mine spoil and serpentine sites. Plant Soil 251: 55-63

Lolkema PC, Vooijs R (1986) Copper tolerance in *Silene cucubalus*. Subcellular distribution of copper and its effects on chloroplasts and plastocyanin synthesis. Planta 167: 30-36

Lorch DW, Schäfer H (1981) Laser microprobe analysis of the intracellular distribution of lead in artificially exposed cultures of *Phymatodocis nordstedtiana* (Chlorophyta). Z Pflanzenphysiol 101: 183-188

Macfie SM, Tarmohamed Y, Welbourn PM (1994) Effects of cadmium, cobalt, copper, and nickel on growth of the green alga *Chlamydomonas reinhardtii*: the influences of the cell wall and pH. Arch Environ Contam Toxicol 27: 454-458

Malone C, Koeppe DE, Miller RJ (1974) Localization of lead accumulated by corn plants. Plant Physiol 53: 388-394

Marienfeld S, Stelzer R (1993) X-ray microanalyses in roots of Al-treated *Avena sativa* plants. J Plant Physiol 141: 569-573

Marienfeld S, Lehmann H, Stelzer R (1995) Ultrastructural investigations and EDX-analyses of Al-treated oat (*Avena sativa*) roots. Plant Soil 171: 167-173

Marmiroli N, Marmiroli M, Gonnelli C, Maestri E, Gabbrielli R (2002) Metal localization by SEM/EDX in a hyperaccumulator and a non-hyperaccumulator species of *Alyssum* living on a serpentine soil in Tuscany. In: Proceedings cost action 837 WG2 workshop on risk assessment and sustainable land management using plants in trace element contaminated soil. Bordeaux, 2002, pp 109-110

Mateo P, Fernandez-Piñas F, Bonilla I (1994) O_2-induced inactivation of nitrogenase as a mechanism for the toxic action of Cd^{2+} on *Nostoc* UAM 208. New Phytol 126: 267-272

McCarthy I, Romero-Puertas MC, Palma JM, Sandalio LM, Corpas FJ, Gómez M, del Río LA (2001) Cadmium induces senescence symptoms in leaf peroxisomes of pea plants. Plant Cell Environ 24: 1065-1073

Miyasaka SC, Hawes C (2001) Possible role of root border cells in detection and avoidance of aluminium toxicity. Plant Physiol 125: 1978-1987

Molas J (2002) Changes of chloroplast ultrastructure and total chlorophyll concentration in cabbage leaves caused by excess of organic Ni(II) complexes. Environ Exp Bot 47: 115-126

Moustakas M, Ouzounidou G, Eleftheriou EP, Lannoye R (1996) Indirect effects of aluminium stress on the function of the photosynthetic apparatus. Plant Physiol Biochem 34: 553-560

Moustakas M, Eleftheriou EP, Ouzounidou G (1997) Short-term effects of aluminium at alkaline pH on the structure and function of the photosynthetic apparatus. Photosynthetica 34: 169-177

Mullins M, Hardwick K, Thurman DA (1985) Heavy metal localisation by analytical electron microscopy in conventionally fixed and freeze-substituted roots of metal tolerant and non-tolerant ecotypes. In: Lekkas TD (ed) Heavy metals in the environment. CEP Consultants, Edinburgh, pp 43-46

Nassiri Y, Mansot JL, Wéry J, Ginsburger-Vogel T, Amiard JC (1997a) Ultrastructural and electron energy loss spectroscopy studies of sequestration mechanisms of Cd and Cu in the marine diatom *Skeletonema costatum*. Arch Environ Toxicol 33: 147-155

Nassiri Y, Wéry J, Mansot JL, Ginsburger-Vogel T (1997b) Cadmium bioaccumulation in *Tetraselmis suecica*: an electron energy loss spectroscopy (EELS) study. Arch Environ Contam Toxicol 33: 156-161

Neumann D, Lichtenberger O, Schwieger W, zur Nieden U (1997a) Silicon storage in selected dicotyledons. Bot Acta 110: 282-290

Neumann D, zur Nieden U, Schwieger W, Leopold I, Lichtenberger O (1997b) Heavy metal tolerance of *Minuartia verna*. J Plant Physiol 151: 101-108

Neumann D, zur Nieden U (2001) Silicon and heavy metal tolerance of higher plants. Phytochemistry 56: 685-692

Ouzounidou G, Eleftheriou EP, Karataglis S (1992) Ecophysiological and ultrastructural effects of copper in *Thlaspi ochroleucum* (Cruciferae). Can J Bot 70: 947-957

Ouzounidou G, Clamporová M, Moustakas M, Karataglis S (1995) Responses of maize (Zea mays L.) plants to copper stress. I. Growth, mineral content and ultrastructure of roots. Environ Exp Bot 35: 167-176

Ouzounidou G, Moustakas M, Eleftheriou EP (1997) Physiological and ultrastructural effects of cadmium on wheat (Triticum aestivum L.) leaves. Arch Environ Contam Toxicol 32: 154-160

Ownby JD (1993) Mechanisms of reaction of hematoxylin with aluminium-treated wheat roots. Physiol Plant 87: 371-380

Panou-Filotheou H, Bosabalidis M, Karataglis S (2001) Effects of copper toxicity on leaves of oregano (Origanum vulgare subsp. Hirtum). Ann Bot 88: 207-214

Pellegrini L, Pellegrini M, Delivopoulos S, Berail G (1991) The effects of cadmium on the fine structure of the brown alga Cystoseira barbata forma repens Zinova et Kaliguna. Br Phycol J 26: 1-8

Percival E, McDowell RH (1981) Algal walls-composition and biosynthesis. In: Tanner W, Loewus FA (eds) Plant carbohydrates II. Extracellular carbohydrates. Encyclopedia of plant physiology, new series, vol 13B. Springer, Berlin Heidelberg New York, pp 277-316

Prevot P, Soyer-Gobillard MO (1986) Combined action of cadmium and selenium on two marine dinoflagellates in culture, Prorocentrum micans Ehrbg. and Crypthecodinium cohnii Biecheler. J Protozool 33: 42-47

Psaras GK, Manetas Y (2001) Nickel localization in seeds of the metal hyperaccumulator Thlaspi pindicum Hausskn. Ann Bot 88: 513-516

Punz WF, Sieghardt H (1993) The response of roots of herbaceous plant species to heavy metals. Environ Exp Bot 33: 85-93

Qureshi JA, Hardwick K, Collin HA (1986) Intracellular localization of lead in a lead tolerant and sensitive clone of Anthoxanthum odoratum. J Plant Physiol 122: 357-364

Rachlin JW, Jensen TE, Baxter M, Jani V (1982) Utilization of morphometric analysis in evaluating response of Plectonema boryanum (Cyanophyceae) to exposure to eight heavy metals. Arch Environ Contam Toxicol 11: 323-333

Rachlin JW, Jensen TE, Warkentine B (1984) The toxicological response of the algae Anabaena flos-aquae (Cyanophyceae) to cadmium. Arch Environ Contam Toxicol 13: 143-151

Rachlin JW, Jensen TE, Warkentine BE (1985) Morphometric analysis of the response of Anabaena flos-aquae and Anabaena variabilis (Cyanophyceae) to selected concentrations of zinc. Arch Environ Contam Toxicol 14: 395-402

Rai LC, Jensen TE, Rachlin JW (1990) A morphometric and X-ray energy dispersive approach to monitoring pH altered cadmium toxicity in Anabaena flos-aquae. Arch Environ Contam Toxicol 19: 479-487

Rascio N, Dalla Vecchia F, Ferretti M, Merlo L, Ghisi R (1993) Some effects of cadmium on maize plants. Arch Environ Contam Toxicol 25: 244-249

Rauser WE (1978) Early effects of phytotoxic burdens of cadmium, cobalt, nickel, and zinc in white beans. Can J Bot 56: 1744-1749

Rauser WE, Ackerley CA (1987) Localization of cadmium in granules within differentiating and mature root cells. Can J Bot 65: 643-646

Rauser WE, Samarakoon AB (1980) Vein loading in seedlings of Phaseolus vulgaris exposed to excess cobalt, nickel and zinc. Plant Physiol 65: 578-583

Rengel Z, Reid RJ (1997) Uptake of Al across the plasma membrane of plant cells. Plant Soil 192: 31-35

Robb J (1981) Early cytological effects of zinc toxicity in white bean leaves. Ann Bot 47: 829-834

Röderer G (1986) On the toxic effects of tetraethyl lead and its derivatives on the Chrysophyte Poterioochromonas malhamensis. VI. Effect on lorica formation, mitosis, and cytokinesis. Environ Res 39: 205-231

Romero-Puertas MC (2002) Metabolismo de especies de oxígeno reactivo en plantas de guisante (*Pisum sativum* L.) y en peroxisomas de hojas en condiciones de estrés por cadmio. PhD thesis, University of Granada, Spain

Rufner R, Barker AV (1984) Ultrastructure of zinc-induced iron deficiency in mesophyll chloroplasts of spinach and tomato. J Am Soc Hort Sci 109: 164-168

Salt DE, Pickering IJ, Prince RC, Gleba D, Dushenkov S, Smith RD, Raskin I (1997) Metal accumulation by aquaculture seedlings of indian mustard. Environ Sci Technol 3: 1636-1644

Salt DE, Prince RC, Baker AJM, Raskin I, Pickering IJ (1999) Zinc ligands in the metal hyperaccumulator *Thlaspi caerulescens* as determined using X-ray absorption spectroscopy. Environ Sci Technol 33: 713-717

Sanità di Toppi L, Fossati F, Musetti R, Mikerezi I, Favali MA (2002) Effects of hexavalent chromium on maize, tomato, and cauliflower plants. J Plant Nutr 25: 701-717

Sarret G, Vangronsveld J, Manceau A, Musso M, D'Haen J, Menthonnex J-J, Hazemann J-L (2001) Accumulation forms of Zn and Pb in *Phaseolus vulgaris* in the presence and absence of EDTA. Environ Sci Technol 35: 2854-2859

Sawidis T, Reiss HD (1995) Effects of heavy metals on pollen tube growth and ultrastructure. Protoplasma 185: 113-122

Schier GA, McQuattie CJ (1995) Effect of aluminum on the growth, anatomy, and nutrient content of ectomycorrhizal and nonmycorrhizal eastern white pine seedlings. Can J For Res 25: 1252-1262

Schützendübel A, Schwanz P, Teichmann T, Gross K, Langenfeld-Heyser R, Godbold DL, Polle A (2001) Cadmium-induced changes in antioxidative systems, hydrogen peroxide content, and differentiation in scots pine roots. Plant Physiol 127: 887-898

Sicko-Goad L, Lazinsky D (1986) Quantitative ultrastructural changes associated with lead-coupled luxury phosphate uptake and polyphosphate utilization. Arch Environ Contam Toxicol 15: 617-627

Sicko-Goad L, Stoermer EF (1979) A morphometric study of lead and copper effects on *Diatoma tenue* var. Elongatum (Bacillariophyta). J Phycol 15: 316-321

Sicko-Goad L, Ladewski BG, Lazinsky D (1986) Synergistic effects of nutrients and lead on the quantitative ultrastructure of *Cyclotella* (Bacillariophyceae). Arch Environ Contam Toxicol 15: 291-300

Silverberg BA (1975) Ultrastructural localization of lead in *Stigeoclonium tenue* (Chlorophyceae, Ulotrichales) as demonstrated by cytochemical and X-ray micro analysis. Phycologia 14: 265-274

Sivaguru M, Baluska F, Volkmann D, Felle HH, Horst WJ (1999) Impacts of aluminum on the cytoskeleton of the maize root apex. Short-term effects on the distal part of the transition zone. Plant Physiol 119: 1073-1082

Smith MA (1983) The effect of heavy metals on the cytoplasmic fine structure of *Skeletonema costatum* (Bacillariophyta). Protoplasma 116: 14-23

Tung G, Temple PJ (1996) Uptake and localization of lead in corn (*Zea mays* L.) seedlings, a study by histochemical and electron microscopy. Sci Total Environ 188: 71-85

Van Steveninck RFM, Van Steveninck ME, Fernando DR, Godbold DL, Horst WJ, Marschner H (1987) Identification of zinc-containing globules in roots of a zinc-tolerant ecotype of *Deschampsia caespitosa*. J Plant Nutr 10: 1239-1246

Van Steveninck RFM, Van Steveninck ME (1991) Microanalysis. In: Hall JL, Hawes C (eds) Electron microscopy of plant cells. Academic, London, pp 415-455

Van Steveninck RFM, Van Steveninck ME, Fernando DR (1993) Heavy-metal (Zn, Cd) tolerance in selected clones of duck weed (*Lemna minor*). In: Randall PJ (ed) Genetic aspects of plant mineral nutrition. Kluwer, Dordrecht, pp 387-396

Vázquez MD, Poschenrieder CH, Barceló J (1987) Chromium VI induced structural and ultrastructural changes in bush bean plants (*Phaseolus vulgaris* L.) Ann Bot 59: 427-438

Vázquez MD, Bennàssar A, Cabot C, Poschenrieder CH, Barceló J (1990) Phytotoxic effects of technetium-99 in beans: influence of cotyledon excision. Environ Exp Bot 30: 271-281

Vázquez MD, Poschenrieder CH, Barceló J (1992a) Ultrastructural effects and localization of low cadmium concentrations in bean roots. New Phytol 120: 215-226

Vázquez MD, Barceló J, Poschenrieder CH, Mádico J, Hatton P, Baker AJM, Cope GH (1992b) Localization of zinc and cadmium in *Thlaspi caerulescens* (Brassicaceae), a metallophyte that can hyperaccumulate both metals. J Plant Physiol 140: 350-355

Vázquez MD, Poschenrieder CH, Barceló J, Baker AJM, Hatton P, Cope HG (1994) Compartmentation of zinc in roots and leaves of the zinc hyperaccumulator *Thlaspi caerulescens* J & C Presl. Bot Acta 107: 243-250

Vázquez MD, Poschenrieder C, Corrales I, Barcelo J (1999) Changes in apoplastic aluminum during the initial growth response to aluminum by roots of a tolerant maize variety. Plant Physiol 119: 435-444

Vesk PA, Nockolds CE, Allaway WG (1999) Metal localization in water hyacinth roots from urban wetland. Plant Cell Environ 22: 149-158

Visviki I, Rachlin JW (1992) Ultrastructural changes in *Dunaliella minuta* follwing acute and chronic exposure to copper and cadmium. Arch Environ Contam Toxicol 23: 420-425

Visviki I, Rachlin JW (1994) Acute and chronic exposure of *Dunaliella salina* and *Chlamydomonas bullosa* to copper and cadmium: effects on ultrastructure. Arch Environ Contam Toxicol 26: 154-162

Vogelei A, Rothe GM (1988) Die Wirkung von Säure und Aluminiumionen auf den Nährelementgehalt und den histologischen Zustand nichtmykorrhizierter Fichtenwurzeln (*Picea abies* [L.] Karst.) Forstw Cbl 107: 348-357

Wagatsuma T, Kaneko M, Hayasaka Y (1987) Destruction process of plant root cells by aluminium. Soil Sci Plant Nutr 33: 161-175

Wang J, Evangelou VP (1995) Metal tolerance aspects of plant cell wall and vacuole. In: Pessarakli M (ed) Handbook of plant and crop physiology. Dekker, New York, pp 695-717

Wellburn AR (1984) Ultrastructural, respiratory and metabolic changes associated with chloroplast development. In: Baker NR, Barber J (ed) Topics in photosynthesis, vol 5. Chloroplast biogenesis. Elsevier, Amsterdam, pp 253-303

Wierzbicka M (1987) Lead accumulation and its translocation barriers in roots of *Allium cepa* L.—autoradiographic and ultrastructural studies. Plant Cell Environ 10: 17-26

Wozny A, Zatorska B, Mlodzianowski F (1982) Influence of lead on the development of lupin seedlings and ultrastructural localization of this metal. Acta Soc Bot Pol 51: 345-351

Wrischer M, Kunst L (1981) Fine structural changes of wheat plastids during cadmium induced bleaching. Acta Bot Croat 40: 79-83

Chapter 10

Water Relations in Heavy Metal Stressed Plants

Ch. Poschenrieder, J. Barceló

Laboratorio de Fisiología Vegetal, Facultad de Ciencias, Universidad Autónoma de Barcelona, 08193 Bellaterra, Spain

10.1 INTRODUCTION

Almost every plant process is affected directly or indirectly by the water supply, and water can be considered as a major factor in the regulation of plant growth (Kramer and Boyer 1995). Therefore, many investigations on plant responses to environmental stresses play considerable attention to water relations from the cell to the whole plant and community level. The influence of excess ions on plant water relations has mainly been investigated in plants exposed to high concentrations of Na^+, Cl^-, and other ions that cause adverse effects in plants at concentrations in the 0.1-1 M range. The interest in water relation studies under salt stress is obvious, because of the significant influence of such high ion concentrations on the osmotic potential of the substrate, and the difficulties for water acquisition by plants under those circumstances. In addition to osmotic stress, ion-specific effects of high salt concentrations on water relations and growth are well documented (Levitt 1980; Montero et al. 1997, 1998).

Ions of heavy metals and Al are toxic to plants at the μM level. The influence of such low concentrations on the substrate's osmotic potential is negligible. This by no means implies that water relationships are unaffected by metal ion toxicity. The primary toxicity mechanism of the different metal ions may be as different as their chemical properties, especially valence, ion radius, and capacity to form organic complexes. Nevertheless, an excess of metal ions in plants can induce a series of effects which present some common characteristics: alteration of plasma membrane properties (Haug and Caldwell 1985; Kennedy and Gonsalves 1987; Ros et al. 1990; de Vos et al. 1991; Ernst 1998), alteration of enzyme activities (van Assche and Clijsters 1990), and inhibition of root growth (Kahle 1993; Punz and Sieghardt 1993; Hagemeyer and Breckle 2002). These early events lead to a large range of secondary effects, such as disturbance of hormone balance, deficiency of essential nutrients, inhibition of photosynthesis, changes in photoassimilate translocation, alteration of water relations, etc., which further enhance the metal-induced growth reduction. Among these, water relations have attracted relatively little attention (Barceló and Poschenrieder 1990; Poschenrieder and Barceló 1999). This is quite surprising in view of the primary toxicity effects of excess ions on membranes and the essential role of water relations for plant growth. Increasing interest in investigations on plant water relations under metal toxicity stress also derives from the fact that plants in metal-enriched habitats frequently suffer from drought stress, mainly because of poor physical soil conditions and shallow root systems. Drought resistance can be considered an important trait in phytoremediation of metal-polluted soils, especially under arid or semiarid climate (Barceló et al. 2001).

10.2 CELL WATER RELATIONS AND CELL EXPANSION GROWTH

Exposure of plants to toxic metal concentrations generally causes a fast inhibition of cell elongation. Cell expansion growth is governed by the dynamic interaction between cell turgor and cell wall extensibility and can be best described by the modified Lockhart's equation (Dale and Sutcliffe 1986):

$$\frac{dV}{dt} = \frac{L'p \cdot \Phi \cdot (\Delta\psi + P - Y)}{\Phi + L'p} \tag{10.1}$$

where $L'p$ is the apparent hydraulic conductance, Φ the extensibility of the wall, $\Delta\psi$ the difference in water potential between the cell and its surroundings, Y the threshold turgor pressure, and P the actual turgor pressure. Any change in these parameters would alter the rate of cell expansion growth.

Extensive membrane damage, as observed in cells of plants stressed by extremely high concentrations of metals that cause peroxidation of membrane lipids, e.g. Cu or Cr(VI) (de Vos et al. 1989; Barceló and Poschenrieder 1997), obviously would result in turgor loss and inhibition of cell expansion. However, there is considerable experimental evidence that prove that metal-induced inhibition of cell expansion can occur at relatively low concentrations without a general loss of membrane integrity. For example, in bean plants exposed to only 3 μM Cd, leaf expansion growth was inhibited after 48 h exposure. The bulk leaf turgor was unaffected, while a decrease in the relative water content was observed (Poschenrieder et al. 1989). These data suggest a Cd-induced decrease of cell wall extensibility and a decrease of the hydraulic conductivity (L_p) in the system.

Only in a few experiments have the influence of toxic metals on the parameters of the Lockhart equation been directly measured. Aluminum has been found to decrease the permeability to water of *Quercus rubra* root cortex cells, while the permeability to nonelectrolytes was increased. High Ca supply had the opposite effect, and the authors conclude that Al induces changes in the lipid packing density of the root cell plasma membrane, leading to decreased values of L_p (Zhao et al. 1987).

Early investigations by Klimashevski and Dedov (1975) have suggested that Al-induced inhibition of root extension growth may be brought about by Al cross-linking of the carboxyl groups of the pectin fraction of cell walls, causing a decrease in cell wall extensibility and cell wall elasticity. Direct measurements of wall extensibility by the Instron technique in maize coleoptiles that had been floated for 4 h on solutions with 100

μM Al (Fig. 10.1) revealed that Al rather increased than decreased total, plastic and elastic extensibility in this system (Barceló et al. 1996). The influence of Cd on wall extensibility of maize coleoptiles has been found to depend on Cd concentration and exposure time (Gunsé et al. 1992). While after 2 h exposure to 50 μM Cd wall extensibility was significantly reduced, longer exposure or higher Cd concentrations increased cell wall extensibility (Fig. 10.1). It is well known that small cells exhibit higher wall extensibility than large ones, and the metal-induced increase in wall extensibility may be a consequence of metal-induced decrease in cell expansion growth. These results do not support the hypothesis that the direct cross-linking of wall polymers by metal ions is responsible for the inhibition of cell expansion. However, results on wall extensibility obtained with in vitro exposure of coleoptiles to metals are not extrapolable to root cells. Moreover, studies performed with tissues or whole organs only provide average values and not data for individual cells.

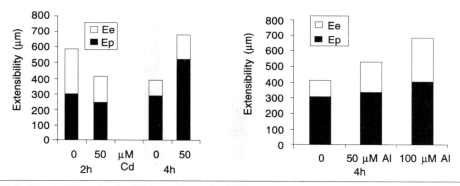

Fig. 10.1 Total (*Et*), plastic (*Ep*) and elastic (*Ee*) cell wall extensibility (μm) of maize coleoptiles floated on Cd- or Al-containing solution (*Et* = *Ep* + *Ee*). Redrawn from Gunsé et al. (1992) and M. Llugany (unpubl. data)

10.2.1 Cell Pressure Probe

The development of cell pressure probe techniques that allow the measurement of water relations in single cells opens the possibility to directly study the influence of toxic metal ions on these parameters. At present there are only a few studies that have used this technique to investigate the influence of metal toxicity on cell water relations. Direct measurement of turgor pressure in root cells of maize varieties exposed to toxic Al concentrations showed that Al either did not affect or increased root cell turgor (Fig. 10.2A; Gunsé et al. 1997). However, Al-tolerant and Al-sensitive varieties exhibited clear differences in the response of the cell membrane hydraulic conductivity (L_{pc}) to Al. While Al increased L_{pc} in the Al-tolerant variety Adour 250, a decrease was observed in Al-sensitive BR 201 F (Fig. 10.2B). In this Al-sensitive variety Al also caused a significant decrease in cell wall elasticity, as shown by the higher values of the cell elastic coefficients (ε_c; Fig. 10.2C). This result clearly demonstrates that Al affects the mechanical properties of root cell walls. As this investigation concerned mature root cells (50 mm from tip), the possible role of Al-induced cell wall stiffening in the inhibition of root cell elongation remains to be established (Gunsé et al. 1997).

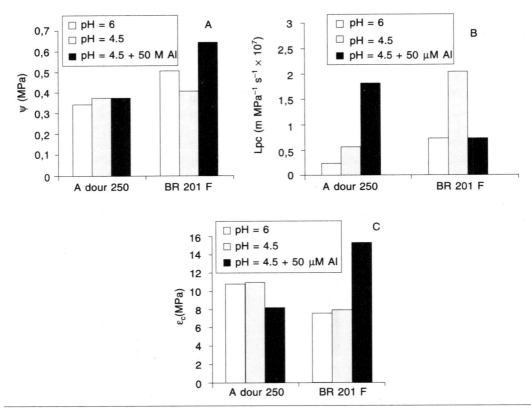

Fig. 10.2 Influence of pH and Al on turgor pressure (ψ_p), hydraulic conductivity (L_{pc}), and elastic coefficient (ε_c) of root cortex cells (third layer) of maize varieties differing in H$^+$ and Al tolerance. (From Gunsé et al. 1997)

10.2.2 Aquaporins

The findings that water channels (aquaporins) are not restricted to animal cells but also occur in plants, and the fact that mercury sulfhydryl reagents act as specific inhibitors of most water channels (Chrispeels and Maurel 1994; Maurel 1997; Tyerman et al. 1999), have stimulated several investigations on the influence of toxic metals on cell water transport. As expected, Hg severely decreases L_{pc} in *Characea* cells (Henzler and Steudle 1995). Also Zn^{2+}, which has less afinity for sulfhydryl groups than Hg, has been reported to decrease L_{pc} in algal cells, probably by blocking water channels (Rygol et al. 1992; Tazawa et al. 1996). The reversibility of the Zn-induced effect by mercaptoethanol strongly suggests that Zn, as Hg, blocks water channels by interaction with the sulfhydryl groups.

The investigations mentioned above clearly demonstrate that toxic metal ions can rapidly affect cell water relations, not by a general destruction of the plasma membrane, but by a rapid decrease in L_{pc}. Studies with *Arabidopsis thaliana* have shown that tonoplast aquaporins are correlated with cell enlargement (Ludevid et al. 1992). However, at the present stage of knowledge, it is not clear to what extent metal-induced inhibition of water transport into cells can be held responsible for metal-induced inhibition of root cell elongation.

10.3 WHOLE PLANT WATER RELATIONS

A plant's water status is the function of three interdependent processes: water uptake, water transport and water loss. Generally, the water movement in the continuous system soil-plant-atmosphere is described by the van den Honert equation:

$$\text{Flow} = \frac{\Psi_{\text{soil}} - \Psi_{\text{root}}}{r_{\text{soil}}} = \frac{\Psi_{\text{root}} - \Psi_{\text{leaf}}}{r_{\text{plant}}} = \frac{\Psi_{\text{leaf}} - \Psi_{\text{air}}}{r_{\text{air}}} \tag{10.2}$$

where Ψ is the water potential and r resistance to water flow.

Under normal water supply conditions, within healthy plants, the movement of water through and out of plants is controlled at the leaf-air interphase, i.e. the stomatal resistance. However, if there is an increase elsewhere in the system, e.g. due to cavitation especially in the rhizosphere or roots, the stomatal resistance is controlled either by the decrease in leaf turgor or, perhaps earlier, by a root emitted hormonal signal (see Sect. 10.3.3.2).

Analyses of the influences of excess heavy metals on plant water relations have to distinguish between effects on water availability in soils, effects on root growth limiting water uptake, and other phytotoxic effects which may influence water relations. Within the limited frame of this chapter, only plant factors can be considered.

10.3.1 Influence of Heavy Metals on Radial Transport in Roots

10.3.1.1 Water Transport Pathways

The passive movement of water across the root to the xylem can occur through different pathways: the apoplastic pathway around the protoplasts and the cell-to-cell (protoplastic) transport that can be divided into a symplastic (via plasmodesmata) and a transcellular (across cell membranes) path. According to the composite membrane model of the root these pathways are arranged in parallel rather than in a serial manner (Steudle 1993; Steudle and Frensch 1996). The relative contribution of these pathways to water transport through roots depends on the driving force of the water flux. In transpiring plants (hydraulic flux) a considerable apoplastic flux occurs, while under root pressure (osmotic driving force) the cell-to-cell path would dominate. Investigations using cell and root pressure probes have found that the relative contribution of the apoplastic and protoplastic pathways for water transport in roots may be different in different plant species. While in barley and bean a substantial cell-to-cell transport seems to occur, in maize and wheat water transport appears to be mainly apoplastic (Steudle 1993). These findings suggest that heavy metals can affect root hydraulic conductivity by multiple mechanisms operating on the apoplastic and/or the symplastic pathway: enhancement of apoplastic barriers, plasma membrane properties, aquaporins, plasmodesmata, metabolic inhibition, etc. Moreover, the consequences of the negative effects of toxic metal ions on root cell hydraulic conductivity for the water transport in whole roots may be different in different species.

The strong inhibitory effect of Hg on the hydraulic conductivity in roots (L_{pr}) has been attributed to a direct blockage of aquaporins by Hg (Maggio and Joly 1995; Carvajal et al. 1996). Mercury at 0.5 mM caused an almost instantaneous decrease of L_{pr} in intact tomato roots. The K^+ concentration in the xylem exudates remained unaffected, so that the rapid decrease in L_{pr} may not be attributed to alterations of the osmotic relations of the entire root (Maggio and Joly 1995). Also, at lower concentrations (50 μM), Hg caused a reduction in sap flow of wheat seedlings (Carvajal et al. 1996). In both cases the inhibitory action of Hg was reversed by reducing agents suggesting that Hg did not cause irreversible damage. However, it must be taken into account that Hg may affect other membrane transport proteins or metabolic events that may affect the osmotic relations of root tissues and water transport (Maurel 1997). Zhang and Tyerman (1999) showed that Hg depolarized the plasma membrane potential of wheat root cells at concentrations similar to that causing decreased hydraulic conductivity. These results indicate that Hg can reduce root hydraulic conductivity not only by blocking water channels, but also through disturbance of ion transport and/or the cell metabolism (Tyerman et al. 1999). Recent investigations in onion roots have shown that Hg moved into the central cortex only in youngest root zones where it reduced the hydraulic conductivity of individual cells (Barrowclough et al. 2000). In older zones of onion and maize roots Hg is localized only in peripheral root cells where it severly reduced turgor of epidermal cells. In maize cortical cells were hardly affected while root conductance was severely reduced. These results suggest that Hg targets are localized in the first root cell layers (Gaspar et al. 2001).

Aluminum, a class A metal with low affinity for sulfhydryl groups, has also been found to reduce cell and root hydraulic conductivities in Al-sensitive maize. In contrast to Hg, however, the hydraulic conductivity of individual root cells was more affected than that of the entire root (Gunsé et al. 1997). This is in line with the hypothesis that in maize roots the apoplastic pathway for water transport is predominant (Jones et al. 1988). At least after long-term exposure to toxic metals, however, anatomical alterations in the root system such as increased suberization and alterations at the hypodermis and endodermis level may contribute to reduced apoplastic water flow.

10.3.1.2 Discussion of Apoplastic Metal Flow to the Xylem

The composite transport model for radial water flow in roots indicates substantial apoplastic transport of water even through the endo- and exodermis (Steudle 2001). This opens the question about the possibility of an apoplastic byflow of nutrient and toxic ions to the xylem without the control of an obligatory symplastic step at the endodermal level. White (2001) recently proposed that important amounts of Ca may be directly moved via the apoplast into the vascular cylinder. White et al. (2002) also provided model calculations supporting the view that Zn hyperaccumulation in *Thlaspi caerulescens* may imply a substantial apoplastic Zn flow to the xylem. According to their calculations taking into account published values for root-to-shoot fresh weight ratios and Zn concentrations in roots and shoots, the influx rate of Zn into the root symplasm is too low to sustain the rate of Zn influx into the xylem. This hypothesis was critizised by Ernst et al. (2002), who mainly questioned the assumption by White et al. on a constant foliar Zn concentration during *Thlaspi caerulescens* growth and provided data that contradict the apparent lack of cation selectivity and competition between cations in the mainly apoplastic route to the xylem emphasized by White's model. Both groups indicate that the operation of a low affinity transport system would largely support the view of symplastic transport. Future investigations into apoplastic and symplastic water and ion transport pathways of hyperaccumulating and non-accumulating metallophytes are clearly necessary.

10.3.2 Xylem Transport

The flow of water in the xylem obeys Hagen Poisseuille's law:

$$\text{Flow rate} = (P_2 - P_1) \cdot \frac{\pi r^4}{8 L \eta} \qquad (10.3)$$

where η is the viscosity of the sap, P_2-P_1 is the pressure gradient along the capillary, L is the length of the pathway and r the radius of the capillary. This means that, if there is a decrease in the pressure difference or of the tracheary diameter, there would be a considerable decrease in xylem transport (Zimmermann 1983).

The magnitude of the pressure gradient along the capillary xylem system is mainly determined by the transpiration rate or, when stomata are closed, by the root pressure. Both factors are influenced by ion toxicity (see Sects. 10.3.3.1 and 10.3.3.3). Moreover, xylem conductivity could be reduced by a metal-induced decrease of the cross-sectional area available for water transport.

Toxic levels of Cd (Lamoreaux and Chaney 1977; Barceló et al. 1988b), Zn (Robb et al. 1980; Paivöke 1983), Al (Bennet et al. 1985) and Cr (Vázquez et al. 1987) have been found to decrease the vessel diameter in diverse plant species. In bean plant stems, 44.5 μM Cd decreased both the vessel radius and the number of vessels, resulting in decreased total vessel area and a more than 50% decrease in the sap flow rate (Barceló et al. 1988b). Other environmental stresses such as drought or freezing also induce the formation of small vessels. However, under those conditions, the decrease in the diameter of the individual conduits may be compensated by an increase in the number of vessels (Aloni 1987). The decrease in both number and size of vessels in Cd-treated plants may be caused by the metal-induced inhibition of cell division in the procambium and cambium, by the inhibition of cell elongation, and by metal-induced alterations of the formation of secondary cell wall thickenings (Barceló et al. 1988b,c). The importance of cytokinins and auxins in the differentiation of both fibers and vessels (Aloni 1987) suggests that Cd may decrease the number and size of these elements through alterations of the hormone balance. However, there is a lack of experimental data on this topic.

Under severe Cd or Zn toxicity, the vessel diameter may be further decreased by the accumulation of amorphous depositions in the conduits, which have been tentatively identified as "lignin-like phenolics", probably derived from cell walls (Fuhrer 1982; Vázquez et al. 1989). Both decreased total vessel area and partial obstruction of vessels by these depositions may account for the lower sap flow rate detected in Cd-stressed plants (Lamoreaux and Chaney 1977; Barceló et al. 1988b). Nevertheless, from the available data it it is not clear to what extent the increased resistance to water flow in the vessels contributes to the general water relations in the whole plants. If important at all, the metal-induced decrease in the xylem capillary radius may only play a role after relatively long metal exposures, while alterations of root and leaf water relations can be observed after a few hours

10.3.3 Effects of Heavy Metals on Transpiratory Water Loss

Early experiments using foliar applications of the metal-containing antitranspirant phenylmercuric acetate (PMA; Davenport et al. 1971), and studies with epidermal peels floating on Pb-, Cd-, Ni-, Tl-, or Al-containing solutions (Bazzaz et al. 1974; Schnabl and Ziegler 1975), have shown that heavy metals may induce stomatal closure. Experiments on excised leaves have demonstrated that metals increase stomatal resistance not only when directly applied to guard or epidermal cells, but also when reaching the leaves via the xylem (Bazzaz et al. 1974; Schnabl and Ziegler 1974; Lamoreaux and Chaney 1978). At that time, the direct effects of metals on stomatal opening were thought to be due to either the metal-induced inhibition of an energy system or the alterations of K^+ fluxes through membranes. More recent studies have shown that aquaporins are present in guard cells (Maurel 1997) and that toxins, that interfere with the polymerization or depolymerization of actin filaments, alter K^+-channel activities in guard cells (Hwang et al. 1997). These observations open new lines of investigations on the possible mechanisms of direct effects of toxic metal ions on stomatal closure. However, in whole plants exposed to substrates with toxic metal concentrations, rather the early effects in the roots may be held responsible for changes in transpiration rates.

Many studies on whole plants with metal supply through roots reveal that the effects of metals on transpiration are quite complex, and depend not only on climatological factors, but also on the metal species and its concentration, the exposure time, and the plant species. Even differences between populations or cultivars of the same species have been observed (Table 10.1).

Table 10.1 Stomatal resistance (r, s cm^{-1}), stomatal conductance (g_s, mmol m^{-2} s^{-1}) and/or transpiration rates (E, mmol H_2O m^{-2} s^{-1}) of plants exposed to different metal concentrations for variable time periods

Metal	Time	Species	r	g_s	E	Reference
Control	50 days	*Chrysanthemum morifolium*	28		0.41	Kirkham (1978)
0.09 μM Cd	50 days	*C. morifolium*	14		0.45	Kirkham (1978)
9 μM Cd	50 days	*C. morifolium*	>30		0.04	Kirkham (1978)
Control	15 days	*Lupinus albus*		0.38		Costa and Spitz (1997)
0.01 μM Cd	15 days	*Lupinus albus*		0.45		Costa and Spitz (1997)
10 μM Cd	15 days	*Lupinus albus*		0.17		Costa and Spitz (1997)
Control	48 h	*Phaseolus vulgaris*	3			Poschenrieder et al. (1989)
3 μM Cd	48 h	*Phaseolus vulgaris*	5			Poschenrieder et al. (1989)
Control	48 h	*Thinopyrum bessarabicum*		420	4.2	Moustakas et al. (1996)
1 mM Al	48 h	*T. bessarabicum*		244	3.0	Moustakas et al. (1996)
Control	120 h	*Phaseolus vulgaris*	18.5			Massot et al. (1994)
370 μM Al	120 h	*P. vulgaris*	36.2			Massot et al. (1994)
Control	28 days	*Triticum aestivum*			4.9	Ohki (1986)
148 μM Al	28 days	*Triticum aestivum*			1.5	Ohki (1986)

(*Contd.*)

Table 10.1 (*Contd.*)

Metal	Time	Species	r	g_s	E	Reference
Control	28 days	*Sorghum bicolor*			1.5	Ohki (1986)
148 μM Al	28 days	*Sorghum bicolor*			3.6	Ohki (1986)
Control	24 h	*Pisum sativum*	16		1.7	Angelov et al. (1993)
500 μM Cu	24 h	*Pisum sativum*	33		0.9	Angelov et al. (1993)
500 μM Cu	4 days	*Pisum sativum*	8.3		2.4	Angelov et al. (1993)
Control	n.a.	*Triticum aestivum*		283	2.7	Moustakas et al. (1997)
Cu-rich soil	n.a.	*Triticum aestivum*		161	2.4	Moustakas et al. (1997)
Control	6 weeks	*Trifolium repens* (Friesland)[a]			2.1	Dueck (1986)
5.5 μM Cu	6 weeks	*Trifolium repens* (Friesland)			1.9	Dueck (1986)
Control	6 weeks	*Trifolium repens* (Brabant)[b]			0.9	Dueck (1986)
5.5 μM Cu	6 weeks	*Trifolium repens* (Brabant)			1.2	Dueck (1986)
Control	6 weeks	*Lolium perenne* (Friesland)			0.8	Dueck (1986)
5.5 μM Cu	6 weeks	*Lolium perenne* (Friesland)			1.7	Dueck (1986)
Control	6 weeks	*Lolium perenne* (Brabant)			1.2	Dueck (1986)
5.5 μM Cu	6 weeks	*Lolium perenne* (Brabant)			0.7	Dueck (1986)
Control	7 days	*Medicago sativa*		130	5.4	Becerril et al. (1989)
0.48 μM Pb	7 days	*Medicago sativa*		42	2.6	Becerril et al. (1989)
Control	14 days	*Phaseolus vulgaris*			0.15	Gunsé (1987)
3.6 μM Cr(VI)	14 days	*Phaseolus vulgaris*			0.21	Gunsé (1987)
27 μM Cr(VI)	14 days	*Phaseolus vulgaris*			0.05	Gunsé (1987)
Control	24 h	*Phaseolus vulgaris*	11			Rauser and Dumbroff (1981)
200 μM Zn	24 h	*Phaseolus vulgaris*	23.5			Rauser and Dumbroff (1981)
400 μM Co	24 h	*Phaseolus vulgaris*	16			Rauser and Dumbroff (1981)
200 μM Ni	24 h	*Phaseolus vulgaris*	15			Rauser and Dumbroff (1981)
Control	5 weeks	*Picea abies*			0.035[c]	Schlegel et al. (1987)
0.1 μM Hg$_{(inorg)}$	5 weeks	*Picea abies*			0.035[c]	Schlegel et al. (1987)
0.01 μM Hg$_{(org)}$	5 weeks	*Picea abies*			0.019[c]	Schlegel et al. (1987)

[a]Friesland population from non-contaminated soil.
[b]Brabant population from Cu-contaminated soil.
[c]mmol H_2O g^{-1} dry weight.
n.a. not available.

10.3.3.1 Metal-Induced Enhancement of Transpiration

Several authors have found that plants exposed to low, only slightly toxic metal concentrations can show higher transpiration rates or stomatal conductances than controls (Kirkham 1978; Dueck 1986; Gunsé 1987; Costa and Morel 1994). Increased transpiration usually was accompanied by increased leaf turgor. Low concentrations of potentially toxic metal ions seem to cause a certain amount of osmotic adjustment, due to the accumulation of soluble sugars (Costa and Spitz 1997). It seems possible that low metal concentrations that do not affect photosynthesis, but are inhibitory to root growth, cause alterations in assimilate partitioning, leading to decreased osmotic potential and turgor maintainance in leaves. There are also several reports on enhanced transpiration after treatments with high metal concentrations (Paul and de Foresta 1981; Angelov et al. 1993). The mechanisms responsible for increased transpiration rates in severely damaged plants seem to be quite different from those leading to increased stomatal conductance and transpiration in plants exposed to low metal supply. Enhanced transpiration in plants stressed by high metal concentrations has been attributed to increased stomatal density, because of the reduction in leaf area (Paul and de Foresta 1981). However, other authors working with growth-reducing concentrations of Cr, Cd, or Zn observed decreased transpiration in spite of increased stomatal density (van Assche et al. 1980; Gunsé 1987; Barceló et al. 1988a,c). It seems likely that in severely intoxicated plants either the onset of premature senescence (van Assche et al. 1984; Vázquez et al. 1989) may be a cause for the complete loss of stomatal control or damage of the cuticle may enhance cuticular transpiration (Greger and Johansson 1992).

10.3.3.2 Metal-Induced Decrease of Transpiration

As a rule, plants exposed to concentrations clearly above the critical toxicity level exhibit increased stomatal resistance and low transpiration rates (Table 10.1). Although direct effects of toxic metals on stomatal guard cells cannot be excluded, the decrease in transpiratory water loss seems mainly a consequence of a water-deficiency stress induced by the toxic metals. Increased levels of proline (Alia and Saradhi 1991; Kastori et al. 1992; Bassi and Sharma 1993a,b; Chen and Kao 1995; Schat et al. 1997) and abscisic acid (ABA; Rauser and Dumbroff 1981; Barceló et al. 1986c; Poschenrieder et al. 1989), the most frequently used biochemical markers for drought stress (Hartung and Davies 1994; Heuer 1994), have been found in plants under metal toxicity stress. Exposure to 76 μM Zn caused a 150% increase in the proline concentration in the fronds of *Lemna minor*. Copper, at a similar concentration (79 μM), was even more effective; the proline concentration was three times higher than in controls (Bassi and Sharma 1993b). Similar results were obtained in Zn- and Cu-stressed sunflower (Kastori et al. 1992) and wheat (Bassi and Sharma 1993a). These results suggest that, at equimolar concentrations, Cu causes a stronger water-deficiency stress than Zn. This is to be expected, taking into acount that Cu is much more toxic to cell membranes than Zn. Schat et al. (1997) have found that metal-tolerant ecotypes of *Silene vulgaris* have constitutively higher proline concentrations than nontolerant ecotypes. However, exposure to high metal concentrations induced an increase in proline concentrations only in the sensitive ecotype. This result strongly indicates that the enhancement of proline concentrations in metal-treated plants is a sign of the intensity of the stress suffered by the plants, and not a means of metal detoxification (Schat et al. 1997). Recent investigations in algae also revealed a good coincidence between proline accumulation and the intensity of metal accumulation and metal-induced stress in the alga *Chlorella vulgaris*. Proline pretreatment of the algae reduced metal-induced K^+ loss and metal-induced lipid peroxidation (Mehta and Gaur 1999). These results support the view that proline acts as a membrane-stabilizing agent under stress conditions rather than a molecule implied in the detoxification of heavy metal cations.

Stomatal closure caused by water-deficiency stress can be brought about by hydropassive (loss of leaf turgor) or hydroactive (ABA-mediated) mechanisms. The increased ABA concentrations in leaves of metal-stressed plants found by several authors indicate that hydroactive stomatal closure is important in metal-induced decrease of transpiration. It is well established that a small reduction in the leaf turgor induces a fast increase in ABA and stomatal closure. However, there is now considerable experimental evidence that water deficiency and Na^+ stress can induce stomatal closure even before a decrease in leaf cell turgor occurs (Blackmann and Davies 1985; Montero et al. 1998). Also in plants exposed to Cd-containing nutrient solution increased stomatal resistance without a reduction in leaf pressure potential has been observed (Poschenrieder et al. 1989). Current evidence, obtained with plants exposed to water deficit or salt, indicates that roots are able to communicate stress situations to the leaves by chemical signals. Among those, ABA seems to play a major role (Davies and Zhang 1991; Montero et al. 1998). Increased transport of root-derived ABA to the leaves may cause stomatal closure before any decrease in leaf turgor and synthesis of leaf ABA occurs. As toxic metal concentrations can rapidly decrease the hydraulic conductivity in roots (see Sect. 10.3.1), it seems likely that also in metal-stressed plants export of ABA from roots to leaves may account for fast stomatal closure without turgor decrease. Unfortunately, there are only a few reports on ABA levels in both roots and leaves of metal-stressed plants. After 48 h exposure to 3 μM Cd, an increase in both root ABA concentrations and stomatal resistance was observed in bean plants, while the enhancement of leaf ABA levels occurred only after 120 h or longer exposure to Cd (Poschenrieder et al. 1989). These results suggest that also in Cd-stressed plants root-derived ABA may play a role in stomatal closure. However, as leaf Cd concentrations were increased, a direct effect of Cd on stomata cannot be excluded.

An Al-induced increase in ABA levels in roots of barley (Kasai et al. 1994) has been observed, but its possible role in stomatal responses of these plants was not reported. The influence of 24 h treatment with 50 μM Al in two zones of root tips of four maize varieties that differed in Al tolerance have shown that Al tended to decrease ABA concentrations in the zones of cell division and cell elongation in Al-tolerant varieties, while in Al-sensitive varieties a small increase in ABA concentrations was detected in the zone of cell elongation The relative ABA concentrations (% of control) in the root tips of the different varieties showed an inverse correlation to the tolerance index based on relative root elongation (Llugany 1994). Unfortunately, data on xylem ABA are not available, and it remains to be established whether these Al-induced changes are mainly related to the regulation of the cell cycle (Müller et al. 1994) and the cell elongation (Saab et al. 1992) in these roots or may also be involved in stress signalling to the shoot.

In addition to ABA as a positive root signal for stomatal closure, also root-derived cytokinins, as a negative signal, may play a role in this process (Davies and Zhang 1991).

Root tips are a major site for cytokinin synthesis and, according to Marschner (1986), all environmental factors, including the supply of mineral elements, which affect root growth are clearly related to changes in the export of cytokinins to the shoot. Taking into account that root tips are a primary site of metal-induced injury, a decrease in cytokinin production in plants exposed to toxic metal concentrations is to be expected. However, there are only a few experimental data on this topic. In bean plants exposed to growth-inhibiting concentrations of Al increased stomatal resistance was observed. However, increased zeatin and dihydrozeatin riboside levels were detected in bean roots after a few minutes of A1 treatment (Massot et al., 2002) and, even after several days of exposure, enhanced citokinin riboside concentrations were detected in roots and leaves (Massot et al., 1994). These results do not support the hypothesis that Al-induced stomatal closure is related to a decrease in cytokinin export from the roots. Unfortunately, no data on the influence of other toxic metals on cytokinin concentrations in plants are available.

10.3.3.3 *Water Use Efficiency*

The negative influence of heavy metal toxicity on photosynthesis is well established (references in Clijsters and van Assche 1985; Prasad 1997). However, the importance of metal-induced stomatal closure for inhibition of photosynthesis is not clear. Some authors have suggested that stomatal closure substantially contributes to the decrease of photosynthesis observed in metal-stressed plants (Carlson et al. 1975), while others indicate that metals act primarily on the metabolic reactions of photosynthesis, so that an increase in stomatal resistance is of secondary importance (van Assche and Clijsters 1983). Moreover, it must be taken into account that a metal-induced inhibition of photosynthesis may increase leaf CO_2 concentrations and cause stomatal closure.

The water use efficiency of metal-stressed plants has been evaluated in different species either in nutrient solution studies or under field conditions. In a comparative study on Pb- and Cd-induced effects in *Medicago sativa*, Becerril et al. (1989) found that at similar leaf metal concentrations Pb caused a drastic reduction of water use efficiency, while Cd inhibited transpiration and CO_2 assimilation to a similar degree. The water use efficiency of Cd-stressed plants was not different from controls. The authors suggest that Cd-induced inhibition of photosynthesis may mainly be due to stomatal closure. Also in Cd-stressed *Picea abies* seedlings decreased CO_2 assimilation was mainly due to stomatal closure (Schlegel et al. 1987). In sunflower, however, Cd inhibited the CO_2 assimilation rate without alterations in the stomatal conductance (di Cagno et al. 2001). Also Zn (van Assche and Clijsters 1983; Schlegel et al. 1987), methyl-Hg (Schlegel et al. 1987), and Pb (Becerril et al. 1989) may affect photosynthesis more than transpiration. Wheat plants grown on Cu-rich soil exhibited considerably lower water use efficiency than plants on fertile soil (Moustakas et al. 1997). A Cu-induced decrease in water use efficiency was also observed in solution-cultured pea plants (Angelov et al. 1993). Aluminum had similar effects on transpiration and net CO_2 assimilation in *Thinopyrum bessarabicum* (Moustakas et al. 1996), while water use efficiency decreased with increasing Al supply in citrus rootstocks as a result of photosynthesis inhibition and enhanced transpiration (Pereira et al. 2000). These results indicate that the relative importance of metal-induced stomatal closure for the inhibition of photosynthesis may not only be different for different metals, but also depends on the plant species and the growth conditions. Most of the studies have analyzed plant responses at only one time sample after relatively long exposure times (several days or weeks). Under these conditions, secondary stress effects, such as metal-induced alteration of mineral nutrition, may contribute to the inhibition of photosynthesis, so that an evaluation of the relative importance of stomatal closure in photosynthesis decline may be impossible. For this purpose, time-dependent, short-term investigations are clearly required.

10.4 RESPONSES TO SIMULTANEOUS DROUGHT AND METAL STRESS

Frequently, plants growing on metal-rich soils under field conditions not only must cope with high metal availability, but also with drought. Transient water shortage due to the weather conditions may even occur in tropical climates. A well-known situation, for example, is the so-called veranico, a short period of drought that in Brazil severely affects crop production on acid, Al-toxic soils. Aluminum toxicity inhibits root growth, and the shallow root system is unable to explore the water reserves of the subsoil. Drought due to unfavorable physical soil conditions has often been observed on both soils disturbed by mining activity (Ernst 1974; Williamson and Johnson 1981) and serpentine soils (Proctor and Woodell 1975; Brooks 1987; Proctor and Nagy 1992).

Although interactions between drought and heavy metal toxicity have been frequently observed, quantitative studies are rare. Moreover, the necessary distinction between soil factors limiting water availability and metal effects in plants that may influence the capacity to regulate water relations has only been made on a few occasions.

10.4.1 Metal-Sensitive Plants

Heavy metal stress can induce in plants a series of events which lead to decreased water loss, i.e. enhanced water conservation: decrease in number and size of leaves, decrease in stomatal size, lower number and diameter of xylem elements, increased stomatal resistance, enhancement of leaf rolling and leaf abscission, higher degree of root suberization (Barceló and Poschenrieder 1990). In metal-sensitive plants these effects mainly seem to be a mere consequence of reduced water uptake and transport, and may not always provide effective mechanisms for improved survival under drought. In well-watered bean plants, for example, Cd supply increased the number of closed stomata and the transpiration rate, but when these plants were exposed to slowly drying substrate, irreversible wilting occurred several days earlier than in drought-stressed controls without Cd supply. In contrast, Cr-stressed bean plants survived drought much better than controls. This was not only due to the Cr-induced stomatal closure, i.e. better water conservation (Gunsé 1987), but also due to enhanced drought tolerance, as indicated by the lower bulk elastic modulus of the leaves (Barceló et al. 1986a). The opposite was true in the Cd-stressed plants; turgor loss occurred at higher relative water contents than in controls, and Cd caused a decrease in leaf cell wall elasticity, indicating that the drought tolerance was substantially decreased by Cd (Barceló et al. 1986b,c, 1988a). Desiccation tolerance in the moss *Tortula ruralis* was also substantially decreased by Cd. This was attributed to additive effects of both stress factors, drought and Cd, on the exhaustion of the antioxidant defence mechanism (Takács et al. 2001).

In metal-sensitive crop plants, synergic interactions have also been observed between drought and Al toxicity (Klimov 1985; Krizek and Foy 1988a; Krizek et al. 1988). These results, in addition to the fact that Al toxicity severely affects crop productivity on acid soils in the tropics, raise the question of the convenience of including drought resistance as a selection factor in breeding programs for Al tolerance. In the open literature, there are only a few reports addressing this point. The influence of Al and drought stress, alone and in combination, has been investigated in barley and sunflower cultivars (Krizek and Foy 1988a; Krizek et al. 1988). The Al-sensitive cultivars were more tolerant to water deficiency than Al-tolerant cultivars. In response to drought alone, the Al-sensitive sunflower cultivar HS-52 increased root development. This suggests that, under drought stress, this cultivar could maintain lower stomatal resistance and higher photosynthesis rates due to more efficient water uptake. However, when exposed to both drought and Al stress, this mechanism was unfavorable, because higher Al concentrations were translocated to the shoots (Krizek and Foy 1988b). This result indicates that, under metal toxicity stress, a plant's ability to increase its root system size in response to drought will only be advantageous if this property is accompanied by mechanisms that enable the plant to tolerate the higher metal uptake. Moreover, it has to be taken into account that many metal-contaminated soils such as mine spoils usually are badly structured, coarse substrates and that the benefits of an increased root-to-shoot area for water stress avoidance depends, among other factors, on soil texture. According to model calculations in a coarse soil rhizosphere, conductance can become so limiting that even a doubling of the root-to-shoot area provides little gain in water availability (Sperry et al. 2002). A deep root system, however would be advantageous for plants on coarse soil under dry climate, where surface drying is predominant.

10.4.2 METALLOPHYTES

Plants that are able to colonize soils with a high metal availability must have efficient mechanisms that allow either exclusion or internal detoxification and compartmentation of the potentially toxic metal species (Ernst et al. 1992; Rengel 1997; Ernst 1998). Drought, due to either or both climate and poor physical soil properties, frequently is an additional stress factor on these soils. This suggests that, in addition to metal tolerance, mechanisms that confer drought avoidance or tolerance may also play an important role for plant performance on metal-rich soils.

Drought tolerance can be brought about by osmotic adjustment leading to lower tissue water potentials without turgor loss. Ernst (1974) found that the osmotic potential in leaves of metallophytes usually is in the range −1.0 to −1.6 MPa, which is similar to the values found in herbaceous plants growing on nearby normal soils. In *Thlaspi alpestre*, a Zn hyperaccumulator species, leaf osmotic potentials down to −3.0 MPa were observed (Ernst 1974). This value is within the range of osmotic potentials observed for species adapted to sandy soils in arid zones and close to the upper limit of osmotic potentials reported for halophytes (Wyn Jones and Storey 1981). The contribution of Zn to the low osmotic potential in this hyperaccumulator has been estimated in a maximum of 10% (Ernst 1974). The constitutively high concentration of organic acids (Tolrà et al. 1996) may also contribute to the low osmotic potential. It has been suggested that Ni may act as an osmoticum in Ni hyperaccumulators (Baker and Walker 1989). However, it is not likely that drought tolerance can provide a general explanation for the adaptive function of metal hyperaccumulation (Baker and Walker 1989; Boyd and Martens 1992).

The observation that many plant species adapted to mine or serpentine soils show morphological differences that resemble xeromorphism (Proctor and Woodell 1975) indicates that mechanisms leading to drought avoidance by water conservation may be important for plants living in metal-rich habitats. However, morphological characteristics are not necessarily correlated with metal tolerance. Morphology and metal tolerance probably are inherited independently, although some genetic coherence may occur (Baker and Dalby 1980). This implies that the degree of drought avoidance displayed by plants adapted to metalliferous soils would largely depend on the force of the selection factor, i.e. drought stress at the site. This may explain, at least in part, the conflicting results reported in the literature.

Water relations and drought avoidance as an adaptive factor in metal-tolerant plants have mainly been investigated comparing serpentinophytes to species or ecotypes from soils with "normal" metal concentrations. In an extensive study on soil and plant water relations on and off a serpentine soil, Hull and Wood (1984) found soil water potentials to be as high or even higher on the serpentine site. Nonetheless, oak species from the serpentine site showed more conservative water use than oak species from the control site. The authors concluded that seasonal soil water potentials do not provide mechanisms for exclusion of oak species from the serpentine site. In contrast, drought was an important factor for structuring the grassland community on a serpentine outcrop in California (Armstrong and Huenneke 1992). In a comparative investigation on water relations of serpentine and sandstone ecotypes of *Bromus hordeaceus* growing on both soil types with and without irrigation, Freitas and Mooney (1996) showed that the serpentine ecotype was better adapted to stress conditions caused by water deficiency and soil texture. These results agree with the view (Proctor and Woodell 1975) that the xeromorphic character of the serpentine species could be a response to other serpentine factors, in addition to drought. Xeromorphism may be considered neither a sufficient nor an indispensable property for all species adapted to metalliferous soils. Probably, a high drought resistance may be a quality which, in addition to the specific metal tolerance strategies, favors plant performance on metalliferous soils, due to preadaptation to other unfavorable edaphic properties occurring on these soils (Macnair 1987). Among those, decreased water availability as well as nutrient factors seem to play a role.

10.5   CONCLUSIONS

Metal toxicity can affect plant water relations at multiple levels. Until recently, most investigations only referred to metal toxicity induced water stress as a secondary effect that occurs relatively late in the complex toxicity syndrome. Many investigations performed after long exposure times with severely intoxicated plants only allowed the observation of the consequences of impaired water uptake and transport. Special attention was paid to the relationship between stomatal effects and photosynthesis. During the last years significant advances in the methods for studying water relation parameters in plant roots have led to a renewed insight into the basic mechanisms of water uptake and transport. The combined use of cell and root pressure probes has changed the former universally accepted view of the root acting as a simple osmometer into the hypothesis of the composite membrane model of the root. Future investigations are required to better establish the relative importance of apoplastic and symplastic pathways not only for radial water transport to the xylem, but also for the transport of nutrients and potentially toxic cations.

Studies at the whole plant level indicate that roots can regulate shoot water relations not only by hydraulic signals, but also by chemical messages. These findings are stimulating intense research into the role of water relations in cell and tissue growth, especially in roots under stress conditions. Alterations of the mechanical properties of cell walls and the impairment of water transport into root cells and through roots may be early events in the metal toxicity stress syndrome. Future experiments should address the possible influence of such effects on elongation of root cells and on signalling to shoots.

Plants that can colonize metal-rich soils must have efficient mechanisms that avoid contact of the potentially toxic metal species with sensitive cell constituents and metabolic reactions. In addition, drought avoidance and tolerance seem to be important traits for plant survival on many metalliferous sites. For both the successful breeding of crop plants that are Al-tolerant and high yielding on acid soils and the development of fast-growing metal-tolerant plants that may be used for phytoremediation of contaminated soils, a better knowledge of the relationship between metal tolerance and drought resistance mechanisms is clearly necessary.

ACKNOWLEDGEMENTS

The work in the author's laboratory is supported by the European Community (ICA4-CT-2000-30017) and the Spanish Government (DGICYTBFI2001-2475-CO2-01)

REFERENCES

Alia, Saradhi PP (1991) Proline accumulation under heavy metal stress. J Plant Physiol 138: 504-508

Aloni R (1987) Differentiation of vascular tissues. Annu Rev Plant Physiol 38: 179-204

Angelov A, Tsonev T, Uzunova A, Gaidardjieva K (1993) Cu^{2+} effect upon photosynthesis, chloroplast structure, RNA and protein synthesis of pea plants. Photosynthetica 28: 341-350

Armstrong JK, Huenneke LF (1992) Spatial and temporal variation in species composition in California grasslands: the interaction of drought and substratum. In: Baker AJM, Proctor J, Reeves RD (eds) The vegetation of ultramafic (serpentine) soils. Intercept Ltd, Andover, Hampshire, pp 213-233

Baker AJM, Dalby DH (1980) Morphological variation between some isolated populations of *Silene maritima* With. in the British Isles with particular reference to inland populations on metalliferous soils. New Phytol 84: 123-138

Baker AJM, Walker PL (1989) Ecophysiology of metal uptake by tolerant plants In: Shaw AJ (ed) Heavy metal tolerance in plants: evolutionary aspects. CRC Press, Boca Raton, Florida, pp 155-193

Barceló J, Poschenrieder CH (1990) Plant water relations as affected by heavy metal stress: a review. J Plant Nutr 13: 1-37

Barceló J, Poschenrieder CH (1997) Chromium in plants. In: Canali S, Tittarelli F, Sequi P (eds) Chromium environmental issues. Franco Angeli Publ, Milano, pp 101-129

Barceló J, Poschenrieder C, Gunsé B (1986a) Water realtions of chromium VI treated bush bean plants (*Phaseolus vulgaris* L. cv Contender) under both normal and water stress conditions. J Exp Bot 37: 178-187

Barceló J, Poschenrieder CH, Andreu I, Gunsé B (1986b) Cadmium-induced decrease of water stress resistance in bush bean plants (*Phaseolus vulgaris* L cv Contender). I. Effects of Cd on water potential, relative water content, and cell wall elasticity. J Plant Physiol 125: 17-25

Barceló J, Cabot C, Poschenrieder CH (1986c) Cadmium-induced decrease of water stress resistance in bush bean plants (*Phaseolus vulgaris* L cv Contender). II. Effects of Cd on endogenous abscisic acid levels. J Plant Physiol 125: 27-34

Barceló J, Poschenrieder CH, Vázquez MD, Gunsé B (1988a) Synergism between cadmium-induced ion stress and drought. In: Öztürk MA (ed) Plants and pollutants in developed and developing countries. Ege Univ Press, Izmir, Turkey, pp 529-544

Barceló J, Vázquez MD, Poschenrieder CH (1988b) Cadmium-induced structural and ultrastructural changes in the vascular system of bush bean stems. Bot Acta 101: 254-261

Barceló J, Vázquez MD, Poschenrieder CH (1988c) Structural and ultrastructural disorders in cadmium-treated bush bean plants (*Phaseolus vulgaris* L.). New Phytol 108: 37-49

Barceló J, Poschenrieder CH, Vázquez MD, Gunsé B (1996) Aluminium phytotoxicity. A challenge for plant scientists. Fertilizer Res 43: 217-223

Barceló J, Poschenrieder C, Lombini A, Llugany M, Bech J, Dinelli E (2001) Mediterranean plant species for phytoremediation. In: Abstracts cost action 837 WG2 workshop on phytoremediation of trace elements in contaminated soils and waters (with special emphasis on Zn, Cd, Pb and As), Madrid, 5-7 April, 23 pp

Barrowclough DE, Peterson CA, Steudle E (2000) Radial hydraulic conductivity along developing onion roots. J Exp Bot 51: 547-557

Bassi R, Sharma SS (1993a) Proline accumulation in wheat seedlings exposed to zinc and copper. Phytochemistry 33: 1339-1342

Bassi R, Sharma SS (1993b) Changes in proline content accompanying the uptake of zinc and copper by *Lemna minor*. Ann Bot 72: 151-154

Bazzaz FA, Carlson RW, Rolfe GL (1974) The effect of heavy metals on plants. I. Inhibition of gas exchange in sunflower by Pb, Cd, and Tl. Environ Pollut 7: 241-246

Becerril JM, González-Murua C, Muñoz-Rueda A, de Felipe MR (1989) Changes induced by cadmium and lead in gas exchange and water relations of clover and lucerne. Plant Physiol Biochem 27: 913-918

Bennet RJ, Breen CM, Fey MV (1985) Aluminium induced changes in the morphology of the quiescent center, proximal meristem and growth region of the root of *Zea mays*. S Afr J Bot 51: 355-362

Blackman PG, Davies WJ (1985) Root to shoot communication in maize plants and the effect of soil drying. J Exp Bot 36: 39-48

Boyd RS, Martens SN (1992) The raison d'être for metal hyperaccumulation by plants. In: Baker AJM, Proctor J, Reeves RD (eds) The vegetation of ultramafic (serpentine) soils. Intercept Ltd, Andover, Hampshire, pp 279-289

Brooks RR (1987) Serpentine and its vegetation. Dioscorides, Portland, Oregon

Carlson RW, Bazzaz FA, Rolfe GL (1975) The effect of heavy metals on plants. II. Net photosynthesis and transpiration of whole corn and sunflower plants treated with Pb, Cd, Ni and Tl. Environ Res 10: 113-120

Carvajal M, Cooke DT, Clarkson DT (1996) Responses of wheat plants to nutrient deprivation may involve the regulation of water-channel function. Planta 199: 372-381

Chen SL, Kao CH (1995) Cd induced changes in proline level and peroxidase activity in roots of rice seedlings. Plant Growth Regul 17: 67-71

Chrispeels MJ, Maurel C (1994) Aquaporins: the molecular basis of facilitated water movement through living plant cells? Plant Physiol 105: 9-13

Clijsters H, van Assche F (1985) Inhibition of photosynthesis by heavy metals. Photosynth Res 7: 31-40

Costa G, Morel JL (1994) Water relations, gas exchange and amino acid content in Cd-treated lettuce. Plant Physiol Biochem 32: 561-570

Costa G, Spitz E (1997) Influence of cadmium on soluble carbohydrates, free amino acids, protein content of in vitro cultured *Lupinus albus*. Plant Sci 128: 131-140

Dale JE, Sutcliffe JF (1986) Water relations of plant cells. In: Steward FC (ed) Plant physiology, a treatise, vol 9. Water and solutes in plants. Academic, London, pp 1-48

Davenport DC, Fisher MA, Hagan GL (1971) Retarded stomatal closure by phenylmercuric acetate. Physiol Plant 24: 330-336

Davies WJ, Zhang J (1991) Root signals and the regulation of growth and development of plants in drying soil. Annu Rev Plant Physiol Plant Mol Biol 42: 55-76

De Vos RCH, Schat H, Voojs R, Ernst WHO (1989) Copper-induced damage to the permeability barrier in roots of *Silene cucbalus*. J Plant Physiol 135: 164-169

De Vos RCH, Schat H, De Wal MAM, Vooijs R, Ernst WHO (1991) Increased resistance to copper-induced damage of the root cell plasmalemma in copper-tolerant *Silene cucubalus*. Physiol Plant 82: 523-528

Di Cagno R, Guidi L, De Gara L, Soldatini GF (2001) Combined cadmium and ozone treatments affect photosynthesis and ascorbate-dependent defences in sunflower. New Phytol 151: 627-636

Dueck TA (1986) The combined influence of sulphur dioxide and copper on two populations of *Trifolium repens* and *Lolium perenne*. In: Dueck TA (ed) Impact of heavy metals and air pollutants on plants. Academisch Proefschrift. Free Univ Press, Amsterdam, pp 102-114

Ernst WHO (1974) Schwermetallvegetation der Erde. Fischer, Stuttgart

Ernst WHO Verkleji JAC, Schat H (1992) Metal tolerance in plants. Acta Bot Neerl 41: 229-248

Ernst WHO (1998) Effects of heavy metals in plants at the cellular and organismic level. In: Schüürmann G, Markert B (eds) Ecotoxicology. Wiley/Spektrum Akad, New York, pp 587-620

Ernst WHO, Assunção AGL, Verkleij JAC, Schat H (2002) How important is apoplastic zinc xylem loading in *Thlaspi caerulescens*. New Phytol 155: 1-7

Freitas H, Mooney H (1996) Effects of water stress and soil texture on the performance of two *Bromus hordaceus* ecotypes from sandstone and serpentine soils. Acta Oecol 17: 307-317

Fuhrer J (1982) Early effects of excess cadmium uptake in *Phaseolus vulgaris*. Plant Cell Environ 5: 263-270

Gaspar M, Sissoëff I, Bousser A, Roche O, Mahé A, Hoarau J (2001) Transient variations of water transfer induced by $HgCl_2$ in excised roots of young maize plants: new data on the inhibition process. Aust J Plant Physiol 28: 1175-1186

Greger M, Johansson M (1992) Cadmium effects on leaf transpiration of sugar beet (*Beta vulgaris*). Physiol Plant 86: 465-473

Gunsé B (1987) Efectos del cromo sobre la nutrición y relaciones hídricas de *Phaseolus vulgaris*. PhD thesis, University of Autónoma de Barcelona, Bellaterra, Spain

Gunsé B, Llugany M, Poschenrieder CH, Barceló J (1992) Growth, cell wall elasticity and plasticity in *Zea mays* L. coleoptiles exposed to cadmium. In: Contaminación: efectos fisiológicos y mecanismos de actuación de contaminantes. ISBN 84-7908-049-3, Alicante University Publishers, Alicante (Spain), pp 179-188

Gunsé B, Poschenrieder CH, Barceló J (1997) Water transport properties of roots and root cortical cells in proton- and Al-stressed maize varieties. Plant Physiol 113: 595-602

Hagemeyer J, Breckle SW (2002) Trace element stress in roots. In: Waisel Y, Eshel A, Kafkafi U (eds) Plant roots: the hidden half, 3rd edn. Dekker, New York, pp 763-785

Hartung W, Davies WJ (1994) Abscisic acid under drought and salt stress. In: Pessarakli M (ed) Handbook of plant stress. Dekker, New York, pp 401-411

Haug A, Caldwell CR (1985) Aluminium toxicity in plants: role of the root plasma membrane and calmoduline. In: St John JB, Berlin E, Jackson PC (eds) Frontiers of membrane research. Beltsville symposium 9. Rowman and Allanheld, Totwa, pp 359-381

Henzler T, Steudle E (1995) Reversible closing of water channels in *Chara* internodes provides evidence for a composite transport model of the plasma membrane. J Exp Bot 46: 199-209

Heuer B (1994) Osmoregulatory role of proline in water- and salt-stressed plants. In: Pessarakli M (ed) Handbook of plant stress. Dekker, New York, pp 363-381

Hull JC, Wood SG (1984) Water relations of oak species on and adjacent to a Maryland serpentine soil. Am Midl Nat 112: 224-234

Hwang JU, Su S, Yi H, Kim J, Lee Y (1997) Actin filaments modulate both stomatal opening and inward K^+-channel activities in guard cells of *Vicia faba* L. Plant Physiol 115: 335-342

Jones H, Leigh RA, Wyn Jones RG, Tomos AD (1988) The integration of whole-root and cellular hydraulic conductivities in cereal roots. Planta 174: 1-7

Kahle H (1993) Response of roots of trees to heavy metals. Environ Exp Bot 33: 99-119

Kasai M, Sasaki M, Yamamoto Y, Maeshima M, Matsumoto H (1994) Possible involvement of abscisic acid in induction of two vacuolar H^+-pump activities in barley roots under aluminum stress. Abstracts of the 5th international symposium on genetics and molecular biology of plant nutrition. Davis, California, p 118

Kastori R, Petrovic M, Petrovic N (1992) Effect of excess lead, cadmium, copper and zinc on water relations in sunflower. J Plant Nutr 15: 2427-2439

Kennedy CD, Gonsalves FAN (1987) The action of divalent zinc, cadmium, mercury, copper and lead on the trans-root potential and H^+ efflux of excised roots. J Exp Bot 38: 800-817

Kirkham M (1978) Water relations of cadmium-treated plants. J Environ Qual 7: 334-336

Klimashevski EL, Dedov VM (1975) Localization of the mechanism of growth inhibiting action of Al^{3+} in elongating cell walls. Fiziol Rast 22: 1040-1046

Klimov SV (1985) Interaction of stress factors: increase of drought effect by the presence of Al^{3+} in the medium. Fiziol Rast 32: 532-538

Kramer PJ, Boyer JS (1995) Water relations of plants and soils. Academic Press, San Diego, 495 pp

Krizek DT, Foy CD (1988a) Role of water stress in differential aluminum tolerance of two barley cultivars grown in an acid soil. J Plant Nutr 11: 351-367

Krizek DT, Foy CD (1988b) Mineral element concentration of six sunflower cultivars in relation to water deficit and aluminum toxicity. J Plant Nutr 11: 409-422

Krizek DT, Foy CD, Wergin WP (1988) Role of water stress in differential aluminum tolerance of six sunflower cultivars grown in an acid soil. J Plant Nutr 11: 387-408

Lamoreaux RJ, Chaney WR (1977) Growth and water movement in silver maple seedlings affected by cadmium. J Environ Qual 6: 201-205

Lamoreaux RJ, Chaney WR (1978) The effect of cadmium on net photosynthesis, transpiration and dark respiration of excised silver maple leaves. Physiol Plant 43: 231-236

Levitt J (1980) Responses of plants to environmental stresses, vol 2. Water, radiation, salt, and other stresses. Academic Press, New York, 607 pp

Llugany M (1994) Respuestas diferenciales de cultivares de *Zea mays* L a la toxicidad por aluminio. PhD thesis University of Autónoma de Barcelona, Spain

Ludevid D, Höfte H, Himelblau E, Chrispeels MJ (1992) The expression pattern of the tonoplast intrinsic protein γ-TIP in *Arabisopsis thaliana* is correlated with cell enlargement. Plant Physiol 100: 1633-1639

Macnair MR (1987) Heavy metal tolerance in plants. A model evolutionary system. Tree 2: 354-359

Maggio A, Joly RJ (1995) Effects of mercuric chloride on the hydraulic conductivity of tomato root systems. Plant Physiol 109: 331-335

Marschner H (1986) Mineral nutrition of higher plants. Academic Press, London

Massot N, Poschenrieder CH, Barceló J (1994) Aluminium-induced increase of zeatin riboside and dihydrozeatin riboside in *Phaseolus vulgaris* L. cultivars. J Plant Nutr 17: 255-265

Massot N, Nicander B, Barceló J, Poschenrieder C, Tillberg E (2002) A rapid increase in cytokinin levels and enhanced ethylene evolution precede Al^{3+} induced inhibition of root growth in bean seedlings. Plant Growth Reg 37: 105-112

Maurel C (1997) Aquaporins and water permeability of plant membranes. Annu Rev Plant Physiol Mol Biol 48: 399-429

Mehta SK, Gaur JP (1999) Heavy-metal-induced proline accumulation and its role in ameliorating metal toxicity in *Chlorella vulgaris*. New Phytol 143: 253-259

Montero E, Cabot C, Barceló J, Poschenrieder C (1997) Endogenous abscisic acid levels are linked to decreased growth of bush bean plants treated with NaCl. Physiol Plant 101: 17-22

Montero E, Cabot C, Barceló J, Poschenrieder CH (1998) Relative importance of osmotic stress and ion-specific effects of ABA-mediated inhibition of leaf expansion growth in *Phaseolus vulgaris*. Plant Cell Environ 21: 54-62

Moustakas M, Ouzounidou G, Eleftheriou EP, Lannoye R (1996) Indirect effects of aluminium on the function of the photosynthetic apparatus. Plant Physiol Biochem 34: 553-560

Moustakas M, Ouzounidou G, Symeonidis L, Karataglis S (1997) Field study of the effects of excess copper on wheat photosynthesis and productivity. Soil Sci Plant Nutr 43: 531-539

Müller ML, Barlow PW, Pilet PE (1994) Effect of abscisic acid on the cell cycle in the growing maize root. Planta 195: 10-16

Ohki K (1986) Photosynthesis, chlorophyll and transpiration responses in aluminium stressed wheat and sorghum. Crop Sci 26: 572-575

Paivöke A (1983) The short-term effects of zinc on the growth anatomy and acid phosphatase of pea seedlings. Ann Bot Fenn 20: 197-203

Paul R, de Foresta E (1981) Effets du cadmium sur la transpiration du plantes. Bull Rech Agron Gembloux 16: 371-378

Pereira WE, de Siqueira DL, Martínez CA, Puiatti M (2000) Gas exchange and chlorophyll fluorescence in four citrus rootstocks under aluminium stress. J Plant Physiol 157: 513-520

Poschenrieder C, Barceló J (1999) Water relations in heavy metal stressed plants. In: Prasad MNV, Hagemayer J (eds) Heavy metal stress in plants. From molecules to ecosystems, Springer, Berlin, pp 207–229

Poschenrieder CH, Gunsé B, Barceló J (1989) Influence of cadmium on water relations, stomatal resistance and abscisic acid content in expanding bean leaves. Plant Physiol 90: 1365-1371

Prasad MNV (1997) Trace metals. In: Prasad MNV (ed) Plant ecophysiology. Wiley, New York, pp 207-249

Proctor J, Nagy L (1992) Ultramafic rocks and their vegetation: an overview. In: Baker AJM, Proctor J, Reeves RD (eds) The vegetation of ultramafic (serpentine) soils. Intercept Ltd, Andover, Hampshire, pp 469-494

Proctor J, Woodell SRJ (1975) The ecology of serpentine soils. Adv Ecol Res 9: 255-366

Punz WF, Sieghardt H (1993) The response of roots of herbaceous plant species to heavy metals. Environ Exp Bot 33: 85-98

Rauser WE, Dumbroff EB (1981) Effects of excess cobalt, nickel and zinc on the water relations of *Phaseolus vulgaris*. Environ Exp Bot 21: 249-255

Rengel Z (1997) Mechanisms of plant resistance to toxicity of aluminium and heavy metals. In: Basra AS, Basra RK (eds) Mechanisms of environmental stress resistance in plants. Harwood Acad Publ, Amsterdam, pp 241-276

Robb J, Busch L, Rauser WE (1980) Zinc toxicity and xylem vessel alterations in white beans. Ann Bot 46: 43-50

Ros R, Cooke DT, Burden RS, James CS (1990) Effects of herbicide MCPA, and the heavy metals, cadmium and nickel on the lipid composition, Mg^{2+}-ATPase activity and fluidity of plasma membranes from rice, *Oryza sativa* (cv Bahia) shoots. J Exp Bot 41: 457-462

Rygol J, Arnold WA, Zimmermann U (1992) Zinc and salinity effects on membrane trasnport in *Chara connivens*. Plant Cell Environ 15: 11-23

Saab IN, Sharp RE, Pritchard J (1992) Effect of inhibition of ABA accumulation on the spatial distribution of elongation in the primary root and mesocotyl of maize at low water potentials. Plant Physiol 99: 26-33

Schat H, Sharma SS, Vooijs R (1997) Heavy-metal-induced accumulation of free proline in a metal-tolerant and a nontolerant ecotype of *Silene vulgaris*. Physiol Plant 101: 477-482

Schlegel H, Godbold DL, Hüttermann A (1987) Whole plant aspects of heavy metal induced changes in CO_2 uptake and water relations of spruce (*Picea abies*) seedlings. Physiol Plant 69: 265-270

Schnabl H, Ziegler H (1974) Der Einfluss des Aluminiums auf den Gasaustausch und das Welken von Schnittpflanzen. Ber Dtsch Bot Ges 87: 13-20

Schnabl H, Ziegler H (1975) Über die Wirkung von Aluminiumionen auf die Stomatabewegung von *Vicia faba*-Epidermen. Z Pflanzenphysiol 74: 394-403

Sperry JS, Stiller V, Hacke UG (2002) Soil water uptake and transport through root systems. In: Waisel Y, Eshel A, Kafkafi U (eds) Plant roots: the hidden half, 3rd edn. Dekker, New York, pp 663-681

Steudle E (1993) Pressure probe techniques: basic principles and application to studies of water and solute relations at the cell tissue and organ level. In: Smith JAC, Griffiths H (eds) Water deficits: plant responses from cell to community. Bios Scientific Publ, Oxford, pp 5-36

Steudle E (2001) The cohesion-tension mechanisms and the acquisition of water by plant roots. Annu Rev Plant Physiol Plant Mol Biol 52: 847-875

Steudle E, Frensch J (1996) Water transport in plants: role of the apoplast. Plant Soil 187: 67-79

Takács Z, Tuba Z, Smirnoff N (2001) Exaggeration of desiccation stress by heavy metal pollution in *Tortula ruralis*: a pilot study. Plant Growth Reg 35: 157-160

Tazawa M, Asai K, Iwasaki N (1996) Characteristics of Hg- and Zn-sensitive water channels in the plasma membrane of *Chara* cells. Bot Acta 5: 388-396

Tolrà RP, Poschenrieder CH, Barceló J (1996) Zinc hyperaccumulation in *Thlaspi caerulescens*. II. Influence on organic acids. J Plant Nutr 19: 1541-1550

Tyerman SD, Bohnert HJ, Maurel C, Steudle E, Smith JAC (1999) Plant aquaporins: their molecular biology, biophysics and significance for plant water relations. J Exp Bot 50: 1055-1071

Van Assche F, Clijsters H (1983) Multiple effects of heavy metals on photosynthesis. In: Marcelle R, Clijsters H, van Pouke M (eds) Effects of stress on photosynthesis. Nijhoff/Junk, The Hague, pp 371-382

Van Assche F, Clijsters H (1990) Effects of metals on enzyme activity in plants. Plant Cell Environ 13: 195-206

Van Assche F, Ceulemans R, Clijsters H (1980) Zinc mediated effects on leaf CO_2 diffusion conductances and net photosynthesis in *Phaseolus vulgaris* L. Photosynth Res 1: 171-180

Van Assche F, Cardinaels C, Put C, Clijsters H (1984) Premature leaf ageing induced by heavy metal toxicity? Arch Int Physiol Biochim 94: PF27-PF28

Vázquez MD, Poschenrieder CH, Barceló J (1987) Chromium VI induced structural and ultrastructural changes in bush bean plants (*Phaseolus vulgaris* L.). Ann Bot 59: 427-438

Vázquez MD, Poschenrieder CH, Barceló J (1989) Pulvinus structure and leaf abscission in cadmium treated bean plants (*Phaseolus vulgaris*). Can J Bot 67: 2756-2764

White PJ (2001) The pathways of calcium movement to the xylem. J Exp Bot 52: 891-899

White PJ, Whiting SN, Baker AJM, Broadley MR (2002) Does zinc move apoplastically to the xylem in roots of *Thlaspi caerulescens*? New Phytol 153: 201-211

Williamson A, Johnson MS (1981) Reclamation of metalliferous mine waste. In: Lepp NW (ed) Effect of heavy metal pollution on plants, vol 2. Applied Science Publ, London, pp 185-212

Wyn Jones RG, Storey R (1981) Betaines. In: Paleg LG, Aspinall D (eds) The physiology and biochemistry of drought resistance in plants. Academic Press, Sydney, pp 171-204

Zhang W-H, Tyerman SD (1999) Inhibition of water channels by $HgCl_2$ in intact wheat root cells. Plant Physiol 120: 849-857

Zhao XJ, Sucoff E, Stadelmann EJ (1987) Al^{3+} and Ca^{2+} alteration of membrane permeability of *Quercus rubra* root cortex cells. Plant Physiol 83: 159-162

Zimmermann MH (1983) Xylem structure and the ascent of sap. Springer, Berlin Heidelberg New York

Chapter 11

Heavy Metals as Essential Nutrients

Zdenko Rengel

Soil Science and Plant Nutrition, Faculty of Natural and Agricultural Sciences, The University of Western Australia, 35 Stirling Highway, Crawley WA 6009, Australia

11.1 INTRODUCTION

Many heavy metals are essential for plants and animals when present in the growing medium in low concentrations (micronutrients: Cu, Zn, Fe, Mn, Mo, Ni, and Co); they become toxic only when a concentration limit is exceeded (in which case the term 'heavy metals' rather than 'micronutrients' is used). This review will concentrate on Fe, Zn and Mn, while the reader is referred to other publications for a more comprehensive coverage of micronutrients (Marschner and Römheld 1991; Marschner 1995; Welch 1995; Reid 2001).

11.2 PHYSIOLOGICAL FUNCTIONS OF MICRONUTRIENTS

Through their involvement in various enzymes and other physiologically active molecules, micronutrients are important for gene expression; biosynthesis of proteins, nucleic acids, growth substances, chlorophyll and secondary metabolites; metabolism of carbohydrates and lipids; stress tolerance, etc. Micronutrients are also involved in structural and functional integrity of various membranes and other cellular components.

11.2.1 Metalloproteins

11.2.1.1 Iron

Iron is required for the functioning of a range of enzymes, especially those involved in oxidation and reduction processes, for synthesis of the porphyrin ring (chlorophyll and heme biosynthesis), reduction of nitrite and sulphate, N_2-fixation (as part of the leghemoghlobin), etc. Iron is a constituent of ferredoxin and takes part in assembling of the thylakoid units, thus having important functions in electron transport as well as in photosynthesis (photosystem I).

11.2.1.2 Zinc

Zinc is required for the activity of various types of enzymes, including dehydrogenases, aldolases, isomerases, transphosphorylases, and RNA and DNA polymerases. In addition, a number of important plant enzymes contain bound Zn; the levels of these enzymes decline sharply under conditions of Zn deficiency (e.g. alcohol dehydrogenase, Moore and Patrick 1988; Cu/Zn superoxide dismutase, Cakmak and Marschner 1988a; Yu et al. 1998, 1999b; carbonic anhydrase, Rengel 1995a; and DNA-dependent RNA polymerases, Falchuk et al. 1976, 1977). Zn-containing enzymes lose their activity upon removal of Zn atoms contained in the functional molecule (e.g. RNA polymerase, Falchuk et al. 1977; carbonic anhydrase, Guliev et al. 1992). Zinc deficiency is therefore associated with an impairment of carbohydrate metabolism and protein synthesis (Pearson and Rengel 1997; Rengel 1999a, 2001).

Zn-metalloproteins forming a specific loop (Zn-finger motif) are regulators of gene expression (DNA-binding transcription factors, Rhodes and Klug 1993; Vallee and Falchuk 1993; Bird et al. 2000). Without them, RNA polymerase cannot complete its function of transcribing genetic information from DNA into RNA. The number of various Zn fingers in cells may be large (e.g. in human cells as much as 1% of the total DNA is suggested to code for Zn fingers, Rhodes and Klug 1993).

A number of genes encoding Zn-finger proteins have been cloned recently (see Takatsuji 1998; Liu et al. 1999). The Zap1 Zn-finger transcriptional activator of *Saccharomyces cerevisiae* controls intracellular zinc homeostasis by functioning not just in DNA binding, but also in sensing of intracellular Zn activity (Bird et al. 2000). Expression of Zn-finger proteins may be tissue-specific. In situ mRNA hybridisation analysis showed that the *DAG1* Arabidopsis gene coding for the Dof Zn-finger protein was expressed in the phloem of all organs of the mother plant, but not in the embryo (Papi et al. 2002).

Carbonic anhydrase (carbonate-lyase, carbonate dehydratase) is a ubiquitous enzyme (found in animals, terrestrial plants, eukaryotic algae, cyanobacteria) and is localised both in the cytosol and in the chloroplasts (Guliev et al. 1992). The gene for carbonic anhydrase has been cloned (e.g. cDNA from the unicellular green algae *Cocomyxa*), and functionally characterised upon overexpression in *Escherichia coli* (Hiltonen et al. 1998). While the mammalian carbonic anhydrase enzyme consists of a single polypeptide chain of about 30 kDa and contains one Zn atom per molecule, carbonic anhydrase in plants can be a dimer, tetramer, hexamer or octamer with a molecular mass ranging from 42 to 250 kDa (Colman 1991) and one Zn atom in every subunit (Guliev et al. 1992). In both mammalian and plant carbonic anhydrase, Zn^{2+} is coordinated to the imidazole rings of the three histidine residues close to the active site of the enzyme (Pocker and Sarkanen 1978). Removal of Zn from the carbonic anhydrase molecule in vitro results in an irreversible loss of catalytic activity (Guliev et al. 1992).

Carbonic anhydrase catalyses the reversible reaction of CO_2 hydration (Guliev et al. 1992; Sültemeyer et al. 1993). Activity of carbonic anhydrase decreases in a number of plant species as a consequence of Zn deficiency; decreased activity of carbonic anhydrase is accompanied by reduced photosynthetic rates (Rengel 1995a, and references therein). The higher photosynthetic rate in some wheat genotypes under Zn deficiency (Fischer et al. 1997) was related to higher CO_2 availability due to a higher carbonic anhydrase activity (Rengel 1995a).

11.2.1.3 *Manganese*

Manganese is a constituent of only two enzymes (water-splitting complex in the photosystem II and Mn-superoxide dismutase), but is required for functioning of a large number of other enzymes. Since Mn is a cofactor for (1) phenylalanine ammonia-lyase that mediates production of cinnamic acid and various other phenolic compounds, and (2) peroxidase involved in polymerisation of cinnamyl alcohols into lignin (Burnell 1988), deficiency of Mn may prevent plants from effectively building up phenolics and lignin content which is considered the primary defence against fungal infection (Burnell 1988; Matern and Kneusel 1988; Rengel et al. 1994; and references therein).

11.2.2 Concentration of Micronutrients in Cellular Compartments

While micronutrients are involved in mediating activities of numerous enzymes (either as direct constituents of the molecule, or as cofactors required for proper functioning of the enzymes), the relationship between the activity of micronutrient-dependent enzymes and the concentration of *total* micronutrients in the tissue is frequently poor. For example, in Cu-deficient tobacco plants, Cu concentration in leaf tissue decreased 20-fold, but the activity of Cu/Zn-superoxide dismutase barely changed compared with control Cu-sufficient plants (Yu et al. 1998). A reason for such a poor relationship may be the absence of suitably sensitive techniques that would allow determination of concentrations of *physiologically active* micronutrients in relevant cell compartments. In contrast, intracellular distribution of macronutrients, like K, in cereal leaf cells is known (Leigh and Tomos 1993).

11.2.2.1 *Intracellular Zn*

Total Zn concentration in roots and leaves is a poor estimate of the plant Zn status (Gibson and Leece 1981; Cakmak and Marschner 1987; Rengel and Graham 1995). Better prediction of the Zn status may come from distinguishing concentration of Zn incorporated into various stable cellular complexes from the concentration representing the labile, loosely bound Zn pool (free Zn^{2+}), which may be responsible for regulating activities of Zn-containing and Zn-dependent enzymes in a particular cell compartment. Direct measurements of labile intracellular Zn^{2+} have been achieved for the first time only recently using cultures of various types of human cells (Zalewski et al. 1994) and the cell-permeant fluorophore zinquin (ethyl [2-methyl-8-*p*-toluenesulphonamido-6-quinolyloxy] acetate), which specifically fluoresces with free or loosely bound intracellular Zn^{2+}. No measurement of ionic activity of loosely bound (free) intracellular Zn^{2+} in plants could be found in the literature.

Plant cells have a finely tuned system for maintaining intracellular Zn^{2+} activity within the optimal range. Some components of this regulatory system have recently been deciphered at the molecular level. Transcription of genes *ZRT1* and *ZRT2* coding for Zn transporters increases under Zn deficiency and is regulated by binding of the Zn-responsive Zap1p transcription regulator whose biosynthesis also increases under Zn deficiency (Zhao et al. 1998).

The vacuole is an important component of intracellular Zn homeostatic mechanism, with Zn being accumulated in the vacuole of Zn-replete yeast cells via Zrc1p and Cot1p transporters. This stored Zn can be mobilised in Zn-deficient cells via another tonoplast-located zinc transporter (Zrt3p) coded for by the gene *ZRT3* that is regulated by the Zap1p transcription regulator (MacDiarmid et al. 2000).

11.2.2.2 *Intracellular Mn*

A ^{31}P-NMR study showed that maize root cells supplied with a range of Mn concentrations in the outside medium maintain low cytosolic Mn^{2+} concentration, with the non-equilibrium distribution between cytosol and the vacuole, where the vacuole acted as a sink for Mn^{2+} (Quiquampoix et al. 1993). This technique should be used more in elucidating distribution of Mn^{2+} among intracellular pools under a variety of conditions.

11.2.2.3 *Oxidative Stress*

Oxidative stress results from deleterious effects of reduced oxygen species such as superoxide and hydrogen peroxide, which can form hydroxyl radicals (OH$^{\bullet}$), the most reactive species known in the field of chemistry. Hydroxyl radicals cause lipid peroxidation, protein denaturation, DNA mutation, photosynthesis inhibition, etc. (Bowler et al. 1992; Foyer et al. 1994; Dat et al. 2000; Kuzniak 2002).

Under physiological conditions, plants produce significant amounts of superoxide and hydrogen peroxide (e.g. in the electron transport that occurs during photosynthetic reactions or during ATP generation in mitochondria; in the process of β-oxidation of fatty acids in glyoxysomes, etc.; Bowler et al. 1992). In order to survive, aerobic organisms have evolved a defence mechanism against oxidative stress. Since hydroxyl radicals are far too reactive to be controlled easily, the defence mechanism is based on elimination of their precursors (superoxide and hydrogen peroxide). Among a number of antioxidative enzymes, the central role in the plant antioxidative mechanism is played by superoxide dismutase (SOD), an enzyme that converts superoxide into hydrogen peroxide, which is then broken down to water by either catalase or various peroxidases.

There are three known types of SOD: Cu/ZnSOD is cytosolic but can be present in the chloroplast stroma as well, MnSOD is located in the mitochondrial matrix, and FeSOD is present in the chloroplast stroma (Bowler et al. 1992; Dat et al. 2000). Cu/ZnSOD is the most abundant form of SOD in higher plant cells and plays an important protective role in various environmental stresses.

Oxidative stress is at the core of a number of environmental stresses, including heavy metal toxicity (Chongpraditnun et al. 1992; Przymusinski et al. 1995; Weckx and Clijster 1996; Prasad et al. 1999; Schickler and Caspi 1999; Dixit et al. 2001; Schutzendubel and Polle 2002) and deficiencies of Zn (Cakmak and Marschner 1988a; Wenzel and Mehlhorn 1995; Cakmak et al. 1997b; Yu et al. 1998, 1999b; Obata et al. 1999; Cakmak 2000), Fe (Iturbe-Ormaetxe et al. 1995), Mn (Kröniger et al. 1995; Yu et al. 1998, 1999a) and Cu (Tanaka et al. 1995; Yu et al. 1998; Quartacci et al. 2001).

Tobacco SOD, particularly Cu/ZnSOD, may be relatively more sensitive to Zn than to Cu deficiency (Yu et al. 1998). Similarly, Vaughan et al. (1982) found that Zn deficiency resulted in a 46% reduction in fresh weight of *Lemna* fronds and a 52% decrease in activity of total SOD, while Cu deficiency caused a 23% reduction in fresh weight and had no apparent effect on activity of total SOD compared with the control. A greater sensitivity to Zn than to Cu deficiency may be due to different roles of Cu and Zn in the SOD enzyme: Cu is catalytic (required for enzyme action), whereas Zn is structural (required for enzyme stability; Vaughan et al. 1982).

The activity of SOD decreases under Zn deficiency in a range of plant species (Cakmak and Marschner 1988a; Wenzel and Mehlhorn 1995; Yu et al. 1998; Cakmak 2000) and in Zn-inefficient wheat genotypes (but not in Zn-efficient ones; Cakmak et al. 1997b; Yu et al. 1999b) because Zn atoms can be released from the Cu/ZnSOD molecules, making them inactive. A similar type of inactivation of Cu/ZnSOD occurred under Cu deficiency (Tanaka et al. 1995).

11.2.3 Membrane Structure and Function

In animal tissues, Zn^{2+} ions have a role in protecting membranes from oxidative and peroxidative damage, thus allowing proper functioning of enzymes bound to the plasma membrane (Bettger and O'Dell 1981). In plant cells, the importance of Zn in maintaining the structural and functional integrity of the plasma membrane has also been recognised. Zinc-deficient plants of various species exhibit high permeability of the plasma membrane, resulting in significant leakage of organic and mineral components from root cells (Welch et al. 1982; Cakmak and Marschner 1988a,b, 1990; Welch and Norvell 1993; Rengel 1995b).

Zinc deficiency results in disturbance of the normal pattern of ion transport across the plasma membrane (Rengel 1995b, 1999b,c; Rengel and Graham 1995, 1996, and references therein). Generally, increased membrane permeability of plant cells is correlated with an increase in peroxidation of membrane components (cf. Dhindsa et al. 1981). Zinc-deficient roots of barley and wheat exhibited membrane leakiness and had lower concentrations of sulfhydryl groups in the root-cell plasma membranes than Zn-sufficient roots (Welch and Norvell 1993; Rengel 1995b); hence, Zn may play a role in the regulation of transmembrane ion fluxes by preventing oxidation of sulfhydryl groups to disulfides in proteins involved in ion-channel gating in the plasma membrane of root cells (Kochian 1993; Welch and Norvell 1993; Rengel 1995b; Welch 1995).

11.2.4 Growth Regulators

The role of micronutrients in biosynthesis and breakdown of growth regulators is not completely understood. Iron is involved in biosynthesis of ethylene; Fe-deficient plants therefore have insufficient ethylene levels. Conversely, in dicots and non-graminaceous monocots, Fe^{3+} reductase activity may be stimulated by ethylene (cf. Romera et al. 1996).

Under suboptimal supply of Mn (either deficient or toxic), the activity of indole-3-acetic acid (IAA) oxidase is high, resulting in inadequate levels of IAA in plant tissues (Marschner 1995). In addition, Zn-deficient plants have low levels of IAA, but it is not clear whether this is a result of decreased biosynthesis or increased degradation (Cakmak et al. 1989).

11.3 PLANT RESPONSES TO MICRONUTRIENT DEFICIENCIES

Plant responses to micronutrient deficiencies have been studied primarily with respect to Fe, Zn and Mn, which are the most common trace metal deficiencies in agriculture. These deficiencies are seldom caused by low soil content, but rather by poor solubility of these micronutrients at neutral to alkaline pH. Plant responses to micronutrient deficiency frequently cause changes in chemistry and biology of the rhizosphere, a layer of soil surrounding roots and varying in thickness between 0.1 and a few millimetres, depending on the length of root hairs (Rengel 1999a, 2001).

11.3.1 Iron

Plants respond to Fe deficiency by employing a range of mechanisms that may be grouped into strategy I (found in dicots and non-graminaceous monocots) and strategy II (in graminaceous monocots; Römheld 1991; Marschner and Römheld 1994; Mori 1994; Welch 1995; Ma and Nomoto 1996; Pearson and Rengel 1997; Crowley and Rengel 1999; Rengel 1999a, 2001).

11.3.1.1 Strategy I

Acidification of the rhizosphere, which occurs primarily at the root apices, increases the solubility of Fe-containing minerals, especially amorphous Fe hydroxide (Jones et al. 1996). A 100-fold increase in Zn and a 1000-fold enhancement in Fe solubility occur for every unit decrease in pH. An increased net extrusion of protons is due to increased activity of the plasma-membrane-bound H^+-ATPase, resulting not only in the rhizosphere acidification, but also in the hyperpolarisation of the plasma membrane and thus an increased driving force for Fe^{2+} uptake into the cell (Rabotti and Zocchi 1994; Brancadoro et al. 1995). Exudation of organic acids, particularly citric and malic (Landsberg 1981; Brancadoro et al. 1995) and caffeic (Olsen et al. 1981), also have a role in Fe-deficiency-induced acidification of the rhizosphere.

In addition to rhizosphere acidification, strategy I plants extrude reducing and/or chelating compounds. Changes in the redox status of the rhizosphere may also be caused by depletion of oxygen by root and microbial respiration in water-saturated soils or at microsites of high microbial activity (Bienfait 1988; Cohen et al. 1997). It is not yet clear which natural compounds function as the primary Fe chelators in the rhizosphere. Siderophores, produced by many rhizosphere microorganisms, are unlikely candidates because the rates of Fe reduction measured for most Fe^{3+}-siderophore complexes are extremely low in comparison with those for synthetic chelates such as EDTA (Crowley and Rengel 1999).

The plasma-membrane-embedded reductase reduces Fe^{3+} in various Fe^{3+}-specific chelates to Fe^{2+} (see Römheld and Marschner 1986; Welch 1995). Fe^{2+} dissociates from the chelate and can then be taken up by a Fe^{2+} transport system in the root-cell plasma membrane. The reductase may also function in reduction of Fe^{3+} in complexes with citrate, malate, heme or fulvic acid.

Recent research has shown that overexpression of root Fe^{3+} reductase activity by *E107* mutant of pea (Welch and La Rue 1990) and *chloronerva* mutant of tomato (Stephan and Grün 1989) was inhibited by ethylene inhibitors, indicating that a large Fe accumulation capacity of these two mutants may be due to a genetic defect in their ability to regulate root ethylene production (Romera et al. 1996), even though the effect might be via the IAA-related changes in ethylene biosynthesis and action.

11.3.1.2 Strategy II

Graminaceous species acquire Fe by releasing phytosiderophores (PS), non-proteinogenic amino acids with a high binding affinity for Fe, and by taking up ferrated PS through a specific transmembrane uptake system (Marschner and Römheld 1994; Mori 1994; Welch 1995; Rengel 2001). An enhanced mobilisation of Fe from a calcareous soil, even as far away from the root surface as 4 mm, demonstrated a high capacity of PS to mobilise Fe (Awad et al. 1994).

An increased exudation of PS under Fe deficiency occurs in a distinct diurnal rhythm, with a peak exudation after the onset of illumination, the light ensuring the continuous supply of assimilates (Römheld 1991; Cakmak et al. 1998). However, it is an increase in temperature during the light period, rather than the onset of light itself, which causes an increase in PS exudation (see Mori 1994).

A number of genes involved in biosynthesis of phytosiderophores and their precursor nicotianamine have been cloned (Mori 1997; Nakanishi et al. 2000; Yamaguchi et al. 2000a,b), including a gene family coding for nicotianamine synthase (Higuchi et al. 1999). Regulation of gene expression and the mosaic of biosynthetic pathways involved in phytosiderophore production are being deciphered at an increasing pace (Itai et al. 2000; Higuchi et al. 2001; Kobayashi et al. 2001). Availability of genes as well as understanding of their induction and regulation will be crucial in future attempts to genetically engineer strategy II crops for increased capacity to synthesise and exude phytosiderophores and thus improve Fe nutrition in environments low in available Fe.

11.3.2 Zinc

Wheat genotypes better adapted to Zn-deficient soils developed longer and thinner roots (a greater proportion of fine roots with diameters ≤ 0.2 mm) than the poorly adapted ones when grown in Zn-deficient soils (Dong et al. 1995), thus resulting in exploration of a larger volume of soil. As Zn ions are transported toward roots by diffusion, an increased root surface area is particularly important because it reduces the distance Zn ions have to travel in soil solution to the root absorption sites (Marschner 1993), hence allowing more efficient scavenging.

Root exudation of PS increases in a range of graminaceous species and genotypes under Zn deficiency (Zhang et al. 1989, 1991; Cakmak et al. 1994, 1996a-c, 1997a, 1998; Walter et al. 1994; Erenoglu et al. 1996, 1999, 2000; Rengel 1997, 1999a-c; Rengel et al. 1998a; Rengel and Römheld 2000; Tolay et al. 2001). However, unequivocal experimental proof that PS play a role in mobilisation and uptake of Zn from Zn-deficient soils has yet to be reported. This is especially important because PS have a greater affinity (up to 2-fold) for Fe than for Zn.

When a range of cereal species was compared, those that were Zn-inefficient (e.g. durum wheat) exuded much less PS in the rhizosphere than species that were comparatively Zn-efficient (bread wheat or rye; Cakmak et al. 1994, 1996b, 1997a; Rengel et al. 1998a; Rengel 1999a-c; Rengel and Römheld 2000). However, there is no correlation between the PS exudation of various bread wheat genotypes and their Zn efficiency, thus casting doubts on a causal role of PS exudation in determining the level of Zn efficiency (Erenoglu et al. 1996). Similarly, there is no relationship between differential Zn efficiency of bread wheat genotypes and their capacity to take up inorganic Zn from the solution (Erenoglu et al. 1999), even when both low- and high-affinity Zn uptake systems are taken into account (Hacisalihoglu et al. 2001). Instead, differential Zn efficiency among bread wheat genotypes appears to be related to differential zinc utilisation efficiency in tissues (Rengel 1995a,b, 1999a-c, 2001; Cakmak et al. 1997b, 1999, 2001; Yu et al. 1999b; Torun et al. 2000).

Zinc deficiency increases root exudation of amino acids, sugars and phenolics in a range of plant species (Zhang 1993; and references therein). The importance of this exudation has not yet been assessed in terms of increasing plant capacity to acquire Zn from soils of low Zn availability.

11.3.3 Manganese

The plant availability of Mn depends on its oxidation state: the oxidised form (Mn^{4+}) is not available to plants, while the reduced form (Mn^{2+}) is. When oxygen is depleted from the growing medium, changes in the redox potential occur; in such a case, NO_3^-, Mn, and Fe serve as alternative electron acceptors for microbial respiration, and are transformed into reduced ionic species. This process increases the solubility and availability of Mn and Fe, but is not under the direct control of the plant.

Manganese availability may also be influenced by the activity of Mn-oxidising and Mn-reducing microorganisms that colonise plant roots (Rengel et al. 1996; Rengel 1997). Mn oxidation $[2Mn^{2+}+O_2+2H_2O \rightarrow 2(Mn^{4+})O_2+4H^+]$ provides energy to support growth of chemolithotrophic bacteria such as Leptothrix, Arthrobacter, Bacillus, Pseudomonas or Hyphomicrobium (Ghiorse 1988). On the other hand, Mn reduction $[(Mn^{4+})O_2+4H^++2e^- \rightarrow Mn^{2+}+2H_2O]$ and Mn reducers are found in many bacterial genera, e.g. Bacillus, Pseudomonas, Clostridium, Micrococcus or Arthrobacter (Ghiorse 1988).

Microorganisms that reduce Mn will increase its availability to plants. Indeed, up to 10-fold greater numbers of Mn-oxidising bacteria (and fewer Mn reducers) have been found on soybean roots in Mn-deficient patches than outside these patches (Huber and McCay-Buis 1993). In addition, altering the ratio of Mn reducers to Mn oxidisers by inoculating wheat with Mn-reducing bacteria increased Mn uptake and improved wheat growth in Mn-deficient soil (Graham 1988; Marschner et al. 1991).

Since availability of Mn is low at neutral to alkaline pH (Tong et al. 1995; Gherardi and Rengel 2001), acidification of the rhizosphere has an important role in mobilising soil Mn (Marschner et al. 1986). pH changes in the rhizosphere depend on the buffering capacity of soils, bulk soil pH, nitrogen sources, and other factors (Tong et al. 1997; and references therein).

The nature and activity of root exudate components that might be involved in mobilisation of Mn are still unclear (for reviews, see Rengel 1997, 1999a, 2000). The importance of organic acids (e.g. malic and citric; Godo and Reisenauer 1980) remains unclear because their effectiveness in forming stable complexes with micronutrients is low at high pH (Jones and Darrah 1994), where Mn deficiency usually occurs. Further research on root exudates effective in mobilising Mn from high-pH substrates is warranted.

11.4 MICRONUTRIENT UPTAKE FROM SOIL AND TRANSPORT IN PLANTS

Deficiency of one micronutrient could substantially facilitate uptake of one or more of other micronutrients (Kochian 1991). The compensatory absorption of micronutrients under deficiency stress has been reported between Mn and Cu or Zn (del Río et al. 1978), between Zn and Fe, Mn or Cu (Rengel and Graham 1995, 1996; Rengel et al. 1998a; Rengel 1999b,c), and between Fe and Mn (Iturbe-Ormaetxe et al. 1995).

A range of genes coding for transporters of various specificity that are involved in uptake of micronutrients have been cloned recently (see Guerinot 2000; Gaither and Eide 2001). Most often, complementation of appropriate yeast mutants deficient in uptake systems for particular nutrients has been used (e.g. Eide et al. 1996; Zhao and Eide 1996a,b; Grotz et al. 1998; Korshunova et al. 1999; Curie et al. 2000; Pence et al. 2000; Thomine et al. 2000; Eckhardt et al. 2001; Vert et al. 2001; Moreau et al. 2002).

11.4.1 Iron

In graminaceous species (strategy II), ferrated PS are transported across the plasma membrane by a transport system which recognises specifically the stereostructure of the Fe^{3+}-PS complex (Oida et al. 1989) in preference to similar complexes with Cu^{2+}, Zn^{2+}, Co^{2+} and Co^{3+} (Ma et al. 1993). However, these metal cations can decrease the rate of uptake of the Fe^{3+}-PS complex by competing with Fe^{3+} for binding to PS in accordance with the stability constants of these metals with PS (Ma and Nomoto 1993).

The undissociated Fe^{3+}-PS complex is taken up by maize and rice roots via an energy-dependent process (von Wiren et al. 1996; and references therein). Von Wiren et al. (1995) found two kinetically distinct components of the Fe^{3+}-PS uptake: (1) a saturable, high-affinity component at low concentrations, and (2) a linear component at higher concentrations, suggesting that either two transporters are present, or that one transporter may assume multiple structural and/or functional forms. Current work in several laboratories around the world is aimed at molecular characterisation of the Fe^{3+}-PS transporter in strategy II plants.

A new Fe-deficiency-induced cDNA, *IDI7*, was isolated from the roots of Fe-deficient barley (Yamaguchi et al. 2002). The regulation by Fe deficiency was restricted to roots only. The transporter has features typical of ATP-binding cassette (ABC) transporters and was situated in the tonoplast (Yamaguchi et al. 2002).

Two genes coding for Fe transporters have been cloned from Arabidopsis (strategy I plant): *AtIRT1* (Eide et al. 1996; Korshunova et al. 1999) and *AtIRT2* (Vert et al. 2001; IRT=iron regulated transporter). Both genes were cloned using functional complementation of a yeast mutant deficient in Fe uptake. These two transporters carry across the plasma membrane not just Fe but also Zn, with the AtIRT1 protein transporting Mn and Cd as well. Replacing the glutamic acid residue at position 103 with alanine, however, resulted in a loss of capacity of the AtIRT1 protein to transport Zn, while replacing aspartic acid with alanine at either position 100 or 136 caused a loss of transport function for Fe and Mn (Rogers et al. 2000). Further work on functional characterisation of parts of the protein sequence is eagerly awaited.

The *AtIRT2* gene is expressed particularly strongly in root tips and root hairs of Fe-deficient Arabidopsis and is up-regulated at the transcriptional level in response to Fe deficiency (Vert et al. 2001). By fusing the *IRT2* promoter to the *uidA* reporter gene, Vert et al. (2001) showed that the *IRT2* promoter is mainly active in the external cell layers of the root subapical zone, thus providing the first tissue localisation of a plant metal transporter.

Using *AtIRT1* as a probe, two similar Fe transporters (*LeIRT1* and *LeIRT2*) have been cloned from Fe-deficient tomato (*Lycopersicon esculentum*; Eckhardt et al. 2001). Expression of these genes was particularly strong in roots, with only *LeIRT1* exhibiting up-regulation by Fe deficiency. These genes complemented the Fe-deficient yeast mutant, and were capable of transporting Mn, Zn and Cu in addition to Fe.

The family of Arabidopsis *AtNRAMP* (natural resistance-associated macrophage protein) genes are up-regulated by Fe deficiency (Curie et al. 2000; Thomine et al. 2000). At least some of these genes are involved in maintaining Fe homeostasis, perhaps through sequestering free cytosolic Fe to the vacuole.

11.4.2 Zinc

Zinc uptake is inversely related to symplasmic Zn concentration; following withdrawal of Zn supply, the ability of plants to take up Zn increases (Rengel and Graham 1996; Rengel and Hawkesford 1997; Rengel et al. 1998a). Zinc deficiency caused an increase in I_{max} (maximum net uptake rate) in some wheat genotypes (Rengel and Wheal 1997; Hart et al. 1988; Rengel et al. 1998a).

In wheat, Zhang et al. (1991) suggested a model in which PS, released across the plasma membrane, mobilise Zn in the apoplasm of root cells, but dissociation of the Zn-PS complex occurs at the plasma membrane, and only Zn is taken up into the cytoplasm. However, recent research has indicated that maize roots can take up both Zn^{2+} and the Zn-PS complex (von Wiren et al. 1996).

The Zn transporters belonging to the ZIP (zinc-induced permease) family have been cloned so far from a range of organisms, including a mammalian Zn efflux transporter that confers resistance to Zn toxicity (Palmiter and Findley 1995), the high- and low-affinity Zn transporters of yeast coded for, respectively, by *ZRT1* and *ZRT2* genes induced under Zn starvation (Zhao and Eide 1996a,b; Eide 1997), as well as Zn transporters from Arabidopsis (Grotz et al. 1998), *Lycopersicon esculentum* (Eckhardt et al. 2001) and *Thlaspi caerulescens* (Pence et al. 2000). The ZIP family has now been divided into four subfamilies (Gaither and Eide 2001).

The ZIP family proteins have eight membrane-spanning regions and can transport a range of cations in addition to Zn (e.g. Mn, Fe, Cd; Guerinot 2000). *AtZIP1* and *AtZIP3* are expressed strongly in Zn-deficient plants (Grotz et al. 1998). The uptake activity of *AtZIP1-3* is dependent on temperature and pH, and is saturable, with K_m similar to values expected in the soil solution in the rhizosphere (Grotz et al. 1998).

Interesting new research has discovered the first *Glycine max* ZIP-type protein transporter (GmZIP1) specifically located in the peribacteroid membrane, an endosymbiotic membrane in *Glycine max* root nodules (Moreau et al. 2002). No expression of the corresponding gene was found in root, shoot or leaf cells. This discovery may open up a new chapter in understanding ion transport in microbe-plant symbioses.

Overexpression of Zn-transporter genes may result in increased uptake of Zn into root cells (e.g. in *T. caerulescens*, Lasat et al. 2000; Pence et al. 2000). In contrast, Zn-transporter genes can also be regulated by protein degradation (e.g. in yeast, Gitan et al. 1998, thus preventing potential Zn toxicity). Potential post-translational regulation of the Arabidopsis Zn transporter *AtZIP1* (Grotz et al. 1998) has been shown in yeast (Guerinot 2000), but work on regulation in higher plants is still lacking.

Net Zn uptake from solutions containing low Zn^{2+} activities was negative, indicating greater efflux than influx (Rengel and Graham 1996). Zinc efflux may be a significant component of net Zn uptake (up to 85% of Zn taken up may be effluxed into external solution; Santa Maria and Cogliatti 1988).

A number of genes coding for transporters mediating efflux of Zn from the cytosol of plant cells have been cloned (e.g. from Arabidopsis; van der Zaal et al. 1999). These genes belong to the CE (cation efflux) family (Maser et al. 2001). Their gene products confer a degree of resistance to heavy metal toxicity (e.g. by sequestering Zn from the cytosol to the vacuole, van der Zaal et al. 1999) and are therefore called MTP (metal tolerance protein) genes (Maser et al. 2001).

11.4.3 Manganese

Little is known about uptake of Mn by plants. Using realistically low Mn activities in a chelate-buffered nutrient solution, Webb et al. (1993) obtained a linear relationship between Mn^{2+} activity in solution and the Mn accumulation rate in barley. Increased complexation of Mn with EDTA in the nutrient solution (around 45% of total Mn present as a complex) increased accumulation of Mn in barley (Laurie et al. 1995), indicating that not only Mn^{2+}, but Mn-EDTA complex as well can be taken up across the plasma membrane.

A promising line of research has been reported for rice (Belouchi et al. 1997), with identification of several genes homologous to the *SMF1* gene in yeast (Supek et al. 1996) that appears to be involved in high-affinity Mn uptake. The *AtNRAMP* family of five genes isolated recently from Arabidopsis (Curie et al. 2000; Thomine et al. 2000) can complement the yeast mutant deficient in the high-affinity Mn uptake system. These genes, however, appear to be non-specific because they transport a range of other divalent metal cations (e.g. Fe, Cd).

11.5 GENOTYPIC DIFFERENCES IN MICRONUTRIENT UPTAKE AND UTILISATION BY PLANTS

Some plant species and genotypes have a capacity to grow and yield well on soils of low fertility (e.g. Graham and Rengel 1993; Gourley et al. 1994; Pearson and Rengel 1997; Rengel 1999a, 2001); these species and genotypes are tolerant to nutrient deficiency (=nutrient-efficient). Efficient genotypes grow and yield well on nutrient-deficient soils by employing specific mechanisms that allow them to gain access to sufficient quantities of nutrients (uptake efficiency) and/or to more effectively utilise nutrients taken up (utilisation efficiency; Sattelmacher et al. 1994).

Uptake efficiency may consist of increased capacity to solubilise non-available nutrient forms into plant-available ones, and/or increased capacity to transport nutrients across the plasma membrane. However, an increased capacity to convert non-available into available nutrients in soil is of greater importance for efficient uptake, especially for nutrients that are transported in soil toward roots by diffusion (Rengel 1999a, 2001).

A given level of phenotypic expression of nutrient efficiency is a result of several mechanisms that may operate at different levels of plant organisation (metabolic, physiological, structural, or developmental). However, elucidation of mechanisms of nutrient efficiency is hampered by inadequate ability to distinguish between causes and effects of nutrient deficiency due to the relatively long time between withdrawal of a nutrient and the time when measurable changes can be observed (Rengel 1999a). It may, however, be assumed that:

1. expression of a certain efficiency mechanism as well as relative effectiveness of various mechanisms are highly dependent on environmental factors;
2. more than one mechanism may be responsible for the level of efficiency in a particular genotype;
3. increased efficiency observed in one genotype compared with another involves additional mechanisms not present in the less efficient genotype; and
4. the effects of differential efficiency mechanisms present in a particular genotype may be additive.

11.5.1   Iron Efficiency

In strategy I plants, the relative importance of various Fe-deficiency responses (enhancement of Fe^{3+} reduction, exudation of protons or reducing/chelating compounds into the rhizosphere, and histological/morphological changes in root growth) is poorly understood due to paucity of comparative measurements of all these components of the Fe-deficiency response. A report by Wei et al. (1997) showed that, under Fe deficiency, Fe-efficient clover had exudation of protons three-fold higher than the Fe-inefficient genotype; in contrast, there was no difference between genotypes in the Fe^{3+} reduction rate, and exudation of reductants was negligible in both genotypes.

In graminaceous species, the rate of release of PS is positively related to Fe efficiency, which decreases in the order: barley >maize >sorghum (Römheld and Marschner 1990) or oat >maize (see Mori 1994). The release of PS and the subsequent uptake of the Fe^{3+}-PS complex are under different genetic control (Römheld and Marschner 1990). Indeed, the Fe-inefficient yellow-stripe mutant of maize (ys1) maintained the rate of PS release similar to that of the Fe-efficient genotype (von Wiren et al. 1994), but Fe inefficiency of ys1 resulted from the mutation that affected a high-affinity uptake component, leading to a decrease in activity and/or number of Fe^{3+}-PS transporters (von Wiren et al. 1995).

11.5.2   Zinc Efficiency

Zinc-efficient genotypes employ a plethora of physiological mechanisms that allow them to withstand the Zn-deficiency stress better than the Zn-inefficient ones. These Zn efficiency mechanisms include:

1. a greater proportion and longer length of fine roots with diameters ≤0.2 mm (Dong et al. 1995);
2. differential changes in the rhizosphere chemistry and biology, including the release of greater amounts of Zn-chelating PS (Cakmak et al. 1994, 1996a-c, 1997a; Walter et al. 1994; Rengel et al. 1998a; Rengel 1999b,c; Rengel and Römheld 2000);

3. an increased I_{max} resulting in increased net Zn accumulation (Cakmak et al. 1996b; Rengel and Graham 1996; Rengel and Wheal 1997; Hart et al. 1998; Rengel et al. 1998a); and

4. more efficient utilisation and compartmentalisation of Zn within cells, tissues and organs (Graham and Rengel 1993), including a greater activity of carbonic anhydrase (Rengel 1995a) and antioxidative enzymes (Cakmak et al. 1997b; Yu et al. 1999b; Cakmak 2000), maintaining sulfhydryl groups in the root-cell plasma membranes in a reduced state (Rengel 1995b), and increased biosynthesis of specific root-cell plasma membrane polypeptides (Rengel and Hawkesford 1997). For a detailed discussion on mechanisms of Zn efficiency, see Rengel (1999, 2001).

11.5.3 Manganese Efficiency

Despite considerable differences among cereal genotypes in tolerance to Mn-deficient soils (eg. Bansal et al. 1991), the mechanisms behind differential Mn efficiency are poorly understood (Graham 1988; Huang et al. 1994, 1996; Rengel et al. 1994; Rengel 1999, 2000). Superior internal utilisation of Mn, a lower physiological requirement, a faster specific rate of Mn absorption, and better root geometry are unlikely to be major contributors to Mn efficiency (Graham 1988). The remaining two possible mechanisms of Mn efficiency are (1) better internal compartmentalisation and remobilisation of Mn, and (2) excretion by roots of Mn-efficient genotypes of greater amounts of substances capable of mobilising insoluble Mn (protons, reductants, Mn-binding ligands, and microbial stimulants; Rengel et al. 1996; Rengel 1997, 2000). The latter mechanism may be of particular importance because genotypic differences in Mn efficiency among barley cultivars were expressed only in the soil-based systems but not in the solution culture (Huang et al. 1994), indicating that differential changes in chemistry and biology of the rhizosphere soil may be responsible for differential ability of barley genotypes to tolerate Mn deficiency. However, recent studies (S. Husted et al., unpubl.) suggest that barley genotypes show differential Mn efficiency even when grown in nutrient solution where rhizosphere gradients in the chemical and/or biological parameters are unlikely to develop.

11.6 ROLE OF RHIZOSPHERE MICROORGANISMS IN MICRONUTRIENT NUTRITION

Different plant species as well as genotypes within a species (see Rengel et al. 1996 for references) differentially influence the quantitative and qualitative composition of microbial populations in the rhizosphere. Soil microorganisms and plant-microbial interactions have an integral role in micronutrient uptake by plants, primarily through the effects on metal solubilisation in the rhizosphere (see reviews by Tinker 1984; Crowley and Gries 1994; Crowley and Rengel 1999).

11.6.1 Iron

In response to Fe deficiency, microorganisms exude metal-chelating siderophores (Crowley et al. 1991), which accumulate at higher concentrations in the rhizosphere than in the bulk soil, and can provide plants with Fe when supplied at concentrations similar to those used for delivery of Fe via synthetic chelating agents (Wang et al. 1993). However, the importance of in situ production of microbial siderophores in the plant rhizosphere remains controversial since siderophores are not readily detected in soil at physiologically relevant concentrations (Crowley and Rengel 1999; Rengel et al. 1999). Because these compounds are subject to microbial degradation, it is likely that they accumulate only transiently at sites of high microbial activity.

The complexes of Fe and pseudomonas-produced pyoverdin-type siderophores are chemically stable and therefore poorly utilised by plants, but can undergo breakdown in soil and thereby indirectly supply Fe to plants (Crowley and Gries 1994). Other siderophores, such as the fungal siderophores, rhizoferrin and rhodotorulic acid, are easily utilised by plants and have been studied for use as micronutrient biofertilisers (Shenker et al. 1992; Crowley and Rengel 1999).

11.6.2 Zinc

Preliminary results show that wheat genotypes differing in Zn efficiency may differentially influence microbial populations in the rhizosphere (Rengel et al. 1996, 1998b; Rengel 1997). Zinc deficiency stimulated growth of fluorescent pseudomonads in the rhizosphere of all wheat genotypes tested, but the effect was particularly obvious for Zn-efficient ones (Rengel et al. 1996; Rengel 1997). A possible causal relationship between increased populations of rhizosphere bacteria and an increased capacity of wheat genotypes to acquire Zn under deficient conditions warrants further research.

11.6.3 Manganese

The genetic control of Mn efficiency may be expressed through the composition of root exudates encouraging a more favourable balance of Mn reducers to Mn oxidisers in the rhizosphere. Greater numbers of Mn-oxidising microorganisms were found in the rhizosphere of Mn-inefficient than in Mn-efficient oat genotypes (Timonin 1946), the latter exuding compounds toxic to Mn-oxidising microorganisms in the rhizosphere (Timonin 1965).

When grown at sufficient Mn supply, no difference in the ratio of Mn reducers and Mn oxidisers was found in the rhizosphere of wheat genotypes differing in Mn efficiency. However, under Mn deficiency, the Mn-efficient Aroona had the ratio of Mn reducers to Mn oxidisers increased three-fold in comparison with the Mn-inefficient genotypes (Rengel et al. 1996). In contrast, microflora in the rhizosphere of other Mn-efficient wheat genotypes (like C8MM) did not show the same response as Aroona. It therefore appears that different mechanisms may underlie the expression of Mn efficiency in different wheat genotypes (Rengel 1997).

11.7 MICRONUTRIENTS IN HUMAN NUTRITION: THE PLANT LINK

Deficiencies of Fe, Zn, I, and vitamin A in human populations are widespread, affecting up to 2 billion people (Graham and Welch 1996; Welch and Graham 1999; Graham et al. 2001). While the traditional remedy of these deficiencies was in the form of food supplements, suitability of agricultural strategies for increasing micronutrient density in grain destined for human consumption is now being assessed as a sustainable, long-term solution (Rengel et al. 1999).

Most crop management practices aim to achieve yields close to the potential yield possible in a given environment. As predicted by the law of diminishing returns, increased additions of micronutrient fertilisers to deficient soils result in a saturable yield crop response. In contrast, concentrations of Zn or Fe in grain increase steadily with an increase in Zn or Fe fertiliser additions (Marschner 1995; Rengel et al. 1999), implying that fertilisation with Zn or Fe in excess of what is required for achieving 90% (or even 100%) of the yield may result in increased concentration of these micronutrients in grain. Agricultural measures would need to be supplemented with appropriate changes in the milling technology to ascertain that increased micronutrient concentrations in some grain parts are passed into the food chain (Rengel et al. 1999).

Remobilisation of nutrients from vegetative tissues into the grain can also be a significant source of micronutrients, but little is known about mechanisms governing such remobilisation (see Pearson and Rengel 1994). However, it is clear that significant differences exist in regulation of loading of various micronutrients into grain (see also Pearson et al. 1995, 1996, 1998, 1999).

11.8 CONCLUSIONS AND FUTURE RESEARCH

Some heavy metals are essential micronutrients and are thus required in low concentrations for plant growth and development. Micronutrients have been studied comparatively less than macronutrients; nevertheless, significant advancements in our understanding of plant responses to micronutrient deficiency have occurred recently. In future research, a greater emphasis should be placed on characterising membrane proteins and studying regulation of expression of corresponding genes involved in transport of micronutrients into the cytoplasm. A better understanding of rhizosphere processes governing availability of micronutrients to plants is needed. A range of agricultural measures should be considered for increasing micronutrient content of grain destined for human consumption.

REFERENCES

Awad F, Römheld V, Marschner H (1994) Effect of root exudates on mobilization in the rhizosphere and uptake of iron by wheat plants. Plant Soil 165: 213-218

Bansal RL, Nayyar VL, Takkar PN (1991) Field screening of wheat cultivars for manganese efficiency. Field Crops Res 29: 107-112

Belouchi A, Kwan T, Gros P (1997) Cloning and characterization of the OsNramp family from *Oryza sativa*, a new family of membrane proteins possibly implicated in the transport of metal ions. Plant Mol Biol 33: 1085-1092

Bettger WJ, O'Dell BL (1981) A critical physiological role of zinc in the structure and function of biomembranes. Life Sci 28: 1425-1438

Bienfait HF (1988) Prevention of stress in iron metabolism of plants. Acta Bot Neerl 38: 105-129

Bird AJ, Zhao H, Luo H, Jensen LT, Srinivasan C, Evans-Galea M, Winge DR, Eide DJ (2000) A dual role for zinc fingers in both DNA binding and zinc sensing by the Zap1 transcriptional activator. EMBO J 19: 3704-3713

Bowler C, van Montagu M, Inze D (1992) Superoxide dismutase and stress tolerance. Annu Rev Plant Physiol Plant Mol Biol 43: 83-116

Brancadoro L, Rabotti G, Scienza A, Zocchi G (1995) Mechanisms of Fe-efficiency in roots of *Vitis* spp. in response to iron deficiency stress. Plant Soil 171: 229-234

Burnell JN (1988) The biochemistry of manganese in plants. In: Graham RD, Hannam RJ, Uren NC (eds) Manganese in soils and plants. Kluwer, Dordrecht, pp 125-137

Cakmak I (2000) Tansley review No. 111. Possible roles of zinc in protecting plant cells from damage by reactive oxygen species. New Phytol 146: 185-205

Cakmak I, Marschner H (1987) Mechanism of phosphorus-induced zinc deficiency in cotton. III. Changes in physiological availability of zinc in plants. Physiol Plant 70: 13-20

Cakmak I, Marschner H (1988a) Enhanced superoxide radical production in roots of zinc-deficient plants. J Exp Bot 39: 1449-1460

Cakmak I, Marschner H (1988b) Increase in membrane permeability and exudation in roots of zinc deficient plants. J Plant Physiol 132: 356-361

Cakmak I, Marschner H (1990) Decrease in nitrate uptake and increase in proton release in zinc deficient cotton, sunflower and buckwheat plants. Plant Soil 129: 261-268

Cakmak I, Marschner H, Bangerth F (1989) Effect of zinc nutritional status on growth, protein metabolism and levels of indole-3-acetic acid and other phytohormones in bean (*Phaseolus vulgaris* L.). J Exp Bot 40: 405-412

Cakmak I, Gülüt KY, Marschner H, Graham RD (1994) Effect of zinc and iron deficiency on phytosiderophore release in wheat genotypes differing in zinc efficiency. J Plant Nutr 17: 1-17

Cakmak I, Öztürk L, Karanlik S, Marschner H, Ekiz H (1996a) Zinc-efficient wild grasses enhance release of phytosiderophores under zinc deficiency. J Plant Nutr 19: 551-563

Cakmak I, Sari N, Marschner H, Ekiz H, Kalayci M, Yilmaz A, Braun HJ (1996b) Phytosiderophore release in bread and durum wheat genotypes difffering in zinc efficiency. Plant Soil 180: 183-189

Cakmak I, Sari N, Marschner H, Kalayci M, Yilmaz A, Eker S, Gülüt KY (1996c) Dry matter production and distribution of zinc in bread and durum wheat genotypes differing in zinc efficiency. Plant Soil 180: 173-181

Cakmak I, Ekiz H, Yilmaz A, Torun B, Köleli N, Gültekin I, Alkan A, Eker S (1997a) Differential response of rye, triticale, bread and durum wheats to zinc deficiency in calcareous soils. Plant Soil 188: 1-10

Cakmak I, Öztürk L, Eker S, Torun B, Kalfa HI, Yilmaz A (1997b) Concentration of zinc and activity of copper/zinc superoxide dismutase in leaves of rye and wheat cultivars differing in sensitivity to zinc deficiency. J Plant Physiol 151: 91-95

Cakmak I, Erenoglu B, Gulut KY, Derici R, Römheld V (1998) Light-mediated release of phytosiderophores in wheat and barley under iron or zinc deficiency. Plant Soil 202: 309-315

Cakmak I, Tolay I, Ozdemir A, Ozkan H, Ozturk L, Kling CI (1999) Differences in zinc efficiency among and within diploid, tetraploid and hexaploid wheats. Ann Bot 84: 163-171

Cakmak O, Ozturk L, Karanlik S, Ozkan H, Kaya Z, Cakmak I (2001) Tolerance of 65 durum wheat genotypes to zinc deficiency in a calcareous soil. J Plant Nutr 24: 1831-1847

Chongpraditnun P, Mori S, Chino M (1992) Excess copper induces a cytosolic copper/zinc-super-oxide dismutase in soybean root. Plant Cell Physiol 33: 239-244

Colman B (ed) (1991) Second international symposium on inorganic carbon utilisation by aquatic photosynthetic organisms. Can J Bot 69: 907-1027

Cohen CK, Norvell WA, Kochian LV (1997) Induction of the root cell plasma membrane ferric reductase: an exclusive role for Fe and Cu. Plant Physiol 114: 1061-1069

Crowley DE, Gries D (1994) Modeling of iron availability in the plant rhizosphere. In: Manthey JA, Crowley DE, Luster DG (eds) Biochemistry of metal micronutrients in the rhizosphere. Lewis Publ, Boca Raton, pp 199-223

Crowley DE, Rengel Z (1999) Biology and chemistry of nutrient availability in the rhizosphere. In: Rengel Z (ed) Mineral nutrition of crops: fundamental mechanisms and implications. Haworth Press, New York, pp 1-40

Crowley DE, Wang YC, Reid CP, Szaniszlo PJ (1991) Mechanisms of iron acquisition from siderophores by microorganisms and plants. Plant Soil 130: 179-198

Curie C, Alonso JM, Le Jean M, Ecker JR, Briat J-F (2000) Involvement of NRAMP1 from *Arabidopsis thaliana* in iron transport. Biochem J 347: 749-755

Curie C, Panaviene Z, Loulergue C, Dellaporta SL, Briat J-F, Walker EL (2001) Maize yellow stripe1 encodes a membrane protein directly involved in Fe(III) uptake. Nature 409: 346-349

Dat J, Vandenabeele S, Vranova E, van Montagu M, Inze D, Van Breusegem F (2000) Dual action of the active oxygen species during plant stress responses. Cell Mol Life Sci 57: 779-795

Del Río LA, Sevilla F, Gómez M, Yanez J, Lopéz-Gorgé J (1978) Superoxide dismutase: an enzyme system for the study of micronutrient interactions in plants. Planta 140: 221-225

Dixit V, Pandey V, Shyam R (2001) Differential antioxidative responses to cadmium in roots and leaves of pea (*Pisum sativum* L. cv. Azad). J Exp Bot 52: 1101-1109

Dhindsa RS, Plumb-Dhindsa P, Thorpe TA (1981) Leaf senescence: correlated with increased levels of membrane permeability and lipid peroxidation, and decreased levels of superoxide dismutase and catalase. J Exp Bot 32: 93-101

Dong B, Rengel Z, Graham RD (1995) Characters of root geometry of wheat genotypes differing in Zn efficiency. J Plant Nutr 18: 2761-2773

Eckhardt U, Buckhout TJ (2000) Analysis of the mechanism of iron assimilation in *Chlamydomonas reinhardtii*: a model system for strategy I plants. J Plant Nutr 23: 1797-1807

Eckhardt U, Marques AM, Buckhout TJ (2001) Two iron-regulated cation transporters from tomato complement metal uptake-deficient yeast mutants. Plant Mol Biol 45: 437-448

Eide D (1997) Molecular biology of iron and zinc uptake in eukaryotes. Curr Opin Cell Biol 9: 573-577

Eide D, Broderius M, Fett J, Guerinot ML (1996) A novel iron-regulated metal transporter from plants identified by functional expression in yeast. Proc Natl Acad Sci USA 93: 5624-5628

Erenoglu B, Cakmak I, Marschner H, Römheld V, Eker S, Daghan H, Kalayci M, Ekiz H (1996) Phytosiderophore release does not relate well with Zn efficiency in different bread wheat genotypes. J Plant Nutr 19: 1569-1580

Erenoglu B, Cakmak I, Römheld V, Derici R, Rengel Z (1999) Uptake of zinc by rye, bread wheat and durum wheat cultivars differing in zinc efficiency. Plant Soil 209: 245-252

Erenoglu B, Eker S, Cakmak I, Derici R, Römheld V (2000) Effect of iron and zinc deficiency on release of phytosiderophores in barley cultivars differing in zinc efficiency. J Plant Nutr 23: 1645-1656

Falchuk KH, Mazus B, Ulpino L, Vallee BL (1976) *Euglena gracilis* DNA dependent RNA polymerase II: a zinc metalloenzyme. Biochemistry 15: 4468-4475

Falchuk KH, Ulpino L, Mazus B, Vallee BL (1977) *E. gracilis* RNA polymerase I: a zinc metalloenzyme. Biochem Biophys Res Commun 74: 1206-1212

Fischer ES, Thimm O, Rengel Z (1997) Zinc nutrition influences gas exchange in wheat. Photosynthetica 33: 505-508

Foyer CH, Descourvieres P, Kunert KJ (1994) Protection against oxygen radicals: an important defence mechanism studied in transgenic plants. Plant Cell Environ 17: 507-523

Gaither LA, Eide DJ (2001) Eukaryotic zinc transporters and their regulation. Biometals 14: 251-270

Gherardi MJ, Rengel Z (2001) Bauxite residue sand has the capacity to rapidly decrease availability of added manganese. Plant Soil 234: 143-151

Ghiorse WC (1988) The biology of manganese transforming microorganisms in soils. In: Graham RD, Hannam RJ, Uren NC (eds) Manganese in soils and plants. Kluwer, Dordrecht. Pp 75-85

Gibson TS, Leece DR (1981) Estimation of physiologically active zinc in maize by biochemical assay. Plant Soil 63: 395-406

Gitan RS, Luo H, Rodgers J, Broderius M, Eide D (1998) Zinc-induced inactivation of the yeast ZRT1 zinc transporter occurs through endocytosis and vacuolar degradation. J Biol Chem 273: 28617-28624

Godo GH, Reisenauer HM (1980) Plant effects on soil manganese availability. Soil Sci Soc Am J 44: 993-995

Gourley CJP, Allan DL, Russelle MP (1994) Plant nutrient efficiency: a comparison of definitions and suggested improvement. Plant Soil 158: 29-37

Graham RD (1988) Genotypic differences in tolerance to manganese deficiency. In: Graham RD, Hannam RJ, Uren NC (eds) Manganese in soils and plants. Kluwer, Dordrecht, pp 261-276

Graham RD, Rengel Z (1993) Genotypic variation in zinc uptake and utilization. In: Robson AD (ed) Zinc in soils and plants. Kluwer, Dordrecht, pp 107-118

Graham RD, Welch RM (1996) Breeding for staple food crops with high micronutrient density. International Food Policy Research Institute, Washington, DC

Graham RD, Welch RM, Bouis HE (2001) Addressing micronutrient malnutrition through enhancing the nutritional quality of staple foods: principles, perspectives and knowledge gaps. Adv Agron 70: 77-142

Grotz N, Fox T, Connolly E, Park W, Guerinot ML, Eide D (1998) Identification of a family of zinc transporter genes from Arabidopsis that respond to zinc deficiency. Proc Natl Acad Sci USA 95: 7220-7224

Guerinot ML (2000) The ZIP family of metal transporters. Biochem Biophys Acta 1465: 190-198

Guliev NM, Bairamov SM, Aliev DA (1992) Functional organization of carbonic anhydrase in higher plants. Sov Plant Physiol 39: 537-544

Hacisalihoglu G, Hart JJ, Kochian LV (2001) High- and low-affinity zinc transport systems and their possible role in zinc efficiency in bread wheat. Plant Physiol 125: 456-463

Hart JJ, Norvell WA, Welch RM, Sullivan LA, Kochian LV (1998) Characterization of zinc uptake, binding, and translocation in intact seedlings of bread and durum wheat cultivars. Plant Physiol 118: 219-226

Higuchi K, Suzuki K, Nakanishi H, Yamaguchi H, Nishizawa NK, Mori S (1999) Cloning of nicotianamine synthase genes, novel genes involved in the biosynthesis of phytosiderophores. Plant Physiol 119: 471-479

Higuchi K, Watanabe S, Takahashi M, Kawasaki S, Nakanishi H, Nishizawa NK, Mori S (2001) Nicotianamine synthase gene expression differs in barley and rice under Fe-deficient conditions. Plant J 25: 159-167

Hiltonen T, Björkbacka H, Forsman C, Clarke AK, Samuelsson G (1998) Intracellular carbonic anhydrase of the unicellular green alga Coccomyxa. Cloning of the cDNA and characterization of the functional enzyme overexpressed in Escherichia coli. Plant Physiol 117: 1341-1349

Huang C, Webb MJ, Graham RD (1994) Manganese efficiency is expressed in barley growing in soil system but not in a solution culture. J Plant Nutr 17: 83-95

Huang C, Webb MJ, Graham RD (1996) Pot size affects expression of Mn efficiency in barley. Plant Soil 178: 205-208

Huber DM, McCay-Buis TS (1993) A multiple component analysis of the take-all disease of cereals. Plant Dis 77: 437-447

Itai R, Suzuki K, Yamaguchi H, Nakanishi H, Nishizawa NK, Yoshimura E, Mori S (2000) Induced activity of adenine phosphoribosyltransferase (APRT) in iron-deficient barley roots: a possible role for phytosiderophore production. J Exp Bot 51: 1179-1188

Iturbe-Ormaetxe I, Moran JF, Arrese-Igor C, Gogorcena Y, Klucas RV, Becana M (1995) Activated oxygen and antioxidant defences in iron-deficient pea plant. Plant Cell Environ 18: 421-429

Jones DL, Darrah PR (1994) Role of root derived organic acids in the mobilization of nutrients from the rhizosphere. Plant Soil 166: 247-257

Jones DL, Darrah PR, Kochian LV (1996) Critical evaluation of organic acid mediated iron dissolution in the rhizosphere and its potential role in root iron uptake. Plant Soil 180: 57-66

Kobayashi T, Nakanishi H, Takahashi M, Kawasaki S, Nishizawa NK, Mori S (2001) In vivo evidence that Ids3 from *Hordeum vulgare* encodes a dioxygenase that converts 2'-deoxymugineic acid to mugineic acid in transgenic rice. Planta 212: 864-871

Kochian LV (1991) Mechanisms of micronutrient uptake and translocation in plants. In: Mortvedt JJ, Fox FR, Shuman LM, Welch RM (eds) Micronutrients in agriculture, 2nd edn. SSSA, Madison, WIS, pp 229-296

Kochian LV (1993) Zinc absorption from hydroponic solutions by plant roots. In: Robson AD (ed) Zinc in soils and plants. Kluwer, Dordrecht, pp 45-57

Korshunova YO, Eide D, Clark WG, Guerinot ML, Pakrasi HB (1999) The IRT1 protein from *Arabidopsis thaliana* is a metal transporter with a broad substrate range. Plant Mol Biol 40: 37-44

Kröniger W, Rennenberg H, Tadros MH, Polle A (1995) Purification and properties of manganese superoxide dismutase from Norway spruce (*Picea abies* L. Karst). Plant Cell Physiol 36: 191-196

Kuzniak E (2002) Transgenic plants: an insight into oxidative stress tolerance mechanisms. Acta Physiol Plant 24: 97-113

Landsberg EC (1981) Organic acid synthesis and release of hydrogen ions in response to Fe-deficiency stress of mono- and dicotyledonous plant species. J Plant Nutr 3: 579-591

Lasat MM, Pence NS, Garvin DF, Ebbs SD, Kochian LV (2000) Molecular physiology of zinc transport in the Zn hyperaccumulator *Thlaspi caerulescens*. J Exp Bot 51: 71-79

Laurie SH, Tancock NP, McGrath SP, Sanders JR (1995) Influence of EDTA complexation on plant uptake of manganese(II). Plant Sci 109: 231-235

Leigh RA, Tomos AD (1993) Ion distribution in cereal leaves: pathways and mechanisms. Philos Trans R Soc Lond B 341: 75-86

Liu LS, White MJ, MacRae TH (1999) Transcription factors and their genes in higher plants—functional domains, evolution and regulation. Eur J Biochem 262: 247-257

Longnecker NE, Robson AD (1993) Distribution and transport of zinc in plants. In: Robson AD (eds) Zinc in soils and plants. Kluwer, Dordrecht, pp 79-91

Ma JF, Nomoto K (1993) Inhibition of mugineic acid-ferric complex uptake in barley by copper, zinc and cobalt. Physiol Plant 89: 331-334

Ma JF, Nomoto K (1996) Effective regulation of iron acquisition in graminaceous plants. The role of mugineic acids as phytosiderophores. Physiol Plant 97: 609-617

Ma JF, Kusano G, Kimura S, Nomoto K (1993) Specific recognition of mugineic acid-ferric complex by barley roots. Phytochemistry 34: 599-603

MacDiarmid CW, Gaither LA, Eide D (2000) Zinc transporters that regulate vacuolar zinc storage in *Saccharomyces cerevisiae*. EMBO J 19: 2845-2855

Marschner H (1993) Zinc uptake from soils. In: Robson AD (ed) Zinc in soils and plants. Kluwer, Dordrecht, pp 59-77

Marschner H (1995) Mineral nutrition of higher plants, 2nd edn. Academic Press, London

Marschner H, Römheld V (1991) Function of micronutrients in plants. In: Mortvedt JJ, Cox FR, Shuman LM, Welch RM (eds) Micronutrients in agriculture, 2nd edn. SSSA, Madison, WIS, pp 297-328

Marschner H, Römheld V (1994) Strategies of plants for acquisition of iron. Plant Soil 165: 261-274

Marschner H, Römheld V, Horst WJ, Martin P (1986) Root induced changes in the rhizosphere: importance for the mineral nutrition of plants. Z Pflanzenernähr Bodenk 149: 441-456

Marschner P, Ascher JS, Graham RD (1991) Effect of manganese-reducing rhizosphere bacteria on the growth of *Gaeumannomyces graminis* var. *tritici* and on manganese uptake by wheat (*Triticum aestivum* L.). Biol Fertil Soils 12: 33-38

Maser P, Thomine S, Schroeder JI, Ward JM, Hirschi K, Sze H, Talke IN, Amtmann A, Maathuis FJ M, Sanders D, Harper JF, Tchieu J, Gribskov M, Persans MW, Salt DE, Kim Sun A, Guerinot ML (2001) Phylogenetic relationships within cation transporter families of Arabidopsis. Plant Physiol 126: 1646-1667

Matern U, Kneusel RE (1988) Phenolic compounds in plant disease resistance. Phytoparasitica 16: 153-170

Moore Jr PA, Patrick Jr WH (1988) Effect of zinc deficiency on alcohol dehydrogenase activity and nutrient uptake in rice. Agron J 80: 882-885

Moreau S, Thomson RM, Kaiser BN, Trevaskis B, Guerinot ML, Udvardi MK, Puppo A, Day DA (2002) GmZIP1 encodes a symbiosis-specific zinc transporter in soybean. J Biol Chem 277: 4738-4746

Mori S (1994) Mechanisms of iron acquisition by graminaceous (strategy II) plants. In: Manthey JA, Crowley DE, Luster DG (eds) Biochemistry of metal micronutrients in the rhizosphere. Lewis Publ, Boca Raton, pp 225-249

Mori S (1997) Reevaluation of the genes induced by iron deficiency in barley roots. Soil Sci Plant Nutr 43: 975-980

Nakanishi H, Yamaguchi H, Sasakuma T, Nishizawa NK, Mori S (2000) Two dioxygenase genes, *Ids3* and *Ids2*, from *Hordeum vulgare* are involved in the biosynthesis of mugineic acid family phytosiderophores. Plant Mol Biol 44: 199-207

Obata H, Kawamura S, Senoo K, Tanaka A (1999) Changes in the level of protein and activity of Cu/Zn-superoxide dismutase in Zn-deficient rice plant, *Oryza sativa* L. Soil Sci Plant Nutr 45: 891-896

Oida F, Ota N, Mino Y, Nomoto K, Sugiura Y (1989) Stereospecific iron uptake mediated by phytosiderophore in gramineous plants. J Am Chem Soc 111: 3436-3437

Olsen RA, Bennett JH, Blume D, Brown JC (1981) Chemical aspects of the Fe stress response mechanism in tomatoes. J Plant Nutr 3: 905-921

Palmiter RD, Findley SD (1995) Cloning and functional characterization of a mammalian zinc transporter that confers resistance to zinc. EMBO J 14: 639-649

Papi M, Sabatini S, Altamura MM, Hennig L, Schäfer E, Costantino P, Vittorioso P (2002) Inactivation of the phloem-specific Dof zinc finger gene *DAG1* affects response to light and integrity of the testa of *Arabidopsis* seeds. Plant Physiol 128: 411-417

Pearson JN, Rengel Z (1994) Distribution and remobilization of Zn and Mn during grain development in wheat. J Exp Bot 45: 1829-1835

Pearson JN, Rengel Z (1997) Mechanisms of plant resistance to nutrient deficiency stresses. In: Basra AS, Basra RK (eds) Mechanisms of environmental stress resistance in plants. Harwood, Amsterdam, pp 213-240

Pearson JN, Rengel Z, Jenner CF, Graham RD (1995) Transport of zinc and manganese to developing wheat grains. Physiol Plant 95: 449-455

Pearson JN, Rengel Z, Jenner CF, Graham RD (1996) Manipulation of xylem transport affects Zn and Mn transport into developing wheat grains of cultured ears. Physiol Plant 98: 229-234

Pearson JN, Rengel Z, Jenner CF, Graham RD (1998) Dynamics of zinc and manganese movement in developing wheat grains. Aust J Plant Physiol 25: 139-144

Pearson JN, Rengel Z, Graham RD (1999) Regulation of zinc and manganese transport into developing wheat grains having different zinc and manganese concentrations. J Plant Nutr 22: 1141-1152

Pence NS, Larsen PB, Ebbs SD, Letham DLD, Lasat MM, Garvin DF, Eide D, Kochian LV (2000) The molecular physiology of heavy metal transport in the Zn/Cd hyperaccumulator *Thlaspi caerulescens*. Proc Natl Acad Sci USA 97: 4956-4960

Pocker Y, Sarkanen S (1978) Carbonic anhydrase: structure, catalytic versatility, and inhibition. Adv Enzymol 47: 149-274

Prasad KVSK, Saradhi PP, Sharmila P (1999) Concerted action of antioxidant enzymes and curtailed growth under zinc toxicity in *Brassica juncea*. Environ Exp Bot 42: 1-10

Przymusinski R, Rucinska R, Gwozdz EA (1995) The stress-stimulated 16 kDa polypeptide from lupin roots has properties of cytosolic Cu:Zn-superoxide dismutase. Environ Exp Bot 35: 485-495

Quartacci MF, Cosi E, Navari-Izzo F (2001) Lipids and NADPH-dependent superoxide production in plasma membrane vesicles from roots of wheat grown under copper deficiency or excess. J Exp Bot 52: 77-84

Quiquampoix H, Loughman BC, Ratcliffe RG (1993) A ^{31}P-NMR study of the uptake and compartmentation of manganese by maize roots. J Exp Bot 44: 1819-1827

Rabotti G, Zocchi G (1994) Plasma membrane-bound H^+-ATPase and reductase activities in Fe-deficient cucumber roots. Physiol Plant 90: 779-785

Reid RJ (2001) Mechanisms of micronutrient uptake in plants. Aust J Plant Physiol 28: 659-666

Rengel Z (1995a) Carbonic anhydrase activity in leaves of wheat genotypes differing in Zn efficiency. J Plant Physiol 147: 251-256

Rengel Z (1995b) Sulfhydryl groups in root-cell plasma membranes of wheat genotypes differing in Zn efficiency. Physiol Plant 95: 604-612

Rengel Z (1997) Root exudation and microflora populations in rhizosphere of crop genotypes differing in tolerance to micronutrient deficiency. Plant Soil 196: 255-260

Rengel Z (1999a) Physiological mechanisms underlying differential nutrient efficiency of crop genotypes. In: Rengel Z (ed) Mineral nutrition of crops: fundamental mechanisms and implications. Haworth Press, New York, pp 227-265

Rengel Z (1999b) Physiological responses of wheat genotypes grown in chelator-buffered nutrient solutions with increasing concentrations of excess HEDTA. Plant Soil 215: 193-202

Rengel Z (1999c) Zinc deficiency in wheat genotypes grown in conventional and chelator-buffered nutrient solutions. Plant Sci 143: 221-230

Rengel Z (2000) Uptake and transport of manganese in plants. In: Sigel A, Sigel H (eds) Metal ions in biological systems. Dekker, New York, pp 57-87

Rengel Z (2001) Genotypic differences in micronutrient use efficiency in crops. Commun Soil Sci Plant Anal 32: 1163-1186

Rengel Z, Graham RD (1995) Wheat genotypes differ in Zn efficiency when grown in chelate-buffered nutrient solution. II. Nutrient uptake. Plant Soil 173: 259-266

Rengel Z, Graham RD (1996) Uptake of zinc from chelate-buffered nutrient solutions by wheat genotypes differing in Zn efficiency. J Exp Bot 47: 217-226

Rengel Z, Hawkesford MJ (1997) Biosynthesis of a 34-kDa polypeptide in the root-cell plasma membrane of Zn-efficient wheat genotype increases upon Zn starvation. Aust J Plant Physiol 24: 307-315

Rengel Z, Wheal MS (1997) Kinetics of Zn uptake by wheat is affected by herbicide chlorsulfuron. J Exp Bot 48: 935-941

Rengel Z, Römheld V (2000) Root exudation and Fe uptake and transport in wheat genotypes differing in tolerance to Zn deficiency. Plant Soil 222: 25-34

Rengel Z, Pedler JF, Graham RD (1994) Control of Mn status in plants and rhizosphere: genetic aspects of host and pathogen effects in the wheat take-all interaction. In: Manthey JA, Crowley DE, Luster DG (eds) Biochemistry of metal micronutrients in the rhizosphere. Lewis Publ/CRC Press, Boca Raton, pp 125-145

Rengel Z, Guterridge R, Hirsch P, Hornby D (1996) Plant genotype, micronutrient fertilisation and take-all infection influence bacterial populations in the rhizosphere of wheat. Plant Soil 183: 269-277

Rengel Z, Römheld V, Marschner H (1998a) Uptake of zinc and iron by wheat genotypes differing in zinc efficiency. J Plant Physiol 152: 433-438

Rengel Z, Ross G, Hirsch P (1998b) Plant genotype and micronutrient status influence colonization of wheat roots by soil bacteria. J Plant Nutr 21: 99-113

Rengel Z, Batten G, Crowley D (1999) Agronomic approaches for improving the micronutrient density in edible portions of field crops. Field Crops Res 60: 27-40

Rhodes D, Klug A (1993) Zinc fingers. Sci Am 268(Feb): 56-65

Rogers EE, Eide DJ, Guerinot ML (2000) Altered selectivity in an Arabidopsis metal transporter. Proc Natl Acad Sci USA 97: 12356-12360

Romera FJ, Welch RM, Norvell WA, Schaefer SC, Kochian LV (1996) Ethylene involvement in the over-expression of Fe(III)-chelate reductase by roots of E107 pea (*Pisum sativum* L. (*brz, brz*)) and *chloronerva* tomato (*Lycopersicon esculentum* L.) mutant genotypes. Biometals 9: 38-44

Römheld V (1991) The role of phytosiderophores in acquisition of iron and other micronutrients in graminaceous species: an ecological approach. Plant Soil 130: 127-134

Römheld V, Marschner H (1986) Evidence for a specific uptake system for iron phytosiderophores in roots of grasses. Plant Physiol 80: 175-180

Römheld V, Marschner H (1990) Genotypic differences among graminaceous species in release of phytosiderophores and uptake of iron phytosiderophores. Plant Soil 123: 147-153

Santa Maria GE, Cogliatti DH (1988) Bidirectional Zn-fluxes and compartmentation in wheat seedling roots. J Plant Physiol 132: 312-315

Sattelmacher B, Horst WJ, Becker HC (1994) Factors that contribute to genetic variation for nutrient efficiency of crop plants. Z Pflanzenernähr Bodenk 157: 215-224

Schickler H, Caspi H (1999) Response of antioxidative enzymes to nickel and cadmium stress in hyperaccumulator plants of the genus *Alyssum*. Physiol Plant 105: 39-44

Schutzendubel A, Polle A (2002) Plant responses to abiotic stresses: heavy metal-induced oxidative stress and protection by mycorrhization. J Exp Bot 53: 1351-1365

Shenker M, Oliver I, Helmann M, Hadar Y (1992) Utilization by tomatoes of iron mediated by a siderophore produced by *Rhizopus arrhizus*. J Plant Nutr 15: 2173-2182

Stephan UW, Grün M (1989) Physiological disorders of the nicotianamine-auxotroph tomato mutant chloronerva at different levels of iron nutrition. II. Iron deficiency response and heavy metal metabolism. Biochem Physiol Pflanzen 185: 189-200

Sültemeyer D, Schmidt C, Fock HP (1993) Carbonic anhydrase in higher plants and aquatic microorganisms. Physiol Plant 88: 179-190

Supek F, Supekova L, Nelson H, Nelson N (1996) A yeast manganese transporter related to the macrophage protein involved in conferring resistance to mycobacteria. Proc Natl Acad Sci USA 93: 5105-5110

Takatsuji H (1998) Zinc-finger transcription factors in plants. Cell Mol Life Sci 54: 582-596

Tanaka K, Takio S, Satoh T (1995) Inactivation of the cytosolic Cu/Zn-superoxide dismutase induced by copper deficiency in suspension-cultured cells of *Marchantia paleacea* var. *diptera*. J Plant Physiol 146: 361-365

Thomine S, Wang R, Ward JM, Crawford NM, Schroeder JI (2000) Cadmium and iron transport by members of a plant metal transporter family in *Arabidopsis* with homology to *Nramp* genes. Proc Natl Acad Sci USA 97: 4991-4996

Timonin MI (1946) Microflora of the rhizosphere in relation to the manganese-deficiency disease of oats. Soil Sci Soc Am Proc 11: 284-292

Timonin MI (1965) Interaction of higher plants and soil microorganisms. In: Gilmore CM, Allen ON (eds) Microbiology and soil fertility. Oregon State University Press, Corvallis, OR, pp 135-138

Tinker PB (1984) The role of microorganisms in mediating and facilitating the uptake of plant nutrients from soil. Plant Soil 76: 77-91

Tolay I, Erenoglu B, Römheld V, Braun HJ, Cakmak I (2001) Phytosiderophore release in *Aegilops tauschii* and *Triticum* species under zinc and iron deficiencies. J Exp Bot 52: 1093-1099

Tong Y, Rengel Z, Graham RD (1995) Effects of temperature on extractable Mn and distribution of Mn among soil fractions. Commun Soil Sci Plant Anal 26: 1963-1977

Tong Y, Rengel Z, Graham RD (1997) Interactions between nitrogen and manganese nutrition of barley genotypes differing in manganese efficiency. Ann Bot 79: 53-58

Torun B, Bozbay G, Gultekin I, Braun HJ, Ekiz H, Cakmak I (2000) Differences in shoot growth and zinc concentration of 164 bread wheat genotypes in a zinc-deficient calcareous soil. J Plant Nutr 23: 1251-1265

Vallee BL, Falchuk KH (1993) The biochemical basis of zinc physiology. Physiol Rev 73: 79-118

Van der Zaal BJ, Neuteboom LW, Pinas JE, Chardonnens AN, Schat H, Verkleij JAC, Hooykaas PJJ (1999) Overexpression of a novel *Arabidopsis* gene related to putative zinc-transporter genes from animals can lead to enhanced zinc resistance and accumulation. Plant Physiol 119: 1047-1055

Vaughan D, Dekock PC, Ord BG (1982) The nature and localisation of superoxide dismutase in fronds of *Lemna gibba* L. and the effects of copper and zinc deficiency on its activity. Physiol Plant 54: 253-257

Vert G, Briat J-F, Curie C (2001) *Arabidopsis IRT2* gene encodes a root-periphery iron transporter. Plant J 26: 181-189

Von Wiren N, Mori S, Marschner H, Römheld V (1994) Iron inefficiency in maize mutant *ys1* (*Zea mays* L. cv. yellow-stripe) is caused by a defect in uptake of iron phytosiderophores. Plant Physiol 106: 71-77

Von Wiren N, Marschner H, Römheld V (1995) Uptake kinetics of iron-phytosiderophores in two maize genotypes differing in iron efficiency. Physiol Plant 93: 611-616

Von Wiren N, Marschner H, Römheld V (1996) Roots of iron-efficient maize also absorb phytosiderophore-chelated zinc. Plant Physiol 111: 1119-1125

Walter A, Römheld V, Marschner H, Mori S (1994) Is the release of phytosiderophores in zinc-deficient wheat plants a response to impaired iron utilization? Physiol Plant 92: 493-500

Wang Y, Brown HN, Crowley DE, Szaniszlo PJ (1993) Evidence for direct utilization of a siderophore, ferrioxamine B, in axenically grown cucumber. Plant Cell Environ 16: 579-585

Webb MJ, Norvell WA, Welch RM, Graham RD (1993) Using a chelate-buffered nutrient solution to establish the critical solution activity of Mn^{2+} required by barley (*Hordeum vulgare* L.). Plant Soil 153: 195-205

Weckx JEJ, Clijsters HMM (1996) Oxidative damage and defense mechanisms in primary leaves of *Phaseolus vulgaris* as a result of root assimilation of toxic amounts of copper. Physiol Plant 96: 506-512

Wei LC, Loeppert RH, Ocumpaugh WR (1997) Fe-deficiency stress response in Fe-deficiency resistant and susceptible subterranean clover: importance of induced H^+ release. J Exp Bot 48: 239-246

Welch RM (1995) Micronutrient nutrition of plants. CRC Crit Rev Plant Sci 14: 49-82

Welch RM, La Rue TA (1990) Physiological characteristics of Fe accumulation in the bronze mutant of *Pisum sativum* L. cultivar Sparkle E107 *brz brz*. Plant Physiol 93: 723-729

Welch RM, Norvell WA (1993) Growth and nutrient uptake by barley (*Hordeum vulgare* L. cv Herta): studies using an N-(2-hydroxyethyl)ethylenedinitrilotriacetic acid-buffered nutrient solution technique. Plant Physiol 101: 627-631

Welch RM, Graham RD (1999) A new paradigm for world agriculture: meeting human needs productive, sustainable, nutritious. Field Crops Res 60: 1-10

Welch RM, Webb MJ, Loneragan JF (1982) Zinc in membrane function and its role in phosphorus toxicity. In: Scaife A (ed) Plant nutrition 1982. Proceedings of the 9th international plant nutrition colloquium. Commonwealth Agricultural Bureau, Farnham Royal, UK, pp 710-715

Wenzel AA, Mehlhorn H (1995) Zinc deficiency enhances ozone toxicity in bush beans (*Phaseolus vulgaris* L. cv. *Saxa*). J Exp Bot 46: 867-872

Yamaguchi H, Nakanishi H, Nishizawa NK, Mori S (2000a) Induction of the *IDI1* gene in Fe-deficient barley roots: a gene encoding a putative enzyme that catalyses the methionine salvage pathway for phytosiderophore production. Soil Sci Plant Nutr 46: 1-9

Yamaguchi H, Nakanishi H, Nishizawa NK, Mori S (2000b) Isolation and characterization of *IDI2*, a new Fe-deficiency-induced cDNA from barley roots, which encodes a protein related to the alpha subunit of eukaryotic initiation factor 2B (elF2B alpha). J Exp Bot 51: 2001-2007

Yamaguchi H, Nishizawa NK, Nakanishi H, Mori S (2002) IDI7, a new iron-regulated ABC transporter from barley roots, localizes to the tonoplast. J Exp Bot 53: 727-735

Yu Q, Osborne LD, Rengel Z (1998) Micronutrient deficiency changes activities of superoxide dismutase and ascorbate peroxidase in tobacco seedlings. J Plant Nutr 21: 1427-1437

Yu Q, Osborne LD, Rengel Z (1999a) Increased tolerance to Mn deficiency in transgenic tobacco overproducing superoxide dismutase. Ann Bot 84: 543-547

Yu Q, Worth C, Rengel Z (1999b) Using capillary electrophoresis to measure Cu/Zn superoxide dismutase concentration in leaves of wheat genotypes differing in tolerance to zinc deficiency. Plant Sci 143: 231-239

Zalewski PD, Millard SH, Forbes IJ, Kapaniris O, Slavotinek A, Betts WH, Ward AD, Lincoln SF, Mahadevan I (1994) Video image analysis of labile zinc in viable pancreatic islet cells using a specific fluorescent probe for zinc. J Histochem Cytochem 42: 877-884

Zhang FS (1993) Mobilisation of iron and manganese by plant-borne and synthetic metal chelators. In: Barrow NJ (ed) Plant nutrition—from genetic engineering to field practice. Kluwer, Dordrecht, pp 115-118

Zhang FS, Römheld V, Marschner H (1989) Effect of zinc deficiency in wheat on the release of zinc and iron mobilizing root exudates. Z Pflanzenernähr Bodenk 152: 205-210

Zhang FS, Römheld V, Marschner H (1991) Diurnal rhythm of release of phytosiderophores and uptake rate of zinc in iron-deficient wheat. Soil Sci Plant Nutr 37: 671-678

Zhao H, Eide D (1996a) The yeast ZRT1 gene encodes the zinc transporter protein of a high-affinity uptake system induced by zinc limitation. Proc Natl Acad Sci USA 93: 2454-2458

Zhao H, Eide D (1996b) The ZRT2 gene encodes the low affinity zinc transporter in *Saccharomyces cerevisiae*. J Biol Chem 271: 23203-23210

Zhao H, Butler E, Rodgers J, Spizzo T, Duesterhoeft S, Eide D (1998) Regulation of zinc homeostasis in yeast by binding of the ZAP1 transcriptional activator to zinc-responsive promoter elements. J Biol Chem 273: 28713-28720

Chapter **12**

Metal Pollution and Forest Decline

A. Hüttermann[1], I. Arduini[2], D.L. Godbold[3]
[1]Forstbotanisches Institut, Abteilung Technische Mykologie, Universität Göttingen, 37077 Göttingen, Germany
[2]Dipartimento di Agronomia e Gestione dell'Agro-Ecosistema, Università degli Studi Pisa, Via S. Michele degli Scalzi 2, 56124 Pisa, Italy
[3]School of Agricultural and Forest Sciences, University of Wales, Bangor, Gwynedd LL57 2UW, UK

12.1 SOURCES AND PATTERN OF HEAVY METAL POLLUTION

Human activities influence the natural distribution and biogeochemical cycling of elements, thus changing the chemical environment where plants develop (Nriagu and Pacyna 1988). Among the elements which could affect plants, heavy metals are not naturally removed or degraded and therefore progressively accumulate in soil or water sediments (Bussotti et al. 1983; Hunter et al. 1987). In forest ecosystems different sources of heavy metal contamination are prevalent. A natural source is the background load from the parent rocks that, on the whole, is rather low (Table 12.1), but may be distributed unevenly. Human activities have resulted in heaps and spoils from mining, short-range emissions from smelters, and long-range emissions from high stacks and lead from automobiles. Special sources of heavy metal contamination may be of local importance, like the treatment of soils with manure from intensive animal husbandry, which may lead to wide-range contamination with copper (Ernst 1985) or polluted rivers that contaminate ecosystems via aerosols (Bussotti et al. 1983).

12.1.1 Short-Range Pollution Owing to Mining Activities

The sources of pollution that are directly connected with mining result in small areas with very different concentrations of heavy metals. Mountain regions with local concentrations of heavy metals are found in Europe, especially in Scandinavia (The Fenno-Scandic Shield) and in the temperate humid zone of central (Table 12.1) and eastern Europe, with a Cambrian mountain ridge stretching from the Ore Mountains in Germany to Romania. In these regions, mining and ore processing activities started several centuries ago, and a very complex emission pattern has evolved.

Table 12.1 Trace elements in main soil-parent rocks (values commonly found, mg/kg, dry-weight basis; Kabata-Pendias et al. 1992)

Element	Magmatic rocks			Sedimentary rocks		
	Mafic	Intermediate	Acid	Argillaceous and shales	Sands tone	Limes, dolomites
Cd	0.13-0.22	0.13	0.09-0.20	0.22-0.30	0.05	0.035
Co	35-50	1.0-10	1-7	11-20	0.3-10	0.1-3.0
Cr	170-200	15-50	4-25	60-100	20-40	5-16
Cu	60-120	15-80	10-30	40	5-30	2-10
Hg	0.0X	0.0X	0.08	0.18-0.40	0.04-0.10	0.04-0.05
Mn	1200-2000	500-1200	350-600	500-850	100-500	200-1000
Mo	1.0-1.5	0.6-1.0	1-2	0.7-2.6	0.2-0.8	0.16-0.40
Ni	130-160	5-55	5-15	50-70	5-20	7-20
Pb	3-8	12-15	15-24	18-25	5-10	3-10
V	200-250	30-100	40-90	100-130	10-60	10-45
Zn	80-120	40-100	40-60	80-120	15-30	10-25

Past mining activities have resulted in short range transported highly concentrated pollution such as mine heaps and emissions from small smelters. Modern smelters with higher stacks have caused medium- and long-range emissions. This complex emission pattern always results in a rather patchy pattern of pollution, an example of which can be cited from a mining site in Poland (Kabata-Pendias et al. 1992).

Depending on human activities and the climatic conditions, regions with high loads of heavy metals are close to those with medium or low loads of pollution. Such patchy patterns of pollution have been found in Germany in the Harz Mountains (Schwarz 1996) and the area around Stolberg (Schneider 1982), in Austria in the Bleiberg, in Carinthia (Sieghardt 1988) and in the Tirol (Göbl and Mutsch 1985), in a lead mining complex in the Pennines, England (Clark and Clark 1981), the damaged forests south of Norilsk in Siberia (Schweingruber and Voronin 1996), and the area of Gyongyosoroszi, Hungary (Horvath and Gruiz 1996) to quote only a few studies.

Near to point sources heavy metals severely influence the growth and regeneration capacity of forest trees (Wotton et al. 1986; Bell and Teramura 1991). Excess heavy metals influence both soil properties and rates of litter decomposition (Bond et al. 1976) and they negatively affect seedlings, whose roots initially develop in the most contaminated layer (Patterson and Olson 1983). This detrimental influence of heavy metal pollution on the soil can go so far that on certain sites no trees have grown for centuries (e.g. in the Stolberg region or the Harz Mountains near Clausthal-Zellerfeld). An idea of the range of heavy metal pollution at given sites is conveyed by Table 12.2, where the published data on pollution in eastern Europe are summarised.

12.1.2 Diffuse and Long-Range Pollution from High Stacks and Automobiles

The change of instrumentation in the mining industry with a trend toward production units with higher capacities and the concern about "clean air" in industrial regions has

Table 12.2 Metal contamination of surface soils of central-eastern Europe (concentrations in mg/kg, dry-weight basis)

Metal	Source of Pollution	Country	Concentration (mg/kg, d.wt)	Reference
Cu	Metal-processing industry	Bulgaria	24-2015	Tchuldziyan and Khinov (1976)
		Poland	72-1710	Kabata-Pendias et al. (1981); Niskavaara et al. (1996)
		Russia, Kola	3-1020	Karczewska (1996)
		Czech		Ustyak and Petrikova (1996)
	Urban gardens and orchards	Poland	12-240	Czamowska (1982)
		Ukraine	50-83	Kiriluk (1980)
	Vineyards Fertilised farmland	Poland	80-1600	Sapek (1980)
Zn	Metal-processing industry	Poland	1665-5660	Faber and Niezgoda (1982); Rybicka (1996); Verner et al. (1996)
		Czech	201-27,600	Rieuwerts and Farago (1996); Ustyak and Petrikova (1996)
	Non-ferric metal mining	Russia	400-4245	Letunova and Krivitskiy (1979)
Cd	Metal-processing industry	Bulgaria	2-5	Petrov and Tsalev (1979)
		Poland	6-270	Faber and Niezgoda (1982) Rybicka (1996); Widera (1980); Greszta et al. (1985)
		Romania	1.4-12	Rauta et al. (1987); Lacatusu et al. (1996)
		Czech		Ustyak and Petrikova (1996)
	Urban gardens	Poland	0.4-5	Czamowska (1982)
	Sludge-treated or fertilised farmland	Hungary	3-6	Vermes (1987)
		Poland	0.4-107	Umifiska (1988)
Pb	Metal-processing industry	Poland	72-3570	Rybicka (1996),
		Kirghizia	3000	Manecki et al. (1981); Petrov and Tsalev (1979); Verner at al. (1996)
		Czech	908-37300	Rieuwerts and Farago (1996);
	Non-ferric metal mining	Kirghizia	21-3044	Niyazova and Letunova (1981); Ustyak and Petrikova (1996)
	Urban gardens	Poland	17-165	Czamowska (1982)

resulted in the establishment of high stacks which can be more than 100 m high. This development has been mainly implemented since the end of the Second World War. Since the filtering techniques in these high stacks were not very effective, at least in the early years, this resulted in long-range transport of pollutants and a concomitant spread of heavy metals over the whole country.

A second source of heavy metal pollution are cars which until the last decade were fuelled with leaded gasoline. In central Europe, the orographic situation of the forests, which mainly grow on hill tops, together with the high filtering effect of the tree canopy, resulted in forests often being subjected to heavier pollutant loads than other ecosystems (Ulrich et al. 1979; Bender et al. 1989). Due to the high affinity of heavy metals for organic matter, even low inputs result in high soil levels over time, especially in the humus layer (Woolhouse 1983; Friedland et al. 1986). In recent decades, an increasing accumulation of heavy metals in the upper soil layer, due to atmospheric deposition, has been observed even far from sources of emissions (Wotton et al. 1986). At such sites, heavy metals appear to contribute to the 'new-type forest decline' (Heale and Ormrod 1982; Godbold and Hüttermann 1985; Godbold 1991), a diffuse disease damaging many forest tree species, especially in northern Europe and North America, that seems to be mainly due to atmospheric pollution. Common symptoms of that disease are crown thinning and an anomalous development of branches, leaves and flowers (Gellini 1989).

12.1.3 Pollution from Marine Aerosols Near Estuaries of Polluted Rivers

An unexpected path for metal emissions into ecosystems was found in recent investigations by Arduini et al. (1994, 1995, 1996), who attempted to assess the impact of heavy metals in the San Rossoré Natural Park in Italy. Heavy metal contamination of this site is due to atmospheric inputs mainly from marine aerosols. Marine pollution results from inputs from the rivers and channels Arno, Serchio and Morto (Bussotti et al. 1983; Table 2).

12.2 THE ROLE OF MYCORRHIZA IN HEAVY-METAL-TREE INTERACTIONS

In all forest ecosystems the fine roots of trees are colonised by mycorrhizas. In boreal forest ecosystems the trees are predominantly ectomycorrhizal, whereas in temperate forest ecosystems many deciduous trees are also colonised by arbuscular mycorrhizas. It has been suggested that mycorrhizas, and in particular ectomycorrhizas, ameliorate heavy metal toxicity (Wilkins 1991). Heavy metal tolerance of the fungus itself is a prerequisite for efficiently protecting the host plant against toxic metals. Fungi, in general, are believed to be highly metal tolerant (Gadd 1993); however, some ectomycorrhizal species have been shown to be sensitive to Cd (Darlington and Rauser 1988; Jongbloed and Borst-Pauwels 1990). Marschner et al. (1999) suggested that the higher tolerance to heavy metals reported for fungi compared with plants is in part due to the different culture methods used for plants and fungi. In field experiments, mycorrhizal formation was strongly suppressed by Cd in oak, pine and spruce seedlings inoculated by mycorrhizal fungi (Dixon 1988; Dixon and Buschena 1988).

In short roots the hyphal mantel of ectomycorrhizas provides a tissue layer between the root and the soil. In a number of cases, mycorrhizal colonisation of roots increased the tolerance of the host plant to metals (Table 12.3). Amelioration of metal toxicity was estimated based on either a smaller decrease in plant biomass in the presence of metals in the ectomycorrhizal plants compared with the non-mycorrhizal plants, or in some cases based on lower concentrations or contents of metals in the shoots of mycorrhizal plants compared with non-mycorrhizal plants.

Table 12.3 Effect of different species and strains of ectomycorrhizas on the metal tolerance of the host tree seedlings

Metal	No effect	Increased metal tolerance	Tree species	Reference
Cd	Scleroderma citrinum Paxillus involutus Suillus luteus NP1	Suillus bovinus P2, S. luteus P2	Pinus sylvestris	Colpaert and Van Assche (1993)
	S. bovinus NP1 S. bovinus P3 Laccaria laccata	P. involutus 533	Picea abies	Jentschke et al. (1999)
Ni	Lactariusrufus L. hibbardae L. proxima	Scleroderma flavidum	Betula papyrifera	Jones and Hutchinson (1986)
	L. rufus P. involutus Cou P. involutus Nau Pisolithus tinctorius Amanita muscaria	P. involutus 533 L. laccata	P. abies	Marschner (1994)
Zn	Thelephora terrestris	P. involutus P. involutus Amanita muscaria	Pinus sylvestris Betula	Colpaert and Van Assche 1992 Brown and Wilkins (1985)

In *Pinus sylvestris* inoculated with a number of mycorrhizal species and strains, lower levels of Cd (Colpaert and Van Assche 1993) and Zn (Colpaert and Van Assche 1992) were found in the needles of mycorrhizal plants inoculated with *Paxillus involutus* and two species of *Suillus* compared with non-mycorrhizal seedlings. However *Thelephora terrestris* increased Zn accumulation in the needles. In *Betula* seedlings, mycorrhization with *Amanita muscaria* and *Paxillus involutus* reduced the Zn contents of the leaves, and in terms of growth improved tolerance to Zn (Brown and Wilkins 1985). In an investigation of the effect of *Scleroderma flavidum*, *Lactarius hibbardae*, *Lactarius rufus* and *Laccaria proxima* on the Cu and Ni tolerance of *Betula papyrifera* (Jones and Hutchinson 1986), only inoculation with *Scleroderma flavidum* increased the Ni tolerance of the seedlings. Although the presence of *Scleroderma flavidum* did not reduce Ni accumulation in the leaves, the growth of the seedlings was improved. However, the Cu tolerance of *Betula papyrifera* was decreased by this fungus. In *Picea abies* (Marschner 1994), Pb tolerance was increased by a strain of *Paxillus involutus* and *Laccaria laccata*, but not by a number of other strains and species.

In *Picea abies* (Jentschke et al. 1999), tolerance to Cd was increased in seedlings inoculated with *Paxillus involutus*, but not in seedlings inoculated with *Laccaria laccata*. In these seedlings, mycorrhizal growth with *Paxillus involutus* was less reduced than non-mycorrhizal seedlings in the presence of Cd. The differences in the ability of these two fungi to ameliorate the toxicity of Cd were not due to decreased Cd accumulation. No changes in the Cd concentrations of the roots or the subcellular distribution of Cd could be found (Winter 1995). Jentschke et al. (1999) suggested that the amelioration of Cd toxicity by *Paxillus involutus* could be due to better P nutrition of the host.

Most mechanisms proposed for amelioration of metal toxicity by ectomycorrhizal fungi involve prevention of metals entering the mycorrhizas or retention of metals in the mycorrhizas restricting metal movement to the plant root cells. It has been suggested that the mechanism by which mycorrhizal fungi ameliorate metal toxicity is by reducing metal uptake into critical locations of the host plant as a result of adsorption of metals in fungal cell walls or intracellular sequestration both in the mantle of the mycorrhizas (Turnau et al. 1993) and the extramatrical mycelium (Denny and Wilkins 1987; Colpaert and Van Assche 1992, 1993; Marschner et al. 1996). Within the hyphae of the extramatrical mycelium and Hartig net a number of binding sites have been suggested. Binding of metals may occur in polyphosphate bodies in the vacuoles (Vare 1991) or in polysaccharide material in the cell walls and vacuoles (Turnau et al. 1996). In addition, metals have also been found associated with cysteine-rich deposits in cytoplasm and vacuoles in mantle cells of *Rhizopogon roseolus* (Turnau et al. 1996). In yeast and filamentous fungi, binding of metals to cysteine-rich metallothioneins and phytochelatins is considered to be an important mechanism in metal tolerance (Gadd 1993). However, in the strain of *Paxillus involutus* shown to ameliorate toxicity of Pb (Marschner 1994) and Cd (Winter 1995), metallothioneins could not be detected (D.A. Thurman, pers. comm.). Additionally, a filter function of the hyphal mantle restricting movements of metals has not been shown experimentally (Galli et al. 1993,1994). Thus there is no strong evidence of intracellular chelation of metals in ectomycorrhizal fungi.

For amelioration of Pb toxicity, Marschner (1994) proposed two mechanisms. In a study with *Laccaria laccata*, Pb was bound extracellularly by organic acids that reduced the toxicity of Pb but did not restrict movement into the plants. In the case of *Paxillus involutus*, Pb was bound and retained in the extramatrical mycelium. Also, for Zn retention in the extramatrical mycelium, the ability to maintain a viable mycelium under metal stress has been strongly linked to amelioration of metal tolerance in the host plant (Colpaert and Van Assche 1992). The ability to retain Zn in the extramatrical mycelium was shown to differ between species despite the fungi having a similar hyphal biomass (Colpaert and Van Assche 1992). There is stronger evidence to suggest that extracellular binding to the extramatrical mycelium may play an important role. However, in a number of cases an increase in metal tolerance was also associated with higher P levels in plants (Marschner 1994, Winter 1995). As suggested by Jentschke et al. (1999), improvement of the mineral nutritional status of the host plant alone may confer a higher resistance to metals.

For ectomycorrhizal fungi large differences have been demonstrated between species in tolerance to single metals, and within a single species to different metals (Rapp and Jentschke 1994, Tamm 1995). For microfungi, Gadd (1993) suggests that tolerance to heavy metals is dependent on intrinsic properties rather than adaptive changes, and that there is little evidence for adaptation of fungi to metals on sites with short or long histories of pollution. In forests, pollution has led to a decrease in species diversity of mycorrhizal fungi (Gadd 1993; Rapp and Jentschke 1994). This suggests a selective elimination of more pollutant-sensitive fungi. Amelioration of metal toxicity by ectomycorrhizal fungi under experimental conditions has been most often achieved using fungi isolated from polluted forests stands (Colpaert and Van Assche 1992, 1993; Marschner 1994; Winter 1995) or sites with high levels of metals in the soil (Colpaert and Van Assche 1993; Jones and Hutchinson 1986), although this is not always the case (Brown and Wilkins 1985). This suggests that under natural conditions some ectomycorrhizal fungi have the potential to ameliorate metal toxicity to the host on polluted sites. However, ectomycorrhizal fungi shown to be tolerant to metals often have no beneficial effect on the metal tolerance of the host plant.

12.3 ASSESSMENT OF THE IMPACT OF HEAVY METAL POLLUTION ON TREES AND FOREST ECOSYSTEMS

Attempts to assess the role of heavy metals in forest decline often require comparison of laboratory studies with field observations. Due to the speciation of metal in the soil solution of forest soils, such comparisons are often fraught with difficulties. Heavy metals have been shown to reduce the growth of trees in a number of investigations (Carlson and Bazzaz 1977; Foy et al. 1978; Kelly et al. 1979; Burton et al. 1984; Breckle 1991). In *Pinus strobus*, *Picea glauca* and *Picea abies*, root growth was inhibited by Cd and Zn (Mitchell and Fretz 1977), in *Pinus strobus* and *Picea glauca* by Ni and Cu (Lozano and Morrison 1982), and in *Picea sitchensis* by Cd, Cu, Ni and Pb (Burton et al. 1984, 1986). Root growth was reduced in *Fagus sylvatica* grown in Pb-contaminated soils (Kahle and Breckle 1986; Kahle 1993). In an attempt to assess the risk of heavy metals in forest ecosystems, Lamersdorf et al. (1991) compared the levels of metals in soil solutions at two sites in Germany with the levels of metals known to inhibit root growth in *Picea abies*. These authors calculated a high risk to forests at these sites due to Pb and a possible risk due to Cd. The levels of heavy metals in the German sites are several times lower than those reported for eastern European forests (Table 12.2), so the impact of heavy metal pollution expected is much higher in the latter.

Arduini et al. (1994, 1995, 1996) carried out an investigation of *Pinus pinea* L., *Pinus pinaster* Ait. and *Fraxinus angustifolia* Vahl. that coexist in the varied habitats of the San Rossoré Natural Park in Italy (Bussotti et al. 1983). In particular, the effects of Cu and Cd on these tree species were investigated. The copper and cadmium uptake, metal distribution in trees and the dose-response patterns varied significantly among species. Considerable differences in the sensitivities of these coexisting species to metals were shown. *Pinus pinea* was more tolerant to cadmium, whereas *F. angustifolia* was highly sensitive to both cadmium and copper. In all species Cu was taken up to a higher degree than cadmium. All species showed a rather linear relation between the Cu-content increase of roots and the rising Cu-supply. This was attributed to be mainly due to passive influx, suggesting that the toxicity threshold had been exceeded at the concentrations used (Fernandes and Henriques 1991) and that membrane selectivity had been lost (Meharg 1993). In contrast, the uptake of Cd did not increase proportionally to its higher availability at the highest Cd concentrations used. Copper is often described as the most toxic of a range of heavy metals in culture solution (Patterson and Olson 1983; Baker et al. 1994) and a higher accumulation of this metal compared with cadmium has been reported in many species (Barbolani et al. 1986; Sela et al. 1988). In contrast, when heavy metals are added to soil, copper binds strongly to many soil components, which considerably reduces its availability to roots, whereas cadmium shows a higher mobility (Smilde 1981; Baker et al. 1994). In the study of Arduini et al. (1994), the inhibition of root growth, which is widely used for testing heavy metal toxicity (Baker et al. 1994), was highly significantly related to the accumulation of copper in roots in all species. This finding suggests that copper exerts a direct toxic effect, perhaps damaging membrane integrity by oxidation and cross-linking of protein thiols or by the peroxidation of the unsaturated fatty acids (Meharg 1993). This is, however, not the case for cadmium, where a low supply enhanced root growth (Arduini et al. 1994).

The findings of Arduini et al. (1994, 1995, 1996) support the idea that heavy metal accumulation in the upper forest soil layers could impair the natural regeneration of forest species, thereby contributing to forest die-back even at relatively low concentrations. Even though a recent survey of cadmium contamination of sites at San Rossoré, where the seeds were collected, are far below the toxicity levels (Bargagli et al. 1986), the load of heavy metals will increase over time (Pantani et al. 1984). This increase over time will become important in plants like *Pinus*, where the vitality of the tap-roots is of extreme importance, allowing them to colonise soils subjected to superficial drying (Sutton 1980; Masetti and Mencuccini 1991). The difference in sensitivity to metals may be important for the natural regeneration and competition between two coexisting species. The response to copper was rather similar among the tested species, whereas marked differences were observed in the cadmium response, which suggest that there could be some innate Cd-tolerance, especially in *P. pinea*. The high sensitivity of *F. angustifolia* to heavy metals could indeed impair the natural regeneration of this species, by affecting seedling survival directly and by decreasing their competitiveness towards the coexisting *Pinus* species.

12.4 AMELIORATION

Except for phytoremediation (see Chaps. 13 and 14) it is impossible to really ameliorate soils that are contaminated with heavy metals in a strict sense. Chemical removal of the heavy metals from soils would mean that the soil would have to be excavated from the stand and treated like any ore from a mine with chemicals which would leach the metals. Since these chemicals are aggressive and would also remove plant nutrients from the soil, the resulting product would very probably be more hostile to plants than the one that originally contained the heavy metals. So it is impractical and economically out of the question to remove heavy metals from contaminated forest soils.

The goal therefore must be to employ strategies that allow the growth of forest trees in stands which are contaminated by heavy metals in spite of their presence. One way is to cover the contaminated sites with uncontaminated fertile soil to provide a substrate for the growth of trees. Apart from being very expensive, this approach is not an optimal solution, since an upward transport of the metals from the spoil to the soil cover may occur and fine roots of trees may even grow through soil covers which are 1 m thick (Borgegard and Rydin 1989a). An alternative approach is to establish conditions in the contaminated soils that allow trees to grow there in spite of the presence of toxic metals. In principle there are several different strategies available which could work in this direction, at least in soils which are not too heavily contaminated: reduction of bioavailability of the heavy metals for planting of trees which are more tolerant to certain heavy metals and establishment of mycorrhizas which might help the trees to survive.

12.4.1 Reduction of Bioavailability

The bioavailability of metal ions is dependent on their solubility in the soil solution, i.e. their general solubility and the stage of equilibrium between the metal cation in its bound form and the free soluble cation. Since the concentration of heavy metal cations in forest soils is usually so low that the solubility behaviour of the metal salts will govern their concentration in the soil solution, the dominating factor for the bioavailability of heavy metal cations in soils is their adsorption to soil structures.

The capacity of a given soil to bind given heavy metals depends on the amount and nature of binding sites in the soil structures and the pH of the soil solution. Generally, it can be stated that the lower the pH value, the more soluble are the metal cations, and the more binding sites which are available in a given soil the lower will be the solubility of the heavy metals. In the case of cadmium ions, the increase in solubility with decreasing pH values starts at a pH of 6.5, in the case of lead and mercury ions at a pH value of 4, ions of arsenic, chromium, nickel, and copper start to dissolve at pH values between these two extremes (cf. Scheffer and Schachtschabel 1989). Thus the pH value of the soil solution in principle is one of the main factors which govern the solubility of heavy metal cations in the soil solution (Gerth and Brümmer 1977; Alloway and Morgan 1985; Tyler et al. 1987; Neite 1989; ; Reddy et al. 1995; Chuan et al. 1996; Kedziorek and Bourg 1996McBride et al. 1997) and its influence on plants under heavy metal stress is well established (e.g. Zhu and Alva 1993). Unfortunately, the acid deposition that has been prevalent all over Europe in recent decades has increased the mobility of heavy metals considerably (e.g. Mannings et al. 1996; Wilson and Bell 1996).

Thus, increasing the pH could be a measure to reduce the bioavailability of heavy metals. This has been shown by Walendzik (1993) for spruce in the western Sudety Mountains. Liming, however, may not always be a good solution since it may increase the rate of nitrogen mineralisation and thus aggravate the NO_3 load in the groundwater (Marschner et al. 1989; Kreutzer 1995; Schuler 1995). The approach of using waste materials like fuel ash (Piha et al. 1995) or sewage sludge (Borgegard and Rydin 1989b) to improve the growth of trees on mine spoils may not always be successful as the authors quoted above have shown. Better results using municipal sewage sludge for establishing sagebrush vegetation on copper mine spoils were reported by Sabey et al. (1990). The simple addition of inorganic fertiliser may not work at all (Borgegard and Rydin 1989b).

Another way to decrease the bioavailability of heavy metals is to increase the binding sites for heavy metal ions in the soil, e.g. by amendment with humic substances or zeolites (Baydina 1996) or expanded clay and porous ceramic material (Figge et al. 1995). When adding organic substances to the soil it is very important to work with water-insoluble material that is not available for rapid degradation by microorganisms (Herms and Brümmer 1984). The authors found that an addition of hay to a soil contaminated with heavy metals increased the solubility of Cu, Cd, and Zn, whereas this effect was not observed for Pb. Amendment of the soil with peat had the opposite effect.

A novel approach was employed by Hüttermann and coworkers applying cross-linked polyacrylates (hydrogels) to metal-contaminated soils. When such a compound (Stockosorb K400) was applied to hydrocultures of scots pine, *Pinus sylvestris*, which contained 1 μM of Pb, two effects were observed: (1) the hydrogel increased the nutrient efficiency of the plants, and (2) the detrimental effect of the heavy metal was completely remediated (Fig. 12.1). Determination of the heavy metal content of the roots revealed that the uptake of the lead was greatly inhibited by the hydrogel (Fig. 12.1). Analysis of the fine roots of 3-year-old spruce which had been grown for one vegetation period in lead-contaminated soil with and with without amendments with the hydrogel showed that amendment of the soil with the cross-linked acrylate did indeed prevent the uptake of the lead into the stele of the fine roots (Hüttermann et al., unpubl. data). The hydrogel acts as a protective gel that inhibits the entrance of the heavy metal into the plant root.

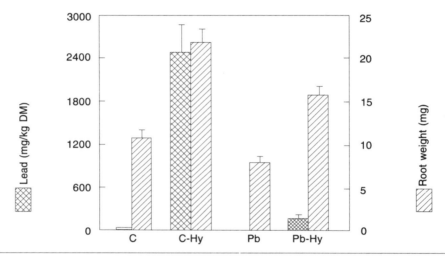

Fig. 12.1 Influence of the presence of hydrogels (Stockosorb K 400) in the culture solution on the root length and lead concentration in the roots of *Pinus sylvestris*. *C* Control, nutrient solution only; *C-Hy* nutrient solution with hydrogel; *Pb* nutrient solution with 1 μM lead per liter; *Pb-Hy* nutrient solution with 1 μM lead per liter, supplemented with hydrogel

12.4.2 Planting of Tolerant Trees

Like any other physiological trait, tolerance against heavy metals exhibits variability depending on the genetic structure of the individual tree. Litzinger (1990) showed that different aspen clones exhibited different degrees of tolerance against cadmium (Table 12.4). Similar differences among different genetic units have been found for other poplar species (Rachwal et al. 1993), sycamore (Dickinson et al. 1992; Watmough and Dickinson 1996), red maple (Watmough and Hutchinson 1997) and willows (Landberg and Greger 1996; Riddell-Black et al. 1997). Watmough et al. (1995) were able to induce metal tolerance in sycamore tissue cultures. Genetic analyses of differently adapted tree populations to heavy metal stress have been performed by Hosius et al. (1996) and Bergmann and Hosius (1996). Although the differences found by these authors between the clones that are most sensitive and most tolerant to heavy metal stress are really remarkable, there is only a very small chance that these tolerant clones of trees will be used for afforestation of metal-polluted sites. Unlike in agriculture, where the seed material usually has to be confined to clones or hybrids, in forestry there are good reasons that prohibit the use of

Table 12.4 Decrease in dry weight (% inhibition) of *Populus tremula* clones exposed to 15 µM Cd for 2 weeks (Litzinger 1990)

Clone	Whole plant	Leaf	Root
4	46.4	50.6	40.0
6	29.1	35.0	0
9	66.7	69.4	57.1
14	48.2	54.9	16.7

single clones for the establishment of stands that are expected to have a rotation time of about a century. A narrow gene pool in such a population would severely reduce the prospects of survival. Therefore, only stands with a short rotation time of less than 15 years can be established with single clones (Table 12.4).

12.4.3 The Use of Mycorrhiza for the Protection of Trees Against Heavy Metals

It is well known that plants with mycorrhizal infections in their roots grow better under stress conditions than plants without these symbionts (e.g. Maronek et al. 1981). Such a stress may have different causes like climate, e.g. the establishment of the tree line in the Alps (Moser 1963; Göbl 1984). Marx (1975, 1976) demonstrated that trees with mycorrhizas grew better on mine tailings and trees with an established symbiosis in their roots were used for afforestation of such sites (Marx et al. 1977; Göbl 1979). Similar results have been reported for plants with arbuscular mycorrhizas growing on mine spoils (Griffioen et al. 1994; Hetrick et al. 1994; Shetty et al. 1994; Noyd et al. 1996). It should be noted in this connection that the reclamation practices used for site preparation have a very high impact on the effectiveness of the establishment of the plants in question (Johnson and McGraw 1988a,b).

More recent and more detailed studies, however, have shown that heavy metal tolerance may vary considerably with different fungal partners (see the literature quoted in Sect. 12.2). Evidence suggests that, while amelioration of metal toxicity to tree seedlings by ectomycorrhizal fungi occurs, it is dependent on the species and strain of the ectomycorrhiza and the metal being considered. Whether amelioration of metal toxicity occurs will depend among other things on the species of ectomycorrhizal fungi colonising the roots and the types and combinations of metals present in the soil.

Use of mycorrhizas which ameliorate metal tolerance may be an integral part of reclamation of some metal-polluted sites such as mine spoils. On such soils mycorrhizas, in combination with soil amendments to lower metal concentrations and increase water holding (like the hydrogels mentioned above), have great potential. In established forest ecosystems, addition of mycorrhizas may be of limited success due to competition from indigenous fungi. However, due to evolution of metal-tolerant tree species, mycorrhizal fungi and novel soil amendments, the tools for biotechnological remediation of some metal-polluted areas may well have been supplied.

REFERENCES

Alloway BJ, Morgan H (1985) The behaviour and availability of Cd, Ni and Pb in polluted soils. In: Assink JW, Brink WJ van den (eds) Contaminated soil. Nijhoff, Dordrecht, pp 101-113

Arduini I, Godbold DL, Onnis A (1994) Cadmium and copper change root growth and morphology of Pinus pinea L. and Pinus pinaster Ait. seedlings. Physiol Planta 92: 675-680

Arduini I, Godbold DL, Onnis A (1995) Influence of copper on root growth and morphology of Pinus pinea L. and Pinus pinaster Ait. seedlings. Tree Physiol 15: 411-415

Arduini I, Godbold DL, Onnis A (1996) Cadmium and copper uptake and distribution Mediterranean tree seedlings. Physiol Plantarum 97: 111-117

Baker AM, Reeves RD, Hajar ASM (1994) Heavy metal accumulation and tolerance in British population of the metallophyte Thalaspi caerulescens J. & C. Presl (Brassicaceae). New Phytol 127: 61-68

Barbolani E, Clauser M, Pantani F, Gellini R (1986) Residual heavy metal (Cu and Cd) removal by Iris pseudacorus. Water Air Soil Pollut 28: 277-282

Bargagli R, D'Amato ML, Iosco FP (1986) Il degrado della vegetazione costiera di San Rossore: possibile incidenza di alcuni elementi in tracce. Atti Soc Tosc Sci Nat Mem Ser B 93: 133-144

Baydina NL (1996) Inactivation of heavy metals by humus and zeolites in industrially contaminated soil. Euras Soil Sci 28: 96-105

Bell R, Teramura AH (1991) Soil metal effects on the germination and survival of Quercus alba L. and Q. prinus L. Environ Exp Bot 31: 145-152

Bender J, Grünhage L, Jäger HJ (1989) Aufnahme und Wirkung von Schwermetallen bei Wald bäumen: Bodenkontaminationsversuche mit Cadmium, Blei und Nickel. Angew Bot 63: 81-93

Bergmann F, Hosius B (1996) Effects of heavy-metal polluted soils on the genetic structure of Norway spruce seedling populations. Water Air Soil Pollut 89: 363-373

Bond H, Lighthart B, Shimabuku R, Russell L (1976) Some effects of cadmium on coniferous forest soil and litter microcosm. Soil Sci 121: 278-287

Borgegard S-O, Rydin H (1989a) Biomass, root penetration and heavy metal uptake in birch in a soil cover over copper tailings. J Appl Ecol 26: 585-595

Borgegard S-O, Rydin H (1989b) Utilization of waste products and inorganic fertilizer in the restoration of iron-mine tailings. J Appl Ecol 26: 1083-1088

Breckle SW (1991) Growth under stress. Heavy metals. In: Waisel Y, Eshel A, Kafkafi U (eds) Plant roots: the hidden half, Dekker, New York, pp 351-373

Brown MT, Wilkins DA (1985) Zinc tolerance of mycorrhizal Betula. New Phytol 99: 101-106

Burton KW, Morgan E, Roig A (1984) The influence of heavy metals upon the growth of Sitka-spruce in South Wales forests. II. Greenhouse experiments. Plant Soil 78: 271-282

Burton KW, Morgan E, Roig A (1986) Interactive effects of cadmium, copper and nickel on the growth of Sitka-spruce and studies of metal uptake from nutrient solutions. New Phytol 103: 549-557

Bussotti F, Rinallo C, Grossoni P, Gellini R, Pantani F, DelPanta S (1983) Degrado della egetazione costiera nella Tenuta di S. Rossore, La Provincia Pisana. Spec. Parco Nat. igliarino S. Rossore Massaciuccoli e aree protette 9: 46-52

Carlson RW, Bazzaz FA (1977) Growth reduction in American sycamore (Platanus occidentalis L.) caused by Pb-Cd interaction. Environ Pollut 12: 243-253

Chuan MC, Shu GY, Liu JC (1996) Solubility of heavy metals in a contaminated soil: effects of redox potential and pH. Water Air Soil Pollut 90: 543-556

Clark RK, Clark SC (1981) Floristic diversity in relation to soil characteristics in a lead mining complex in the Pennines, England. New Phytol 87: 799-815

Colpaert JV, Van Assche JA (1992) Zinc toxicity in ectomycorrhizal Pinus sylvestris. Plant Soil 143: 201-211

Colpaert JV, Van Assche JA (1993) The effects of cadmium on ectomycorrhizal *Pinus sylvestris*. New Phytol 123: 325-333

Czamowska K (1982) Heavy metal contents of surface soils and plants in urban gardens. Symposium on Environmental Pollution, Nov 12, Plock (in Polish)

Darlington AB, Rauser WE (1988) Cadmium alters the growth of the ectomycorrhizal fungus *Paxillus involutus*: a new growth model accounts for changes in branching. Can J Bot 66: 225-229

Denny HJ, Wilkins DA (1987) Zinc tolerance in Betula spp. I. Effect of external concentration of zinc on growth and uptake. New Phytol 106: 517-524

Dickinson NM, Turner AP, Watmough SA, Lepp NW (1992) Acclimation of trees to pollution stress: cellular metal tolerance traits. Ann Bot 70: 569-572

Dixon RK (1988) Response of ectomycorrhizal Quercus rubra to soil cadmium, nickel and lead. Soil Biol Biochem 20: 555-559

Dixon RK, Buschena CA (1988) Response of ectomycorrhizal *Pinus banksiana* and *Picea glauca* to heavy metals in soil. Plant Soil 105: 265-271

Ernst WHO (1985) Schwermetallimmissionen—ökophysiologische und populations genetische Aspekte. Düsseldorfer Geobot Kolloq 2: 43-57

Faber A, Niezgoda J (1982) Contamination of soils and plants in the vicinity of the zinc and lead smelter. 1. Soils (in Polish). Roczniki Gleboznawcze 33: 93-107

Fernandes JC, Henriques FS (1991) Biochemical, physiological, and structural effects of excess copper in plants. Bot Rev 57: 246-273

Figge DAH, Hetrick BAD, Wilson GWT (1995) Role of expanded clay and porous ceramic amendments on plant establishment in mine spoils. Environ Pollut 88: 161-165

Foy CD, Chaney RL, White C (1978) The physiology of metal toxicity in plants. Ann Rev Plant Physiol 29: 511-566

Friedland AJ, Johnson AH, Siccama TG (1986) Zinc, Cu, Ni and Cd in the northeastern United States. Water Air Soil Pollut 29: 233-243

Gadd GM (1993) Interactions of fungi with toxic metals. Tansley review No 47. New Phytol 124: 25-60

Galli U, Meier M, Brunhold C (1993) Effects of cadmium on nonmycorrhizal and mycorrhizal Norway spruce seedlings *Picea abies* (L.) Karst and its ectomycorrhizal fungus *Laccaria laccata* (Scop. Ex Fr.) Bk. & BR.: sulphate reduction, thiols and distribution of the heavy metal. New Phytol 125: 837-843

Galli U, Schüepp H, Brunhold C (1994) Heavy metal binding by mycorrhizal fungi. Physiol Plant 92: 364-368

Gellini R (1989) Inquinamento e condizioni di efficienza del bosco. Econ Montana - Linea Ecol Anno XXI 6: 11-22

Gerth J, Brümmer G (1977) Quantitäts-Intensitäts-Beziehungen von Cd, Zn und Ni in Böden unterschiedlichen Stoffbestands. Mitt Dtsch Bodenkund Ges 29: 555-566

Göbl F (1979) Erfahrungen in der Anwendung von Mykorrhiza-Impfmaterial. I. Zirbe. Central Gesamte Forstwes 96: 30-43

Göbl F (1984) Forstliche Mykorrhizaforschung in Österreich. Allg Forstz 94: 318-319

Göbl F, Mutsch F (1985) Schwermetallbelastung von Wäldern in der Umgebung eines Hüttenwerkes in Brixlegg/Tirol. I. Untersuchung der Mykorrhiza und Humusauflage. Central Gesamte Forstwes 102: 28-40

Godbold DL (1991) Die Wirkung von Aluminium und Schwermetallen auf Picea abies Sämlinge. Sauerländers, Frankfurt

Godbold DL, Hüttermann A (1985) Effect of zinc, cadmium and mercury on root elongation of *Picea abies* (Karst.) seedlings, and the significance of these metals to forest die-back. Environ Pollut 38: 375-381

Greszta J, Braniewski S, Chrzanowska E (1985) Heavy metals in soils and plants around a zinc smelter. In: Kabata-Pendias A (ed) 3rd National Conference on Effects of Trace Metal Pollution on Agricultural and Environmental Quality. 2: 58-61 (in Polish) IUNG, Pulawy

Griffioen WAJ, Ietswaart JH, Ernst WHO (1994) Mycorrhizal infection of an *Agrostis capillaris* population on a copper contaminated soil. Plant Soil 158: 83-89

Heale EL, Ormrod DP (1982) Effects of nickel and copper on *Acer rubrum, Cornus stolonifera, Lonicera tatarica* and *Pinus resinosa*. Can J Bot 60: 2674-2681

Herms U, Brümmer G (1984) Einflußgrößen der Schwermetallöslichkeit und bindung in Böden.Z Pflanzenernähr Bodenkd 147: 400-424

Hetrick BAD, Wilson GWT, Figge DAH (1994) The influence of mycorrhizal symbiosis and fertilizer amendments on establishment of vegetation in heavy metal mine spoil. Environ Pollut 86: 171-179

Horvath B, Gruiz K (1996) Impact of metalliferous ore mining activity on the environment in Gyongyosoroszi, Hungary. Sci Total Environ 184: 215-227

Hosius B, Bergmann F, Hattemer HH (1996) Physiologische und genetische Anpassung von Fichtensämlingen verschiedener Provenienz an schwermetall-kontaminierte Böden. Forstarchiv 108-114

Hunter BA, Johnson MS, Thompson DJ (1987) Ecotoxicology of copper and cadmium in a contaminated grassland ecosystem. I. Soil and vegetation contamination. J Appl Ecol 24: 573-586

Jentschke G, Winter S, Godbold DL (1999) Ectomycorrhizas and cadmium toxicity in Norway spruce seedlings. Tree Physiol 19: 23-30

Johnson NC, McGraw A-C (1988a) Vesicular-arbuscular mycorrhizae in taconite tailings. I. Incidence and spread of endogonacenous fungi following reclamation. Agric Ecos Environ 21: 135-142

Johnson NC, McGraw A-C (1988b) Vesicular-arbuscular mycorrhizae in taconite tailings. II. Effects of reclamation practices. Agric Ecos Environ 21: 143-152

Jones MD, Hutchinson TC (1986) The effect of mycorrhizal infection on the response of *Betula papyrifera* to nickel and copper. New Phytol 102: 429-442

Jongbloed RH, Borst-Pauwels GWFH (1990) Differential response of some ectomycorrhizal fungi to cadmium *in vitro*. Acta Bot Neerl 39: 241-246

Kabata-Pendias A, Bolibrzuch E, Tarlowski P (1981) Impact of a copper smelter on agricultural environments. Rocz Glebozn 32: 207-214

Kabata-Pendias A, Dudka S, Chlopecka A, Gawinowska T (1992) Background levels and environmental influences on trace metals in soils of the temperate humid zone of Europe. In: Adriano DC (ed) Biogeochemistry of trace metals. Lewis, Boca Raton, pp 61-84

Kahle H (1993) Response of roots of trees to heavy metals. Environ Exp Bot 33: 99-119

Kahle H, Breckle SW (1986) Wirkungen ökotoxischer Schwermetalle auf Buchenjungwuchs. Statuskolloquium "Luftverunreinigungen and Waldschäden" Düsseldorf, MURL, pp 84-90

Karczewska A (1996) Metal species distribution in top- and sub-soil in an area affected by copper smelter emissions. Appl Geochem 11: 35-42

Kedziorek MAM, Bourg A CM (1996) Acidification and solubilisation of heavy metals from single and dual-component model solids. Appl Geochem 11: 299-304

Kelly JM, Parker GR, McFee WW (1979) Heavy metal accumulation and growth of seedlings of five forest species as influenced by soil cadmium levels. J Environ Qual 8: 361-364

Kiriluk VP (1980) Accumulation of copper and silver in chernozems of vineyards. In: Vlasyuk PA (ed) Microelements in the environment. Naukova Dumka, Kiyev, p 76 (in Russian)

Kreutzer K (1995) Effects of forest liming on soil processes. Plant Soil 168/169: 447-470

Lacatusu R, Rauta C, Carstea S, Ghelase I (1996) Soil-plant-man relationships in heavy metal polluted areas in Romania. Appl Geochem 11: 105-108

Lamersdorf N, Godbold DL, Knoche D (1991) Risk assessment of some heavy metals for the growth of Norway spruce. Water Air Soil Pollut 57/58: 535-543.

Landberg T, Greger M (1996) Differences in uptake and tolerance to heavy metals in Salix from unpolluted and polluted areas. Appl Geochem 11: 175-180

Letunova SV, Krivitskiy VA (1979) Concentration of zinc in biomass of soil microflora in south Urals copper-zinc subregion of biosphere (in Russian). Agrokhimiya 6: 104-111

Litzinger M (1990) Untersuchungen über die Cadmium toleranz verschiedener Aspenklone (Populus tremula cv. Ahle). Diplom, University of Göttingen, Germany

Lozano F, Morrison IK (1982) Growth and nutrition of white pine and white spruce seedlings in solutions of various nickel and copper concentrations. J Environ Qual 11: 437-441

Manecki A, Klapyta Z, Schejbal-Chwastek M, Skowrofiski A, Tarkowski J, Tokarz M (1981) The effect of industrial pollutants of the atmosphere on the geochemistry of natural environment of the Niepolomice Forest. PAN Miner Trans 71: 58 (in Polish)

Mannings S, Smith S, Bell JNB (1996) Effect of acid deposition on soil acidification and metal mobilisation. Appl Geochem 11: 139-143

Maronek DM, Hendrix JW, Kiermann J (1981) Mycorrhizal fungi and their importance in horticultural crop production. Hort Rev 3: 172-213

Marschner P (1994) Einfluß der Mykorrhizierung auf die Aufnahme von Blei bei Fichtenkeimlingen, PhD thesis, University of Göttingen, Germany

Marschner B, Stahr K, Rengen M (1989) Potential hazards of lime application in a damaged pine forest ecosystem in Berlin, Germany. Water Air Soil Pollut 48: 45-57

Marschner P, Godbold DL, Jentschke G (1996) Dynamics of lead accumulation in mycorrhizal and non-mycorrhizal Norway spruce (Picea abies (L.) Karst.) Plant Soil 178: 239-24

Marschner P, Klam A, Jentschke G, Godbold DL (1999) Aluminum and lead tolerance in ectomycorrhizal fungi. J Soil Sci Plant Nutrition 162: 281-286

Marx DH (1975) Mycorrhizae and establishment of trees on strip-mined land. Ohio J Sci 75: 288-297

Marx DH (1976) Use of specific mycorrhizal fungi on tree roots for forestation of disturbed land. In: Proceedings of the Conference on Forestation of Disturbed Surface Areas. Birmingham, Al, pp 47-65

Marx DH, Bryan WC, Cordell CE (1977) Survival and growth of pine seedlings with Pisolithus ectomycorrhizae after two years in reforestation sites in North Carolina and Florida. For Sci 23: 363-373

Masetti C, Mencuccini M (1991) Régénération naturelle du pin pignon (Pinus pinea L.) dans la Pineta Granducale di Alberese (Parco Naturale della Maremma, Toscana, Italie). Ecol Medit 17: 103-118

McBride M, Sauve B, Hendershot W (1997) Solubility control of Cu, Zn, Cd and Pb in contaminated soils. Eur J Soil Sci 48: 337-346

Meharg AA (1993) The role of plasmalemma in metal tolerance in angiosperms. Physiol Plant 88: 191-198

Mitchell CD, Fretz TA (1977) Cadmium and zinc toxicity in white pine, red maple and Norway spruce. J Am Soc Hortic Sci 1021: 81-84

Moser M (1963) Die Bedeutung der Mykorrhiza bei Aufforstungen unter besonderer Berücksichtigung der Hochlagen. In: Rawald W, Lyr H (eds) Mykorrhiza. Fischer Verlag

Neite H (1989) Zum Einfluß von pH und organischem Kohlenstoffgehalt auf die Löslichkeit von Eisen, Blei, Mangan und Zink in Waldböden. Z Pflanzenernähr Bodenkd 152: 441-445

Niskavaara H, Reimann C, Chekushin V (1996) Distribution and pathways of heavy metal and sulphur in the vicinity of the copper-nickel smelters in Nikel and Zapoljarnij, Kola Peninsula, Russia as revealed by different sample media. Appl Geochem 11: 25-34

Niyazova GA, Letunova SV (1981) Microelement accumulation by soil microflora under the conditions of the Sumsaraky lead-zinc biogeochemical province in Kirghizya. Ekologiya 5: 89-100 (in Russian)

Noyd RK, Pfleger FL, Norland MR (1996) Field responses to added organic matter, arbuscular mycorrhizal fungi, and fertilizer in reclamation of taconite iron ore tailing. Plant Soil 179: 89-97

Nriagu, JO, Pacyna JM (1988) Quantitative assessment of worldwide contamination of air, water and soils by trace metals. Nature 333: 134-139

Pantani F, Cellini P, DelPanta S, Bussotti F (1984) Sulla deposizione acida nell'area della tenuta di San Rossore (Pisa). Inf Bot Ital 16: 182-191

Patterson WA III, Olson JJ (1983) Effects of heavy metals on radical growth of selected woody species germinated on filter paper, mineral and organic soil substrates. Can J For Res 13: 233-238

Petrov II, Tsalev DL (1979) Atomic absorption methods for determination of soil arsenic based on arsine generation. Pochvozn Agrokhim 14: 20-25 (in Bulgarian)

Piha MI, Vallack HW, Michael N, Reeler BM (1995) A low input approach to vegetation establishment on mine and coal ash wastes in semi-arid regions. II. Lagooned pulverized fuel ash in Zimbabwe. J Appl Ecol 32: 382-390

Rachwal L, de Temmerman LO, Istas JR (1993) Differences in the accumulation of heavy metals in poplar clones of various susceptibilities to air pollution. Arbor-Kornickie 7: 101-111

Rapp C, Jentschke G (1994) Acid deposition and ectomycorrhizal symbiosis: field investigations and causal relationships. In: Godbold DL, Hüttermann A (eds) The effect of acid rain on forest processes. Wiley, New York, pp 181-230

Rauta C, Carstea S, Mihailescu A (1987) Influence of some pollutants on agricultural soils in Romania. Arch Ochr Srodowiska 1/2: 33-37

Reddy KJ, Wang L, Gloss SP (1995) Solubility and mobility of copper, zinc and lead in acidic environments. Plant Soil 171: 53-58

Riddell-Black D, Pulford ID, Stewart C (1997) Clonal variation in heavy metal uptake by willow. Aspects Appl Biol 49: 327-334

Rieuwerts J, Farago M (1996) Heavy metal pollution in the vicinity of a secondary lead smelter in the Czech Republic. Appl Geochem 11: 17-23

Rybicka EH (1996) Impact of mining and metallurgical industries on the environment in Poland. Appl Geochem 11: 3-9

Sabey BR, Pendleton RL, Webb BL (1990) Effect of municipal sewage sludge application on growth of two reclamation shrub species in copper mine spoils. J Environ Qual 19: 580-586

Sapek B (1980) Copper behavior in reclaimed peat soil of grassland. Rocz Nauk Roln 8OF: 13-39 (in Polish)

Scheffer F, Schachtschabel P (1989) Lehrbuch der Bodenkunde. Enke, Stuttgart

Schneider FK (1982) Untersuchung über den Gehalt an Blei und anderen Schwermetallen in den Böden und Halden des Raumes Stolberg (Rheinland). Schweizerbarth, Stuttgart

Schuler G (1995) Waldkalkung als Bodenschutz. Allg Forst Zschr 50: 430-433

Schwarz T (1996) Waldböden im Bereich des Forstamtes Grund (Oberharz) als Indikator für atmosphärisch eingetragene Schadstoffe unter besonderer Berücksichtigung der kleinräumigen Elementverteilungsmuster im Umfeld eines mittelalterlichen Verhüttungsplatzes. Diplom, University of Göttingen, Germany

Schweingruber FH, Voronin V (1996) Eine dendrochronologisch-bodenchemische Studie aus dem Waldschadengebiet Norilsk, Sibirien und die Konsequenzen für die Interpretation grossflächiger Kronentaxationsinventuren. Allg Forst Jagdztg 167: 53-67

Sela M, Tel-Or E, Fritz E, Hüttermann A (1988) Localization and toxic effects of cadmium, copper, and uranium in *Azolla*. Plant Physiol 88: 30-36

Shetty KG, Hetrick BAD, Figge DAH, Schwab AP (1994) Effects of mycorrhizae and other soil microbes on revegetation of heavy metal contaminated mine spoil. Environ Pollut 86: 181-188

Sieghardt H (1988) Schwermetall- und Nahrelementgehalte von Pflanzen und Bodenproben schwermetallhaltiger Halden im Raum Bleiberg in Karnten (Osterreich). II. Holzpflanzen Z Pflanzenernaehr Bodenkd 151: 21-26

Smilde KW (1981) Heavy-metal accumulation in crops grown on sewage sludge amended with metal salts. Plant Soil 62: 3-14

Sutton RF (1980) Root system morphogenesis. NZ J For Sci 10: 264-292

Tamm PCF (1995) Heavy metal tolerance by ectomycorrhizal fungi and metal amelioration of Pisolithus tinctorius. Mycorrhiza 5: 181-187

Tchuldziyan H, Khinov G (1976) On the chemistry of copper pollution of certain soils. Poch-vozn Agrokhim 11: 41-46

Turnau K, Kottke I, Oberwinkler F (1993) Paxillus involutus—Pinus sylvestris mycorrhizae from heavily polluted forest. 1. Element localization using electron energy loss spectroscopy and imaging. Bot Acta 106: 213-219

Turnau K, Kottke I, Drexheimer J (1996) Toxic elements filtering in Rhizopogon roseolus—Pinus sylvestris mycorrhizas collected from calamine dumps. Mycol Res 100: 16-22

Tyler G, Berggren D, Bergkvist B, Falkengren-Grerup U, Folkeson L, Ruhling A (1987) Soil acidification and metal solubility in forests of southern Sweden. In: Hutchinson TC, Meema KM (eds) Effects of atmospheric pollutants on forests, wetlands, and agricultural-ecosystems. Springer, Berlin Heielberg New York, pp 347-359

Ulrich B, Mayer R, Khanna PK (1979) Deposition von Luftverunreinigungen und ihre Auswirkungen in Waldökosystemen im Solling. Sauerländer, Frankfurt

Umifiska R (1988) Assessment of hazardous levels of trace elements to health in contaminated soils of Poland. Inst. Medycyny Wsi. Warsaw, p 188 (in Polish)

Ustyak S, Petrikova V (1996) Heavy metal pollution of soils and crops in northern Bohemia. Appl Geochem 11: 77-80

Vare H (1991) Aluminium polyphosphate in the ectomycorrhizal fungus Suillus variegatus (Fr) O. Kunze as revealed by energy dispersive spectrometry. New Phytol 116: 663-668

Vermes L (1987) Results of research work and status of regulation of heavy metal contamination concerning sewage sludge land application in Hungary. Arch Ochr Srodowiska 2: 21-32

Verner JF, Ramsey MH, Helios-Rybicka E, Jedrzejczyk B (1996) Heavy metal contamination of soils around a Pb-Zn smelter in Bukowno, Poland. Appl Geochem 11: 11-16

Walendzik RJ (1993) Deterioration of forest soils in the western Sudety mountains (Poland) and attempts at its limitation. Sylwan 137: 29-38 (in Polish)

Watmough SA, Dickinson NM (1996) Variability of metal resistance in Acer pseudoplatanus L. (sycamore) callus tissue of different origins. Environ Exp Bot 36: 293-302

Watmough SA, Hutchinson TC (1997) Metal resistance in red maple (Acer rubrum) callus cultures from mine and smelter sites in Canada. Can J For Res 27: 693-700

Watmough SA, Gallivan CC, Dickinson NM (1995) Induction of zinc and nickel resistance in Acer pseudoplatanus L. (sycamore) callus cell lines. Environ Exp Bot 35: 465-473

Widera S (1980) Contamination of the soil and assimilative organs of the pine tree at various distances from the source of emission. Arch Ochr Srodowiska 3/4: 141-146 (in Polish)

Wilkins DA (1991) The influence of sheathing (ecto-)mycorrhizas of trees on the uptake and toxicity of heavy metals. Agric Ecos Environ 35: 245-260

Wilson MJ, Bell N (1996) Acid deposition and heavy metal mobilization. Appl Geochem 11: 133-137

Winter S (1995) Der Einfluß von Cadmium auf das Wachstum von Fichten am Beispiel der Mykorrhizapilze Laccaria laccata und Paxillus involutus. MSc thesis, University of Göttingen, Germany

Woolhouse HW (1983) Toxicity and tolerance in the responses of plants to metals. In: Lange OL, Nobel PS, Osmond CB Ziegler H (eds) Physiological plant ecology III. Responses to the chemical and biological environment. Encyclopedia of plant physiology, new series, vol 12C. Sringer, Berlin Heidelberg New York, pp 245-300

Wotton DL, Jones DC, Phillips SF (1986) The effect of nickel and copper deposition from a mining and smelting complex on coniferous regeneration in the boreal forest of northern Manitoba. Water Air Soil Pollut 31: 349-358

Zhu B, Alva AK (1993) Effect of pH on growth and uptake of copper by single citrumelo seedlings. J Plant Nutr 16: 1837-1845

Chapter 13

Root and Rhizosphere Processes in Metal Hyperaccumulation and Phytoremediation Technology

Walter W. Wenzel[1], Enzo Lombi[2], Domy C. Adriano[3]

BOKU - University of Natural Resources and Applied Life Sciences, Vienna, Institute of Soil Science, Gregor Mendel Strasse 33, 1180 Vienna, Austria
[2]Adelaide Laboratory, PMB2 Glen Osmond, SA 5064 Australia
[3]Savannah River Ecology Laboratory, University of Georgia, Aiken, SC 29802, USA

13.1 INTRODUCTION

According to most legislative schemes, a soil may require remediation if a certain concentration of one or more heavy metals is exceeded in a designated part (topsoil, subsoil) of the soil profile. A multitude of remediation technologies has been developed for cleanup of heavy metal polluted soils (Iskandar and Adriano 1997; Pierzynski 1997). Classic methods, such as excavation, thermal treatment and chemical soil washing, are typically expensive and destructive.

Recently, the potential role of higher terrestrial plants in remediation of metal-polluted soils has been studied by an increasing number of scientists from various disciplines, including plant and soil sciences. Comprehensive reviews of the emerging phytoremediation technologies and the underlying fundamental processes are available (Baker et al. 1994; Cunningham and Ow 1996; Entry et al. 1996; Wenzel et al. 1999; McGrath et al. 2002).

A focal point of soil-plant interactions is the microecosystem surrounding the plant roots, the rhizosphere (Hiltner 1904). This microecosystem is characterized by rather characteristic physical, chemical, and biological conditions as opposed to the bulk soil. These are created by the plant roots and their microbial associations. These rhizosphere-related biogeochemical processes are unstable because they are considerably influenced by edaphic and climatic conditions. The edaphic influence is in turn modified by the physical, mineralogical, chemical, and biological features of the soil. Due to the limited areal extension of the rhizosphere, special tools and techniques are required to study its characteristics and processes (Brown and Ul-Haq 1984; Youssef and Chino 1988; Zoyas et al. 1997).

The role of rhizosphere processes in metal tolerance (Ryan et al. 1995; Pellet et al. 1995 1997) and phytoremediation has been considered by several authors (Stomp et al. 1994; Entry et al. 1996; Wenzel et al. 1999), and only recently addressed experimentally (e.g. Bernal et al. 1994; McGrath et al. 1997; Wenzel et al. 2003a).

The mechanisms of the uptake of metals and metalloids have been investigated for some time. However, only recently has this process been studied in hyperaccumulator plants that are the most promising plant candidates to be used in phytoremediation. In these very particular plants the mechanisms of metal uptake and their regulation are significantly different from common plants.

The focus of this chapter is centered on the role of rhizosphere biogeochemical processes and mechanisms of metal/metalloid uptake in hyperaccumulator plants and their application to emerging phytoremediation technology. In Section 13.2, important aspects of the fate of metals in contaminated soils are highlighted. In Section 13.3 a summary of phytoremediation processes and technologies is presented. A more detailed picture of the rhizosphere as a microenvironment and the biogeochemical processes involved is presented in Section 13.4. Special attention is given to the interaction of rhizosphere processes with metals. Root and rhizosphere processes, including root-soil interactions as well as physiological and molecular mechanisms responsible for uptake of metals and metalloids in hyperaccumulator plants and their relevance for phytoremediation of metal-contaminated soils, are discussed by compiling the latest devlopments in this emerging field of technology and research, and identifying gaps in knowledge and research needs.

13.2 PHYTOREMEDIATION AND METAL HYPERACCUMULATION

Traditional technologies for the remediation of metal-contaminated soils such as excavation and subsequent ex situ solidification, vitrification, and soil washing, or in situ encapsulation, attentuation, volatilization and electrokinetics (Pierzynski 1997) are time-consuming and expensive. In addition, these technologies can generate hazardous waste (Wenzel et al. 1999). Bioremediation based on microbial processes has been employed to remediate soils polluted with organic toxicants. The use of fungi and bacteria to remediate metal-contaminated soils has potential for success, but needs scaling up from laboratory tests to field application (Iskandar and Adriano 1997).

In general, non-destructive in situ techniques are considered to be superior to other remedial actions. Where applicable, soil cleanup is preferable relative to immobilization and containment techniques. An evaluation of different approaches for soil remediation has to include not only cost and time required to achieve the remedial goals, but should also address safety for workers and short- and long-term environmental risks. The sustainability of a remedial action may be measured on the basis of renewable versus non-renewable energy and material input and the volume and safety of the residuals. From this point of view, in situ technologies that could remove pollutants from soils and mixed wastes and concentrate them in small volumes primarily based on biological processes would be required.

Phytoremediation, i.e. the utilization of green plants in the remediation of contaminated soils, wastes, and waters, can play a major role in achieving these goals through their indigenous processes and as a catalyst in enhancing other natural processes (e.g. microbial activities) at a relatively low level of financial and technical input (Chaney 1983; Cunningham and Berti 1993; Raskin et al. 1994; Wenzel et al. 1999). The concept of phytoremediation is primarily based on the utilization of solar energy via photosynthesis and transpiration.

Five fundamental processes by which plants can be used to remediate contaminated soils, sediments, and waters have been identified (Table 13.1).

Table 13.1 Fundamental processes involved in phytoremediation of contaminated soils and waste materials

Process	Effect on pollutant[a]	Target pollutants[b]
Phytostabilization	C	**HM**, MO, HA, **RA**, OR
Phytoimmobilization	C	**HM**, MO, HA
Phytoextraction	R	**HM**, **MO**, **HA**, **RA**, OR
Phytovolatilization	R	HM, **MO**, HA, **OR**
Phytodegradation	R	**OR**

[a] *C* Containment; *R* removal from polluted soil.

[b] *HM* Heavy metals; *MO* metalloids; *HA* halides; *RA* radionuclides; *OR* organic pollutants; bold symbols indicate the primary target pollutants.

Phytostabilization is a containment process using pollutant-tolerant plants to mechanically stabilize polluted soils to prevent bulk erosion and airborne transport to other environments. In addition, leachability of pollutants may be reduced due to higher evapotranspiration rates relative to bare soils. *Phytoimmobilization* processes prevent the movement and transport of dissolved contaminants using plants to decrease the mobility of pollutants in soils (Wenzel et al. 1999). *Phytoextraction* removes both metallic and organic constituents from soil by direct uptake into plants and translocation to aboveground biomass. *Phytodegradation* results in uptake and plant-internal degradation and/or plant-assisted microbial degradation of organic contaminants in the rhizosphere (Walton and Anderson 1990; Anderson and Coats 1994; Cunningham and Lee 1995). *Phytovolatilization* processes involve specialized enzymes that can transform, degrade and finally volatilize contaminants in the plant-microbe-soil system (Terry et al. 1992; Schnoor et al. 1995; Terry 1995; Meagher and Rugh 1996).

Because of the plant-microbe interactions involved, such as the association of fungi and bacteria with plant roots, i.e. in the rhizosphere, plant-based cleanup technologies have been referred to as plant-microbe treatment systems (Wenzel et al. 1999).

The idea of using plants for the cleanup of metal-laden soils initially emerged from the discovery of plant species that accumulate metals at unusually high concentrations in their shoots (Chaney 1983; Baker et al. 1994). Metal hyperaccumulator plants have a number of characteristics that differentiate them from other plants. Firstly, hyperaccumulator plants are extremely tolerant to elevated external concentrations of the metals and metalloids that they accumulate. Secondly, the concentration of metals in these plants is 100- to 1000-fold higher than those of non-hyperaccumulator plants growing on the same soils. Thirdly, they have an enhanced translocation of metals and metalloids from roots to shoots that results in the majority of these elements being accumulated in the aboveground biomass.

Metal hyperacumulator plants have been deemed for use in phytoextraction and represent ideal model mechanisms to study genetic, molecular, physiological and root/rhizosphere processes responsible for metal uptake, tolerance and accumulation in terrestrial plants.

13.3 THE RHIZOSPHERE

13.3.1 The Rhizosphere as a Soil-Plant Microenvironment

The term "rhizosphere" was first proposed by Hiltner (1904) to designate the zone of enhanced microbial abundance in and around the roots. The studies of Rovira in the 1960s improved our understanding on the role of root exudates in the rhizosphere (Rovira and Bowen 1966; Rovira 1969).

The soil-root interface is generally called the rhizoplane. The rhizosphere is represented by few millimeters of soil surrounding the plant roots and influenced by their activities. Because the rhizosphere is characterized by steep gradients of microbial abundance and chemical characteristics, the boundary between rhizosphere and bulk soil is not accurately determined (Fig. 13.1). The rhizosphere can be defined as a highly dynamic, solar/plant-driven *microenvironment* which is characterized by feedback loops of interactions between plant root processes, soil characteristics, and the dynamics of the associated microbial populations. In short, it is characterized by very dynamic biogeochemical processes.

Plant roots represent highly dynamic systems that explore the rooting medium, typically soil, to stabilize the plant mechanically and to take up water and mineral nutrients. As a consequence, water and nutrients become depleted earlier in the rhizosphere relative to the bulk soil.

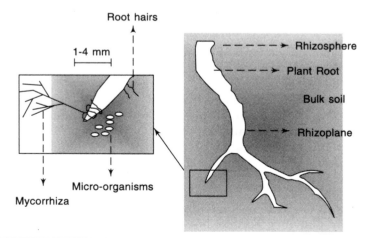

Fig. 13.1 Schematic presentation of the rhizosphere as a microenvironment showing the major structural components

In turn, the plant roots provide *structural elements* such as the rhizoplane and act as a continuous source of energy and materials that create specific conditions in the rhizosphere soil. Considerable amounts of organic compounds released into the rhizosphere provide the *substrate* and the *energy source* for specific microbial populations. These microbial associations include non-symbiotic and symbiotic organisms such as mycorrhizae. The microbial populations are an essential part of the rhizosphere and affect the rhizosphere soil by their various activities such as water and nutrient uptake, exudates, and biological transformations.

Both plant roots and microbes exude inorganic and organic compounds, e.g. protons, HCO_3^-, and various functional groups possessing acidifying, chelating and/or reductive power. These compounds can trigger a range of chemical reactions and biological transformations in the rhizosphere. In this sense, plant roots introduce directly, or indirectly via the enhanced activities of microorganisms, *"reactivity"* into the rhizosphere system, resulting in altered solubility, speciation and state (phase) of chemical elements and compounds relative to the bulk soil. Due to continued accelerated input and output of energy, materials, and chemical reactivity, the rhizosphere soil is far from equilibrium, thus quickly differentiating from non-rhizosphere soil. Even relatively static soil properties, such as mineralogy (Hinsinger et al. 1992), may be affected within a relatively short time. The effects on soil characteristics are discussed in more detail in Section 13.3.2.

Using ^{14}C-labelling techniques, the amount of organic C input to the rhizosphere soil by plant roots has been estimated to be in the range 10-40% of the total net C assimilation of arable crops (Whipps and Lynch 1983; Whipps 1984; Helal and Sauerbeck 1986; Keith et al. 1986; Liljeroth et al. 1990; Gregory and Atwell 1991; Martin and Merckx 1992). This C input may include:

- water-insoluble materials such as cell walls, sloughed materials, and mucilage (Cheng et al. 1994).
- water-soluble organic exudates such as sugars, organic acids, and amino acids.

Plant roots also release inorganic ions and compounds at considerable rates, primarily including:

- protons
- CO_2/HCO_3^-

13.3.2 Biogeochemical Processes in the Rhizosphere

As roots grow and penetrate the soil, they may increase the bulk density of the rhizosphere soil and thus cause physical changes (Hoffman and Barber 1971). Schreiner et al. (1997) indicated that root pressure exerted on the soil was responsible for the formation of soil aggregates. Roots, exudates, and hyphae of mycorrhizal fungi can bind together soil particles, making the soil more resistant to erosion (Miller and Jastrow 1990). A relationship between mycorrhizal fungi and the stability of large soil aggregates has been reported by several authors (Tisdall and Oades 1982; Burns and Davis 1986; Schreiner et al. 1997).

Changes in pH that occur in the rhizosphere and release of organic compounds that complex metals can cause dissolution or precipitation of minerals. Hinsinger et al. (1993) have demonstrated that rape can acidify its rhizosphere to pH values <4.5 and mobilize the structural elements of phyllosilicates, in particular Al. Organic acids released by plant roots or microbes can act as a driving force of mineral weathering (Lundström 1994). Jones and von Kochian (1996) reported that citric acid released from the roots can dissolve $Al(OH)_3$. Transformation of minerals in the rhizosphere has been demonstrated by Hinsinger et al. (1992, 1993) and Hinsinger and Jallard (1993), who reported rapid vermiculization of phlogopite in the rhizosphere of *Lolium multiflorum* and *Brassica napus*. Gobran et al. (1997) found a significantly lower amount of amphiboles and of interlayered minerals in the rhizosphere of Norway spruce compared with the bulk soil.

In conclusion, the rhizosphere is characterized by accelerated rates of mineral transformation and weathering, along with depletion of nutrients. Increased aggregate stability and resistance against soil erosion can contribute to the success of phytostabilization of polluted soils.

13.3.2.1 *Their Influence on Chemical Properties of Soils*

Soil pH typically displays microheterogeneity (Davey and Conyers 1988) that is partly caused by root activities (Pijnenborg et al. 1990). Soil pH near the root surface may differ considerably from that of the bulk soil (Riley and Barber 1969; Nye 1981; Blanchard and Lipton 1986; Dormaar 1988; Pijenenborg et al. 1990). Changes in rhizosphere pH can modify the solubility and thus the mobility and bioavailablity of micro- and macronutrients, pollutant metals, and cause shifts in the structure, abundance and activity of microbial populations. Such pH changes are induced by imbalanced plant uptake of cation and anion equivalents in order to maintain electroneutrality at root surfaces (Breteler 1973; Hedley et al. 1982). Changes in pH may also be due to exudation of organic acids by plant roots and microorganisms in the rhizosphere. It is generally accepted that the dominant factor influencing the pH of the rhizosphere is the nitrogen form taken up by the plant. Uptake of NH_4^+ results in a NH_4^+/NO_3^- uptake ratio >1, hence causing a net release of protons from the roots to maintain electrical neutrality. The result is acidification of the rhizosphere. On the other hand, the uptake of NO_3^- primarily results in a net release of HCO_3^-, with a corresponding increase in rhizosphere pH (Nye 1981; Rygicwycz et al. 1984; Gijsman 1990). Smiley (1974) found differences of up to 2.2 pH units in the rhizosphere of wheat plants when a different source of N (NH_4^+ or NO_3^-) was used. Riley and Barber (1971) demonstrated that fertilization of soybean with NH_4^+ decreased rhizosphere pH. HCO_3^- accumulation and the increases in pH were related to the NO_3^- concentration in the soil solution. Olsthoorn et al. (1991) investigated the effects of Douglas-fir seedlings grown in sand and found that the pH was typically larger in the rhizosphere than in the bulk soil. These findings were explained by the predominant uptake of NO_3^-.

Rhizosphere pH patterns show regularities not only normal to, but also along the rhizoplane (Römheld and Marschner 1984; Häussling et al. 1985; Jaillard et al. 1996). Differences in rhizosphere pH among root tip, cell division zone, and cell elongation zone have been observed (Marschner and Römheld 1983; Pilet et al. 1983; Blanchard and Lipton 1986). Schaller and Fischer (1985) found that the pH in the rhizosphere of maize plants displayed differences of up to 4 units between the root tip and the root hair zone. Gollany and Schumaker (1993) found that the patterns of pH distribution in the rhizosphere of soybean, maize, sorghum, oats and barley varied among these species.

Rhizosphere interactions exert considerable influence on the acquisition of micro- and macronutrients by plants (Dommergues and Krupta 1978; Rovira et al. 1983; Curl and

Truelove 1986; Marschner et al. 1986; Mench and Martin 1991). Root exudates may influence nutrient solubility and uptake indirectly through their effects on pH, microbial activity, and physical properties of the rhizosphere soil, and directly by chelation, precipitation, and oxidation-reduction reactions of nutrients with root exudates (Uren and Reisenauer 1988).

13.3.2.2 *Their Influence on Biological Properties of Soils*

Microbial populations in the rhizosphere have access to a continuous flow of organic substrates derived from the roots. Some of these C compounds (e.g. sugars) are lost in small quantities by the root purely as a result of passive diffusion along the large concentration gradient between the root cytosol and the soil solution, over which the plants appear to have only limited control (Jones and Darrah 1995). Other compounds (e.g. organic acids, phytosiderophores), however, are often excreted in large quantities in direct response to environmental stress (Marschner 1995). Due to this continuous input of substrate into the rhizosphere, microbial biomass and activities are generally larger than in the bulk soil. The number of microorganisms in the rhizosphere is typically one order of magnitude larger than in in non-rhizosphere soil; the microbial community is more diverse, active, and synergistic than in non-rhizosphere soil (Anderson and Coats 1994). Rhizodeposition does not occur uniformly across the root surface, but is concentrated at the root tip and at the zones of lateral branch formation (McDougall and Rovira 1970; McCully and Canny 1985; Trofymov et al. 1987). In the rhizosphere of fast growing roots, the number of microorganisms has been reported to increase from apical to basal zones along the root axis (Schönwitz and Ziegler 1986; Bowen and Rover 1991). In contrast, Smith and Paul (1990) concluded from a number of rhizosphere studies that the microbial biomass decreases from the apical to the basal zones. These experimental findings are supported by mathematical modeling of the effect of dissolved and solid-phase organic carbon on the dynamics of microbial populations around roots (Darrah 1991a,b). Because it is likely that the concentration of organic carbon decreases with distance from the root surface, microbial biomass is thought to decrease also with distance from the rhizoplane (Darrah 1991a,b). However, the extent of the effect of rhizodeposition on microbial biomass is not yet well established. While Jensen and Sorensen (1994) proposed that any effect was strictly limited to the rhizoplane, Youssef and Chino (1989a) demonstrated that microbial biomass and activity were considerably increased up to a distance of 4 mm from the rhizoplane. Helal and Sauerbeck (1986) reported a doubling of the microbial biomass C even in root-free soil adjacent to a densely rooted soil zone.

A schematic presentation of the distribution of microbes at the rhizoplane is shown in Fig. 13.2. It should be noted that, although supported by most studies of the rhizosphere, this generalization does not apply to all cases.

Patterns of enzyme activities in the rhizosphere vary considerably with time, and among different soils and plant species (Dormaar 1988). Tarafdar and Jungk (1987) reported that dehydrogenase activity often increased towards the rhizoplane, while phosphatase activities decreased. Opposing results were obtained by Dormaar (1988). Utilizing destructive (Häussling and Marschner 1989) and non-destructive (Dinkelaker and Marschner 1992) methods, it has been demonstrated that phosphatase activities increase from the bulk to the rhizosphere and rhizoplane soil. This may be due to predominant association of

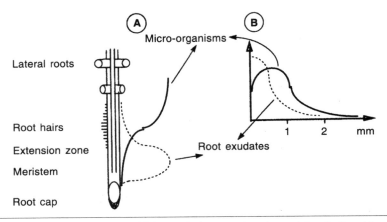

Fig. 13.2 Distribution patterns of microbial populations and root exudates in the rhizosphere *A* along the rhizoplane and *B* perpendicular to the rhizoplane (compiled from Römheld 1991 and Marschner 1995)

phosphatase with roots rather than with microorganisms in non-myccorhizal systems (Ridge and Rovira 1971; McLachlan 1980).

13.3.3 Rhizosphere Processes Influencing Plant-Metal/Metalloid Interactions

Metal cation solubility is expected to increase in zones acidified by the plant roots, and where root exudates with chelating power are released (cf. Sect. 13.2). The predominant form in which N is absorbed by roots strongly affects the pH. Moreover, there is increasing evidence that dissolved root exudates increase the solubility of metals in the rhizosphere depending on plant species and cultivars (Kawai et al. 1988; Jolley and Brown 1989; Römheld and Marschner 1990).

Studies of metal interactions with rhizosphere processes have primarily focused on iron acquisition by plants and aluminum toxicity. Even though these elements are not pollutant metals sensu stricto, these well-studied elements can serve as a model for other metals. A summary of our knowledge of the fate of Fe in the rhizosphere is presented below.

Two different types of root response to *iron* deficiency have been identified in plant species. Strategy I is employed by dicotyledenous and monocotyledenous species, except of the family of the *Graminaceae* (grasses). In this case the components involved in Fe aquisition (Marschner and Römheld 1994) are:

- plasma membrane bound inducible reductase ("turbo");
- enhanced net excretion of protons; and
- in many cases, enhanced release of reductants/chelators.

The role of the latter component is controversial and certainly the least important for strategy I plants (Chaney and Bell 1987). Ohkawi and Sugahara (1997) studied the capacity of chickpea roots to acidify the rhizosphere in response to iron deficiency. They found that the exudation of protons mediated by a non-specific H^+/cation antiport was considerably larger than that of carboxylic acid.

Strategy II of root response to Fe deficiency is confined to grasses and is characterized by enhanced production and release of phytosiderophores into the rhizosphere and subsequent transport of chelated Fe by a specific uptake system at the root surface. The chemical nature of phytosiderophores can differ among plant species and even cultivars, but typically one dominant phytosiderophore is characteristic for a certain genotype (Kawai et al. 1988; Römheld and Marschner 1990). The release of phytosiderophores is typically located in apical root zones (Marschner et al. 1987) and occurs as a flush for only a few hours at the beginning of the light period (Takagi et al. 1984; Marschner et al. 1986). Within the same species, Fe-efficient cultivars are able to produce phytosiderophores while Fe-inefficient ones are not. This has been demonstrated for oat cultivars (Brown et al. 1991; Mench and Fargues 1994), while some other authors have postulated low efficiency of phytosiderophores due to their rapid degradation by microorganisms (Crowley et al. 1988). However, Schönwitz and Ziegler (1986) showed that only a small part of the rhizoplane is colonized by microbes. A spatial differentiation between sites of siderophore release and microbial activity has been identified in indigenous rhizosphere soils (Römheld 1991). The release of phytosiderophores decreases from apical to basal root zones (Marschner et al. 1987), but an opposite trend is expected for the activity of microorganisms (Uren and Reisenauer 1988). Different patterns of phytosiderophore concentrations and microbial activity normal to the rhizoplane are also evident (see Sect. 13.4.2.4).

Concentrations of phytosiderophores up to molar levels have been calculated for nonsterile conditions in the rhizosphere of Fe-deficient barley plants (Römheld 1991). Awad et al. (1994) found enhanced Fe mobilization up to a distance of 4 mm from wheat roots, and concluded that phytosiderophores are highly effective in Fe acquisition.

Siderophores released by microbes have been proposed to serve as Fe sources for Fe-efficient plants from both groups (Bienfait 1989). For instance, iron transport systems for microbial siderophores have been postulated to exist in oat (Crowley et al. 1988, 1991) and maize roots (Ganmore-Neumann et al. 1992).

Jones and Darrah (1996) evaluated the potential role of iron dissolution triggered by organic acids in root iron uptake. They concluded that citrate and malate may contribute significantly to the Fe demand of strategy I plants through the formation of plant-available organic-Fe^{3+} complexes in the rhizosphere, while proteinaceous amino acids may be less important.

The fate of other *metals* in the rhizosphere has been less well studied. Phytosiderophores form chelates not only with Fe, but also with Zn, Cu, and Mn, and therefore trigger the solubilization of these metals in calcareous soils (Treeby et al. 1989; Römheld 1991). It has been shown that grasses release phytosiderophores also upon Zn deficiency (Cakmak et al. 1994; Walter et al. 1994; Zhang et al. 1989). Cakmak et al. (1996) demonstrated that enhanced release of phytosiderophores upon Zn deficiency is associated with Zn efficiency in wheat genotypes. Wiren et al. (1996) proposed two pathways for the uptake of Zn from Zn phytosiderophores in grasses: (1) via the transport of the free Zn cation, and (2) the uptake of non-dissociated Zn phytosiderophore complexes.

Mench and Martin (1991) collected dissolved root exudates from two tobacco species (*Nicotiana tabacum*, *N. rustica*) and corn (*Zea mays*) to extract metals from two soils. They observed increased extractability of Mn and Cu, while Ni and Zn were not affected. Root exudates of the tobacco species enhanced the solubility of Cd. The amount of Cd extracted by the root exudates was in the range of Cd uptake in these plants grown on soil (*N. tabacum* >*N. rustica* >*Zea mays*). Therefore, increased solubility of Cd in the rhizosphere due to the release of organic exudates in the apical root zone is considered a source of Cd accumulation in tobacco.

Youssef and Chino (1989b) found smaller concentrations of dissolved Cu and Zn in the rhizosphere soil to be associated with larger pH. Neng-Chang and Huai-Man (1992) studied the fate of Cd in wheat rhizosphere. Rhizosphere pH was changed by wheat roots dependent on the soil conditions; the extractability of Cd by 0.1 M $CaCl_2$ was negatively correlated with pH. The changes in rhizosphere pH were closely related to the uptake balance between anions and cations. Thus, physiologically alkaline fertilizers were recommended to limit plant uptake of toxic metal cations.

Liao et al. (1993) investigated the distribution of Si, Fe, and Mn in rice rhizosphere of paddy soils. They identified particular zones of accumulation and depletion across the rhizosphere due to interactions between roots and soil processes associated with flooding. Doyle and Otte (1997) found that As and Zn accumulated in the rhizosphere of wetlands plants relative to non-rhizosphere soil, providing evidence that the iron oxide plaque formed due to oxidative conditions around the roots acts as a sink for metals.

On the other hand, roots can facilitate reduction of chelated metals by specific plasma membrane reductase enzymes. As described above, a typical response of dicotyledonous plants to Fe-deficiency is the induction of root Fe(III) chelate reductase activity, which appears to be effective also for the reduction of Cu(II) (Yi and Guerinot 1996). There is also evidence that ferric-chelate reductase may be involved in Cu and Mn uptake by plants (Welch et al. 1993).

Yet, the fate of the metalloid arsenic in the rhizosphere has been hardly explored. Only recently, Fitz and Wenzel (2002) developed a conceptual model of As speciation and interactions in the soil-soil solution-plant system. Based on the available literature, they identified iron oxides as the major sorbent in well-aerated soils. Expected major controls of arsenic solubility/bioavailability in the rhizosphere that can be influenced by plant roots include: redox potential, pH, exudation of organic acids, competition with phosphorus for uptake and sorption sites.

13.4 ROLE OF RHIZOSPHERE PROCESSES IN HYPERACCUMULATION AND PHYTOREMEDIATION OF METAL-CONTAMINATED SOILS

13.4.1 Overview

Based on a literature review of metal interactions with plant-microbial associations in the rhizosphere (Sect. 13.3.3), we can derive a flowchart presenting the major processes involved that may be beneficial for the rhizosphere-based phytoremediation technologies (Fig. 13.3).

13.4.2 Rhizosphere Processes Relevant for Phytostabilization/ Immobilization

Phytostabilization: The importance of plant roots to mechanically stabilize the soil and reduce wind and water erosion is well established. In addition root exudates, such as

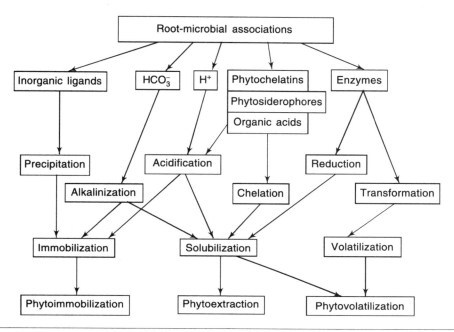

Fig. 13.3 Possible effects of root-microbial associations on the biogeochemical processes influencing the fate of metals and metalloids in the rhizosphere and their implication for phytoremediation technologies

sugars and mucilage, can increase soil structure and improve soil biological activity. Hyphae of mycorrhizal fungi can mechanically bind together soil particles, making the soil more resistant to erosion (Miller and Jastrow 1990). Therefore, *phytostabilization* technologies are strictly related to the development and activity of roots and can hardly be improved without knowledge about rhizosphere processes.

Phytoimmobilization of metal cations may be supported by alkalinization of the rhizosphere soil due to release of HCO_3^-. However, oxyanions of metals (e.g. Mo, Cr) and metalloids (e.g. As, Se) may be immobilized by release of protons into the rhizosphere. For Pb, the formation of a Pb-phosphate mineral has been identified to cause precipitation and thus immobilization. Accordingly, the accumulation of inorganic ligands (e.g. phosphate) in the rhizosphere may be effective in phytoimmobilization (Fig. 13.3).

Rhizosphere research has already emphasized the role of root activity to reduce the mobility and/or leachability of pollutants in soils. Neng-Chang and Huai-Man (1992) pointed out that Cd extractability in the rhizosphere was controlled by the pH, and the changes in the pH in the rhizosphere were closely related to the uptake balance between anions and cations. The use of fertilizer able to increase soil pH seems to be useful in remediation programs to reduce the mobility of toxic metals in the soil. Salt et al. (1995) found that in the rhizosphere of *Brassica juncea* the mobile and less toxic Cr(VI) was reduced to the less mobile and toxic Cr(III). In the rhizosphere of *Agrostis capillaris*, a Pb/Zn-tolerant plant, precipitation of a lead-phosphate mineral, pyromorphite, has been identified as a possible mechanism of phytoimmobilization (Cotter-Howells et al. 1994). The metalloid arsenic can be immobilized in the rhizosphere of marsh plants through fixation in the iron plaque that forms around roots due to oxygen leaking via aerenchyma (Doyle and Otte 1997). This process might offer a strategy to control arsenic solubility and leaching in reduced soils in which the more soluble and toxic As(III) dominates.

Although not presented in Fig. 13.3, mycorrhizal associations may be a focal point in the development of phytoimmobilization technologies. Mycorrhizae can immobilize metals and constrain their translocation to shoots (Leyval et al. 1997).

13.4.3 Rhizosphere Processes Relevant for Phytoextraction

Phytoextraction could benefit from rhizosphere processes primarily through plant-microbe induced solubilization prior to uptake by the plant (Fig. 13.3). Based on information available primarily from plant nutrition related research, it may be expected that excretion of protons and exudation of organic compounds carrying acidic functional groups, such as organic acids, can decrease the pH and thus solubilize metal cations. Metal cations may also be solubilized through the formation of metal-organic chelates with phytosiderophores and organic acids released by plant roots or microbes. Reductive enzymes may play a role in the solubilization of As, Cu and Mn, while Cr can be immobilized. Metalloids such as Se and As may be solubilized upon alkalinization due to excretion of carboxylic acid by roots or microbes. In the following sections we discuss recent findings with emphasis on metal-hyperaccumulating plants.

13.4.3.1 *The Role of Rhizosphere Acidification in Hyperaccumulation/Phytoextraction*

McGrath et al. (1997) found that the Zn hyperaccumulator *Thlaspi caerulescens* did not cause a significant decrease in rhizosphere pH, suggesting that the increased metal uptake was not related to pH-controlled changes in metal lability. For seven soils differing in soil properties and the level of Zn pollution, Knight et al. (1997) found pH changes ranging between −0.3 and +0.9 units in the rhizosphere of *Thlaspi caerulescens*. Most other studies revealed even slightly increased pH in the rhizosphere of various hyperaccumulator plants (Table 13.2). Overall, it is now well established that-in contrast to expectations from previous research on plant nutrition—root-induced changes in pH are not involved in metal hyperaccumulation.

However, rhizosphere pH can, to some extent, be controlled by using different N fertilizers. Bernal et al. (1994) compared the changes in pH and redox potential in the rhizosphere of Ni-accumulating *Alyssum murale* with a crop plant (*Raphanus sativus*). They found that pH changes were regulated mainly by the source of nitrogen. The acidification and reducing activity of the hyperaccumulator was lower than that of the crop plant. Therefore, they concluded that the different uptake of the two species could be rather regulated by differential exudation of chelating compounds. The efficiency of manipulating rhizosphere pH by adding acidifying $(NH_4)_2SO_4$ has been tested by Puschenreiter et al. (2001). Their results indicate that nitrification of the ammonium source occurred in spite of the addition of an inhibitor, rendering the amendment ineffective in terms of its acidification capacity. Similarly, the addition of elemental sulfur to various phytoremediation crops in a field experiment turned out to be inefficient at decreasing rhizosphere pH (Kayser et al. 2000). Summarized, there appears to be little scope for

Table 13.2 pH changes observed in the rhizosphere of metal-hyperaccumulating plants

Initial pH	pH change	pH method	Contami-nants	Soil name	Plant	Reference
7.3	0	Soil solution	Zn (1722)	Arras	*Thlaspi caerulescens*	Knight et al. (1997)
6.6	0.9	Soil solution	Zn (400)	Avon	*Thlaspi caerulescens*	Knight et al. (1997)
7.5	−0.1	Soil solution	Zn (383)	Bonn	*Thlaspi caerulescens*	Knight et al. (1997)
5.8	0.8	Soil solution	Zn (329)	Bordeaux	*Thlaspi caerulescens*	Knight et al. (1997)
6.9	0.6	Soil solution	Zn (3259)	Felton	*Thlaspi caerulescens*	Knight et al. (1997)
7.3	−0.3	Soil solution	Zn (2166)	Sala	*Thlaspi caerulescens*	Knight et al. (1997)
6.1	0.4	Soil solution	Zn (210)	Woburn	*Thlaspi caerulescens*	Knight et al. (1997)
6.44	−0.21	Water (1:2.5)	Zn (1317), Cd (16.5)	Arras	*Thlaspi caerulescens*	McGrath et al. (1997)
6.48	−0.3	Water (1:2.5)	Zn (340), Cd (0.9)	Bonn	*Thlaspi caerulescens*	McGrath et al. (1997)
7.36	−0.29	Water (1:2.5)	Zn (155), Cd (2.9)	Woburn	*Thlaspi caerulescens*	McGrath et al. (1997)
6.05	0.41	Micro-electrodes	Zn (285)	—	*Alyssum murale*	Bernal et al. (1994)
6.5	0.2	Soil solution	Zn (500), Cd (20)	—	*Thlaspi caerulescens*	Luo et al. (2000)
6.5	0	Water (1:2.5)	Pb (4800), Zn (2190), Cd (15.4)	Arnoldstein	*Thlaspi caerulescens*	Puschenreiter et al. (2001)
6.5	0.1	Water (1:2.5)	Pb (4800), Zn (2190), Cd (15.4)	Arnoldstein	*Thlaspi goesingense*	Puschenreiter et al. (2001)
6.37	0.39	Water (1:2.5)	Ni (2580), Cr (1910)	Redlschlag	*Thlaspi goesingense*	Wenzel et al. (2003a)
7.24	0.24	0.01 M CaCl$_2$ (1:2.5)	Pb (12300), Zn (2710), Cd (19.7)	Arnoldstein	*Thlaspi goesingense*	Puschenreiter et al. (2001)
7.54	0.19	0.01 M CaCl$_2$ (1:2.5)	Zn (714), Cd (6,83), Cu (232)	Untertie-fenbach	*Thlaspi goesingense*	Puschenreiter et al. (2001)
6.61	−0.29	Water (1:2.5)	Zn (20.5)	Woburn	*Thlaspi caerulescens*	Whiting et al. (2001b)
7.29	0.11	0.01 M CaCl$_2$ (1:2.5)	As (2270)	St. Margareten	*Pteris vittata*	Fitz et al. (2003).
7.37	−0.02	0.01 M CaCl$_2$ (1:2.5)	As (2270)	St. Margareten	*Pteris cretica*	Fitz et al.. unpubl.

decreasing rhizosphere pH in phytoremediation crops, last but not least owing to the typically high pH buffer capacity of many contaminated soils (Wenzel et al. 2000).

13.4.3.2 Accessibility of Non-labile Metal Pools by Hyperaccumulating Plants in Relation to Root/Microbial Activities

Two different approaches have been used to assess so-called labile metal pools and their change upon uptake by metal hyperaccumulator species: Isotopic techniques (Hutchinson et al. 2000) and chemical extraction using neutral salt solutions that act via cation exchange (McGrath et al. 1997; Wenzel et al. 2003a). Even though in the following discussion we use the term "labile" for both approaches, it should be noted that they are likely to address different pools.

Based on their experiments using isotopic dilution techniques (determining L and E values), Hutchinson et al. (2000) suggested that the hyperaccumulator *Thlaspi caerulescens* was accessing Cd from the same labile (=isotopically exchangeable) metal pool as the non-accumulator *Lepidium heterophyllum*. They concluded that *T. caerulescens* did not mobilize Cd from non-labile pools via root exudates or alteration of rhizosphere pH. Moreover, they assumed that this could imply significant restrictions in metal phytoavailability even to hyperaccumulator species. However, the isotopically exchangeable Cd pools in the study of Hutchinson et al. (2000) ranged between 19.6 and 52.5 mg Cd kg^{-1} soil. These values corresponded well to Cd extractable by 1 M CaCl$_2$, being several magnitudes of order larger than 1 M NH$_4$NO$_3$-extractable Cd pools in highly contaminated soils found in other studies (Puschenreiter et al. 2001). As stated by McGrath et al. (2001), under these conditions the L value may be unable to detect subtle differences between plants in metal utilization.

Using rhizobags, McGrath et al. (1997) studied the rhizosphere of the hyperaccumulator *T. caerulescens* and the non-accumulator *T. ochroleucum*. *T. caerulescens* was more effective at decreasing the concentration of labile Zn than *T. ochroleucum*. In contrast to the work of Hutchinson et al. (2000), mass balances comparing metal accumulation in plants with the corresponding reduction in the labile (NH$_4$NO$_3$-extactable) metal pool in soil provide an indication that hyperaccumulator plants may forage from non-labile metal fractions (McGrath et al. 1997; Puschenreiter et al. 2001; Wenzel et al. 2003a; Whiting et al. 2001a, b). The reduction of the labile metal pool in the rhizosphere generally accounted only for 2-50% of the amount accumulated in the plants (Table 13.3). Therefore, hyperaccumulators seemed to be able to solubilize metals/metalloids from less soluble fractions unless the buffer power of the experimental soils listed in Table 13.3 was large enough to replenish the labile metal pool at a rate sufficient to explain the uptake in the hyperaccumulator plants. As it has been repeatedly shown that pH changes in the rhizosphere of hyperaccumulators plants are typically too small to explain additional metal solubilization (Table 13.2, Sect. 13.4.3.1), it was suggested that root or microbial exudates could be involved (Knight et al. 1997; McGrath et al. 1997). Evidence for increased root and/or microbial exudation being involved in Ni solubilization in the rhizosphere of *T. goesingense* growing indigenously on a serpentine site was recently presented by Wenzel et al. (2003a). In a co-cultivation experiment, Whiting et al. (2001b) showed that the hyperaccumulator *T. caerulescens* did not enhance Zn availability to the excluders *T. arvense* and *Festuca rubra*. They concluded that *T. caerulescens* did not strongly mobilize Zn in the rhizosphere.

Table 13.3 Changes in the labile metal fraction in the rhizosphere of metal-hyperaccumulating plants

Metal	Initial LF	LF change	% change	Method of LF measurement	% of plant uptake explained by change of LF	Contaminants	Soil name	Plant	Reference
Zn	1.05	-0.1	-10	0.05 M NH_4NO_3	<10	Zn (1317)	Arras	*Thlaspi caerulescens*	McGrath et al. (1997)
Zn	0.03	0.08	267	0.05 M NH_4NO_3	<10	Zn (340)	Bonn	*Thlaspi caerulescens*	McGrath et al. (1997)
Zn	0.5	0.05	10	0.05 M NH_4NO_3	<10	Zn (155)	Woburn	*Thlaspi caerulescens*	McGrath et al. (1997)
Ni	7.72	-2.66	-34	0.05 M NH_4NO_3		Ni (2580), Cr (1910)	Redlschlag	*Thlaspi goesingense*	Wenzel et al. (2003a)
Zn	2.63	-0.68	-26	0.05 M NH_4NO_3	17	Pb (4800), Zn (2190), Cd (15.4)	Arnoldstein	*Thlaspi caerulescens*	Puschenreiter et al. (2001)
Zn	2.46	-0.36	-15	0.05 M NH_4NO_3	2	Pb (4800), Zn (2190), Cd (15.4)	Arnoldstein	*Thlaspi goesingense*	Puschenreiter et al. (2001)
Cd	0.05	-0.01	-25	0.05 M NH_4NO_3	2	Pb (4800), Zn (2190), Cd (15.4)	Arnoldstein	*Thlaspi caerulescens*	Puschenreiter et al. (2001)
Cd	0.03	0.02	87	0.05 M NH_4NO_3	—	Pb (4800), Zn (2190), Cd (15.4)	Arnoldstein	*Thlaspi goesingense*	Puschenreiter et al. (2001)
Zn	6.01	-2.6	-43	0.05 M NH_4NO_3	29	Pb (12300), Zn (2710), Cd (19.7)	Arnoldstein	*Thlaspi goesingense*	Puschenreiter et al. (2001)
Zn	0.78	-0.33	-42	0.05 M NH_4NO_3	26	Zn (714), Cd (6,83), Cu (232)	Untertiefenbach	*Thlaspi goesingense*	Puschenreiter et al. (2001)

(Contd.)

Table 13.3 (*Contd.*)

Metal	Initial LF	LF change	% change	Method of LF measurement	% of plant uptake explained by change of LF	Contaminants*	Soil name	Plant	Reference
Cd	0.09	0.01	11	0.05 M NH_4NO_3	—	Pb (12300), Zn (2710), Cd (19.7)	Arnoldstein	*Thlaspi goesingense*	Puschenreiter et al. (2001)
Cd	0.08	−0.03	−38	0.05 M NH_4NO_3	300	Zn (714), Cd (6,83), Cu (232)	Untertiefenbach	*Thlaspi goesingense*	Puschenreiter et al. (2001)
Pb	2.03	−0.46	−23	0.05 M NH_4NO_3	7	Pb (12300), Zn (2710), Cd (19.7)	Arnoldstein	*Thlaspi goesingense*	Puschenreiter et al. (2001)
Cu	0.4	0.07	18	0.05 M NH_4NO_3	—	Zn (714), Cd (6,83), Cu (232)	Untertiefenbach	*Thlaspi goesingense*	Puschenreiter et al. (2001)
Zn	0.23	−0.1	−43	0.05 M NH_4NO_3	24	Zn (35)	Woburn	*Thlaspi caerulescens* (Prayon)	Whiting et al. (2001a)
Zn	0.23	−0.1	−43	0.05 M NH_4NO_3	50	Zn (35)	Woburn	*Thlaspi caerulescens* (Cloughwood)	Whiting et al. (2001a)
Zn	0.21	−0.16	−76	0.05 M NH_4NO_3	33	Zn (20.5)	Woburn	*Thlaspi caerulescens*	Whiting et al. (2001b)
Zn	0.46	0.03	7	0.05 M NH_4NO_3	—	Zn (150)	Woburn	*Thlaspi caerulescens*	Whiting et al. (2001b)
As	3.674	−1.127	−31	0.05 M $(NH_4)_2SO_4$	9	As (2270)	St. Margareten	*Pteris vittata*	Fitz et al. (2003)
As	3.834	−0.754	−20	0.05 M $(NH_4)_2SO_4$	14	As (2270)	St. Margareten	*Pteris cretica*	Fitz et al., unpubl.

* The number is parantheses refers to the total metal concentration in soil ($mg\text{-}kg^{-1}$)

However, considering the competition for metal uptake in the shared rhizosphere and only 33% of Zn accumulation in *T. caerulescens* being explained by the decrease in the labile (1 M NH_4NO_3-extractable) metal pool, metal mobilization via root activities cannot be ruled out.

Among the compounds that have been proposed to participate in metal chelation in hyperaccumulator species are citrate (Lee et al. 1977) and free histidine (Krämer et al. 1996). However, Salt et al. (2000) have shown in a hydroponic experiment that the release of citrate and histidine did not appear to play an important role in Ni hyperaccumulation in *T. goesingense*. They suggested that there could be other Ni-chelating compounds released from roots which could not be detected. As root exudates from *T. caerulescens* collected in hydroponic experiments did not significantly enhance Zn and Cd solubility when added to soil, Zhao et al. (2001) concluded that root exudates from *T. caerulescens* are not involved in hyperaccumulation. However, root exudation in hydroponic conditions may considerably differ from that of soil-grown plants (Marschner 1995). This is particularly true if activities exerted by rhizosphere microbes are considered. Experiments comparing Zn uptake in *T. caerulescens* grown in sterile conditions and inoculated with bacteria seem to provide indication of bacterially mediated dissolution of Zn from the non-labile phase in soil (Whiting et al. 2001c). However, these authors compared previously autoclaved, inoculated treatments with non-autoclaved soil, which likely explains some of the differences observed. More research is required to establish such microbial effects using proper methodology and to establish their longevity in real-world conditions. Inoculation with mycorrhiza enhanced Cd, Pb and Zn accumulation in metal-accumulating *Salix caprea* substantially relative to sterile control soil (Sommer et al. 2002), indicating that microbes indeed may be involved in metal accumulation in certain plant species.

Whiting et al. (2001a) compared two populations of *T. caerulescens* in terms of Zn accumulation and corresponding changes in the labile (1 M NH_4NO_3-extractable) Zn pool in the rhizosphere. Their results showed that *T. caerulescens* sequestered Zn from non-labile pools, but it was not possible to assign this to either root activities or a passive process. The authors concluded that dissolution of non-labile Zn might have been due to high root density in pot experiments where the soil volume available for roots is limited.

However, a mass balance calculated from the data of Wenzel et al. (2003a) demonstrates that even in field conditions the decrease of labile Ni in *T. goesingense* rhizosphere explains only a small proportion of Ni accumulation in the plants, whereas uptake in excluder species at the same site corresponded fairly well with the observed changes in the labile pool.

In conclusion, the role of root activities, in particular of metal-chelating exudates, remains unclear, showing the need for further research on this aspect.

13.4.3.3 *Root Processes in Hyperaccumulator Plants*

Root morphology: Soil contamination by heavy metals is often non-uniform with both vertical and horizontal heterogeneity in the distribution of metals. Most plants preferentially develop their root systems in uncontaminated soil zones and in doing so reduce their metal uptake. It has been reported that the Zn and Cd hyperaccumulator *T. caerulescens* does exactly the opposite: it actively forages for metals. Schwartz et al. (1999) studied the distribution of *T. caerulescens* roots in rhizoboxes filled with uncontaminated soil with inclusions of contaminated soil. Roots preferentially grew in the Zn-contaminated patches. Similarly, Whiting et al. (2000) investigated the response of *T. cearulescens* roots to Zn and Cd. In their experiment rhizoboxes were filled with homogeneous soil (either contaminated or not) or juxtaposed control and metal-enriched soil (Zn and Cd in the form of oxides). Plants allocated approximately 70% of their root biomass and assimilated ^{14}C into roots developing in the metal-enriched soil. In contrast the non-accumulator *T. arvense* showed reduced root allocation in Zn-enriched soil.

These results indicate that root morphology plays a very important role in metal hyperaccumulation. However, the physiological and molecular mechanisms involved in the preferential root development in metal-rich areas are still unknown. Also, the root response to metals in hyperaccumulator plants other than *T. caerulescens* has not been thoroughly investigated.

Root uptake of metals and metalloids: In recent years, root uptake of metals and metalloids in hyperaccumulator plants has been the subject of intensive investigation. The mechanism of uptake in hyperaccumulator plants is probably element- and species-specific. Most of the information available regards the uptake of Cd and Zn by *T. caerulescens*. It was shown that in this plant Zn is taken up via a high-affinity Zn transport system and that the maximum influx velocity (V_{max}) was 4.5-fold larger in *T. caerulescens* than in the non-hyperaccumulator *T. arvense* (Lasat et al. 1996). More recently, this mechanism has been investigated at the molecular level. Genes encoding for Zn transporters, *ZNT1* and *ZNT2*, were recently cloned from *T. caerulescens* by Pence et al. (2000) and Assunção et al. (2001). Compared with *T. arvense*, *ZNT1* was found to be expressed at much higher levels in both roots and shoots of *T. caerulescens*. Also, a down-regulation of the *ZNT1* expression occurred at a much higher concentration of Zn supply (50 μM) for *T. caerulescens* than for *T. arvense* (1 μM; Pence et al. 2000). The results suggest that an alteration in the regulation of Zn transport by Zn status plays a role in enhanced Zn uptake and Zn hyperaccumulation in *T. caerulescens*.

The mechanism of Cd hyperaccumulation in *T. caerulescens* has been studied by comparing at the physiological and molecular level two ecotypes of this species that are similar in terms of Zn uptake but which differ significantly in terms of Cd accumulation. The two ecotypes were named Ganges (high Cd accumulation) and Prayon (low Cd accumulation; Lombi et al. 2000). Kinetic studies revealed a marked difference in Cd influx, but little difference in Zn influx, between the Ganges and Prayon ecotypes (Lombi et al. 2001a). The Ganges ecotype showed a clear saturable component in the low Cd concentration range, but this was much less evident in the Prayon ecotype. The K_m values were not significantly different between the two ecotypes but the maximum influx rate (V_{max}) for Cd was 5-fold higher in the Ganges than in the Prayon ecotype. In addition, at an equimolar concentration, Zn inhibited Cd uptake by Prayon but not by Ganges. From the physiological evidence, it was postulated that a transport system with a high affinity for Cd was highly expressed in the Ganges ecotype (Lombi et al. 2001a). Recent results indicated that a Fe-induced transporter (TcIRT1-G) is overexpressed in Ganges in comparison to the Prayon ecotype and may be responsible for the differences in Cd uptake observed between these ecotypes of *T. caerulescens* (Lombi et al. 2002).

Despite the fact that Ni hyperaccumulator plants represent more than three-quarters of the known metal hyperaccumulator species, Ni uptake by these plants is still hardly understood. The uptake of Ni was studied in *Thaspi goesingense* by Krämer et al. (1997) who found that the rate of uptake and root-to-shoot translocation were similar to those of the non-Ni hyperaccumulator *T. arvense* (at least when Ni was not toxic to *T. arvense*). It was suggested that Ni tolerance could be sufficient to induce Ni hyperaccumulation when plants were grown in the presence of large concentrations of available Ni. This conclusion is significantly different from that observed in the case of Cd and Zn hyperaccumulation, where overexpression of high-affinity transport systems seems to be responsible for the enhanced uptake of these metals. However, Ni uptake has not been comprehensively investigated in other Ni hyperaccumulator species.

Hyperaccumulation of As has been very recently reported to occur in ferns (Ma et al. 2001; Francesconi et al. 2002; Zhao et al. 2002a). In all plants investigated, it has been reported that arsenate is taken up via the P transport system (Meharg and Hartley-Whitaker 2002). Wang et al. (2002) conducted a series of physiological experiments with the As hyperaccumulator *P. vittata* to investigate P/As competition for plant uptake and the effect of P deficiency on As uptake. They concluded that, similarly to what has been observed in other plants, arsenate is taken up via P transporters.

The studies quoted above suggest that high-affinity transport systems may be key components in metal hyperaccumulation. In Table 13.4 the maximum influx (V_{max}) and the K_m (which are a measure of the affinity of a transport system for an element) for Zn, Cd, As and Ni in hyperaccumulator and normal plants are summarized. Presently, it is unknown whether the main difference in terms of metal uptake between hyperaccumulator and normal plants is due to differences in the regulation of common transport systems or to the presence of novel transport systems with high specificity for metals in hyperaccumulator plants.

Table 13.4 Maximum influx (V_{max}) and apparent K_m values (concentration in solution where the influx is half of the V_{max}) for different metals in hyperaccumulator (H) and non-hyperaccumulator (control - C) plants

Species	Hyperacc./ Control	Metal	V_{max} (nmol g^{-1} h^{-1})	K_m (μM)	Reference
T. caerulescens	H	Zn	270	8	Lasat et al. (1996)
T. arvense	C	Zn	60	6	
T. caerulescens (Ganges)	H	Zn	176	0.30	Lombi et al. (2001a)
T. caerulescens (Prayon)	H	Zn	239	0.92	
T. caerulescens (Ganges)	H	Cd	160	0.18	
T. caerulescens (Prayon)	H (C)	Cd	33	0.26	
T. caerulescens (Ganges)	H	Cd	143	0.45	Zhao et al. (2002b)
P. vittata	H	As	130	0.52	Wang et al. (2002)
T. goesingense	H	Ni	286	15.7	Puschenreiter et al., unpubl.

13.4.3.4 Rhizosphere Manipulation Using Chelates

For the success of phytoextraction technologies, large biomass production and rates of metal uptake and translocation into shoots are critical to achieve reasonable metal extraction rates. Artificial chelates such as EDTA have been tested to enhance metal phytoavailability in soil and subsequent uptake and translocation in shoots (Blaylock et al. 1997; Huang et al. 1997; Cooper et al. 1999; Kayser et al. 2000; Puschenreiter et al. 2001; Wenzel et al. 2003b).

Reported extraction rates of Pb by chelate-induced phytoextraction inferred from short-term experiments range from 265 (Huang and Cunningham 1996) to 0.4 kg Pb ha^{-1} $year^{-1}$ (Cooper et al. 1999); for instance, an extraction rate of 180 kg Pb ha^{-1} $year^{-1}$ was calculated based on shoot concentrations of *Brassica juncea* grown on a soil previously amended with 1200 mg kg^{-1} $PbCO_3$ (Blaylock et al. 1997), corresponding to approximately 5000 kg Pb ha^{-1} if a bulk density of 1200 kg m^{-3} and a remediation depth of 0.3 m are assumed. To decrease the Pb concentration in this soil to 600 mg kg^{-1}, it would take 14 years. In contrast to the success stories published for Pb (Blaylock et al. 1997; Epstein et al. 1999) and U extraction (Huang et al. 1998) by Indian mustard and corn (*Zea mays* L. Fiesta; Huang et al. 1997), most other reports, including other metals such as Cd, Cu and Zn, in particular when based on long-term contaminated soils and field experiments, are less encouraging (Cooper et al. 1999; Kayser et al. 2000; Grcman et al. 2001; Lombi et al. 2001b; Römkens et al. 2002; Wenzel et al. 2003b). Lower extraction rates are related to lower metal concentrations in the soil (Kayser et al. 2000; Wenzel et al. 2003b), to toxicity problems leading to yield reduction (Cooper et al. 1999; Lombi et al. 2001b), and differences in the performance between greenhouse and field studies (Wenzel et al. 2003b). Leaching of metal-laden seepage towards the groundwater represents a major problem in the application of chelate-assisted phytoextraction (Luo et al. 2000; Grcman et al. 2001; Lombi et al. 2001b; Sun et al. 2001; Römkens et al. 2002; Wenzel et al. 2003b). This is consistent with the known biostability of EDTA (Hong et al. 1999; Nörtemann 1999).

Given these severe limitations of chelate-assisted phytoextraction, further efforts to develop phytoextraction should focus on natural, continuous technologies using high-biomass perennial plants capable of accumulating metals in leaves, such as willows, poplars (Zn, Cd) or ferns (As).

13.4.4 Rhizosphere Processes Relevant for Phytovolatilization

Phytovolatilization has been related to specific enzymes produced by rhizospheric microorganisms. Enzyme-mediated transformation of Hg, Se and possibly other elements such as As to volatile compounds is the main process involved in this phytoremediation technology (Terry et al. 1992; Terry and Zayed 1994; Terry 1995, 1996).

Azaizeh et al. (1997) studied selenium volatilization in the rhizosphere and bulk soil of a constructed wetland. They demonstrated that Se volatilization from rhizosphere soil cultures is generally higher than from bulk soil cultures. They concluded that microbial populations in the rhizosphere appear to be more effective in volatilizing Se than those present in non-rhizosphere soil.

Several reports indicated that rhizosphere bacteria, and not fungi, may be responsible for Se volatilization in the rhizosphere (Zayed and Terry 1994; de Souza et al. 1999a). In fact, rhizosphere bacteria facilitated 35% of plant Se volatilization and 70% of Se accumulation in plants (de Souza et al. 1999a,b). A heat-labile proteinaceous compound, not yet identified, produced by the root-bacteria interaction may be responsible for this enhanced plant Se volatilization in the presence of bacteria.

Therefore, rhizosphere processes are considered essential for enhancing microbially mediated volatilization of pollutant metals and metalloids.

13.5 ENHANCEMENT OF ROOT-RHIZOSPHERE PROCESSES IN PHYTOREMEDIATION AND RESEARCH NEEDS

The density and depth to which plant roots penetrate the soil is critical to potential application of any phytoremediation technology. Enhancement of root biomass and morphology (i.e. root density, depth, etc.) is therefore required in any phytoremediation technology. Plants having dense and deep roots that can exploit larger volumes of contaminated soils should have larger surface areas to stabilize contaminated soils, intercept and uptake the contaminants, or facilitate the microbial volatilization of metalloids in the rhizosphere. Fibrous roots offer a large surface area for contaminant absorption (as required in phytoextraction) and plant-microbe interactions (as important in phytovolatilization). Deep-rooted plants with high transpiration rates, such as hybrid poplar, have access and can be used to remediate soils at depths of as much as 6 m (Wenzel et al. 1999). Willow species such as *Salix caprea* that have been identified to accumulate substantial amounts of Cd and Zn (Sommer et al. 2002) hold promise for use in phytoextraction.

Strategies to enlarge root biomass and to enhance root-rhizosphere associations, inluding mycorrhizae, have been reviewed by several authors (Stomp et al. 1994; Cunningham at al. 1996; Entry et al. 1996). Genetic engineering seems to provide a tool to modify root morphology and/or to identify and clone relevant enzyme-controlling genes into more deep-rooted plants. However, as demonstrated by the recent discovery of arsenic-accumulating fern species (Ma et al. 2001; Francesconi et al. 2002), additional screening combined with traditional breeding also offers a rich resource for enhancing phytoextraction and associated rhizosphere technology (Baker and Whiting 2002; Fitz and Wenzel 2002). *Agrobacterium rhizogenes* can mediate DNA transfer from bacterium to plant cells. A wild-type *A. rhizogenes* strain with high efficiency of transformation has been used to obtain trees with larger than normal root masses (Stomp et al. 1994); these trees are being evaluated for their growth rates, phenotype stability through time, and their ability to take up and translocate Pb. Preliminary results indicate that the growth rate is enhanced by the larger root biomass. Increased shoot growth rates may increase the transpiration rates and potentially the uptake of contaminants in the tree. Accordingly, phytoextraction and plant-internal degradation and volatilization could be enhanced using this approach.

Modification of root morphology could change microbial associations in the rhizosphere qualitatively and quantitatively (Cunningham et al. 1996). For example, a mature scots pine has been estimated to have 50,000 m of roots with 5 million root tips. These root tips could support the growth of 5 kg bacteria. *A. rhizogenes* transformation could increase the number of root tips two- to fourfold and result in enhanced microbially mediated remediation rates (Stomp et al. 1994). This is supported by findings that uptake of Cd from sewage sludge by *A. rhizogenes* transformed *Calystegia pepium* plants is larger than by non-transformed specimens (Tepfer et al. 1989).

Because the management of soil microorganisms including mycorrhizal fungi is a pre-requisite for the success of any soil restoration programs (Haselwandter and Bowen 1996; Haselwandter 1997), studies of the relationships between roots and microorganisms in the rhizosphere are needed.

Inoculation with specific mycorrhizal fungi has been considered to increase the uptake of nutrients and pollutants by phytoextractor species (Entry et al. 1994, 1996). Rogers and Williams (1986) found that inoculation of *Melilotus officinalis* and *Sorghum sudanense* with vesicular arbuscular mycorrhizae increased the uptake of ^{137}Cs. However, results of green-house studies on the effect of mycorrhizae on contaminant uptake cannot be directly translated to field conditions because of different environmental conditions and time scale (short-term greenhouse vs long-term field conditions). Plant physiological and soil conditions in the field typically differ greatly from those obtained in greenhouse experiments (Entry et al. 1996).

Therefore, better control of root-rhizosphere associations and their dynamics will be a key to improve phytoremediation technologies. Because of the large number of factors influencing microbial associations in the rhizosphere, e.g. soil conditions and climate, their control will be a difficult task. As root exudates are among the most important components of plant-microorganism interactions, they have been considered as a logical controlling point from which to start (Cunningham et al. 1996). As the contribution to metal accumulation of complexing compounds exuded by plant roots still remains un-clear, further research should focus on this important topic. Success will largely depend on employing/developing appropriate experimental approaches, including improved rhizobox designs (Wenzel et al. 2001), tools that allow for the collection of exudates from soil-grown plants, and more sensitive analytical techniques for the measurement of organic acids.

Phytoimmobilization: Plants may be identified and/or engineered that exude compounds capable of immobilizing contaminants using redox processes or precipitation of insoluble compounds in the rhizosphere. An example is the precipitation of Pb-P compounds in roots and the rhizosphere of *Agrostis capillaris* (Cotter-Howells et al. 1994, 1999). Yet, little progress has been made towards application of these processes in the remediation of metal-contaminated soil.

Phytoextraction: A better understanding of the physio-biochemical processes involved in hyperaccumulation is needed before target genes can be identified for transfer into high biomass plants. Recently, progress has been made in this direction with the observation that Ni accumulation by Ni-hyperaccumulating *Alyssum lesbiacum* is associated with Ni chelation by histidine in both the rhizosphere and within the plant (Smith et al. 1995). The addition of histidine to the root zone soil of *A. montanum* also caused an increase in both the concentration and flux of Ni within the xylem sap of this non-Ni-accumulating species, confirming the important role of this compound in Ni hyperaccumulation by *Alyssum* spp.

An immediate strategy is to improve pollutant accumulation by plants by manipulating biochemical/physiological mechanisms relevant to this process. A possible approach is to employ transgenic techniques (i.e. genetic engineering), to impart into the host plant desirable features from a hyperaccumulator species. Research efforts have been concentrating on Cd, Zn, Pb, Ni, Hg and Se, but may be extended to other contaminants, e.g. As,

Cu, Cs, Sr, V, and organics. Breeding of hyperaccumulator species to improve their biomass and other features relevant to crop management has been considered as an alternative strategy (Li et al. 1996), but is probably less effective in relation to the phytoextraction capacity that can be achieved with this approach within a reasonable time.

In particular, the main challenge includes improving the ease with which pollutants are mobilized in the soil, taken up by the roots and translocated to other plant parts. To enhance pollutant mobilization, transgenic plants may be developed that exude ligands selective for specific metals into the rhizosphere (Ma and Nomoto 1996). The identification of an increasing number of genes encoding for metal transporters in hyperaccumulator plants may provide the genetic resources to increase the phytoextraction potential of plants (Zhao and Eide 1996).

Even though the use of microbially hardly degradable chelates such as EDTA to enhance metal uptake in phytoremediation crops is limited by the risk of leaching, there may be some scope for developing techniques in which degradable chelates are applied.

The success of phytoextraction technology will also depend on a move from our current paradigm of remedial targets based on total metal concentrations towards the concept of "bioavailable contaminant stripping" (BCS), as proposed by Hamon and McLaughlin (1999). Even under optimal conditions, in most cases phytoextraction would take at least a few decades to clean up metal-polluted soil to currently accepted target values. BCS offers a viable alternative by targeting the extraction of only the most labile metal pools that pose a risk to the environment via leaching or uptake in organisms. However, before BCS can be implemented in regulations, thorough research, including experimental and modeling (Schnepf et al. 2002; Puschenreiter et al., unpubl.) approaches, is required to establish the long-term fate of the remaining non-labile metal fractions. Studying the dissolution kinetics of metal-bearing phases (e.g. Fe oxides, layer silicates) in the soil will be a key to understand/predict the potential of metal remobilization after termination of the phytoextraction exercise (Wenzel et al. 2003a).

Phytovolatilization: Volatilization, along with degradation and extraction of contaminants, is an inherent process in the phytoremediation of organically polluted soils (Schnoor et al. 1995). However, phytovolatilization as applied to inorganic contaminants is still in an early stage of development. Phytovolatilization has some potential to remediate soil polluted with Hg, Se, B, and possibly other elements. Recent efforts have concentrated on developing transgenic phytovolatilizator species with increased capacities for volatilizing Hg (Meagher and Rugh 1996) and Se (Terry 1995, 1996). Future work will focus on generating libraries of mutant merA genes involved in volatilization of Hg, and identifying mutants that reduce toxic metal ions such as Pb^{2+}, Cu^{2+}, and Cd^{2+} (Meagher and Rugh 1996). For Se, recent efforts include screening of hypervolatilizator species (Terry et al. 1992), testing and genetic engineering of these plants. Previous results indicate that the conversion of selenate to volatile Se species may be rate-limited by several enzymes in the Se volatilization pathway. Using *Brassica juncea*, genetic engineering is being employed to develop transgenic plants with increased activities of potentially rate-limiting enzymes (Terry 1996).

Mechanisms of tolerance, such as the well-studied Al-tolerance, are fundamental in almost every kind of phytoremediation and biodegradation process localized in the rhizosphere. An improvement in our knowledge on the rhizosphere process that regulates tolerance of plants to pollutants and solubility/uptake of toxic substances is urgently required to solve the environmental hazard associated with polluted soils.

ACKNOWLEDGEMENTS

The senior author was supported by grants from the University of Agricultural Sciences Vienna—BOKU (BOKU Priority Area Project 16) and from the Austrian Bundesministerium für Bildung, Wissenschaft und Kultur (grant # GZ 38.038/1-VIII/A/4/2000).

REFERENCES

Anderson TA, Coats JR (eds) (1994) Bioremediation through rhizosphere technology. American Chemical Society, Washington, DC

Assunção AGL, De Costa Martins P, De Folter S, Vooijs R, Schat H, Aarts MGM (2001) Elevated expression of metal transporter genes in three accessions of the metal hyperaccumulator *Thlaspi caerulescens*. Plant Cell Environ 24: 217-226

Awad F, Römheld V, Marschner H (1994) Effect of root exudates on mobilization in the rhizosphere and uptake of iron by wheat plants. Plant Soil 165: 213-218

Azaizeh HA, Gowthaman S, Terry N (1997) Microbial selenium volatilization in rhizosphere and bulk soils from a constructed wetland. J Environ Qual 26: 666-672

Baker AJM, Whiting SN (2002) In search of the holy grail—a further step in understanding hyperaccumulation? New Phytol 155: 1-4

Baker AJM, McGrath SP, Sidoli CMD, Reeves RD (1994) The possibility of *in situ* heavy metal decontamination of polluted soils using crops of metal-accumulating plants. Resources Conserv Recycl 11: 41-49

Bernal MP, McGrath SP, Miller AJ, Baker AJM (1994) Comparison of the chemical changes in the rhizosphere of the nickel hyperaccumulator *Alyssum murale* with the non-accumulator Raphanus sativus. Plant Soil 164: 251-259

Blaylock MJ, Salt DE, Dushenkov S, Zakharova O, Gussman C, Kapulnik Y, Ensley BD, Raskin I (1997) Enhanced accumulation of Pb in Indian mustard by soil-applied chelating agents. Environ Sci Technol 31: 860-865

Bienfait F (1989) Prevention of stress in iron metabolism of plants. Acta Bot Neerl 38: 105-129

Blanchard RW, Lipton DS (1986) The pe and pH in alfalfa seedlings rhizosphere. Agron J 78: 216-218

Bowen GD, Rover AD (1991) The rhizosphere, the hidden half of the hidden half. In: Waisel Y, Eshel A, Kafkafi U (eds) Plant roots: the hidden half. Dekker, New York, pp 641-669

Breteler H (1973) A comparison between ammonium and nitrate nutrition of young sugar-beet plants grown in nutrient solutions at constant acidity. I. Production of dry matter, ionic balance and chemical composition. Nether J Agric Sci 21: 227-244

Brown DA, Ul-Haq A (1984) A porous membrane-root culture technique for growing plants under controlled soil condition. Soil Sci Soc Am J 48: 692-695

Brown JC, von Jolley D, Lytle M (1991) Comparative evaluation of iron solubilizing substances (phytosiderophores) released by oats and corn: iron efficient and inefficient plants. Plant Soil 130: 157-163

Burns RG, Davis JA (1986) The microbiology of soil structure. Biol Agric Hortic 3: 95-113

Cakmak I, Gülüt KY, Marschner H, Graham RD (1994) Effect of zinc and iron deficiency on phytosiderophores release in wheat genotypes different in zinc efficiency. J Plant Nutr 17 :1-17

Cakmak I, Sari N, Marschner H, Ekiz H, Kalayei M, Yilmaz A, Braun HJ (1996) Phytosiderophore release in bread and durum wheat genotypes differing in zinc efficiency. Plant Soil 180: 183-189

Chaney RL (1983) Plant uptake of inorganic waste constituents. In: Parr JF, March PD, Kla JM (eds) Land treatment of hazardous wastes. Noyes Data Corp. Park Ridge, NJ

Chaney RL, Bell PF (1987) Complexity of iron nutrition: lessons for plant-soil interaction research. J Plant Nutr 10: 963-994

Cheng W, Coleman DC, Caroll CR, Hoffman CA (1994) Investigating short-term carbon flows in the rhizospheres of different plant species, using isotopic trapping. Agron J 86: 782-788

Cooper EM, Sims JT, Cunningham SD, Huang JW, Berti WR (1999) Chelate-assisted phytoextraction of lead from contaminated soil. J Environ Qual 28: 1709-1719

Cotter-Howells JD, Champness PE, Charnock JM, Pattrick RAD (1994) Identification of pyromorphite in mine-waste contaminated soils by ATEM and EXAFS. Eur J Soil Sci 45: 393-402

Cotter-Howells JD, Champness PE, Charnock JM (1999) Mineralogy of lead-phosphorus grains in the roots of Agrostis capillaris L. by ATEM and EXAFS. Min Mag 63: 777-789

Crowley DE, Reid CPP, Szaniszlo PJ (1988) Utilization of microbial siderophores in iron acquisition by oat. Plant Physiol 87: 680-685

Crowley DE, Wang YC, Reid CPP, Szaniszlo PJ (1991) Mechanisms of iron acquisition from siderophores by microorganisms and plants. Plant Soil 130: 179-198

Cunningham SD, Berti WR (1993) Remediation of contaminated soils with green plants: an overview. In Vitro Cell Dev Biol 29P: 207-212

Cunningham SD, Lee CR (1995) Phytoremediation: plant-based remediation of contaminated soils and sediments. Bioremediation: science and applications. SSSA special publication 43

Cunningham SD, Ow DW (1996) Promises and prospects of phytoremediation. Plant Physiol 110: 715-719

Cunningham SD, Anderson TA, Schwab AP, Hsu FC (1996) Phytoremediation of soils contaminated with organic pollutants. Adv Agron 56: 55-114

Curl EA, Truelove B (1986) The rhizosphere. Springer, Berlin Heidelberg New York

Dalal RC (1997) Soil organic phosphorus. Adv Agron 29: 83-117

Darrah PR (1991a) Models of the rhizosphere. I. Microbial population dynamics around a root releasing soluble and insoluble carbon. Plant Soil 133: 187-199

Darrah PR (1991b) Models of the rhizosphere. II. A quasi three-dimensional simulation of the microbial population dynamics around a growing root releasing soluble exudates. Plant Soil 138: 147-158

Davey BG, Conyers MK (1988) Determining the pH of acid soils. Soil Sci 146: 141-150

Delhaize E, Ryan PR, Randall PJ (1993) Aluminium tolerance in wheat. II. Aluminium stimulated excretion of malic acid from root apices. Plant Physiol 103: 695-702

De Souza MP, Chu D, Zhao M, Zayed AM, Ruzin SE, Schichnes D, Terry N (1999a) Rhizosphere bacteria enhance selenium accumulation and volatilisation by Indian mustard. Plant Physiol 119: 563-573

De Souza MP, Huang CPA, Chee N, Terry N (1999b) Rhizosphere bacteria enhance the accumulation of selenium and mercury in wetland plants. Planta 209: 259-263

Dinkelaker B, Marschner H (1992) *In vivo* demonstration of acid phosphatase activity in the rhizosphere of soil-grown plants. Plant Soil 144: 199-205

Dommergues YR, Krupta SV (1978) Interaction between non-pathogenic soil microorganisms and plants. Elsevier, Amsterdam

Dormaar JF (1988) Effect of plant roots on chemical and biochemical properties of surrounding discrete soil zones. Can J Soil Sci 68: 233-242

Doyle MO, Otte ML (1997) Organism-induced accumulation of iron, zinc and arsenic in wetland soils. Environ Pollut 96: 1-11

Entry JA, Rygiewicz PT, Emmingham WH (1994) ^{90}Sr uptake by *Pinus ponderosa* and *Pinus radiata* seedlings inoculated with ectomycorrhizal fungi. Environ Pollut 86: 201-206

Entry JA, Vance NC, Hamilton MA, Zabowski D, Watrud LS, Adriano DC (1996) Phytoremediation of soil contaminated with low concentrations of radionuclides. Water Air Soil Pollut 88: 167-176

Epstein A, Gussman C, Blaylock M, Yermiyahu U, Huang J, Kapulnik Y, Orser C (1999) EDTA and Pb-EDTA accumulation in Brassica juncea grown in Pb-amended soil. Plant Soil 208: 87-94

Fitz WJ, Wenzel WW (2002) Arsenic transformations in the soil-rhizosphere-plant system: fundamentals and potential application to phytoremediation. J Biotechnol 99: 259-278

Fitz, WJ, Wenzel WW, Zhang H, Nurmi J, öllensperger G, Stipek K, Fischerova Z, Sscheiget P, Ma LQ, Stingeder G (2003). Rhizosphere characteristics of the arsenic hyperaccumulator pteris vittata L. and application in phytoextraction, Environ Sci. Technot, in Press.

Francesconi K, Visoottiviseth P, Sridockhan W, Goessler W (2002) Arsenic species in an hyperaccumulating fern, *Pityrogramma calomelanos*: a potential phytoremediator. Sci Total Environ 284: 27-35

Ganmore-Neumann R, Bar Yosef B, Shanzer A, Libman J (1992) Enhanced iron uptake by synthetic siderophores in corn roots. J Plant Nutr 15: 1027-1037

Gijsman AJ (1990) Rhizosphere pH along different root zones of Douglas-fir (*Pseudotsuga menziesii*) as effected by sources of nitrogen. In: Van Beusichem ML (ed) Plant nutrition—physiology and application. Kluwer, Dordrecht, pp 45-51

Gobran GR, Clegg S, Courchesne F (1997) Rhizospheric effects on nutrient availability in forest soils: processes and feedbacks. Extended abstracts of the 4th international conference on the biogeochemistry of trace elements, June 23-26, 1997, Berkely. pp 355-356

Gollany HT, Schumaker TE (1993) Combined use of colorimetric and microelectrode methods for evaluating rhizosphere pH. Plant Soil 154: 151-159

Grcman H, Velikonja-Bolta S, Vodnik D, Kos B, Lestan D (2001) EDTA enhanced heavy metal phytoextraction: metal accumulation, leaching and toxicity. Plant Soil 235: 105-114

Gregory PJ, Atwell BJ (1991) The fate of carbon in pulse-labeled crops of barley and wheat. Plant Soil 136: 205-213

Hamon RE, McLaughlin JM (1999) Use of the hyperaccumulator *Thlaspi caerulescens* for bioavailable contaminant stripping. In: Wenzel WW, Adriano DC, Alloway B, Doner HE, Keller C, Lepp NW, Mench M, Naidu R, Pierzynski GM (eds) Proceedings of the 5th international conference on biogeochemistry of trace elements. ICOBTE, Vienna, pp 908-909

Haselwandter K (1997) Soil microorganisms, mycorrhiza, and restoration ecology. In: Urbanska KM, Webb NR, Edwards PJ (eds) Restoration ecology and sustainable development. Cambridge University Press, Cambridge

Haselwandter K, Bowen GD (1996). Mycorrhizal relations in trees for agroforestry and land rehabilitation. For Ecol Manage 81: 1-17

Häussling M, Marschner H (1989) Organic and inorganic soil phosphate and acid phosphatase activity in the rhizosphere of 80-year-old Norway spruce (*Picea abies*) trees. Biol Fertil Soil 8: 128-133

Häussling M, Leisen E, Marschner H, Römheld V (1985) An improved method for non-destructive measurements of the pH at the root-soil interface (rhizosphere). J Plant Physiol 117: 371-375

Hedley MJ, Nye PH, White RH (1982) Plant induced changes in the rhizosphere of rape (Brassica napus. var. Emerald) seedlings. I. pH changes and the increase in P concentration in the soil solution. New Phytol 91: 19-29

Helal HM, Sauerbeck D (1986) Effects of plant roots on carbon metabolism of soil microbial biomass. Z Pflanzenernaehr Bodenkd 149: 181-188

Hiltner L (1904) Über neuere Erfahrungen und Probleme auf dem Gebiet der Bodenbakteriolgie unter besonderer Berücksichtigung der Gründüngung und Brache. Arb Dtsch Landwirt Ges 98: 59-78

Hinsinger P, Jaillard B (1993) Root-induced release of interlayer potassium and the vermiculization of phlogopite as related to potassium depletion in the rhizosphere of ryegrass. J Soil Sci 44: 525-534

Hinsinger P, Jaillard B, Dufey J (1992) Rapid weathering of a trioctahedral mica by the roots of ryegrass. Soil Sci Soc Am J 56: 997-982

Hinsinger P, Elsass F, Jaillard B, Robert M (1993) Root-induced irreversible transformation of a trioctahedral mica in the rhizosphere of rape. J Soil Sci 44: 535-545

Hoffmann WF, Barber SA (1971) Phosphorus uptake by wheat (*Triticum aestivum*) as influenced by ion accumulation in the rhizocylinder. Soil Sci 112: 256-262

Hong PKA, Li C, Banerji SK, Regmi T (1999) Extraction, recovery and biostability of EDTA for remediation of heavy metal-contaminated soil. J Soil Contam 8: 81-103

Huang JW, Cunningham SD (1996) Lead phytoextraction: species variation in lead uptake and translocation. New Phytol 134: 75-84

Huang JW, Chen J, Berti WB, Cunningham SD (1997) Phytoremediation of lead-contaminated soils: role of synthetic chelates in lead phytoextraction. Environ Sci Technol 31: 800-805

Huang JW, Blaylock MJ, Kapulnik Y, Ensley BD (1998) Phytoremediation of uranium-contaminated soils: role of organic acids in triggering uranium hyperaccumulation in plants. Environ Sci Technol 32: 2004-2008

Hutchinson JJ, Young SD, McGrath SP, West HM, Black CR, Baker AJM (2000) Determining uptake of "non-labile" soil cadmium by *Thlaspi caerulescens* using isotopic dilution techniques. New Phytol 146: 453-460

Iskandar IK, Adriano DC (1997) Remediation of soils contaminated with metals—a review of current practices in the USA. In: Iskandar IK, Adriano DC (eds) Remediation of soils contaminated with metals. Science Reviews, Northwood, UK

Jaillard B, Ruitz L, Arvieu JC (1996) pH mapping in transparent gel using color indicator videodensitometry. Plant Soil 183: 85-95

Jensen LS, Sorensen J (1994) Microscale fumigation-extraction and substrate-induced respiration methods for measuring microbial biomass in barley rhizosphere. Plant Soil 162: 151-161

Jolley VD, Brown JC (1989) Iron efficient and inefficient oats. I. Differences in phytosiderophores release. J Plant Nutr 12: 423-435

Jones DL, Darrah PR (1995) Influx and efflux of organic acids across the soil root interface of Zea mays L. and its applications in rhizosphere C flow. Plant Soil 173: 103-109

Jones DL, Darrah PR (1996) Re-sorption of organic compounds by roots of Zea mays L. and its consequence in the rhizosphere. III. Characteristics of sugar influx and efflux. Planta Soil 178: 153-160

Jones DL, von Kochian L (1996) Aluminium-organic acid interactions in acid soils. I. Effect of root-derived organic acid on the kinetics of Al dissolution. Plant Soil 182: 221-228

Kawai S, Takagi S, Sato Y (1988) Mugineic acid-family phytosiderophores in root-secretions of barley, corn and sorghum varieties. J Plant Nutr 11: 633-642

Kayser A, Wenger K, Keller A, Attinger W, Felix HR, Gupta SK, Schulin R (2000) Enhancement of phytoextraction of Zn, Cd, and Cu from calcareous soil: the use of NTA and sulfur amendments. Environ Sci Technol 34: 1178-1183

Keith H, Oades JM, Martin JK (1986) Input of carbon to soil from wheat plants. Soil Biol Biochem 18: 445-449

Kirlew PW, Bouldin DR (1987) Chemical properties of the rhizosphere in acid sub-soil. Soil Sci Soc Am J 51: 128-132

Knight B, Zhao FJ, McGrath SP, Shen ZG (1997) Zinc and cadmium uptake by the hyperaccumulator Thlaspi caerulescens in contaminated soils and its effects on the concentration and chemical speciation of metals in soil solution. Plant Soil 197: 71-78

Krämer U, Cotter-Howells JD, Charnock JM, Baker AJM, Smith JAC (1996) Free histidin as a metal chelator in plants that hyperaccumulate nickel. Nature 379: 635-638

Krämer U, Smith RD, Wenzel WW, Raskin I, Salt DE (1997) The role of metal transport and tolerance in nickel hyperaccumulation by *Thlaspi goesingense* Halacsy. Plant Physiol 115: 1641-1650

Lasat MM, Baker AJM, Kochian LV (1996) Physiological characterisation of root Zn^{2+} absorption and translocation to shoots in Zn hyperaccumulator and nonaccumulator species of *Thlaspi*. Plant Physiol 112: 1715-1722

Lee J, Reeves RD, Brooks RR, Jaffré T (1977) Isolation and identification of a citrato-complex of nickel from nickel-accumulating plants. Phytochemistry 16: 1503-1505

Leyval C, Turnau K, Haselwandter K (1997) Effect of heavy metal pollution on mycorrhizal colonization and function: physiological, ecological and applied aspects. Mycorrhiza 7: 139-153

Li YM, Chaney RL, Angle JS, Chen KY, Kerschner BA, Baker AJM (1996) Genotypical differences in zinc and cadmium hyperaccumulation in Thlaspi caerulescens. Agron Abstr 1996:27

Liao ZW, Wang JL, Liu ZY (1993) Si, Fe and Mn distributions in rice (Oryza sativa L.) rhizosphere of red earths and paddy soils. Pedosphere 3: 1-6

Liljeroth JA, van Veen JA, Miller HJ (1990) Assimilate translocation to the rhizosphere of two wheat lines and subsequent utilization by rhizosphere microorganisms at two nitrogen concentration. Soil Biol Biochem 22: 1015-1021

Lombi E, Zhao FJ, Dunham SJ, McGrath SP (2000) Cadmium accumulation in populations of *Thlaspi caerulescens* and *Thlaspi goesingense*. New Phytol 145: 11-20

Lombi E, Zhao FJ, McGrath SP, Young S, Sacchi GA (2001a) Physiological evidence for a high affinity cadmium transporter in a *Thlaspi caerulescens* ecotype. New Phytol 149: 53-60

Lombi E, Zhao FJ, Dunham SJ, McGrath SP (2001b) Phytoremediation of heavy metal contaminated soils: natural hyperaccumulation versus chemically-enhanced phytoextraction. J Environ Qual 30: 1919-1926

Lombi E, Tearall KL, Howarth JR, Zhao FJ, Hawkesford MJ, McGrath SP (2002) Influence of iron status on cadmium and zinc uptake by different ecotypes of the hyperaccumulator *Thlaspi caerulescens*. Plant Physiol 128: 1359-1367

Lundström US (1994) Significance of organic acids for weathering and the podolization process. Environ Int 20: 21-30

Luo YM, Christie P, Baker AJM (2000) Soil solution Zn and pH dynamics in non-rhizosphere soil and in the rhizosphere of *Thlaspi caerulescens* grown in a Zn/Cd-contaminated soil. Chemosphere 41: 161-164

Ma JF, Nomoto K (1996) Effective regulation of iron acquisition in graminaceous plants. The role of mugineic acids as phytosiderophores. Physiol Plant 97: 609-617

Ma LQ, Komar KM, Tu C, Zhang WH, Cai Y, Kennelley ED (2001) A fern that hyperaccumulates arsenic. Nature 409:579

Marschner H (1995) Mineral nutrition of higher plants. Academic Press, London

Marschner H, Römheld V (1983) *In vivo* measurement of root-induced pH changes at the soil-root interface: effect of plant species and nitrogen source. Z Pflanzenrnaehr Bodenkd 111: 241-251

Marschner H, Römheld V (1994) Strategies of plants for acquisition of iron. Plant Soil 165: 261-274

Marschner H, Römheld V, Kissel M (1986) Different strategies in higher plants in mobilization and uptake of iron. J Plant Nutr 9: 695-713

Marschner H, Römheld V, Kissel M (1987) Localization of phytosiderophores release and iron uptake along intact barley roots. Physiol Plant 71: 157-162

Martin JK, Merckx R (1992) The partitioning of photosynthetically fixed carbon within the rhizosphere of mature wheat. Soil Biol Biochem 24: 1147-1156

McCully ME, Canny MJ (1985) Localization of translocated ^{14}C in roots and root exudates of field grown maize. Physiol Plant 65: 380-392

McDougall BM, Rovira AD (1970) Site of exudation of ^{14}C-labelled compounds from wheat roots. New Phytol 69: 999-1003

McGrath SP, Shen ZG, Zhao FJ (1997) Heavy metals uptake and chemical changes in the rhizosphere of Thlaspi caerulescens and Thlaspi ochroleucum grown in contaminated soils. Plant Soil 188: 153-159

McGrath SP, Zhao FJ, Lombi E (2001) Plant and rhizosphere processes involved in phytoremediation of metal-contaminated soils. Plant Soil 232: 207-214

McGrath SP, Zhao FJ, Lombi E (2002) Phytoremediation of metals, metalloids and radionuclides. Adv Agron 75: 1-56

McLachlan KD (1980) Acid phosphatase activity of intact roots and phosphorus nutrition in plants. I. Assay conditions and phosphatase activity. Aust J Agric Res 31: 429-440

Meagher RB, Rugh C (1996) Phytoremediation of mercury and methyl mercury pollution using modified bacterial genes. IBC conference 5/8/96— abstract and outline

Meharg AA, Hartley-Whitaker J (2002) Arsenic uptake and metabolism in arsenic resistant and nonresistant plant species. New Phytol 154: 29-44

Mench M, Martin E (1991) Mobilization of cadmium and other metals from two soils by root exudates of *Zea mays* L, *Nicotiana tabacum* L, and *Nicotiana rustica* L. Plant Soil 132: 187-196

Mench MJ, Fargues S (1994) Metal uptake by iron-efficient and inefficient oats. Planta Soil 165: 227-233

Miller RM, Jastrow J (1990) Hierarchy of root and mycorrhizal interactions with soil aggregation. Soil Biol Biochem 22:579

Neng-Chang C, Huai-Man C (1992) Chemical behavior of cadmium in wheat rhizosphere. Pedosphere 2: 363-371

Nörtemann B (1999) Biodegradation of EDTA. Appl Microbiol Biotechnol 51: 751-759

Nye PH (1981) pH changes across the rhizosphere induced by roots. Plant Soil 61: 7-26

Ohkawi Y, Sugahara K (1997) Active extrusion of protons and exudation of carboxylic acids in response to iron deficiency by roots of chickpea (*Cicer arietinum* L.) Plant Soil 189: 49-55

Olsthoorn AFM, Keltjens WG, van Baren B, Hopman MCG (1991) Influence of ammonium on fine root development and rhizosphere pH of Douglas-fir seedlings in sand. Plant Soil 133: 75-81

Pellet DM, Grunes DL, Kochian LV (1995) Organic acids exudation as an aluminium-tolerance mechanism in maize (*Zea mays*). Planta 196: 788-795

Pellet DM, Papernik LA, Jones DL, Darrah PR, Grunes DL, Kochian LV (1997) Involvement of multiple aluminium exclusion mechanisms in aluminium tolerance in wheat. Plant Soil 192: 63-68

Pence NS, Larsen PB, Ebbs SD, Letham DLD, Lasat MM, Garvin DF, Eide D, von Kochian L (2000) The molecular physiology of heavy metal transporter in the Zn/Cd hyperaccumulator *Thlaspi caerulescens*. Proc Natl Acad Sci USA 97: 4956-4960

Pierzynski GM (1997) Strategies for remediating trace-element contaminated sites. In: Iskandar IK, Adriano DC (eds) Remediation of soils contaminated with metals. Science Reviews, Northwood, UK

Pijnenborg JWM, Lie TA, Zehender AJB (1990) Nodulation of lucerne (*Medicago sativa* L.) in an acid soil: pH dynamics in the rhizosphere of seedlings growing in rhizotrons. Plant Soil 126: 161-168

Pilet PE, Versef JM, Mayor G (1983) Growth distribution and surface pH patterns along maize roots. Planta 164: 96-100

Puschenreiter M, Stöger G, Lombi E, Horak O, Wenzel WW (2001) Phytoextraction of heavy metal contaminated soils with *Thlaspi goesingense* and *Amaranthus hybridus*: rhizosphere manipulation using EDTA and ammonium sulfate. J Plant Nutr Soil Sci 164: 615-621

Raskin I, Kumar PBAN, Dushenkow S, Salt DE (1994) Bioconcentration of heavy metals by plants. Curr Opin Biotechnol 5: 285-290

Ridge EH, Rovira AD (1971) Phosphatase activity of intact young wheat roots under sterile and non-sterile condition. New Phytol 70: 1017-1026

Riley D, Barber SA (1969) Bicarbonate accumulation and pH changes at the soybean (Glycine max) root-soil interface. Soil Sci Soc Am Proc 33: 905-908

Riley D, Barber SA (1971) Effect of ammonium and nitrate fertilization on phosphorus uptake as related to root-induced pH changes at the root-soil interface. Soil Sci Soc Am Proc 35: 301-306

Rogers RD, Williams SE (1986) Vesicular-arbuscular mycorrhiza: influence on plant uptake of cesium and cobalt. Soil Biol Biochem 18: 371-376

Römheld V (1991) The role of phytosiderophores in acquisition of iron and other micronutrients in graminaceous species: an ecological approach. Plant Soil 130: 127-134

Römheld V, Marschner H (1984) Plant induced pH changes in the rhizosphere of "Fe-efficient" soybean and corn cultivars. J Plant Nutr 7: 623-630

Römheld V, Marschner H (1990) Genotypical differences among graminaceous species in release of phytosiderophores and uptake of ironphytosiderophores. Plant Soil 123: 147-153

Römkens P, Bouwman L, Japenga J, Draaisma C (2002) Potentials and drawbacks of chelate-enhanced phytoremediation of soils. Environ Pollut 116: 109-121

Rovira AD (1969) Plant root exudates. Bot Rev 35: 35-57

Rovira AD, Bowen GD (1966) The effects of microorganisms upon plant growth. II. Detoxification of heat sterilized soils by fungi and bacteria. Plant Soil 14: 199-214

Rovira AD, Bowen GD, Foster RC (1983) The significance of rhizosphere microflora and mycorrhizas in plant nutrition. In: Läuchli A, Bieleski RL (eds) Inorganic plant nutrition. Encyclopedia of plant physiology, new edition, vol 15a. Springer, Berlin Heidelberg New York, pp 61-93

Ryan PR, Delhaize E, Randall PJ (1995) Malate efflux from root apices and tolerance to aluminium are highly correlated in wheat. Aust J Plant Physiol 22: 531-536

Rygiewicz PT, Bledsoe CS, Zasoski RJ (1984) Effects of ectomycorrhizae and solution pH on [^{15}N] nitrate uptake by coniferous seedlings. Can J For Res 14: 885-892

Salt DE, Blaylock M, Kumar NPBA, Dushenkov V, Ensley BD, Chet I, Raskin I (1995) Phytoremediation: a novel strategy for the removal of toxic metals from the environment using plants. Biotechnology 13: 468-474

Salt DE, Kato N, Krämer U, Smith RD, Raskin I (2000) The role of root exudates in nickel hyperaccumulation and tolerance in accumulator and nonaccumulator species of Thlaspi. In: Terry N, Bañuelos G (eds) Phytoremediation of contaminated soil and water. Lewis Publ, Boca Raton, pp 189-200

Schaller G, Fisher WR (1985) pH-Änderungen in der Rhizosphäre von Masiund Erdnusswurzeln. Z Pflanzenernaehr Bodenkd 148: 306-320

Schnepf A, Schrefl T, Wenzel WW (2002) The suitability of pde-solvers in rhizosphere modeling, exemplified by three mechanistic rhizosphere models. J Plant Nutr Soil Sci 165: 713-718

Schnoor JL, Licht LA, McCutcheon SC, Wolfe NL, Carreira LH (1995) Phytoremediation of organic and nutrient contaminants. Environ Sci Technol 29: 318-323

Schönwitz R, Ziegler H (1986) Quantitative and qualitative aspects of a developing rhizosphere micoflora and hydroponically grown maize seedlings. Z Pflanzenernaehr Bodenkd 149: 623-634

Schreiner RP, Mihara KL, McDaniel H, Bethlenfalvay GJ (1997) Mycorrhizal fungi influence plant and soil functions and interactions. Plant Soil 188: 199-209

Schwartz C, Morel JL, Saumier S, Whiting SN, Baker AJM (1999) Root development of the zinc-hyperaccumulator plant Thlaspi caerulescens as affected by metal origin, content and localization in soil. Plant Soil 208: 103-115

Smiley RW (1974) Rhizosphere pH as influenced by plants, soils, and nitrogen fertilizers. Soil Sci Soc Am Proc 38: 795-799

Smith JL, Paul EA (1990) The significance of soil microbial biomass estimation. In: Bollag JM, Stotzsky G (eds) Soil biochemistry. Dekker, New York, pp 357-396

Smith JA, Krämer U, Baker AJM (1995) Role of metal transport and chelation in nickel hyperaccumulation in the genus *Alyssum*. Abstract book of the 14th annual symposium on current topics in plant biochemistry, physiology and molecular biology, University of Missouri, pp 11-12

Sommer P, Burguera G, Wieshammer G, Strauss J, Ellersdorfer G, Wenzel WW (2002) Effects of mycorrhizal associations on the metal uptake by willows from polluted soils: implication for soil remediation by phytoextraction. Mitteil Österr Bodenkundl Gesellsch 66: 113-119

Stomp A-M, Han K-H, Wilbert S, Gordon MP, Cunningham SD (1994) Genetic strategies for enhancing phytoremediation. Ann NY Acad Sci 721: 481-492

Sun B, Zhao FJ, Lombi E, McGrath SP (2001) Leaching of heavy metals from contaminated soils using EDTA. Environ Pollut 113: 111-120

Takagi S, Nomoto K, Takemoto T (1984) Physiological aspect of mugineic acid, a possible phytosiderophore of graminaceous plants. J Plant Nutr 7: 469-477

Tarafdar JC, Jungk A (1987) Phosphatase activity in the rhizosphere and its relation to the depletion of soil organic phosphorus. Biol Fertil Soil 3: 199-204

Tepfer D, Metzger L, Prost R (1989) Use of roots transformed by Agrobacterium rhizogenes in rhizosphere research: applications in studies of cadmium assimilation from sewage sludges. Plant Mol Biol 13: 295-302

Terry N (1995) Can plants solve the Se problem? Abstract book of the 14th annual symposium on current topics in plant biochemistry, physiology and molecular biology, University of Missouri, pp 63-64

Terry N (1996) The use of phytoremediation in the clean-up of selenium polluted soils and waters. IBC conference 5/8/96—abstract and outline

Terry N, Zayed AM (1994) Selenium volatilization in plants. In: Frankenberger WT Jr, Benson S (eds) Selenium in the environment. Dekker, New York, pp 343-367

Terry N, Carlson C, Raab TK, Zayed AM (1992) Rates of selenium volatilization among crop species. J Environ Qual 21: 341-344

Tisdall JM, Oades JM (1982) Organic matter and water stable aggregation in soils. J Soil Sci 33: 141-163

Treeby M, Marschner H, Römheld V (1989). Mobilization of iron and other micronutrients from a calcareous soil by plant-borne microbial and synthetic metal chelators. Plant Soil 114: 217-226

Trofymov JA, Coleman DC, Cambardella C (1987) Rate of rhizodeposition and ammonium depletion in the rhizosphere of axenic oat roots. Plant Soil 97: 333-344

Uren NC, Reisenauer HM (1988) The role of root exudates in nutrient acquisition. Adv Plant Nutr 3: 79-114

Walter A, Römheld V, Marschner H, Mori S (1994) Is the release of phytosiderophores in zinc-deficient plants a response to impaired iron utilization? Physiol Plant 92: 493-500

Walton B, Anderson TA (1990) Microbial degradation of trichloroethylene in the rhizosphere: potential application to biological remediation of waste sites. Appl Environ Microb 56: 1012-1016

Wang J, Zhao FJ, Meharg AA, Raab A, Feldmann J, McGrath SP (2002) Mechanisms of arsenic hyperaccumulation in *Pteris vittata*: uptake kinetics, interactions with phosphate, and arsenic speciation. Plant Physiol 130: 1552-1561

Welch RM, Norvell WA, Schaefer SC, Shaff JE, von Kochian L (1993) Induction of iron(III) and copper(II) reduction in pea (*Pisum sativum* L.) roots by Fe and Cu status: does the root-cell plasmalemma Fe(III) chelate reductase perform a general role in regulating cation uptake. Planta 190: 555-561

Wenzel WW, Adriano DC, Salt DE, Smith R (1999) Phytoremediation: a plant-microbe based system. In: Adriano DC, Bollag J-M, Frankenberger WT Jr, Sims RC (eds) Remediation of contaminated soils. SSSA Spec Monogr 37: 457-510

Wenzel WW, Puschenreiter M, Horak O (2000) Role and manipulation of the rhizosphere in soil remediation/revegetation. In: Luo YM, McGrath SP, Cao ZH, Thao FJ, Chen YX, Xu JM (eds) Proceedings of the international conference on soil remediation, SOILREM, 15-19 Oct 2000, China, pp 176-180

Wenzel WW, Wieshammer G, Fitz WJ, Puschenreiter M (2001) Novel rhizobox design to assess rhizosphere characteristics at high spatial resolution. Plant Soil 237: 37-45

Wenzel WW, Bunkowski M, Puschenreiter M, Horak O (2003a) Rhizosphere characteristics of indigenously growing nickel hyperaccumulator and excluder plants on serpentine soil. Environ Pollut 123: 131-138

Wenzel WW, Unterbrunner R, Sommer P, Sacco P (2003b) Chelate-assisted phytoextraction using canola (*Brassica napus* L.) in outdoor pot and field-lysimeter experiments. Plant Soil 249: 83-96

Whipps JM (1984) Environmental factors affecting the loss of carbon from the roots of wheat and barley seedlings. J Exp Bot 35: 767-733

Whipps JM, Lynch JM (1983) Substrate flow and utilization in the rhizosphere of cereals. New Phytol 95: 605-623

Whiting SN, Leake JR, McGrath SP, Baker AJM (2000) Positive responses to Zn and Cd by roots of the Zn and Cd hyperaccumulator *Thlaspi caerulescens*. New Phytol 145: 199-210

Whiting SN, Leake JR, McGrath SP, Baker AJM (2001a) Zinc accumulation by *Thlaspi caerulescens* from soils with different Zn availability: a pot study. Plant Soil 236: 11-18

Whiting SN, Leake JR, McGrath SP, Baker AJM (2001b) Assessment of Zn mobilization in the rhizosphere of *Thlaspi caerulescens* by bioassay with non-accumulator plants and soil extraction. Plant Soil 237: 147-156

Whiting SN, de Souza MP, Terry N (2001c) Rhizosphere bacteria mobilize Zn for hyperaccumulation by *Thlaspi caerulescens*. Environ Sci Technol 35: 3144-3150

Wiren N, Marshner H, Römheld V (1996) Roots of iron-efficient maize also absorb phytosiderophore-chelated Zn. Plant Physiol 111: 1119-1125

Yi Y, Guerinot ML (1996) Genetic evidence for induction of root Fe(III) chelate reductase activity is necessary for iron uptake under iron deficiency. Plant J 10: 835-844

Youssef RA, Chino M (1988) Development of a new rhizobox system to study the nutrient status in the rhizosphere. Soil Sci Plant Nutr 34: 461-465

Youssef RA, Chino M (1989a) Root induced changes in the rhizosphere of plants. I. Changes in relation to the bulk soil. J Soil Sci Plant Nutr 35:4 61-468

Youssef RA, Chino M (1989b) Root-induced changes in the rhizosphere of plants. II. Distribution of heavy metals across the rhizosphere in soils. J Soil Sci Plant Nutr 35: 609-621

Zhang F, Römheld V, Marschner H (1989) Effect of zinc deficiency on the release of zinc and iron mobilizing root exudates. Z Pflanzenrmaehr Bodenkd 152: 205-210

Zhao FJ, Hamon R, McLaughlin MJ (2001) Root exudates of the hyperaccumulator *Thlaspi caerulescens* do not enhance metal mobilization. New Phytol 151: 613-620

Zhao FJ, Dunham SJ, McGrath SP (2002a) Arsenic hyperaccumulation by different fern species. New Phytol 156: 27-31

Zhao FJ, Hamon R, Lombi E, McLaughlin MJ, McGrath SP (2002b) Characteristics of cadmium uptake in two contrasting ecotypes of the hyperaccumulator *Thlaspi caerulescens*. J Exp Bot 53: 535-543

Zhao H, Eide D (1996) The yeast ZRT1 gene encodes the zinc transporter protein of a high affinity system induced by zinc limitation. Proc Natl Acad Sci USA 93: 2455-2458

Zayed AM, Terry N (1994). Selenium volatilization in roots and shoots: effects of shoot removal and sulfate level. J Plant Physiol 143: 8-14

Zoyas AKN, Loganathan P, Hedley MJ (1997) A technique for studying rhizosphere processes in tree crops: soil phosphorus depletion around camelia (*Camelia japonica* L.) roots. Plant Soil 190: 253-269

Chapter 14

Phytoremediation of Metals and Radionuclides in the Environment: The Case for Natural Hyperaccumulators, Metal Transporters, Soil-Amending Chelators and Transgenic Plants

M.N.V. Prasad

Department of Plant Sciences, University of Hyderabad, Hyderabad 500046 AP, India

14.1 INTRODUCTION

All compartments of the environment, viz. air, water and soil, are polluted by a variety of metals including radionuclides that can interfere with biogeochemical cycles. Heavy metals and radionuclides are considered as major environmental pollutants and are considered to be cytotoxic, mutagenic and carcinogenic (Hadjiliads 1997). Phytoremediation has been accepted widely both in developed and developing nations for its potential to clean up polluted and contaminated sites (Figs 14.1-14.3).

This technology is currently gaining considerable importance owing to its potential for applications to real-world ecosystems. Use of plants to remove metals from the environment has received increasing attention in recent years (Bell et al. 1988; Berstein 1992; Anderson et al. 1993; Brown et al. 1994; Raskin et al. 1994, 1997; Baker et al. 1995; Brown 1995; Cunningham et al. 1995, Dodds-Smith et al. 1995; Salt et al. 1995, 1998; Adler 1996; Azadpour and Matthews 1996; Cunningham and Ow 1996; Kadlec and Knight 1996; Ow 1996; Raskin 1996; Banuelos et al. 1997a,b; Bishop 1997; Chaney et al. 1997; Dushenkov et al. 1997a,b, 1999; Entry et al. 1997; Huang et al. 1997, 1998; McIntyre and Lewis 1997; Watanabe 1997; Brooks 1998; Heaton et al. 1998; Lasat et al. 1998b; Vangronsveld and Cunningham 1998; Glass 1999, 2000; Greger and Lindberg 1999; Prasad and Freitas 1999; Saxena et al. 1999; Kadlec et al. 2000; Kaltsikes 2000; Negri and Hinchman 2000; Raskin and Ensley 2000; Terry and Bañuelos 2000; Wise et al. 2000; Guerinot and Salt 2001). Compared with existing physical and chemical methods of soil remediation, the use of

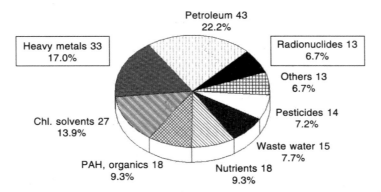

Fig. 14.1 In a report from the year 2000, the USA Environmental Protection Agency (EPA) lists about 166 ongoing bioremediation filed research projects. Heavy metals and radionuclides represent about 25% of this activity supporting the hypothesis that bioremediation is a feasible technology to clean up the environment

plants is cost-effective and less disruptive to the environment (Raskin et al. 1994, 1997; Saxena et al. 1999).

Phytoremediation can be defined as the use of plants, including trees and grasses, to remove, destroy or sequester hazardous contaminants from media such as soil, water and air. It is being investigated and/or used commercially to treat a variety of contaminants in a number of different scenarios. Researchers have found that plants can be used to treat metals and radionuclides. Plant species are selected for phytoremediation based on their potential to accumulate metals, their growth rates and yield, and the depth of their root zone. Conventional technologies suitable for water and soil remediation used in situ and ex situ are: (1) soil flushing, (2) pneumatic fracturing, (3) solidification/stabilization, (4) vitrification, (5) electrokinetics, (6) chemical reduction/oxidation, (7) soil washing, and (8) excavation, retrieval and off-site disposal. These technologies are cost-prohibitive and the processes often generate secondary waste (Cunningham and Ow 1996).

Heavy metals are conventionally defined as elements with metallic properties (ductility, conductivity, stability as cations, ligand specificity, etc.) and atomic number >20. The most common heavy metal contaminants are Cd, Cr, Cu, Hg, Pb, and Zn. Metals are natural components in soil. Contamination, however, has resulted from industrial activities, such as mining and smelting of metalliferous ores, electroplating, gas exhaust, energy and fuel production, fertilizer and pesticide application, and generation of municipal waste (Kabata-Pendias 2001). Concentration ranges and regulatory limits for several major metal contaminants are shown in Table 14.1.

Toxicity can lead to metal mobilization in runoff water and subsequent deposition into nearby bodies of water. Furthermore, bare soil is more susceptible to wind erosion and spreading of contamination by airborne dust. In such situations, the immediate goal of remediation is to reclaim the site by establishing a vegetative cover to minimize soil erosion and pollution spread.

Metal-contaminated soils are notoriously hard to remediate. Current technologies resort to soil excavation and either landfilling or soil washing followed by physical or chemical separation of the contaminants. The cost of soil remediation is highly variable and depends on the contaminants of concern, soil properties, and site conditions. Cost

Table 14.1 Some toxic metals in soil—concentration ranges and regulatory guidelines

Metal	Soil concentration range[a] (mg kg^{-1})	Regulatory limits (mg kg^{-1})
Pb	1.00-6900	600
Cd	0.10-345	100
Cr	0.05-3950	100
Hg	<0.01-1800	270
Zn	150.00-5000	1500

[a]New Jersey Department of Environmental Protection, nonresidential direct contact soil cleanup criteria 1996.

Table 14.2 Costs for the treatment of metal-contaminated soils (Glass 1999)

Treatment	Cost ($/ton)	Additional factors/expenses
Vitrification	75-425	Long-term monitoring
Landfilling	100-500	Transport/excavation/monitoring
Chemical treatment	100-500	Recycling of contaminants
Electrokinetics	20-200	Monitoring
Phytoextraction	5-40	Monitoring

estimates associated with the use of several technologies for the cleanup of metal-contaminated soil are shown in Table 14.2.

Cleaning of metal-contaminated soils via conventional engineering methods can be prohibitively expensive (Salt et al. 1995). The costs estimated for the remediation of sites contaminated with heavy metals, and heavy metals mixed with organic compounds, are also shown in Table 14.2.

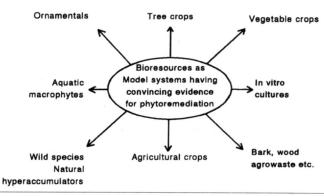

Fig. 14.2 Biodiversity with convincing evidence to clean up the metal-contaminated environment

Soil remediation is needed to eliminate risk to humans or the environment from toxic metals (Hadjiliads 1997). Livestock and wildlife have suffered from Se poisoning (Banuelos et al. 1997a,b). In addition, soil contamination with Zn, Ni and Cu caused by mine wastes and smelters is known to be phytotoxic to sensitive plants (Chaney et al. 1999). One of the greatest concerns for human health is caused by Pb contamination. Exposure to Pb can

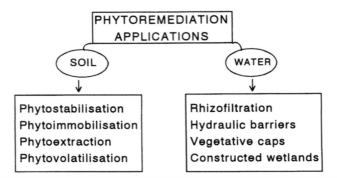

Fig. 14.3 Different phytoremediation applications to soil and water polluted/contaminated with metals and radionuclides

occur through multiple pathways, including inhalation of air and ingestion of Pb in food, water, soil or dust. Excessive Pb exposure can cause seizures, mental retardation and behavioral disorders. The danger of Pb is aggravated by low environmental mobility even under high precipitations.

In soil, metals are associated with several fractions: (1) in soil solution, as free metal ions and soluble metal complexes, (2) adsorbed to inorganic soil constituents at ion-exchange sites, (3) bound to soil organic matter, (4) precipitated such as oxides, hydroxides, carbonates, and (5) embedded in the structure of the silicate minerals. Soil sequential extractions are employed to isolate and quantify metals associated with different fractions (Tessier et al. 1979). For phytoextraction to occur, contaminants must be bioavailable (ready to be absorbed by roots). Bioavailability depends on metal solubility in soil solution. Only metals associated with fractions 1 and 2 (above) are readily available for plant uptake. Some metals, such as Zn and Cd, occur primarily in exchangeable, readily bioavailable form. Others, such as Pb, occur as a soil precipitate, a significantly less bioavailable form.

To grow and complete the life cycle, plants must acquire not only macronutrients (N, P, S, Ca, and Mg), but also essential micronutrients such as Fe, Zn, Mn, Ni, Cu, and Mo. Plants have evolved highly specific mechanisms to take up, translocate, and store these nutrients. For example, metal movement across biological membranes is mediated by proteins with transport functions. In addition, sensitive mechanisms maintain intracellular concentrations of metal ions within the physiological range. In general, the uptake mechanism is selective, plants preferentially acquiring some ions over others. Ion uptake selectivity depends on the structure and properties of membrane transporters. These characteristics allow transporters to recognize, bind and mediate the transmembrane transport of specific ions. For example, some transporters mediate the transport of divalent cations, but do not recognize mono- or trivalent ions. Many metals such as Zn, Mn, Ni and Cu are essential micronutrients. In common nonaccumulator plants, accumulation of these micronutrients does not exceed their metabolic needs (<10 ppm). In contrast, metal hyperaccumulator plants can accumulate exceptionally high amounts of metals (in the thousands of ppm). Since metal accumulation is ultimately an energy-consuming process, one would wonder what evolutionary advantage does metal hyperaccumulation give to these species?

Recent studies have shown that metal accumulation in the foliage may allow hyperaccumulator species to evade predators including caterpillars, fungi and bacteria (Boyd and Martens 1994; Pollard and Baker 1997). Hyperaccumulator plants not only accumulate high levels of essential micronutrients, but can also absorb significant amounts of nonessential metals, such as Cd. The mechanism of Cd accumulation has not been elucidated. It is possible that the uptake of this metal in roots is via a system involved in the transport of another essential divalent micronutrient, possibly Zn^{2+}. Cadmium is a chemical analog of the latter, and plants may not be able to differentiate between the two ions (Chaney et al. 1997).

14.2 NATURAL METAL HYPERACCUMULATORS

Plants that hyperaccumulate metals have attracted the attention of scientists all over the world. This work expanded during 1970-1980 by the contributions of the late Professor Robert Brooks of Massey University in New Zealand and Professor Alan Baker of the University of Sheffield, UK (currently working for the University of Melbourne, Australia). Brooks (1998) defined hyperaccumulators as those plants that contain more than 1000 mg/kg (0.1% of dry weight) of Co, Cu, Cr, Pb or Ni, or more than 10,000 mg/kg (1.0% of dry weight) of Mn or Zn in their dry matter. Hyperaccumulators have often been isolated from nature in areas of high contamination or high metal concentration (Baker and Brooks 1989; Homer et al. 1991; Vázquez et al. 1992, 1994; Boyd et al. 1994; Brown 1995; Varennes et al. 1996; Krämer et al. 1997; Robinson et al. 1997a,b; Brooks 1998; Huang et al. 1998; Lasat et al. 2000; Pence et al. 2000; Ma et al. 2001). Among the earliest-identified terrestrial hyperaccumulating species are members of the genus *Thlapsi*, which are known to accumulate Zn, Cd and Pb, and the *Alyssum* genus, which are Ni accumulators. Specific examples include *Thlaspi caerulescens* (pennycress), a Zn accumulator, *Armeria maritima*, a Pb accumulator, and two species isolated in Africa as Cu and Co accumulators, *Aeolanthus biformifolius* and *Haumaniastrum katangense* (Raskin et al. 1994). More recent research has identified other hyperaccumulators, notably *Brassica juncea* (Indian mustard) and *Brassica nigra*, Pb accumulators. A research team at the University of Guelph reported that lemon-scented geraniums (*Pelargonium* sp. 'Frensham') accumulated large amounts of Cd, Pb, Ni and Cu from soil in greenhouse experiments (Saxena et al. 1999).

The commercial potential of hyperaccumulators derived from their rates of accumulation of metals, combined with their growth (i.e. biomass accumulation) rates: multiplying the metal accumulation rate (gm metal per kilogram of plant tissue) by the growth rate (kilograms of tissue per hectare per year), yields a value for metal removal from soil (gm or kg of metal per hectare per year). It has been estimated that this rate must be at least several hundred, and perhaps as much as 1000 kg/ha/year, to be commercially viable, and, even then, remediation may take 15-20 years depending on initial metal concentrations and the depth of the contamination.

As noted above, the threshold for defining a naturally occurring plant as a hyperaccumulator is either 0.1 or 1.0% of dry weight, depending on the metal in question. There are numerous reports in the literature of natural or selected species being able to accumulate larger concentrations of metal, in the range 2.0-4.0% of dry weight. It has been estimated that accumulations of 1.0-2.0% might yield metal removal rates of 200-1000 kg/

ha/year, depending on the growth rate of the plants, and thus be sufficient for commercial use, albeit over a 20-year period (Cunningham et al. 1995). However, the biomass production of several of the hyperaccumulating plants is not satisfactory due to metal toxicity. Another theoretical calculation based on the accumulation rates of *Thlaspi* provided an estimate of 125 kg/ha/year for removal of zinc, which would require 16 years to remediate a "typical contaminated" site. These forecasts have left the impression that natural hyperaccumulators may not fit into the commercial remediation time frames, and that genetic/molecular improvement of plants is needed.

A major drawback of phytoextraction using natural hyperaccumulators is that the rate of natural uptake of metals by plants is very slow, and may not be fast enough to enable commercial use in remediation. The two primary reasons for this are the slow growth rates exhibited by most naturally ocurring metal hyperaccumulators and the limited solubility of metals in soils (i.e. the high affinity of metal ions for soil particles). The latter problem is being investigated through the use of chelators or other surfactants in the soil, or even combination of phytoremediation with other in situ techniques like electroosmosis, to stimulate metal mobility in the soil. The former problem is being solved by the selection or creation of new plant varieties, including the use of genetic engineering, which might be used to introduce biochemical traits that enchance hyperaccumulation. Such efforts are directed towards the use of artificially improved or selected varieties to improve uptake of the metal contaminant (Cunningham and Ow 1996). Another strategy to enhance metal uptake by plants, being pursued by a number of research groups, is the use of symbiotic root-dwelling fungi. These fungi, called mycorrhizae, naturally live on the roots of many types of plants, and benefit their plant host by improving the plant's ability to take up water and certain types of nutrients, by effectively increasing the surface area of the root available for such uptake (Wenzel et al. 1999)

Interest in phytoremediation has grown significantly following the identification of metal-hyperaccumulating plant species (Figs. 14.4-14.8). Hyperaccumulators are conventionally defined as species capable of accumulating metals at levels 100-fold greater than those typically measured in common nonaccumulator plants. Thus, a hyperaccumulator will concentrate more than: 10 ppm Hg; 100 ppm Cd; 1000 ppm Co, Cr, Cu, and Pb;

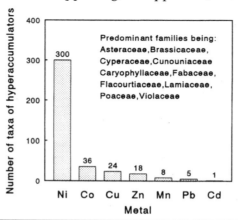

Fig. 14.4 Taxa of various angiospermous families that hyperaccumulate metals

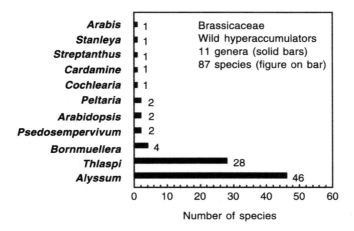

Fig. 14.5 Wild Brassicaceae that hyperaccumulate metals

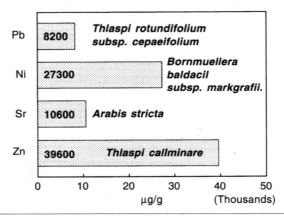

Fig. 14.6 Selected examples of Brassicaceae hyperaccumulation of lead, nickel, selenium, strontium and zinc

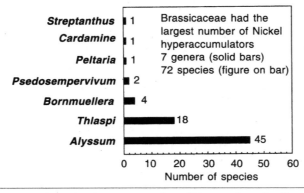

Fig. 14.7 Brassicaceae has the largest number (72 species distributed in 7 genera) of nickel hyperaccumulators

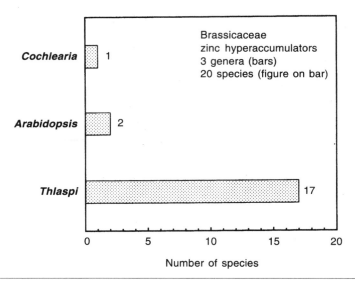

Fig. 14.8 Zinc hyperacccumulators in Brassicaceae represent 20 species distributed in 3 genera

10,000 ppm Ni and Zn. To date, approximately 400 plant species from at least 45 plant families have been reported to hyperaccumulate metals (Tables 14.3-14.5). Most hyperaccumulators bioconcentrate Ni, about 30 absorb either Co, Cu, and/or Zn, even fewer species accumulate Mn and Cd, and there are no known natural Pb hyperaccumulators (Reeves and Baker 1999).

Table 14.3 Several metal hyperaccumulator species and their bioaccumulation potential

Plant species	Metal	Leaf content (ppm)	Reference
Thlaspi caerulescens	Zn:Cd	39,600:1800	Reeves and Brooks (1983); Baker and Walker (1990)
Ipomea alpina	Cu	12,300	Baker and Walker (1990)
Haumaniastrum robertii	Co	10,200	Brooks (1977)
Astragalus racemosus	Se	14,900	Beath et al. (1937)
Sebertia acuminata	Ni	25% by wt dried sap	Jaffre et al. (1976)

Possibly, the best-known metal hyperaccumulator is *Thlaspi caerulescens* (alpine penny-cress). While most plants show toxicity symptoms at Zn accumulation of about 100 ppm, *T. caerulescens* was shown to accumulate up to 26,000 ppm without showing any injury (Brown et al. 1995b). It is possible that hyperaccumulator plants have a higher require-ment for metals such as Zn than nonaccumulator species. In support of this, many hyperaccumulators, including *T. caerulescens,* have been shown to colonize metal-rich soils such as calamine soil (soil enriched in Pb, Zn, and Cd). Because of this ability, considerable efforts have been directed to identify hyperaccumulator plants endemic to metal-rich soils (Baker and Proctor 1990).

Table 14.4 Plants used in phytoremediation (the list is not exhaustive; Glass 1999; Palmer et al. 2001)

Plant name	Function in phytoremediation
Alyssum	Nickel accumulator
Amaranthus retroflexus	Accumulator of ^{137}Cs
Armoracia rustica	Hairy root cultures remove heavy metals
Armeria maritima	Lead accumulator
Atriplex prostrata	Removes salt from soil
Azolla pinnata	Accumulator of lead, copper, cadmium and iron
Brassica canola	Remediates ^{137}Cs-contaminated soil
B. juncea	Hyperaccumulator of metals
Cannabis sativa	Hyperaccumulator of metals
Cardamonopsis hallerii	Hyperaccumulator of metals
Ceratophyllum demersum	Metal accumulator
Datura innoxia	Barium accumulator
Eucalyptus sp.	Removes sodium and arsenic
Eichhornia crassipes	Accumulator of lead, copper, cadmium and iron
Helianthus annus	Accumulator of lead and uranium; removes ^{137}Cs and ^{90}Sr in hydroponic reactors
Hydrocotyle umbellata	Accumulator of lead, copper, cadmium and iron
Kochia scoparia	Removes ^{137}Cs and other radionuclides
Lemna minor	Accumulator of lead, copper, cadmium and iron
Phaseolus acutifolius	Accumulator of ^{137}Cs
Salix sp.	Phytoextraction of heavy metals, waste water and leachate
Tamarisk	Removes sodium and arsenic
Typha sp.	Selenium volatilization
Thlaspi sp.	Accumulator of zinc, cadmium and lead

Ecological studies have revealed the existence of specific plant communities, endemic floras, which have adapted on soils contaminated with elevated levels of Zn Cu, and Ni. Different ecotypes of the same species may occur in areas uncontaminated by metals. To plants endemic to metal-contaminated soils, metal tolerance is an indispensable property. In comparison, in related populations inhabiting uncontaminated areas, a continuous gradation between ecotypes with high and low tolerance usually occurs. Plants have evolved several effective mechanisms for tolerating high concentrations of metals in soil. In some species, tolerance is achieved by preventing toxic metal uptake into root cells. These plants, coined excluders, have little potential for metal extraction. Such an excluder is "Merlin", a commercial variety of red fescue (*Festuca rubra*), used to stabilize erosion-susceptible metal-contaminated soils. A second group of plants, accumulators, does not prevent metals from entering the root. Accumulator species have evolved specific mechanisms for detoxifying high metal levels accumulated in the cells. These mechanisms allow bioaccumulation of extremely high concentrations of metals. In addition, a third group of plants, termed indicators, shows poor control over metal uptake and transport processes.

Table 14.5 Selected field/laboratory investigations of phytoremediation (Wenzel et al. 1999)

Location	Application	Contaminants	Medium	Plant
Chevron Corp., USA	Phytoremediation	Arsenic	Sediment/soil, groundwater, wetlands	Lettuce, maple, Typha, water lily
USDA-ARS, MD, USA; Viridian Environmental, TX, USA; Environmental Consultancy, Sheffield, UK; OSU, OR, USA	Phytomining	Nickel	Soil	*Alyssum murale, A. corsicum*
North of France	Phytoremediation	Zinc, cadmium	Soil	*Cardaminopsis halleri*
Western part of USA	Photovolatilization	Selenium	Water	Wetland plant community
Univ. Georgia, GA, USA	Phytoremediaiation	Mercury-based media	Agar	Poplar
USDA-ARS, MD, USA	Phytoremediation	Zinc, cadmium	Soil	*Thlaspi caerulescens*
Stockholm, Sweden	Phytoextraction	Cadmium	Soil	*Salix viminalis, T. caerulescens, Alyssum murale*
Switzerland	Phytoextraction	Cadmium	Soil	Tobacco
Heidelberg, Germany	Phytoextraction	Cadmium, zinc	Soil	*Brassica juncea, B. napus*
California, USA	Photovolatilization	Selenium	Drainage water	*Salicornia*
Hungary (former galvanization plant)	Phytoextraction	Cadmium, copper, nickel, zinc, chromium	Soil	*Brassica juncea, B. napus, Cannabis sativa, Beta vulgaris, Sinapisalba, Amaranthus hypochondriacus, Raphanus sativus*
Globe, AZ, USA	Phytoremediation	Copper	Soil	*Prosopis, Eucalyptus*
Northeast Portugal	Phytoremediation	Nickel	Soil	*Quercus ilex*
Greenhouse pot expt.	Phytostabilization	Cadmium	Soil amended with lime	*Brassica juncea, Agropyron elongatum*
Stockholm, Sweden	Phytoremediation	Cadmium, copper, lead, zinc	Storm water ditch, sewage wetlands	*Phalaris rundinaceae, Typha latifolia, Scirpus sylvaticus*
Ukraine	Rhizofiltration	^{137}Cs, ^{90}Sr	Water	*Helianthus annus, Brassica juncea, Pisum sativum*
Savannah River, SC, USA	Phytoremediation	Nickel	Hydroponic bioreactor	Hybrid poplar cuttings DN5 and NM6 (*Populus nigra x P. deltoides*), *P. maximowiczii*

In these plants, the extent of metal accumulation reflects metal concentration in the rhizospheric soil. Indicator species have been used for mine prospecting to find new ore bodies (Raskin et al. 1994).

14.3 MECHANISMS OF METAL TRANSPORT AND DETOXIFICATION

Because of their charge, metal ions cannot move freely across the cellular membranes, which are lipophilic structures. Therefore, ion transport into cells must be mediated by membrane proteins with transport functions, generically known as transporters. Transmembrane transporters possess an extracellular binding domain to which the ions attach just before the transport, and a transmembrane structure that connects extracelluar and intracellular media. The binding domain is receptive only to specific ions and is responsible for transporter specificity. The transmembrane structure facilitates the transfer of bound ions from extracellular space through the hydrophobic environment of the membrane into the cell. These transporters are characterized by certain kinetic parameters, such as transport capacity (V_{max}) and affinity for ions (K_m). V_{max} measures the maximum rate of ion transport across the cellular membranes. K_m measures transporter affinity for a specific ion and represents the ion concentration in the external solution at which the transport rate equals $V_{max}/2$. A low K_m value, high affinity, indicates that high levels of ions are transported into the cells even at low external ion concentration. By studying these kinetic parameters, K_m and V_{max}, plant biologists can gain insights into the specificity and selectivity of the transport system.

It is important to note that of the total amount of ions associated with the root, only a part is absorbed into cells. A significant ion fraction is physically adsorbed at the extracellular negatively charged sites (COO-) of the root cell walls. The cell-wall-bound fraction cannot be translocated to the shoots and, therefore, cannot be removed by harvesting shoot biomass (phytoextraction). Thus, it is possible for a plant exhibiting significant metal accumulation into the root to express a limited capacity for phytoextraction. For example, many plants accumulate Pb in roots, but Pb translocation to shoots is very low. In support of this, Blaylock and Huang (1999) concluded that the limiting step for Pb phytoextraction is the long-distance translocation from roots to shoots.

Binding to the cell wall is not the only plant mechanism responsible for metal immobilization into roots and subsequent inhibition of ion translocation to the shoot. Metals can also be complexed and sequestered in cellular structures (e.g. vacuole) becoming unavailable for translocation to the shoot (Lasat et al. 1998a). In addition, some plants, coined excluders, possess specialized mechanisms to restrict metal uptake into roots. However, the concept of metal exclusion is not well understood (Peterson 1983).

Metal uptake and accumulation in plants involves five major steps: (1) a metal fraction is sorbed at the root surface, (2) bioavailable metal moves across the cellular membrane into root cells, (3) a fraction of the metal absorbed into roots is immobilized in the vacuole, (4) intracellular mobile metal crosses cellular membranes into root vascular tissue (xylem), and (5) metal is translocated from the root to aerial tissues (stems and leaves).

Uptake of metals into root cells, the point of entry into living tissues, is a step of major importance for the process of phytoextraction. However, for phytoextraction to occur, metals must also be transported from the root to the shoot. Movement of metal-containing sap from the root to the shoot, termed translocation, is primarily controlled by two processes: root pressure and leaf transpiration. Following translocation to leaves metals can be reabsorbed from the sap into leaf cells.

Although micronutrients such as Zn, Mn, Ni and Cu are essential for plant growth and development, high intracellular concentrations of these ions can be toxic. To deal with this potential stress, common nonaccumulator plants have evolved several mechanisms to control the homeostasis of intracellular ions. Such mechanisms include regulation of ion influx (stimulation of transporter activity at low intracellular ion supply, and inhibition at high concentrations) and extrusion of intracellular ions back into the external solution. Metal hyperaccumulator species, capable of taking up metals in the thousands of ppm, possess additional detoxification mechanisms. For example, research has shown that in *T. goesingense*, a Ni hyperacccumulator, high tolerance was due to Ni complexation by histidine that rendered the metal inactive (Krämer et al. 1996, 1997). Sequestration in the vacuole has been suggested to be responsible for Zn tolerance in the shoots of the Zn-hyperaccumulator *T. caerulescens* (Lasat et al. 1996, 1998a). Several mechanisms have been proposed to account for Zn inactivation in the vacuole including precipitation as Zn-phytate, and binding to low molecular weight organic acids (Salt et al. 1999). Complexation to low molecular weight organic compounds (<10 kDa) was also shown to play a role in tolerance to Ni. Cadmium, a potentially toxic metal, has been shown to accumulate in plants, where it is detoxified by binding to phytochelatins a family of thiol (SH)-rich peptides (Rauser 1999, 2000). Metallothioneins (MTs), identified in numerous animals and, more recently, in plants and bacteria (Kägi 1991), are also compounds (proteins) with heavy metal binding properties (Tomsett et al. 1992).

The hyperaccumulator *Thlaspi caerulescens* and the related nonaccumulator *T. arvense* differ in their transcriptional regulation of ZNT1, a member of the ZIP family of membrane transporters (TC 2.A.5.1.1; Saier 2000). When expressed in *S. cerevisiae*, ZNT1 is capable of conferring uptake of Cd^{2+} and Zn^{2+} (Lasat et al. 2000; Pence et al. 2000) and

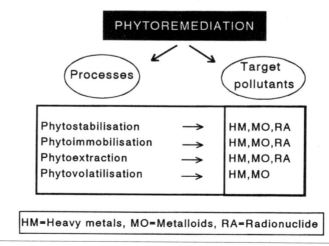

Fig. 14.9 Basic processes involved in the phytoremediation of metal, metalloids and radionuclides (see also Chap. 13)

Fig. 14.10 Uptake of metals from soil by movement up into plant shoots (aerial part) and retainment in the underground part (root system) or uniformly distributed in shoot and root systems

both expression of ZNT1 and root zinc uptake rates are elevated in *T. caerulescens*, when compared with *T. arvense*.

Zinc-mediated downregulation of ZNT1 transcript levels in the hyperaccumulator occurs at about 50-fold higher external metal concentrations compared with the nonhyperaccumulator. In several nickel hyperaccumulators, metal exposure elicits a large and dose-dependent increase in the concentrations of free histidine, which can act as a specific chelator able to detoxify Ni^{2+} and which enhances the rate of nickel translocation from the rooting medium into the xylem for transport into the shoot via the transpiration stream (Krämer et al. 1996).

In the shoots of hyperaccumulating plants, metal detoxification is achieved by both metal chelation and subcellular compartmentalization into the vacuole and the apoplast (Vázquez et al. 1992) and by sequestration within specific tissues, e.g. in the epidermis or in trichomes. The plant detoxification systems remain to be characterized at the molecular level. The generation and analysis of crosses between hyperaccumulators and related nonhyperaccumulators will be one key tool in identifying the genes responsible for the metal hyperaccumulator phenotype.

Based on a preliminary genetic analysis of a number of F2 progeny from crosses between the cadmium- and zinc-tolerant zinc hyperaccumulator *Arabidopsis halleri* ssp. *halleri* (L.) and the closely related, nontolerant, nonaccumulator *A. lyrata* ssp. *petraea* (L.), it was postulated that only a small number of major genes are involved in both zinc hyperaccumulation and zinc tolerance in *A. halleri*.

14.4 PHYTOEXTRACTION

Bioconcentration of heavy metals by plants is well established and today a large number of higher plants are known to exhibit different strategies when exposed to heavy metals. Plants growing in metal-contaminated and polluted terrestrial ecosystems take up toxic heavy metals and complex them in their parts, thereby helping in environmental decontamination. (Figs 14.9-14.10)

Phytoextraction is the use of plants to absorb contaminants from the soil into plant roots, in many cases eventually to be translocated and concentrated in shoots or other aboveground biomass (Nandakumar et al. 1995; Salt et al. 1995; Huang and Cunniungham 1996; Ebbs and Kochian 1997; Ebbs et al. 1997; Huang et al. 1997; Banuelos et al. 1999). Phytoextraction is used for the recovery of metals (Brooks 1998). This technology can also

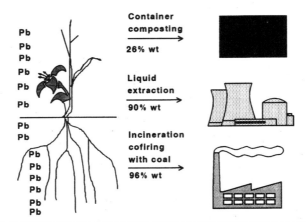

Fig. 14.11 Phytoextraction of lead—feasibility of commercialization. Cofiring with coal/incineration and composting are the two main methods of concentration methods. A subsample of harvested phytomass is burnt with coal under typical electrical power plant combustion conditions. The third method, liquid extraction, is a seperation method. Another fraction of phytomass is composted in containers and the third portion is subjected to liquid extraction using chelation agents. Hyperaccumulators of metals contain leachable metals that would recontaminate the environment thereby requiring some type of post-phytoremediation treatment. Cofiring with coal reduces the lead-contaminated mass by about 90 wt% by concentrating the lead into small fly ash particles. Container composting reduces the lead-contaminated material by about 26 wt%, while extraction using chelating agents removes over 90 wt% of the lead in two sequential batch extractions

be used for inorganic contaminants like selenium (Banuelos and Meeks 1990). In order for plant roots to take up metals from soil, the soil-bound metal must first be solubilized (i.e. freed from the organic or inorganic soil components to which they are bound). This can be accomplished by secretion of metal-chelating molecules or metal reductases by the plant roots (Raskin et al. 1994). Once solubilized, the metal ions may enter roots via extracellular or intracellular pathways, with the intracellular pathways generally requiring the presence of an ion channel or a metal transport protein in the plasma membrane of the root cells. It has been speculated that some of these channels may be nonspecific so that they can be used by different metals (Raskin et al. 1994). Once metal ions reach the root, they can either be stored in vacuoles in the root, often in chelated form, or they can be transported to the shoots: it has been reported that some hyperaccumulating species accumulate metals both in roots and shoots.

Phytoextraction of lead, arsenic, uranium and other minerals is being commercialized in the United States (Fig. 14.11). The EPA (US Environmental Protection Agency) estimates that there are more than 30,000 sites throughout the US that require environmental treatment. Heavy metals comprise a particularly difficult component of this problem, because many metal compounds resist chemical breakdown and because soil excavation and removal is expensive. Depending on site conditions and metal concentrations, solar-powered phytoremediation can cost as little as 5% of alternative treatment methods.

Sites for phytoremediation include thousands of government and private firing ranges, as well as industrial facilities used by primary and secondary metal manufacturers, scrap metal recyclers, paint manufacturers, battery recycling and production companies, chemical and petrochemical manufacturers, automobile manufacturers, utility companies, transportation companies, mining companies, and landfill operators.

A variety of different metals can be taken up by natural hyperaccumulators. The most readily bioavailable metals include Cd, Ni, Zn, As, Se and Cu, while more moderately bioavailable metals are Co, Mn and Fe. Pb, Cr and U are generally not considered very bioavailable, but all are capable of phytoremediation to some extent: Pb can be taken up by certain plants, but availablility is enhanced by the use of chelators; Cr and U are best remediated by rhizofiltration (Glass 1999).

A potentially emerging environmental application of plants is "phytomining" using hyperaccumulating plants to remove precious metals or other industrially important metals from naturally occurring locations where conventional mining would not be economically feasible. Recently, several companies have seriously discussed the commercial application of phytomining. The use of microorganisms (e.g. sulfur-oxidizing bacteria) has long been contemplated for this same purpose, and in fact "biomining" is used to a significant extent for mining of copper.

Several companies and research groups are pursuing phytomining strategies (Brooks 1998). It was demonstrated that *Brassica juncea* and other plant species accumulated as much as 20 ppm of gold in greenhouse pot experiments, when plants were supplemented with the solubilizing agent ammonium thiocyanate. It is reported that certain commercial companies are gaining economic benefits from phytoextraction, not only by recovering extracted metals from plant biomass, but also in using the biomass for energy generation (Glass 2000).

14.5 PHYTOSTABILIZATION

Phytostabilization relies on plants, or compounds they secrete, to stabilize low levels of contaminants that are present in soils (e.g. by absorption or precipitation), to prevent them from mobilizing or leaching in a manner that would endanger public health (Salt et al. 1995; Cunningham and Ow 1996). Possible mechanisms might include sequestering the contaminant in or on cell wall lignins ("lignification"), absorption of contaminants to soil humus, via plant or microbial enzymes ("humification"), or other mechanisms whereby the contaminant is sequestered in the soil, e.g. by binding to organic matter. Phytostablilization is primarily applicable to metal contamination, and might best be used near the end of remediation by traditional means, or after site closure, for those sites where it is acceptable under regulatory guidelines to leave a nonbioavailabe portion of the contaminant remaining in the soil (e.g. a risk-based regulatory approach). The term is also used to refer to the use of a vegetative cover, e.g. at a landfill, to prevent contaminants from leaching into groundwater or surface waters.

14.6 PHYTOVOLATILIZATION

This method is a specialized form of phytoextraction that can be used only for those contaminants that are highly volatile (Terry and Zayed 1998). Inorganic contaminants like mercury or selenium, once taken up by the plant roots, can be converted into nontoxic forms and volatilized into the atmosphere from the roots, shoots, or leaves. For examples, Se can be taken up by plants of the *Brassica* genus and other wetland plants, and converted (e.g. by methylation to the volatile dimethyl selenium) into nontoxic forms which are volatilized by the plants: field testing has shown this to be a potentially effective method (Hansen et al. 1998). A similar mechanism can be exploited for Hg, although there are no naturally occurring plants that can accomplish this: the goal here is to engineer bacterial genes for mercury reduction into plants, and here too laboratory experiments are highly encouraging (Rugh et al. 1996).

Chemical properties of a small number of pollutant trace elements—mercury and selenium—allow the use of phytovolatilization. Main difference: instead of accumulating inside the plant, the trace element is enzymatically transformed into a less toxic, volatile compound and is subsequently released into the atmosphere (Rugh et al. 1996; Pilon-Smits et al. 1999). Both phytovolatilization and phytoextraction can also serve to treat contaminated waters, in which the bioavailability of trace elements is generally higher than in soils. In this case, accumulation of the trace element in root biomass (rhizofiltration), followed by the removal of entire plants, is sufficient (Dushenkov et al. 1996, 1997a,b, 1999). Metal hyperaccumulator plants occur on metal-rich soils and accumulate metals in their aboveground tissues to concentrations between one and three orders of magnitude higher than surrounding "normal" plants grown at the same site (Baker and Brooks 1989). Hyperaccumulation has been confirmed for (1) cadmium (up to 0.2% Cd in shoot dry biomass), (2) cobalt (up to 1.2% Co in shoot dry biomass), (3) nickel (up to 3.8% Ni in shoot dry biomass), (4) zinc (up to 4% Zn in shoot dry biomass; Baker and Brooks 1989), (5) selenium (up to 0.4% Se in shoot dry biomass, and (6) metalloid arsenic (up to 0.75% As in shoot dry biomass; Ma et al. 2001).

Metal hyperaccumulators are highly attractive model organisms, because they have overcome major physiological bottlenecks limiting metal accumulation in shoots and metal tolerance. Chemical transformation of a trace element into a less toxic, volatile compound is a very effective strategy for detoxification, because the potentially harmful element is removed from the tissues. In mercury-contaminated soils and sediments, microbial activity results in the conversion of toxic Hg(II) into organomercurials, e.g. the highly toxic methylmercury (CH_3Hg^+). Mercury-resistant bacteria have been isolated which are able to transform organomercurials and Hg(II) into significantly less toxic elemental mercury. Methylmercury is converted to the less toxic Hg(II) by organomercurial lyase, encoded by the gene *merB*. A second enzyme, encoded by *merA*, catalyzes the reduction of Hg(II) to elemental mercury using NADPH as the electron donor. Under ambient conditions, elemental mercury enters the global biogeochemical cycle upon volatilization (Bizily et al. 1999).

Rugh et al. (1996) transferred the pathway of mercury volatilization into plants. The nucleotide sequence of a bacterial *merA* gene had to be modified in order to allow for high-level expression in plants. *Arabidopsis thaliana* expressing *merA* under the control of a constitutive cauliflower mosaic virus 35S promoter germinated and developed on agarose media containing 50 and 100 μM $HgCl_2$, concentrations that completely inhibited germination of wild-type seeds. The MerA plants showed a significantly higher tolerance to Hg^{2+} and volatilized Hg. Their tolerance to methylmercury was unchanged. They were also more tolerant to Au^{3+}. The MerB plants were significantly more tolerant to methylmercury and other organomercurials. They effectively converted highly toxic methylmercury to Hg^{2+}, which is 100 times less toxic to plants. To study the effects of both, the MerA and Merb plants were crossed and F1 generation was selfed. The F2-MerAMerB double transgenics showed the highest tolerance to organic mercury (10 mkM).

MerAMerB plants were shown to volatilize elemental mercury when supplied with organic mercury. Submicromolar concentrations of highly toxic organomercurials abolish germination of both wild-type and merA-expressing A. thaliana (Bizily et al. 1999).

The combined expression of *merA* and *merB* in a high biomass plant could be a promising step towards the generation of an improved mercury phytoremediator plant. Modified *merA* genes were then introduced into higher biomass plants. Tobacco transformants expressing a modified *merA* gene were able to develop and flower on soils containing up to 500 ppm Hg(II), but mercury removal from soil substrates has yet to be determined (Heaton et al. 1998).

It might be advantageous to use trees in phytovolatilization, because of their large root systems, long life span and extensive production of litter, which may serve to enhance metal availability in the soil (Rugh et al. 1996). Mercury volatilization was ten-fold higher in *merA*-expressing transgenic yellow poplar plantlets when compared with the wild type. The transgenics volatilized Hg(0) at an average rate of approximately 1 μg g^{-1} tissue day^{-1} when grown in agarose media containing 10 μM $HgCl_2$ (Rugh et al. 1996). It has been reported that some plants are able to volatilize selenium at low rates, either as dimethyl selenide or as dimethyl diselenide (Frankenberger and Engberg 1998). The enzymes involved have not been identified so far. Selenium volatilization involves prior conversion of selenate into organic forms of selenium. Volatilization of any other trace elements from plants appears to require the introduction of microbial genes.

14.7 RHIZOFILTRATION

Rhizofiltration can be defined as the use of plant roots to absorb, concentrate, and/or precipitate hazardous compounds, particularly heavy metals or radionuclides, from aqueous solutions (Salt et al. 1995; Dushenkov at al. 1996, 1997a,b, 1999). Experimental evidence showing nonlinear kinetics of disappearance of metals from solution suggests that several different mechanisms, of differing speeds, operate simultaneously. Surface absorption by the roots, the fastest and often the most prevalent mechanism, most likely depends on physicochemical processes (e.g. ion exchange, chelation) and can even take place on dead roots.

Rhizofiltration is the use of plant roots to accumulate metals from water. Hydroponically cultivated plants rapidly remove heavy metals from water and concentrate them in the roots and shoots. Harvested plants containing heavy metals can be disposed of or treated to recycle the metal. Today scientists have identified select species of plants demonstrating high biomass production and metal removal capacity for a wide variety of metals. Rhizofiltration has many of the benefits of other phytoextraction techniques, including low cost and minimal environmental disruption. A continuous flow system circulates the contaminated water through specially designed plant containment units. Periodically, older plants are harvested and replaced.

In its reliance on surface absorption as the primary mechanism for removing metals from wastestreams, rhizofiltration is related to the process known as biosorption, in which microbial, fungal or other biomass, living or dead, is used to absorb large quantities of materials such as heavy metals. In addition to surface absorption, other, slower mechanisms underlying rhizofiltration may also occur: these might include biological processes (intracellular uptake, deposition in vacuoles, and translocation to the shoot), or precipitation of the metal from solution by plant exudates (the slowest mechanism of the three).

There are several naturally occurring plant species with a natural propensity to accumulate high concentrations of metal. The desired species are those where root surface absorption is the primary, or fastest mechanism: in those plants where a substantial amount of shoot translocation occurs, more of the plant biomass becomes contaminated with the metal or radionuclide and must therefore be disposed as hazardous (or radioactive) waste, thus negatively affecting economics and efficiency. Rhizofiltration is believed to be effective (and perhaps most economically attractive) for dilute concentrations of contaminants in large volumes of water, and this feature may make it especially attractive for radionuclide contamination (Salt et al. 1995).

Rhizofiltration can be practiced in situ (e.g. on surface waters), but more likely would be operated in an aboveground artificial flow-through reactor, akin to hydroponics, and could thus be applicable to groundwater or industiral wastestreams. At the conclusion of the process, plant roots and any other tissue containing metals can be harvested for disposal or metal recovery: any uncontaminated shoots can, in some cases, be used to regenerate new roots. Another alternative rhizofiltration approach involves the use of unicellular eukaryotic algae such as *Chlamydomanas reinhardtii*. Certain of these species are tolerant to heavy metals and express low levels of phytochelatins, which are metal-binding polypeptides, and because they have high surface area to volume ratios may be suitable to take up large amounts of metals from aqueous solutions. One recent example of a research project along these lines is the work of Richard Sayre of Ohio State University, in which *C. reinhardtii* was transformed to express metallothionein, another metal-binding protein, and these engineered cells showed enhanced uptake of Cd in experimental solutions (Cai et al. 1999). Today there are several companies conducting research on blue-green algae for environmental decontamination or food-processing purposes.

Aquatic metal hyperaccumulators have also been known for some time. These include *Datura innoxia*, a Ba accumulator, and *Eichhornia crassipes* (water hyacinth), *Hydrocotyle umbellata* (pennywort), *Lemma minor* (duckweed) and *Azolla pinnata* (water velvet), which can take up Pb, Cu, Cd, Fe and Hg from aqueous solutions (Dushenkov et al. 1996; Raskin and Ensley 2000). Recently, investigators have turned to the common sunflower, *Helianthus annuus*, and *B. juncea* as accumulators of Pb, U and other metals from aqueous wastestreams.

14.8 CONSTRUCTED WETLANDS (REED BEDS)

Phytoremediation is a natural extension of what has historically been the most common use of plants to treat aqueous wastestreams: constructed wetlands, in which man-made ecosystems, including organic soils, microbial fauna, algae and vascular plants, work together to remove metals from the wastestreams. Sometimes known as reed beds or managed wetlands, these systems have been used to treat a variety of wastestreams, including municipal and industrial wastewater and metal-contaminated waters such as acid mine drainage. The use of such systems, sometimes called "land treatment", dates back several centuries, and was the major method of human waste treatment prior to the advent of artificial engineered systems.

Wetland systems have been used commercially for many years; however, because they rely on the combined efforts of the total ecosystem (i.e. consortia of plants in concert with microbes, algae, etc., as well as photochemical effects) rather than on the action of the plants themselves, it is generally not appropriate to consider use of wetlands to be phytoremediation per se.

Currently, commercial uses of wetlands and creation of specialized plant varieties for use in constructed wetlands are emerging as market opportunities for phytoremediation. A number of commercial enterprenuers have publicized their interest in developing such products. Vegetation caps and constructed wetlands can also be included in the overall classification of the phytoremediation approaches. However, results obtained so far with various phytoremediation approaches show that phytoextraction, rhizofiltration and phytostabilization methods hold more promise as successful commercial technologies (Albers and Camardese 1993; Otte et al. 1995; Kadlec and Knight 1996; Vymazal 1996; Azaizeh et al. 1997; Crites et al. 1997; Carbonell et al. 1998; Cole 1998; Hansen et al. 1998; Kadlec et al. 2000; Odum 2000).

Polluted water from acid mine drainage (AMD) that normally contains high levels of iron, aluminum and acid is a major contaminant of the mining industry, particularly coal mining. Acid mine drainage comes from pyrite or iron sulfide, a mineral associated with coal mining. Also of concern is acid rock drainage (ARD), the term used to describe leachate, seepage or drainage that has been affected by the natural oxidation of sulfur minerals contained in rock which is exposed to air and water. The constituents of AMD wastestreams will depend on the type of metal being mined, with the most common being Cd, Cu, Pb, and Zn. AMD is regulated in the US by the EPA under the Clean Water Act, and discharges must be treated to reduce metal concentrations to specified levels before the water can be discharged into a public waterway.

Constructed wetlands, consisting of a variety of aquatic plants, have long been used to treat AMD. Such wetlands are often constructed over a layer of limestone to provide neutralizing capacity as well. Influent waters, with high metal concentrations and low pH, flow through aerobic and anaerobic zones of the wetland. The metals are removed by the combined actions of filtration, ion exchange, adsorption, and precipitation, although microbial sulfate reduction/precipitation may be the most prevalent mechanism. Initially, the most commonly used plant was *Sphagnum*, which helps bioaccumulate the metals, but today other plants such as free-floating macrophytes and rooted emergent aquatic macro-phytes (e.g. cattails) are used more often because they can better withstand extremes in pH and metal concentration and are easier to grow and transport. Contaminants that are treated using wetlands include Fe, Mn, Cu, Ni and Al, and efficiencies of greater than 90% are achievable (although subsequent treatment by chemical precipitation may still be necessary to meet regulatory endpoints).

There are at least 200-300 constructed wetlands for AMD treatment in the Appalachain region of the US alone, and are common in Canada for this purpose as well. Each 15 m^2 of constructed wetlands can treat 1 L/min flow of contaminated water. Typical flows at AMD sites can range from 100 gallons/min to several thousands of gallons/min. Al, Cd, Cr, Cu and Zn were all reduced by 98% or more, Fe by 99%, Pb by 94% and Ni by 84%, although Mn removal was only 9-44%.

Constructed wetlands with reed beds and floating-plant systems have been common for the treatment of various types of wastewaters for many years. This strategy is currently gaining importance globally and expanding to address contaminated/polluted soils and water bodies. Bioconcentration of trace metals by aquatic macrophytes is of special concern to human welfare and for environmental protection and conservation (Kadlec et al. 2000).

In wetland ecosystems a wide variety of processes ranging from physicochemical and biological operate which can provide a suitable situation for removal of metals. For example, in the case of acidic metal rich mine drainage, the principal processes include oxidation of dissolved metal ions and subsequent precipitation of metal hydroxides, bacterial reduction of sulfate and precipitation of metal sulfides, the co-precipitation of metals with iron hydroxides, the adsorption of metals onto precipitated hydroxides, the adsorption of metals onto organic or clay substrates, and finally metal uptake by growing macrophytes.

Natural wetland ecosystems are inherently complex. Hence, for the purpose of treatment of metal-contaminated waters, it is advantageous to construct separate tanks within the treatment system with each tank designated to perform a particular function maximally (occasionally more than one would be beneficial). The design of wetlands constructed for the treatment of metal-contaminated waters attempts to identify and optimize the key processes which promote the removal of a specific targeted metal. Alternatively, this also includes suppression of potentially interfering and competing processes.

Treatment of wastewaters/natural waters containing a single metal such as iron can be achieved using a constructed wetland designated to optimize only one of the possible processes. For example, removal of iron involves precipitation of iron hydroxide in an aerobic environment. In contrast, if the water contains a mixture of metals, e.g. iron and zinc, in high concentrations, the constructed wetland may have to adapt different strategies like application of aerobic and anaerobic processes. An aerobic environment promotes the precipitation of aluminum and iron hydroxides and co-precipitation of arsenic. An anaerobic situation promotes the reduction of sulfates and the consequent precipitation of sulfides primarily for copper, cadmium and zinc.

The precipitation of hydroxides is regulated by pH and the availability of oxygen that can be ensured by (1) construction of shallow wetlands with a maximal depth of about 3 m water, (2) organic detritus to be minimized as it demands oxygen for decomposition; it is preferable to use a large inorganic substrate, (3) designing the landscape into ridges and gullies to ensure continual mixing of the water within the system so as to prevent stratification of water into oxygen-rich and oxygen-depleted zones, (4) incorporating cascades at the point of influent to promote oxygenation of air, and (5) utilizing reed beds comprising *Phragmites australis* (common reed), *Typha latifolia* (cattail), etc., which have the ability to transfer oxygen to the root zone.

Revegetation of mine tailings is a challenging task in view of the fact that the metal mine tailings are usually very poor in nutrients, rich toxic metal content and have low capacity to retain water. Furthermore, wind erosion of mine tailings poses a serious environmental problem. All these problems could be averted if tailings were revegetated with wetlands using metal-tolerant wetland plants. Wetlands have been used and constructed for the treatment of metal-contaminated water in recent years indicating that wetland plants that hyperaccumulate metals can tolerate elevated metal concentrations.

Glyceria fluitans (floating sweetgrass) is an amphibious plant that was found growing in the tailings pond of an abandoned lead/zinc mine in Glendalough, Co. Wicklow, Ireland. Greenhouse experiments demonstrated that *G. fluitans* could grow in sand culture treated with high concentration zinc sulfate solution. Further research confirmed that two populations of *G. fluitans*, one from a metal-contaminated and the other from a noncontaminated site, could be grown successfully on mine tailings with a high zinc content. *G. fluitans* and two other wetland plants, *Phragmites australis* and *Typha latifolia*, have since been grown on both alkaline and acidic zinc mine tailings in field conditions under fertilized and nonfertilized conditions. Research findings obtained so far indicate that *G. fluitans* can be easily established on zinc mine tailings. It appears also to have a very low nutrient requirement thus keeping fertilizer costs to a minimum during rehabilitation of mine tailings.

Wetlands have been constructed in Ireland for the passive treatment of tailings water originating from a lead/zinc mine. Water originating from mine tailings (slags) is often characterized by high metal and sulfate concentrations compared with background levels. Conventional methodology of tailing water treatment involves chemical treatment, which is a costly procedure requiring intensive chemical and labour input.

Therefore, more recently, both constructed and natural wetlands have been utilized for metal removal and wastewater quality control. Wetlands with their diversified macrophytes are known to retain substances such as metals from water passing through them. Aquatic macrophytes encompassing many common weeds enable cost-effective treatment and remediation technologies for wastewaters contaminated with inorganics and organics. Acid mine drainage (AMD) significantly impairs the quality of aquatic ecosytems. The exposure and oxidation of iron sulfide from coal mining results in AMD. AMD is a serious water pollution problem in the Appalachian region of the US. In the US, estimates by the US Bureau of Mines indicate that about 20000 km of streams and rivers are impaired by AMD. Therefore, in recent years, several low-cost preventive and passive technologies have been developed that utilize biological and natural chemical processes to remediate the contaminated mine waters without chemical treatment. In this regard, constructed wetlands and assemblage, anoxic limestone drains (ALD) and successive alkalinity-producing systems (SAPS) have potential, and promising results have been obtained in several instances.

14.9 CHELATE-ENHANCED PHYTOREMEDIATION

The chemistry of metal interactions with the soil matrix is central to the phytoremediation concept. In general, sorption to soil particles reduces the activity of metals in the system. Thus, the higher the cation-exchange capacity (CEC) of the soil, the greater the sorption and immobilization of the metals. In acidic soils, metal desorption from soil-binding sites into solution is stimulated due to H^+ competition for binding sites. Soil pH affects not

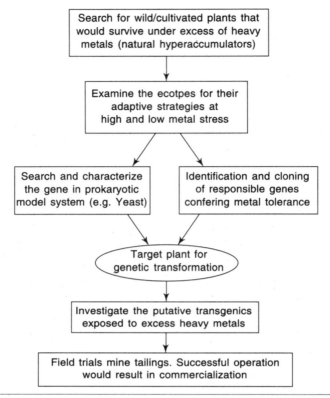

Fig. 14.12 Scheme for genetic engineering and production of transgenics having tolerance and metal accumulation ability for use in phytoremediation

only metal bioavailabilty, but also the very process of metal uptake into roots. This effect appears to be metal-specific. For example, in *T. caerulescens*, Zn uptake in roots showed a small dependence on pH, whereas uptake of Mn and Cd was more dependent on soil acidity (Brown et al. 1995a).

Use of soil amendments such as synthetic (ammonium thiocyanate) and natural zeolites has yielded promising results (Krämer et al. 1996; Blaylock et al. 1997; Huang et al. 1997; Churchmann et al. 1999; Zorpas et al. 1999). EDTA, NTA, citrate, oxalate, malate, etc., have been used as chelators for rapid mobility and uptake of metals from contaminated soils by plants. Use of synthetic chelators significantly increased Pb and Cd uptake and translocation from roots to shoots facilitating phytoextraction of the metals from low-grade ores (Anderson et al. 1998; Table 14.6).

Synthetic cross-linked polyacrylates, hydrogels, have protected plant roots from heavy metal toxicity and prevented the entry of toxic metals into roots. Application of synthetic and natural zeolites on a large scale may not be a practical solution due to exorbitant costs (Churchmann et al. 1999; Zorpas et al. 1999). However, low-cost chelators are applied to the soil surface through irrigation at specific stages of plant growth (See Chap. 13 for additional information).

Table 14.6 Chelator-facilitated phytoremediation of metals: regulatory processes (Huang et al. 1997; Laperche et al. 1997; Vassil et al. 1998; Churchmann et al. 1999; Zorpas et al. 1999; Matsumoto 2002)

Bioaccumulation is limited by:
— Plant growth rate
— Rate of uptake
Biostabilization
— Valency transformation
— Cr^{6+} (highly soluble) $\Leftrightarrow$ Cr^{3+} (highly insoluble)
— Extracellular precipitation
Phosphonation of lead (Pb) into e.g. 'chloropyromorphite'
Volatilization
$Hg^{2+} \Leftrightarrow Hg^0$
Rate of uptake and bioavailability
— Bioavailability is influenced by:
Contaminant mobility
— Species and/or genus of organism
Mobility is dependent
— particulate $\Leftrightarrow$ dissolved equilibrium
$$p \Leftrightarrow d$$
and this p $\Leftrightarrow$ d relationship is affected by
— pH
— Alkalinity
— Eh (redox potential of the soil)
— Type of soil - sandy-loam, brown-earth, podzol, clay
— OM concentration especially DOCs such as HA and FAs
OM: Organic molten; DOC: dissolved organic carbon; HA: humic acid; FA: Fulvic acid
— The salinity, hardness and ionic strength of the surface soil water
— Microorganism interactions
— Complexation
What is a complex?
Complex=Metal+Ligand
— $M2^+ + L \Leftrightarrow ML$
— $M2^+ + 2L \Leftrightarrow ML_2$
What is a chelate?
— A ligand that can bond with the central atom at two or more points
— e.g. EDTA, citrate
Classification of complexes
Inorganic (e.g. Cl^-, NH_3)
Organic
— large e.g. HA, FAs   *(Contd.)*

Table 14.6 (*Contd.*)

— small e.g. RCOO⁻, ROH

Inert exhibit strong ligand-central atom bonds

Labile exhibit weak ligand-central atom bonds

Influence of chelates on bioavailability

    Hamper bioavailability

    ⇓ reactivity and uptake.

    ⇓ lability of complex and ⇓ rate of metals release for uptake by organism

    ⇑ physical size of the complex=⇓ speed of transport across membranes

    Assist bioavailability

    Desorb contaminants from soil particles

    Alters chemical/physical properties=⇑ bioavailability

    Alters size, HA vs small organics

    Dissolution of precipitates, e.g. Fe and Mn oxides

Types of chelates used in phytoremediation

    EDTA (ethylenediaminetetraacetic acid)

      Indian mustard plant

      $[Pb]_{shoot} \propto [EDTA]_{soil}$

      1.5% Pb in shoots.

    HEDTA (n-hydroxy-EDTA)

    Alyssum, hyperaccumulator of Ni, naturally produce

      Citrate

      Malate

      Histidine

      Ni (histidine, malate) bi-ligand complex

Chelate-enhanced phytoremediation: associated problems

    Isolation of nontargeted contaminants

    Compartmentalization of toxic substances in plants

    Application of chelate solution transport of chelated metals

      — to deeper subsoils

      — too quickly into root zone ⇒ lethal exposure

Metal chelators produced by plants, viz. organic acids and amino acids, phytin, metallothioneins (class II and III), metal-sequestering proteins/peptides, and cellular complexation processes, and understanding their physiology and biochemistry, would help in achieving goals in this field (Cobbet and Goldsbrough 1999; Cobbet 2000a,b; Prasad and Strzalka 2002).

For some toxic metals such as Pb, a major factor limiting the potential for phytoextraction is limited solubility and bioavailability for uptake into roots. One way to induce Pb solubility is to decrease soil pH (McBride 1994). Following soil acidification, however, mobilized Pb can leach rapidly below the root zone. In addition, soluble ionic lead has little propensity for uptake into roots. The use of specific chemicals, synthetic chelates, has been shown to dramatically stimulate the potential for Pb accumulation in plants. These compounds prevent Pb precipitation and keep the metal as soluble chelate-Pb complexes available for uptake into roots and transport within plants. For example, addition of EDTA (ethylenediaminetetraacetic acid) at a rate of 10 mmol/kg soil stimulated Pb accumulation in shoots of maize up to 1.6% (Blaylock et al. 1997). In a subsequent study, Indian mustard exposed to Pb and EDTA in hydroponic solution was able to accumulate more than 1% Pb in dry shoots (Vassil et al. 1998). Another synthetic chelator, HEDTA (hydroxyethylethylenediaminetriacetic acid), applied at 2.0 g/kg soil contaminated with 2500 ppm Pb, increased Pb accumulation in shoots of Indian mustard from 40 to 10,600 ppm (Huang and Cunningham 1996). Accumulation of elevated Pb levels is highly toxic and can cause plant death. Because of the toxic effects, it is recommended that chelates be applied only after a maximum amount of plant biomass has been produced. Prompt harvesting (within 1 week of treatment) is required to minimize the loss of Pb-laden shoots. Blaylock et al. (1997) indicated that, in addition to Pb, chelate-assisted phytoextraction is applicable to other metals. These authors indicated that application of EDTA also stimulated Cd, Cu, Ni, and Zn phytoaccumulation. It has been shown that chelate ability to facilitate phytoextraction is directly related to its affinity for metals. For example, EGTA [ethylenebis(oxyethylenetrinitrilo)tetraacetic acid] has a high affinity for Cd^{2+}, but does not bind Zn^{2+}. EDTA, HEDTA, and DTPA (diethylenetriaminepentaacetic acid) are selective for Zn. In fact, zinc binding by DTPA is so strong that plants cannot use Zn from this complex and potentially suffer from Zn deficiency.

14.10 IMPROVING PHYTOREMEDIATING PLANTS AND PRODUCTION OF TRANSGENICS THAT HYPERACCUMULATE METALS

It has been suggested that phytoremediation would rapidly become commercially available if metal-removal properties of hyperaccumulator plants, such as *T. caerulescens*, could be transferred to high-biomass producing species, such as Indian mustard (*Brassica juncea*) or maize (*Zea mays*; Brown et al. 1995b). Biotechnology has already been successfully employed to manipulate metal uptake and tolerance properties in several species (Fig. 14.12). For example, in tobacco (*Nicotiana tabacum*), increased metal tolerance has been obtained by expressing the mammalian metallothionein, metal-binding proteins, and genes (Lefebvre et al. 1987; Maiti et al. 1991). Possibly, the most spectacular application of biotechnology for environmental restoration has been the bioengineering of plants capable of volatilizing mercury from soil contaminated with methylmercury. Methylmercury, a strong neurotoxic agent, is biosynthesized in Hg-contaminated soils. To detoxify this toxin, transgenic plants (*Arabidopsis* and tobacco) were engineered to express bacterial genes *merB* and *merA*. In these modified plants, *merB* catalyzes the protonolysis of the carbon-mercury bond with the generation of Hg^{2+}, a less mobile mercury species. Subsequently, *merA* converts Hg(II) to Hg(0), a less toxic, volatile element which is released into the atmosphere (Rugh et al. 1996; Heaton et al. 1998). Although regulatory concerns restrict the use of plants modified with *merA* and *merB*, this research illustrates the tremendous potential of biotechnology for environmental restoration. In an effort to address regulatory concerns related to phytovolatilization of mercury, Bizily et al. (1999) demonstrated that plants engineered to express *MerBpe* (an organomercurial lyase under the control of a plant promoter) may be used to degrade methylmercury and subsequently remove ionic mercury via extraction. Despite recent advances in biotechnology, little is known about the genetics of metal hyperaccumulation in plants. In particular, the heredity of relevant plant mechanisms, such as metal transport and storage (Lasat et al. 2000) and metal tolerance (Ortiz et al. 1995), must be better understood. Recently, Chaney et al. (1999) proposed the use of traditional breeding approaches for improving metal hyperaccumulator species and possibly incorporating significant traits such as metal tolerance and uptake characteristics into high-biomass-producing plants. Partial success has been reported in the literature. For example, in an effort to correct for the small size of hyperaccumulator plants, Brewer et al. (1997) generated somatic hybrids between *T. caerulescens* (a Zn hyperaccumulator) and *Brassica napus* (canola) followed by hybrid selection for Zn tolerance. High biomass hybrids with superior Zn tolerance were recovered. These authors have also advocated a coordinated effort to collect and preserve germplasm of accumulator species.

Initially, phytoremediation trials were performed using plants known to accumulate metals and/or to possess metal tolerance—*Silene vulgaris* (Moench) Garcke L., Brassicaceae plants *Brassica oleracea* and *Raphanus sativus* and metal hyperaccumulators like *Thlaspi caerulescens* and *Alyssum* L. spp. (Brown et al. 1994).

Metal hyperaccumulators were most efficient at metal removal in these field trials. In order to clean up a moderately contaminated soil, 6 and 130 croppings would be needed for zinc and cadmium, respectively. In pot trials, a low rate of biomass production, common to most hyperaccumulators, was shown to limit zinc removal from a contaminated soil by *T. caerulescens*, whereas high biomass nonaccumulator Brassica crops were more effective (Ebbs et al. 1997). For phytoextraction to become a viable technology, dramatic improvements would be required in either hyperaccumulator biomass yield or nonaccumulator metal accumulation (Chaney et al. 1999). Plants able to tolerate and accumulate several metals are required: polluted soils often contain high levels of several contaminant trace elements. Soils polluted with arsenic, cadmium, lead or mercury provide major targets for remediation. To date, no plants have been identified which reproducibly hyperaccumulate lead or mercury. In most naturally occurring tolerant plants studied to date, tolerance to arsenic or lead appears to be based on exclusion from the plant (Macnair et al. 1992; Stomp et al. 1994; Raskin 1996; Heaton et al. 1998; Cai et al. 1999; Arisi et al. 2000).

The development of a phytoremediation technology for some trace elements is thus likely to require the transfer of genes into plants across species borders. Although we know little about the molecular basis of trace element detoxification and hyperaccumulation in plants, a number of trace element detoxification systems from bacteria and yeast have been characterized genetically and functionally at the molecular level—for the detoxification of metals (Raskin and Ensley 2000).

Despite the difficulty in predicting the effects of microbial genes in a complex multicellular organism like a plant, the successful introduction of a modified bacterial mercuric ion-reductase gene into yellow poplar (*Liriodendron tulipifera*) and *Nicotiana tabacum* demonstrates that bacterial genes may be extremely valuable in phytoremediation (Heaton et al. 1998; Rugh et al. 1996, 1998). There are a number of processes that limit the performance of plants in phytoremediation.

These are:

1. the availability of contaminant trace elements in the soil for uptake by plant roots,
2. the rate of uptake of contaminants into plant roots,
3. the extent of tolerance or the rate of chemical transformation into less toxic, possibly volatile compounds, and
4. the translocation of trace elements from roots into shoots.

Several transporters implicated in the uptake of divalent nutrient cations like Ca^{2+}, Fe^{2+} or Zn^{2+} appear to be able to transport other divalent cations. For example, heterologous expression in yeast of IRT1 from *A. thaliana*, an iron-repressed transporter in the ZIP family of metal transporters, suggests a broad-range specificity of transport for Cd^{2+}, Fe^{2+}, Mn^{2+}, Zn^{2+} and possibly other divalent cations (TC 2.A.5.1).

The expression of IRT1 is strongly induced in plants under conditions of iron deficiency and is repressed in iron-replete plants. This correlates well with the finding that cadmium uptake is enhanced in iron-deficient pea seedlings. However, these transporters are tightly regulated at both the transcriptional and post-transcriptional level, and there are no reports to date on plants engineered to overexpress transporters of the ZIP family. Tobacco plants engineered to contain increased amounts of NtCBP4 protein (TC 1.A.1.5.1), a putative cyclic-nucleotide and calmodulin-regulated cation channel in the plasma membrane, displayed an increased sensitivity to lead, a 1.5- to 2.0-fold shoot accumulation of lead and an increased nickel tolerance.

Yeast cells expressing the wheat LCT1 cDNA (TC 9.A.20.1.1), encoding a low-affinity cation transporter, were hypersensitive to Cd^{2+} and accumulated increased amounts of cadmium (Clemens et al. 1999). Plants overexpressing *AtNramp3* (TC 2.A.55), a member of the Nramp family of metal transporters, were hypersensitive to Cd^{2+}, but enhanced cadmium accumulation was not observed.

In phytoremediation, it has to be considered that a transporter capable of transporting a specific contaminant metal cation might primarily transport other competing cations, like Ca^{2+} or Zn^{2+}, under natural soil conditions, should the latter ions be present in large excess. Therefore, it is desirable to better understand what governs the specificity of membrane transporters, in order to generate mutated transporters with altered specificities.

Understanding the regulation of ZIP family members in *T. caerulescens* and analyzing Arabidopsis mutants with altered metal responses will also help to identify novel target genes and strategies for the generation of plants with enhanced metal uptake. Detoxification of trace elements in phytoremediation, tolerance to elevated concentrations of trace elements either in the rooting medium or in plant tissues is desirable. A number of transgenic approaches were aimed at reducing the concentration of the potential toxin, for example, a free metal cation, in the cytoplasm. Three major strategies followed are (1) chemical transformation and/or volatilization, (2) over-production of metal chelators or compounds which bind the trace element, and (3) efflux from the cytoplasm, e.g. through transport into the vacuole.

To date there are numerous examples that have been demonstrated to have the potential for phytoremediation; e.g. Pb, Ni, Zn, Al, Se, Au and As. Arazi et al. (1999) have described a tobacco plasma membrane calmodulin-binding transporter that confers Ni^{2+} tolerance and Pb^{2+} hypersensitivity. Transgenic plants with the ZAT coding sequence exhibited increased Zn resistance and accumulation in the roots at high Zn concentrations. Furthermore, ZIP genes that confer Zn uptake activities in yeast have also recently been described (Pence et al. 2000). Moffat (1999) reported the characterization of two Arabidopsis mutants that were resistant to high levels of aluminum. The genes have yet to be cloned but one of the mutants, on chromosome 1, secretes organic acids to bind Al in the soil before it enters the plant. The second mutant, mapped to chromosome 4, increased the flux of hydrogen outside the root, changing the pH, which transformed the Al^{3+} ions into aluminum hydroxides and aluminum precipitates. These forms are incapable of entering the plant via the roots. Rate-limiting steps in selenium assimilation and volatilization have been deduced in Indian mustard. TP sulfurylase was determined to be involved in selenate reduction and, when overexpressed in Indian mstard, conferred Se accumulation, tolerance and volatilization. The Arabidopsis transgenic plants with mer operon (see Sect. 14.9) have conferred tolerance to gold. Schmöger et al. (2000) demonstrated that phytochelatins can be coupled to arsenic. This has important implications for the glutathione, phytochelatin synthetase pathway transgenics.

14.11 CHEMICAL TRANSFORMATION

Selenate, the oxoanion of the element selenium, is taken up and metabolized by higher plants because of its chemical similarity to sulfate. Thus, growth on soils contaminated with selenate results in the formation of excess amounts of selenocysteine and selenomethionine. These are incorporated into proteins of sensitive plants instead of cysteine and methionine, rendering the affected proteins nonfunctional. In an approach to increase selenium assimilation by plants, the plastidic *A. thaliana* APS1 cDNA encoding ATP sulfurylase was expressed in *B. juncea* under the control of a 35S promoter (Pilon-Smits et al. 1999). Transgenic plants exhibited a slightly increased tolerance to selenate when compared with wild-type controls and accumulated approximately two-fold higher concentrations of selenium in their shoots.

Enhanced sulfur assimilation in these transgenic plants resulted in an increase in glutathione concentrations by approximately 100 and 30% in shoots and roots, respectively, suggesting that ATP sulfurylase might also be an interesting target for the phytoremediation of other metals, especially cadmium.

14.11.1 Metal Binding or Chelation as an Intrinsic Principle of Phytoremediation

Metallothioneins (MTs) are low molecular weight proteins with a high cysteine content and a high affinity for binding metal cations such as those of cadmium, copper and zinc. Metallothionein-like proteins have been implicated in metal homeostasis in cyanobacteria, yeasts, animals and plants. Metallothioneins from animal sources were introduced into plants in a transgenic approach, mainly to reduce metal accumulation in shoots by trapping the metal in the roots. Expression of a mammalian MT in *Nicotiana tabacum* L. under the control of a constitutive promoter was able to reduce the translocation of cadmium into the shoot. Following exposure to a low cadmium concentration (0.02 μM) in the rooting medium, leaf cadmium concentrations were 20% lower in the transgenics than in wild-type plants. However, under field conditions, a consistent difference between transgenic and control plants could not be observed in either leaf cadmium content or plant growth.

These results demonstrate that trace element uptake observed on non-soil substrates under glasshouse or growth chamber conditions may not be extrapolated to predict the performance of transgenic plants on soil substrates or under field conditions.

14.11.2 Overexpression of MTs as a Means to Increase Cadmium Tolerance

Plants overexpressing mammalian MTs were reported to be unaffected by concentrations of 100-200 μM cadmium, whereas growth of *N. tabacum* control plants was severely inhibited at external cadmium concentrations of 10 μM (Misra and Gedamu 1989).

Transformants of *Brassica oleracea* expressing the yeast metallothionein gene CUP1 tolerated 400 μM cadmium, whereas wild-type plants were unable to grow at concentrations above 25 μM cadmium in a hydroponic medium. Transformants grown at 50 μM cadmium accumulated 10-70% higher concentrations of cadmium in their upper leaves than did nontransformed plants grown at 25 μM cadmium. This indicates that the enhanced tolerance observed in the transgenic plants was unlikely to be a consequence of excluding cadmium from the leaves.

The concept of using plants to clean up contaminated environments is not new. About 300 years ago, plants were proposed for use in the treatment of wastewater. At the end of the nineteenth century, *Thlaspi caerulescens* and *Viola calaminaria* were the first plant species documented to accumulate high levels of metals in leaves. Members of the genus *Astragalus* were capable of accumulating selenium up to 0.6% in dry shoot biomass. Despite subsequent reports claiming hyperaccumulators, the existence of plants hyperaccumulating metals other than Cr, Ni, Mn, Se and Zn has been questioned and requires additional confirmation (Salt et al. 1995). The idea of using plants to extract metals from contaminated soil, and the first field trial on Zn and Cd phytoextraction, were conducted by Baker et al. (1994). In the last decade, extensive research has been conducted to investigate the biology of metal phytoextraction. Despite significant success, our understanding of the plant mechanisms that allow metal extraction is still emerging. In addition, relevant applied aspects, such as the effect of agronomic practices on metal removal by plants, are largely unknown. It is conceivable that maturation of phytoextraction into a commercial technology will ultimately depend on the elucidation of plant mechanisms and application of adequate agronomic practices. Natural occurrence of plant species capable of accumulating extraordinarily high metal levels makes the investigation of this process particularly interesting.

14.12 ORNAMENTALS AND VEGETABLE CROPS

14.12.1 *Canna* x *generalis*

Canna x *generalis* is an important ornamental cultivated in urban landscapes. Hydroponic cultures of this plant treated with lead for 1 month suggest that this plant is a suitable candidate for phytoextraction of lead as the plant produces appreciable quantities of biomass (Figs. 14.13 and 14.14; Trampczynska et al. 2001)

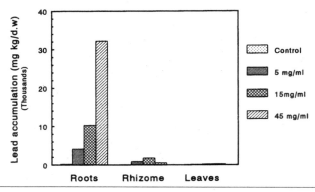

Fig. 14.13 Lead accumulation in *Canna* x *generalis*

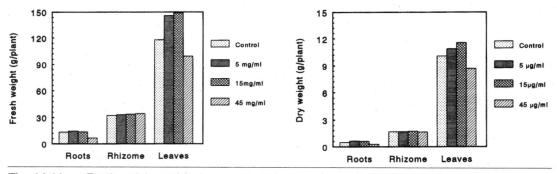

Fig. 14.14 **a** Fresh weight and **b** dry weight of *Canna* x *generalis* treated with lead

14.12.2 Scented Geranium

Scented geranium was identified as one of the most efficient metal hyperaccumulator plants. In a greenhouse study, young cuttings of scented geranium (*Pelargonium* sp. "Frensham"), grown in artificial soil and fed different metal solutions, were capable of taking up large amounts of three major heavy metal contaminants (e.g. Pb, Cd and Ni) in a relatively short time. These plants were capable of extracting from the feeding solution and stocking in their roots amounts of lead, cadmium and nickel equivalent to 9, 2.7 and 1.9% of their dry weight material, respectively. With an average root mass of 0.5-1.0 g in dry weight, scented geranium cuttings could extract 90 mg of Pb, 27 mg of Cd and 19 mg of Ni from the feeding solution in 14 days. If these rates of uptake could be maintained under field conditions, scented geranium should be able to clean up heavily contaminated sites in less than 10 years. For example, a phytoremediation lead cleanup program consisting of 16 successive croppings of scented geranium planted at a density of 100 plants m^{-2} over the summer could easily remove up to 72 g of lead m^{-2} year^{-1}. In our estimates, scented geranium would extract 1000-5000 kg of lead per ha^{-1} year^{-1}. Thus, these reported figures are close to Cunningham and Ow's (1996) estimations of metal removal rates of 200-1000 kg ha^{-1} year^{-1} for plants capable of accumulating 1.0-2.0% metal. Thus, if scented geranium is planted in soil where the lead contamination is 1000 mg kg^{-1} of soil, which is the acceptable limit for the province of Ontario (Canada), it could clean up the soil completely in 8 years. Scented geranium also has the ability to survive on soils containing one or more metal contaminants (either individually or in combination) and on soils contaminated with a mixture of metal and hydrocarbons (up to three metal-hydrocarbon contaminated soils, >3% total hydrocarbon in combination with several metal contaminants). Genetic transformation (improved transgenic plants) and soil amendments may accelerate phytoremediation and would be more realistic (Saxena et al. 1999).

14.12.3 Amaranthaceae

The dominant leaf vegetable producing species, viz. *Amaranthus spinosus, Alternanthera philoxeroides* and *A. sessiles*, grow on sewage sludge. The transfer factor (TF) for several metals was high. TF is calculated as: metal content in plant part (dry wt.)/metal content in soil (dry.wt). The concentration of metals is invariably high in leaf tissue. Thus, it is possible to use these species to restore the biosolid and sewage sludge contaminated sites. *Alternanthera philoxeroides* was used for removal of lead and mercury from polluted waters (Prasad 2001b, and references therein). It is also possible to supplement the dietary requirement of human food with Zn and Fe as these are essential nutrients and the plant species are edible. However, there is a need to monitor the metal transfer factor through the food chain.

14.13 PHYTOREMEDIATION OF RADIONUCLIDES

Radionuclides are deposited on the soil surface through mining, milling, and drilling for oil (Negri and Hinchman 2000). Natural weathering and chemical leaching can also release radionuclides into the soil (Dushenkov et al. 1998). Physical and biological nutrient cycles can distribute radionuclides throughout soil and water (Entry et al. 1997). Once radionuclides are deposited on the soil surface, they eventually are incorporated into the soil structure. Radionuclides can then absorb onto soil particles, bind with soil constituents, or remain in the soil solution. Soil constituents, such as organic matter and oxides, can temporarily bind radionuclides, allowing their release under specific environmental conditions (Negri and Hinchman 2000). The amount of bound radionuclides is usually linearly correlated with the amount of time that has passed since the radionuclide deposition. If radionuclides are adsorbed or bound in the soil, then they are not bioavailable to plants, microorganisms, or soil invertebrates. If radionuclides remain in the soil solution, though, then they are bioavailable to soil biota and plants. Radionuclide bioavailability mostly depends on the type of radionuclide deposition, the time of deposition, and the soil characteristics (Dushenkov et al. 1999). Radionuclides ^{137}Cs and plutonium tightly bind to soil particles, decreasing their bioavailability (Negri and Hinchman 2000).

Four common radionuclides are ^{137}Cs, ^{90}Sr (strontium), ^{234,235,238}U (uranium), and $^{238-241}$Pu (plutonium). Cesium-137 and ^{90}Sr come from fission by-products while Pu is released from nuclear weapon testing and nuclear fuel facilities (Negri and Hinchman 2000). Uranium is released from nuclear fuel cycles, but is the only one of the four radionuclides that occurs naturally (Negri and Hinchman 2000). If any of these radionuclides are in the soil solution and bioavailable, they can pose a risk to environmental and human health. Radionuclides cannot be degraded, so they have the potential to accumulate in plant species and increase in concentration as they make it higher in the food chain (Entry et al. 1997). For example, a plant that has accumulated ^{137}Cs is then eaten by a deer, which humans hunt for food. Humans then become exposed to the radionuclides as they ingest the contaminated food (Entry et al. 1997). Humans can also be directly exposed to the radionuclide through contact in soil or water. Some human health effects of radionuclide contamination are forms of cancer and genetic mutations (Entry et al. 1997).

Where radionuclide concentrations in soil warrant remediation, two traditional soil treatments are usually used, viz. soil excavation and soil washing. Unfortunately, both these processes are cost prohibitive. Therefore, the alternative method is phytoremediation. If radionuclide contamination occurs in water or soil, rhizofilteration and phytoextraction are feasible.

In order for phytoremediation of radionuclides to be successful, a few criteria have to be met. The most important is that the radionuclides be spread throughout a large area and present in low-level concentrations (Entry et al. 1997, 1999; Negri and Hinchman 2000). The radionuclides must be bioavailable in water or soil solution in order for plants to take them up into their roots. The plants themselves must also be tolerant of the radionuclides when they are accumulated into their biomass. The best plants for phytoremediation are those that have an extensive root system (rhizofiltration) and adequate aboveground biomass (phytoextraction). A plant's bioaccumulation coefficient factor (CF) determines if radionuclides remain in the roots or are translocated from roots to shoots. The CF is the concentration of radionuclides in dry plant tissue/concentration of radionuclides in solution. Uranium has a CF range of 0.01 to 0.0001 for roots, so it is accumulated in the shoots (Pilon-Smits and Pilon 2000). This is similar for Pu (Pilon-Smits and Pilon 2000). Cesium-137, though, has a shoot CF range of 0.01 to 1.0 and a root CF range of 38 to 165, which indicates that ^{137}Cs mostly remains in the roots (Negri and Hinchman 2000). Strontium-90 is taken up into the roots and is translocated to aboveground biomass depending on the plant species (Table 14.7).

Additions of soil amendments have been studied to determine their enhancement of radionuclide bioavailability in soil. Mycorrhizal fungi or other microbial root associations and chelating agents are two such amendments (Entry et al. 1997). Mycorrhizal fungi increase the surface area of plant roots, allowing roots to acquire more nutrients, water, and therefore more available radionuclides in soil solution. Entry et al. (1999) found that the mycorrhizal fungi *Glomus mosseae* and *G. intraradices* increase the aboveground biomass of bahia lovegrass (*Eragrostis bahiensis*), Johnsongrass (*Sorghum halepense*), and switchgrass (*Panicum virgatum*) while also increasing the accumulation of ^{137}Cs in a relatively short period of time. Synthetic chelating agents form complexes with bound radionuclides, releasing them into soil solution (Entry et al. 1997; Ensley 2000). The complexes prevent the radionuclides from binding to other soil constituents, making them more

Table 14.7 Chelating agents with associated plants that have enhanced the bioavailability and accumulation of radionuclides (Hossner et al. 1998)

Chelating Agents	Radionuclide	Bioavailable radionuclide	Plants	Reference
Nitric acid	^{137}Cs	20.7%		Dushenkov et al. 1999
Ammonium salt	^{137}Cs			Dushenkov et al. 1999
Ammonium nitrate	^{137}Cs	25%	Cabbage, tepary beans, Indian mustard, reed canary grass, *Amaranthus retroflexus* (redroot pigweed)	Negri and Hinchman 2000
Ammonium chloride	^{137}Cs		Cabbage, tepary beans, Indian mustard, reed canary grass *Amaranthus retroflexus* (redroot pigweed)	Negri and Hinchman 2000
Ammonium acetate	^{137}Cs			Dushenkov et al. 1999
Monovalent cations (K^+, NH_4^+, Rb^+, Na^+, and Cs^+)	^{137}Cs			Dushenkov et al. 1999
Citric acid. Agent concentration applied >10 mmol kg^{-1}	U		*Brassica juncea, B. chinensis, B. narinosa* plus *Amaranthus* spp.	Huang et al. 1998
Natural organic matter, dissolved organic ompounds	^{137}Cs, U, Pu, ^{90}Sr			Negri and Hinchman 2000
Microbial excudates	^{137}Cs, U, Pu, ^{90}Sr			Negri and Hinchman 2000
Sulfur	Pu			

bioavailable for plant uptake. Chelates have the greatest efficiency when they have high affinity for the radionuclides of interest (Salt et al. 1998). Table 14.7 lists chelating agents that have successfully increased radionuclide bioavailability in soil. It also lists the associated plant species that have accumulated radionuclides due to their increased bioavailability. In a greenhouse study, ammonium nitrate and ammonium chloride increased the uptake of ^{137}Cs into cabbage, tepary beans, Indian mustard (*Brassica juncea*), and reed canary grass, but, unfortunately, ^{137}Cs also decreased the plants' biomass production. In a field trial, though, ammonium yielded no visible effects on biomass production of Redroot pigweed (*Amaranthus retroflexus*; Negri and Hinchman 2000). Citric acid enhances uranium hyperaccumulation in plants, is biodegradable, and is easily obtained as an industrial by-product (Huang et al. 1998). When citric acid was applied to soil with *Brassica juncea, B. chinensis, B. narinosa* and *Amaranthus* species, it resulted in shoot U-concentrations that were a 1000 times greater than if these plants were not treated with citric acid (Huang et al. 1998; Table 14.8).

Water sources need less enhancement of radionuclide availability. Radionuclides have fewer constituents to bind to in water than in soil. Rhizofiltration is most effective when plant roots have large densities and radionuclides are available for plant uptake (Entry et al. 1999). Plant root densities can be enhanced by providing adequate nutrients and optimum environmental growth conditions. Two plants that have proven the most successful at rhizofiltration are sunflower (*Helianthus annuus*) and water hyacinth (*Eichornia crassippes*). Both have been found to accumulate significant percentages of radionuclides (^{137}Cs, U, ^{90}Sr) within a few hours to a few days (Negri and Hinchman 2000). A pond near the Chernobyl nuclear reactor was phytoremediated with sunflowers and their roots accumulated up to eight times more ^{137}Cs than timothy or foxtail, yielding a bioaccumulation coefficient of 4900-8600 (Negri and Hinchman 2000). At a contaminated wastewater site in Ashtabula, Ohio, 4-week-old sunflowers were able to remove more than 95% of uranium in 24 h (Dushenkov et al. 1996, 1997a,b, 1999). Water hyacinth grown in a water source with a pH of 9 accumulated high concentrations of ^{90}Sr with approximately 80-90% confined to the roots (Negri and Hinchman 2000). Table 14.8 lists extra examples of plant species, separated into trees, grasses and forbs, that have been found to accumulate radionuclides into their biomass. Even though the CF of ^{137}Cs suggests that these radionuclides mostly remain in the roots, *Amaranth* cultivars produce high aboveground biomass, which accumulates high concentrations of ^{137}Cs (up to 3000 Bq kg^{-1}; Dushenkov et al. 1998). Becqueral (Bq) is a measurement of decay per second. Maximum ^{137}Cs concentration is reached after 35 days of growth (Dushenkov et al. 1998). Legume and *Umbelliferae* family species accumulate the highest levels of ^{90}Sr (Negri and Hinchman 2000). Spiderworts (*Tradescantia bracteata*) have been found to act as biological indicators of radionuclides because their stamens are usually blue or blue-purple in color, but, when exposed to radionuclides, their stamens turn a pink color. Black spruce trees accumulate U in their twigs while oak and juniper trees accumulate U in their roots (Negri and Hinchman 2000). Pine seedlings of *Pinus radiata* and *P. ponderosa* accumulate 1.5-4.5% of ^{90}Sr in their shoots (Negri and Hinchman 2000).

Once plants have accumulated radionuclides to their maximum concentrations, the plants must be harvested to remove the radionuclides from the site. Plants used for rhizofiltration are completely removed from the water, since radionuclides collect in their

Table 14.8 Tree, grass, and forb species able to accumulate radionuclides

Tree species	Radionuclide	Grass/Forb species	Radionuclide	Reference
Red maple (Acer rubrum)	^{137}Cs, ^{238}Pu, ^{90}Sr	Tall fescue (Festuca arundinacea)	^{137}Cs	Entry et al. 1997
Liquidambar stryaciflua	^{137}Cs, ^{238}Pu, ^{90}Sr	F. rubra	^{137}Cs	Entry et al. 1997
Tulip tree (Liriodendron tulipifera)	^{137}Cs, ^{238}Pu, ^{90}Sr	Perennial ryegrass (Lolium perenne)	^{137}Cs	Entry et al. 1997
Coconut palm (Cocos nucifera)	^{137}Cs	White clover (Trifolium repens)	^{137}Cs	Entry et al. 1997
Montery pine (Pinus radiata)	^{137}Cs, ^{90}Sr	Chickweed (Cerastium fontanum)	^{137}Cs	Entry et al. 1997
Ponderosa pine (P. ponderosa)	^{137}Cs, ^{90}Sr	Bentgrass (Agrostis plant communities)	^{137}Cs	Entry et al. 1997
Forest redgum (Eucalyptus tereticornis)	^{137}Cs, ^{90}Sr	Redroot (Amaranthus retroflexus cv. belozernii, aureus, and Pt-95)	^{137}Cs	Entry et al. 1997; Dushenkov et al. 1999
Black spruce (Picea mariana)	U	Beet, quinoa, and Russian thistle (Chenopodiaceae)	^{137}Cs, ^{90}Sr	Negri and Hinchman 2000
Oak (Quercus)	U	Umbelliferae and Legume family (a)	^{90}Sr	Negri and Hinchman 2000
Juniper (Juniperus)	U	Spiderwort (Tradescantia bracteata)*		Negri and Hinchman 2000

*Biological indicator of radionuclides

roots. Plants used for phytoextraction have their shoots mowed and collected, leaving the roots to resprout more shoots. The harvested biomass can be either incinerated/combusted at high temperatures to oxidize the radionuclides into ash or composted to concentrate the radionuclides into smaller biomass before being disposed of in a radioactive-waste facility (Entry et al. 1997, 1999; Huang et al. 1998).

Soils with bioavailable concentrations of U and Pu, if low, can have citric acid incorporated into the soil to release U into soil solution and sulfate to release Pu. Switchgrass and Johnsongrass seeds, inoculated with *Glomus mosseae* mycorrhizal fungi, and *Brassica juncea* seeds would be broadcast onto the soil. Tree species, such as ponderosa pine and juniper or oak, could also be transplanted into the soil if radionuclides are present deep in the landfill. The plants' shoots and soil would be monitored on a monthly basis for radionuclide concentrations. After 2 months, spiderwort would be planted to serve as an indicator of radionuclides. The plants would be harvested and their shoots incinerated in the proper facilities to collect the radionuclides in ash. The ash would be disposed of in a hazardous-waste landfill. The trees would be completely removed after they had reached maximum radionuclide accumulation. They would also be incinerated and properly disposed of.

Phytoremediation could be commercially successful if adequate knowledge on mechanisms of metal transport is aquired. "Hyperaccumulators like *Thlaspi* are a marvellous model system for elucidating the fundamental mechanisms of—and ultimately the genes that control—metal hyperaccumulation. These plants possess genes that regulate the amount of metals taken up from the soil by roots and deposited at other locations within the plant. There are a number of sites in the plant that could be controlled by different genes contributing to the hyperaccumulation trait. These genes govern processes that can increase the solubility of metals in the soil surrounding the roots as well as the transport proteins that move metals into root cells. From there, the metals enter the plant's vascular system for further transport to other parts of the plant and are ultimately deposited in leaf cells.

Thlaspi accumulates metals in its shoots at astoundingly high levels. A typical plant may accumulate about 100 parts per million (ppm) zinc and 1 ppm cadmium. *Thlaspi* can accumulate up to 30,000 ppm zinc and 1500 ppm cadmium in its shoots, while exhibiting few or no toxicity symptoms". A normal plant can be poisoned with as little as 1000 ppm of zinc or 20-50 ppm of cadmium in its shoots. The research also suggests an approach for economically recovering these metals. Zinc and cadmium are metals that can be removed from contaminated soil by harvesting the plant's shoots and extracting the metals from them (Lasat et al. 1998a,b, 2000).

Soils contaminated with uranium can be remediated by adding organic acid citrate which increases the solubility of uranium and its bioavailability for plant uptake and translocation. Specific agronomic practices and plant species to remediate soils contaminated with radioactive cesium or cesium-137 have been identified (Lasat et al. 1997, 1998b). It was discovered that the ammonium ion was most effective in dissolving cesium-137 in soils. This treatment increased the availability of cesium-137 for root uptake and significantly stimulated radioactive cesium accumulation in plant shoots (Lasat et al. 1997, 1998b).

14.14 COMMERCIALIZATION OF PHYTOREMEDIATION

The potential global phytoremediation market figures have been projected by Glass (1999). Plants use photosynthetic energy to extract ions from the soil and to concentrate them in their biomass, according to nutritional requirements. When present at elevated levels, contaminants which are essential or nonessential trace elements are able to enter higher plants by virtue of their chemical similarity to (other) nutrient ions. AsO_4^{3-} or Cd^{2+} can enter roots through the uptake systems for PO_4^{3-} or Fe^{2+}/Ca^{2+}, respectively (Clemens 2001).

Phytoextraction is aimed at exploiting the nutrient acquisition system of plants in order to achieve maximum accumulation of heavy metals and other contaminants in the aboveground tissues. Aboveground biomass is then harvested, thereby removing the pollutant from the site in a small number of successive growth periods. Plant material can be ashed and possibly be recycled in metal smelting, or deposited in specialized dumps. When grown at a contaminated site a plant used in phytoextraction is required: to accumulate large amounts of one or several trace elements in the shoot, to exhibit a high rate of biomass production, and to develop an extensive root system.

A comprehensive analysis of phytoremediation markets was published by Glass (1999). The author indicated that the estimated 1999 phytoremediation market was twofold greater than 1998 estimates. This growth has been attributed to an increased number of companies offering services, particularly companies in the consulting engineering sector, and to growing acceptance of the technology. An estimate of 1999 US phytoremediation markets related to a variety of contaminated media and contaminants of concern is shown in Table 14.9.

Table 14.9 Estimated 1999 US phytoremediation markets for metals and radionuclides (Glass 1999)

Landfill leachate	$5-8 million
Metals in soil	$4.5-6 million
Inorganics in wastewater	$2-4 million
Inorganics in groundwater	$2-3 million
Metals in groundwater	$1-2 million
Radionuclides	$0.5-1 million
Metals in wastewater	$0.1-0.2 million

Estimated revenues for 1999 and 2000 were slightly lower than what had been previously projected, largely due to slower commercialization of the technology for the cleanup of metal- and radionuclide-contaminated sites (Glass 1999). The second largest market for phytoremediation was identified in Europe, although the European market was estimated to be 10-fold smaller than the US market (Glass 1999).

14.15 ADVANTAGES AND LIMITATIONS OF PHYTOREMEDIATION

Phytoremediation of metals and radionuclides has many advantages over the traditional treatments. First, in phytoremediation the soil is treated in situ, which does not cause further disruption to the soil dynamics. Secondly, once plants are established, they remain for consecutive harvests to continually remove the contaminants. In a single growing season, it is possible to grow and harvest multiple crops (Huang et al. 1998). The plants also provide soil nutrients and stabilization by reducing wind and water erosion. Thirdly, phytoremediation reduces the time workers are exposed to the radionuclides (Negri and Hinchman 2000). Finally, phytoremediation can be used as a long-term treatment that can provide an affordable way to restore radionuclide-contaminated areas (Dushenkov et al. 1999).

When the concept of phytoextraction was reintroduced (approximately two decades ago), engineering calculations suggested that a successful plant-based decontamination of even moderately contaminated soils would require crops able to concentrate metals in excess of 1-2%. Accumulation of such high levels of heavy metals is highly toxic and would certainly kill the common nonaccumulator plant. However, in hyperaccumulator species, such concentrations are attainable. Nevertheless, the extent of metal removal is ultimately limited by plant ability to extract and tolerate only a finite amount of metals. On a dry weight basis, this threshold is around 3% for Zn and Ni, and considerably less for more toxic metals, such as Cd and Pb. The other biological parameter that limits the potential for metal phytoextraction is biomass production. With highly productive species, the potential for biomass production is about 100 tons fresh weight/hectare. The values of these parameters limit the annual removal potential to a maximum of 400 kg metal/ha/year. It should be mentioned, however, that most metal hyperaccumulators are slow growing and produce little biomass. These characteristics severely limit the use of hyperaccumulator plants for environmental cleanup. Even though phytoremediation is a more ecologically and environmentally sound alternative, it does have some disadvantages.

1. Removal of radionuclides through plants can take more time than traditional treatments. For ^{137}Cs and ^{90}Sr removal, it can take from 5 to 20 years to reach full remediation (Entry et al. 1997, 1999).
2. The cost of phytoremediating radionuclides and their disposal has not been fully worked out (Negri and Hinchman 2000). Estimates only exist for phytoremediation of soils and water. Soils run about $25 to $100 per ton and water runs from $0.60 to $6 per 1000 gallons (Glass 2000).
3. Phytoremediation of radionuclides may pose risk factors to the surrounding biotic community. Therefore, these risks must be taken into consideration and can be reduced by harvesting plants before they set seed, selecting plants less palatable to grazers, and chosing plants that are wind pollinated rather than by insects (Entry et al. 1997). Finally, some plants that are successful at rhizofiltration in water sources cannot be applied to phytoextraction in soils. Corn, sunflower, and Indian mustard have about two to three times more ^{137}Cs accumulation in the roots as in the aboveground biomass (Dushenkov et al. 1999).

14.16 CONCLUSIONS

There are certain limitations to the use of bioresources and biodiversity in the upcoming technology of bioremediation. These need to be addressed in laboratory and field research carefully before implementing on the site of remediation. To a considerable extent these include: limited regulatory acceptance, long duration of time, potential contamination of the vegetation and food chain, and often extreme difficulty in establishing and maintaining vegetation on contaminated sites, e.g. mine tailings with high levels of residual metals. For metal contaminants, plants show potential for phytoextraction (uptake and recovery of contaminants into aboveground biomass), filtering metals from water onto root systems (rhizofiltration), or stabilizing waste sites by erosion control and evapotranspiration of large quantities of water (phytostabilization). After the plants have been allowed to grow for some time, they are harvested and either incinerated or composted to recycle the metals. This procedure may be repeated as necessary to bring soil contaminant levels down to allowable limits. If plants are incinerated, the ash must be disposed of in a hazardous-waste landfill.

Having considered and accepted the rationale of metal accumulation and degradation of organics by plants an obvious question that arises is: to what extent and what kind of pollutants can be cleaned up by plants in aquatic and terrestrial ecosystems? This question is not only of academic significance, but also the underlying scientific principles will have wider management implications for sustainable global development (Vangronsveld and Cunningham 1998; Glass 1999; Saxena et al. 1999; Wenzel et al. 1999; Kaltsikes 2000; Terry and Bañuelos 2000; Barcelo and Poschenrieder 2002). Plant-metal interactions in the rhizosphere, metal bioavailability for uptake by roots, effect of soil microorganisms and rhizosphere exudates on metal uptake (see Chap. 13 for details), and the development of appropriate agronomic practices for phytoextraction would lead to the success of phytoremediation.

Conventional remediation technologies (dredging, capping, ploughing, soil flushing, pneumatic fracturing, vitrification, electrokinetics, chemical conversions, soil/sediment washing and excavation, retrieval and off-site disposal) for soil and water (in situ and ex situ) are exorbitantly expensive and, hence, phytoremediation has tremendous potential not only to lockup the toxicants, but also for the recovery of useful components.

ACKNOWLEDGEMENTS

Thanks are due to funding agencies, viz. the Ministry of Environment and Forests, Department of Science and Technology, Council of Scientific and Industrial Research, New Delhi, and the Goverment of India, who supported trace metal research in the author's laboratory.

REFERENCES

Adler T (1996) Botanical cleanup crews. Sci News 150: 42-43

Albers PH, Camardese M (1993) Effects of acidification on metal accumulation by aquatic plants and invertebrates: wetlands, ponds, and small lakes. Environ Toxicol Chem 12: 969-976

Alcantara E, Barra R, Benlloch M, Ginhas A, Jorrin J, Lopez JA, Lora A, Ojeda MA, Pujadas A, Requejo R, Romera J, Sancho ED, Shilev S, Tena M (2000) Phytoremediation of a metal contaminated area in southern Spain. Intercost workshop, Sorrento, pp 121-123

Anderson CWN, Brooks RR, Stewart RB, Simcock R (1998) Harvesting a crop of gold in plants. Nature 395: 553-554

Anderson TA, Guthrie EA, Walton BT (1993) Bioremediation in the rhizosphere, Environ Sci Technol 27: 2530-2626

Arazi T, Sunkar T, Kaplan B (1999) A tobacco plasma membrane calmodulin-binding transporter confers Ni^{2+} tolerance and Pb^{2+} hypersensitivity in transgenic plants. Plant J 20: 171-182

Arisi ACM, Mocquot B, Lagriffoul A, Mench M, Foyer CH, Jouanin L (2000) Responses to cadmium in leaves of transformed poplars overexpressing γ-glutamylcysteine synthetase. Physiol Plant 109: 143-149

Azadpour A, Matthews JE (1996) Remediation of metal-contaminated sites using plants. Remed Summer 6(3): 1-19

Azaizeh HA, Gowthaman S, Terry N (1997) Microbial selenium volatilization in rhizosphere and bulk soils from a constructed wetland. J Environ Qual 26: 666-672

Bae W, Chen W, Mulchandani A, Mehra RK (2000) Enhanced bioaccumulation of heavy metals by bacterial cells displaying synthetic phytochelatins. Biotechnol Bioeng 70: 518-524

Baker AJM, Brooks RR (1989) Terrestrial higher plants which hyperaccumulate metal elements. A review of their distribution, ecology, and phytochemistry. Biorecovery 1: 81-126

Baker AJM, Proctor J (1990) The influence of cadmium, copper, lead, and zinc on the distribution and evolution of metallophytes in the British Isles. Plant Syst Evol 173: 91-108

Baker AJM, Walker PL (1990) Ecophysiology of metal uptake by tolerant plants. In: Shaw AJ (ed) Heavy metal tolerance in plants: evolutionary aspects. CRC Press, Boca Raton

Baker AJM, McGrath SP, Sidoli CMD, Reeves RD (1995) The potential for heavy metal decontamination. Mining Environ Manage 3(3): 12-14

Banuelos GS, Meek DW (1990) Accumulation of selenium in plants grown on selenium-treated soil. J Environ Qual 19(4): 772-777

Banuelos GS, Ajwa HA, Zambrzuski S (1997a) Selenium-induced growth reduction in *Brassica* land races considered for phytoremediation. J Ecotoxicol Environ Saf 36(3): 282

Banuelos GS, Ajwa HA, Mackey B, Wu LL, Cook C, Akohoue S, Zambrzuski S. (1997b) Evaluation of different plant species used for phytoremediation of high soil selenium. J Environ Qual 26: 639-646

Banuelos GS, Shannon MC, Ajwa H, Draper JH, Jordahl J, Licht L (1999) Phytoextraction and accumulation of boron and selenium by poplar (*Populus*) hybrid coles. Int J Phytochem 1: 81-96

Barcelo J, Poschenrieder C (2002) Fast root growth responses, root exudates, and internal detoxification as clues to the mechanisms of aluminium toxicity and resistance: a review. Environ Exp Bot 48: 75-92

Beath OA, Eppsom HF, Gilbert CS (1937) Selenium distribution in and seasonal variation of vegetation type occurring on seleniferous soils. J Am Pharm Assoc 26: 394-405

Bell JNB, Minski MJ, Grogan HA (1988) Plant uptake of radionuclides. Soil Use Manage 4(3): 76-84

Berstein EM (1992) Scientists using plants to clean up metals in contaminated soil. NY Times 141: C4

Bishop J (1997) Phytoremediation: a new technology gets ready to bloom. Environ Solutions 10(4): 29

Bizily SP, Rugh CL, Summers AO, Meagher RB (1999) Phytoremediation of methylmercury pollution: merB expression in *Arabidopsis thaliana* confers resistance to organimercurials. Proc Natl Acad Sci USA 96: 6808-6813

Blaylock MJ, Huang JW (1999) Phytoextraction of metals. In: Raskin I, Ensley BD (eds) Phytoremediation of toxic metals: using plants to clean up the environment. Wiley, New York, pp 53-70

Blaylock MJ, Salt DE, Dushenkov S, Zakharova O, Gussman C, Kapulnik Y, Ensley BD, Raskin I (1997) Enhanced accumulation of Pb in Indian mustard by soil-applied chelating agents. Environ Sci Technol 31: 860-865

Boyd RS, Martens SN (1994) Nickel hyperaccumulated by *Thlaspi montanum* var. *montanum* is acutely toxic to an insect herbivore. Oikos 70: 21-25

Boyd RS, Shaw JJ, Martens SN (1994) Nickel hyperaccumulation in *S. Polygaloids* (Brassicaceae) as a defense against pathogens. Am J Bot 81: 294-300

Brewer EP, Saunders JA, Angle JS, Chaney RL, McIntosh MS (1997) Somatic hybridization between heavy metal hyperaccumulating *Thlaspi caerulescens* and canola. Agron Abstr 1997:154

Briat JF, Lebrun M (1999) Plant responses to metal toxicity. CR Acad Sci Paris/Life Sci 322: 43-54

Brooks RR (1977) Copper and cobalt uptake be *Haumaniastrum* species. Plant Soil 48: 541-544

Brooks RR (1998) Plants that hyperaccumulate heavy metals. CAB International, Wallingford, UK

Brown KS (1995) The green clean: the emerging field of phytoremediation takes root. Bio Sci 45: 579-582

Brown SL, Chaney RL, Angle JS, Baker AJM (1994) Phytoremediation potential of *Thlaspi caerulescens* and bladder campion for zinc and cadmium contaminated soil. J Environ Qual 23: 1151-1157

Brown SL, Chaney RL, Angle JS, Baker AM (1995a) Zinc and cadmium uptake by hyperaccumulator *Thlaspi caerulescens* and metal tolerant *Silene vulgaris* grown on sludge amended soils. Environ Sci Technol 29: 1581-1585

Brown SL, Chaney RL, Angle JS, Baker AM (1995b) Zinc and cadmium uptake by hyperaccumulator *Thlaspi caerulescens* grown in nutrient solution. Soil Sci Am J 59: 125-133

Brown TA, Shrift A. (1982) Selenium: toxicity and tolerance in higher plants. Biol Rev Cambridge Philos Soc 57: 59-84

Cai XH, Bown C, Adhiya J, Traina SJ, Sayre RT (1999) Growth and heavy metal binding properties of transgenic *Chlamydomonas* expressing a foreign metallothionein. Int J Phytorem 1: 53-65

Carbonell AA, Aarabi MA, DeLaune RD, Gambrell RP, Patrick WH Jr (1998) Bioavailability and uptake of arsenic by wetland vegetation: effects on plant growth and nutrition. J Environ Sci Health A33: 45-66

Chaney RL, Malik M, Li YM, Brown SL, Brewer EP, Angle JS, Baker AJM (1997) Phytoremediation of soil metals. Curr Opin Biotechnol 8: 279

Chaney RL, Li YM, Angle JS, Baker AJM, Reeves RD, Brown SL, Homer FA, Malik M, Chin M (1999) Improving metal hyperaccumulator wild plants to develop commercial phytoextraction systems: approaches and progress. In: Terry N, Bañuelos GS (eds) Phytoremediation of contaminated soil and water. CRC Press, Boca Raton

Churchmann GJ, Slade PG, Rengasamy P, Peter P, Wright M, Naidu R (1999) Use of fine grained minerals to minimize the bioavaliability of metal contaminants. Environmental impacts of metals. Int Workshop, Tamil Nadu Agricultural University, Coimbatore, India, pp 49-52

Clemens S (2001) Molecular mechanisms of plant metal tolerance and homeostasis. Planta 212: 475-486

Clemens S, Kim EJ, Neumann D, Schroeder JI (1999) Tolerance to toxic metals by a gene family of phytochelatin synthases from plants and yeast. EMBO J 18: 3325-3333

Cobbet CS (2000a) Phytochelatins and their roles in heavy metal detoxification. Plant Physiol 123: 825-832

Cobbet CS (2000b) Phytochelatins biosynthesis and function in heavy-metal detoxification. Curr Opin Plant Biol 3: 211-216

Cobbett CS, Goldsbrough PB (1999) Mechanisms of metal resistance: phytochelatins and metallothioneins. In: Raskin I, Ensley BD (eds) Phytoremediation of toxic metals: using plants to clean up the environment. Wiley, New York, pp 247-269

Cole S (1998) The emergence of treatment wetlands. Environ Sci Technol 32(9): 218A-223A

Comis D (1996) Green remediation: using plants to clean the soil. J Soil Water Conserv 51: 184-187

Crites RW, Dombeck GD, Williams CR (1997) Removal of metals and ammonia in constructed wetlands. Water Environ Res 69: 132

Cunningham SD, Ow DW (1996) Promises and prospects of phytoremediation. Plant Physiol 110: 715-719

Cunningham SD, Berti WR, Huang JW (1995) Phytoremediation of contaminated soils. Trends Biotechnol 13: 393-397

Dodds-Smith ME, Payne CA, Gusek JJ (1995) Reedbeds at wheal jane. Mining Environ Manage 3(3): 22-24

Dushenkov V, Nanda Kumar, PBA, Motto H, Raskin I (1996) Rhizofiltration: the use of plants to remove heavy metals from aqueous streams. Environ Sci Technol 29: 1239-1245

Dushenkov S, Vasudev D, Kapulnik Y, Gleba D, Fleisher D, Ting KC, Ensley B (1997a) Removal of uranium from water using terrestrial plants. Environ Sci Technol 31: 3468-3474

Dushenkov S, Kapulnik Y, Blaylock M, Sorochisky B, Raskin I, Ensley B (1997b) Phytoremediation: a novel approach to an old problem. Stud Environ Sci 66: 563

Dushenkov S, Mikheev A, Prokhnevsky A, Ruchko M, Sorochinsky B (1998) Phytoremediation of radiocesium-contaminated soil in the vicinity of Chernobyl, Ukraine. Environ Sci Technol 33: 469-475

Dushenkov S, Mikheev A, Prokhnevsky A, Ruchko M, Sorochinsky B (1999) Phytoremediation of radiocesium-contaminated soil in the vicinity of Chernobyl, Ukraine. Environ Sci Technol 33: 469-475

Ebbs SD, Kochian LV (1997) Toxicity of zinc and copper to *Brassica* species: implications for phytoremediation. J Environ Qual 26: 776-781

Ebbs SD, Kochian LV (1998) Phytoextraction of zinc by oat (*Avena sativa*), barley (*Hordeum vulgare*), and Indian mustard (*Brassica juncea*). Environ Sci Technol 32: 802-806

Ebbs SD, Lasat MM, Brandy DJ, Cornish J, Gordon R, Kochian LV (1997) Heavy metals in the environment. Phytoextraction of cadmium and zinc from a contaminated soil. J Environ Qual 26: 1424-1430

Ensley BD (2000) Rationale for use of phytoremediation. In: Raskin I, Ensley BD (eds) Phytoremediation of toxic metals: using plants to clean up the environment. Wiley-Interscience, New York, chap 1

Entry JA, Watrud LS, Reeves M (1999) Accumulation of ^{137}Cs and ^{90}Sr from contaminated soil by three grass species inoculated with mycorrhizal fungi. Environ Pollut 104: 449-457

Entry JA, Watrud LS, Manasse RS, Vance NC (1997) Phytoremediation and reclamation of soils contaminated with radionuclides. Phytoremediation of soil and water contaminants, chap 22. American Chemical Society, New York

Frankenberger WR, Engberg RA (eds) (1998) Environmental chemistry of selenium. Dekker, New York, 736 pp

Glass DJ (1999) US and international markets for phytoremediation, 1999-2000. DJ Glass Associates Inc, Needham, MA, USA, 266 pp

Glass DJ (2000) Economic potential of phytoremediation. Raskin I, Ensley BD (eds) Phytoremediation of toxic metals: using plants to clean up the environment, chap 2. Wiley-Interscience Publ, New York

Greger M, Lindberg T (1999) Use of willow in phytoremediation. Int J Phytorem 1: 115-123

Guerinot ML, Salt DE (2001) Fortified foods and phytoremediation. Two sides of the same coin. Plant Physiol 125: 164-167

Hadjiliads ND (ed) (1997) Cytotoxicity, mutagenic and carcinogenic potential of heavy metals related to human environment. NATO ASI series 2. Environment, vol 26. Kluwer, Dordrecht, pp 629

Hansen D, Duda PJ, Zayed A, Terry N (1998) Selenium removal by constructed wetlands: role of biological volatilization. Environ Sci Technol 32: 591-597

Heaton ACP, Rugh CL, Wang N, Meagher RB (1998) Phytoremediation of mercury- and methylmercury-polluted soils using genetically engineered plants. J Soil Contam 7: 497-510

Homer FA, Reeves RD, Brooks RR, Baker AJM (1991) Characterization of the nickel-rich extract from the nickel hyperaccumulator *Dichapetalum gelonioides*. Phytochemistry 30: 2141-2145

Hossner LR, Loeppert, RH, Newton RJ, Szaniszlo PJ, Moses Attrep J (1998) Literature review: phytoaccumulation of chromium, uranium, and plutonium in plant systems. Amarillo National Resource Center for Plutonium

Huang JW, Cunningham SD (1996) Lead phytoextraction: species variation in lead uptake and translocation. New Phytol 134: 75

Huang JW, Chen J, Berti WR, Cunningham SD (1997) Phytoremediation of lead contaminated soil: role of synthetic chelates in lead phytoextraction. Environ Sci Technol 31: 800-805

Huang JW, Blaylock MJ, Kapulnik Y, Ensley BD (1998) Phytoremediation of uranium-contaminated soils: role of organic acids in triggering uranium hyperaccumulation in plants. Environ Sci Technol 32: 2004-2008

Jaffre T, Brooks RR, Lee J, Reeves RD (1976) *Sebertia acuminata*: a nickel-accumulating plant from New Caledonia. Science 193: 579-580

Kabata-Pendias A (2001) Trace elements in the soil and plants. CRC Press, Boca Raton

Kadlec RH, Knight RL (1996) Treatment wetlands. Lewis Publ, Boca Raton

Kadlec RH, Knight RL, Vymazal J, Brix H, Cooper P, Habert R (2000) Constructed wetlands for pollution control. Control processes, performance, design and operation. IWA Publ, London

Kägi JHR (1991) Overview of metallothioneins. Methods Enzymol 205: 613-623

Kaltsikes PJ (2000) Phytoremediation—state of the art in Europe, an international comparison. Agricultural University of Athens, COST action 837, first workshop

Krämer U, Cotter-Howells JD, Charnock JM, Baker AJM, Smith JAC (1996) Free histidine as a metal chelator in plants that accumulate nickel. Nature 373: 635-638

Krämer U, Smith R, Wenzel WW, Raskin I, Salt DE (1997) The role of metal transport and tolerance in nickel hyperaccumulation by *Thlaspi goesingense* halacsy. Plant Physiol 115: 1641

Laperche V, Logan TJ, Gaddam P, Traina SJ (1997) Effect of apatite amendments on plant uptake of lead from contaminated soil. Environ Sci Technol 31: 2745-2753

Lasat MM, Baker AJM, Kochian LV (1996) Physiological characterization of root Zn^{2+} absorption and translocation to shoots in Zn hyperaccumulator and nonaccumulator speciesof *Thlaspi*. Plant Physiol 112: 1715-1722

Lasat MM, Norvell WA, Kochian LV (1997) Potential for phytoextraction of ^{137}Cs from a contaminated soil. Plant Soil 195: 99-106

Lasat MM, Baker AJM, Kochian LV (1998a) Altered Zn compartmentation in the root symplasm and stimulated Zn absorption into the leaf as mechanisms involved in Zn hyperaccumulation in *Thlaspi caerulescens*. Plant Physiol 118: 875-883

Lasat MM, Fuhrmann M, Ebbs SD, Cornish JE, Kochian LV (1998b) Phytoremediation of a radiocesium-contaminated soil: evaluation of cesium-137 bioaccumulation in the shoots of three plant species. J Environ Qual 27: 165-169

Lasat MM, Pence NS, Garvin DF, Ebbs SD, Kochian LV (2000) Molecular physiology of zinc transport in the Zn hyperaccumulator *Thlaspi caerulescens*. J Exp Bot 51: 71-79

Lefebvre DD, Miki DBL, Laliberte JF (1987) Mammalian metallothioneins functions in plants. Bio/Technol 5: 1053-1056

Ma LQ, Komar KM, Tu C, Zhang W, Cai Y, Kennelley ED (2001) A fern that hyperaccumulates arsenic. Nature 409: 579

Macnair MR, Cumbes QJ, Meharg AA (1992) The genetics of arsenate tolerance in Yorkshire fog, Holcus lanatus L. Heredity 69: 325-335

Maiti IB, Wagner GJ, Hunt AG (1991) Light inducible and tissue specific expression of a chimeric mouse metallothionein cDNA gene in tobacco. Plant Sci 76: 99-107

Matsumoto H (2002) Metabolism of organic acids and metal tolerance in plants exposed to aluminum In: Prasad MNV, Strzalka K (eds) Physiology and biochemistry of metal toxicity and tolerance in plants, Kluwer, Dordrecht, pp 95-109

McBride MB (1994) Environmental chemistry of soils. Oxford University Press, New York, pp 336-337

McIntyre T, Lewis GM (1997) Advancement of phytoremediation as an innovative environmental technology for stabilization, remediation, or restoration of contaminated sites. J Soil Contam 6: 227

Meharg AA, Macnair MR (1991) Uptake, accumulation, and translocation of arsenate in arsenate-tolerant and non-tolerant Holcus lanatus L. New Phytol 117: 225-231

Misra S, Gedamu L (1989) Heavy metal tolerant transgenic Brassica napus L. and Nicotiana tabacum L. plants. Theor Appl Genet 78: 161-168

Moffat AS (1999) Engineering plants to cope with metals. Science 285: 369-370

Nandakumar PBA, Dushenkov S, Salt DE, Raskin I (1994) Crop Brassicas and phytoremediation—a novel environmental technology. Cruciferae Newslett 16: 18-19

Nandakumar PBA, Dushenkov S, Motto H, Raskin I (1995) Phytoextraction: the use of plants to remove heavy metals from soils. Environ Sci Technol 29: 1232-1238

Negri CM, Hinchman RR (2000) The use of plants for the treatment of radionuclides. Raskin I, Ensley BD (eds) Phytoremediation of toxic metals: using plants to clean up the environment, chap 8. Wiley-Interscience, New York

Odum HT (2000) Heavy metals in the environment. Using wetlands for their removal. CRC Press, Boca Raton, p 401

Ortiz DF, Ruscitti T, McCue KF, Ow DV (1995) Transport of metal-binding peptides by HMT1, a fission yeast ABC-type B vacuolar membrane protein. J Biol Chem 270: 4721-4728

Otte ML, Kearns CC, Doyle MO (1995) Accumulation of arsenic and zinc in the rhizosphere of wetland plants. Bull Environ Contam Toxicol 55: 154-161

Ow DW (1996) Heavy metal tolerance genes: prospective tools for bioremediations. Res Conserv Recycl 18: 135-149

Palmer CE, Warwick S, Keller W (2001) Brassicaceae (Cruciferae) family, plant biotechnology and phytoremediation. Int J Phytoremed 3: 245-287

Pence NS, Larsen PB, Ebbs SD, Letham DL, Lasat MM, Garvin DF, Eide D, Kochian LV (2000) The molecular physiology of heavy metal transport in the Zn/Cd hyperaccumulator Thlaspi caerulescens. Proc Natl Acad Sci USA 97: 4956-4960

Peterson PJ (1983) Adaptation to toxic metals. In: Robb DA, Pierpoint WS (eds) Metals and micronutrients: uptake andutilization by plants. Academic Press, London, pp 51-69

Pilon-Smits E, Pilon M (2000) Breeding mercury-breathing plants for environmental cleanup. Trends Plant Sci 5: 235-236

Pilon-Smits EAH, Hwang S, Lytle CM, ZhuYL, Tay JC, Bravo RC, Chen Y, Leustek T, Terry N (1999) Overexpression of ATP sulfurylase in Indian mustard leads to increased selenate uptake, reduction, and tolerance. Plant Physiol 119: 123-132

Pollard JA, Baker AJM (1997) Deterence of herbivory by zinc hyperaccumulation in Thlaspi caerulescens (Brassicacea). New Phytol 135: 655-658

Prasad MNV (1999) Metallothioneins and metal binding complexes in plants. In: Prasad MNV, Hagemeyer J (eds) Heavy metal stress in plants: from molecules to ecosystems. Springer, Berlin Heidelberg New York, pp 51-72

Prasad MNV (2001a) Metals in the environment—analysis by biodiversity. Dekker, New York

Prasad MNV (2001b) Bioremediation potential of Amaranthaceae. In: Leeson A, Foote EA, Banks MK, Magar VS (eds) Phytoremediation, wetlands, and sediments. Proceedings of the 6th international in situ and on-site bioremediation symposium, vol 6(5). Battelle Press, Columbus, OH, pp 165-172

Prasad MNV, Freitas H (1999) Feasible biotechnological and bioremediation strategies for serpentine soils and mine spoils. Electr J Biotechnol 2: 35-50. Website: http://ejb.ucv.cl or http://www.ejb.org

Prasad MNV, Hagemeyer J (1999) Heavy metal stress in plants—from molecules to ecosystems. Springer, Berlin Heidelberg New York

Prasad MNV, Strzalka K (eds) (2002) Physiology and biochemistry of metal toxicity and tolerance in plants. Kluwer, Dordrecht, 460pp

Raskin I (1996) Plant genetic engineering may help with environmnetal cleanup. Proc Natl Acad Sci USA 93: 3164-3166

Raskin I, Ensley BP (2000) Phytoremediation of toxic metals—using plants to clean up the environment. Wiley, New York

Raskin I, Nanda Kumar PBA, Dushenkov S, Salt DE (1994) Bioconcentration of heavy metals by plants. Curr Opin Biotechnol 5: 285-290

Raskin I, Smith RD, Salt DE (1997) Phytoremediation of metals: using plants to remove pollutants from the environment. Curr Opin Biotechnol 8: 221-226

Rauser WE (1999) Structure and functions of metal chelators produced by plants: the case for organic acids, amino acids, phytin, and metallothioneins. Cell Biochem Biophys 31: 19-46

Rauser WE (2000) The role of thiols in plants under metal stress, In: Brunold C, Rennenberg H, De Kok LJ (eds) Sulfur nutrition and sulfur assimilation in higher plants. Haupt, Bern, pp 169-183

Reeves RD, Baker AJM (1999). Metal-accumulating plants. In: Raskin I, Ensley BD (eds) Phytoremediation of toxic metals: using plants to clean up the environment. Wiley, New York, pp 193-229

Reeves RD, Brooks RR (1983) European species of *Thlaspi* L. (Cruciferae) as indicators of nickel and zinc. J Geochem Explor 18: 275-283

Robinson BH, Chiarucci A, Brooks RR, Petit D, Kirkman JH, Gregg PEH, de Dominicis V (1997a) The nickel hyperaccumulator plant Alyssum bertolonii as a potential agent for phytoremediation and phytomining of nickel. J Geochem Explor 59(2): 75

Robinson BH, Brooks RR, Howes AW, Kirkman JH, Gregg PEH (1997b) The potential of the high-biomass nickel hyperaccumulator *Berkheya coddii* for phytoremediation and phytomining. J Geochem Explor 60(2): 115-126

Rugh CL, Wilde HD, Stack NM, Thompson DM, Summers AO, Meagher RB (1996) Mercuric ion reduction and resistance in transgeneic *Arabidopsis thaliana* plants expressing a modified bacterial merA gene. Proc Natl Acad Sci USA 93: 3182-3187

Rugh CL, Gragson GM, Meagher RB (1998) Toxic mercury reduction and remediation using transgenic plants with modified bacterial genes. HortScience 33: 618-621

Saier MH JR (2000) A functional-phylogenetic classification system for transmembrane solute transporters. Microbiol Mol Biol Rev 64: 354-411

Salt DE, Thurman DA, Sewell AK (1989) Copper phytochelatin of *Mimulus guttatus*. Proc R Soc Lond B 236: 79-89

Salt DE, Blaylock M, Nanda Kumar PBA, Dushenkov S, Ensley BD, Chet I, Raskin I (1995) Phytoremediation: a novel strategy for the removal of toxic metals from the environment using plants. Biotechnology 13: 468-474

Salt DE, Smith RD, Raskin I (1998) Phytoremediation. Annu Rev Plant Physiol Plant Mol Biol 49: 643-668

Salt DE, Prince RC, Baker AJM, Raskin I, Pickering IJ (1999) Zinc ligands in the metal hyperaccumulator *Thlaspi caerulescens* as determined using X-ray absorption spectroscopy. Environ Sci Technol 33: 713-717

Samecka-Cymerman A, Kempers AJ (1996) Bioaccumulation of heavy metals by aquatic macrophytes around Wroclaw, Poland. Ecotoxicol Environ Saf 35: 242

Sanità di Toppi L, Prasad MNV, Ottonello S (2002) Metal chelating peptides and proteins in plants. In: Prasad MNV, Strzalka K (eds) Physiology and biochemistry of metal toxicity and tolerance in plants. Kluwer, Dortrecht, pp 59-93

Saxena PK, KrishnaRaj S, Dan T, Perras MR, Vettakkorumakankav NN (1999) Phytoremediation of metal contaminated and polluted soils. In: Prasad MNV, Hagemeyer J (eds) Heavy metal stress in plants: from molecules to ecosystems. Springer, Berlin Heidelberg New York, pp 305-329

Schäfer HJ, Haag-Kerwer A, Rausch T (1998) cDNA cloning and expression analysis of genes encoding GSH synthesis in roots of the heavy-metal accumulator *Brassica juncea* L.: evidence for Cd-induction of a putative mitochondrial γ-glutamylcysteine synthetase isoform. Plant Mol Biol 37: 87-97

Schmöger MEV, Oven M, Grill E (2000) Detoxification of arsenic by phytochelatins in plants. Plant Physiol 122: 793-801

Stomp AM, Han KH, Wilbert S, Gordon MP, Cunningham SD (1994) Genetic strategies for enhancing phytoremediation. Ann NY Acad Sci 721: 481-491

Terry N, Bañuelos G (2000) Phytoremediation of contaminated soil and water. Lewis Publ, Boca Raton

Terry N, Zayed AM (1998) Phytoremediation of selenium. In: Frankenberger WT, Engberg RA (eds) Environmental chemistry of selenium. Dekker, New York

Tessier A, Campbell PGC, Bisson M (1979) Sequential extraction procedure for the speciation of particulate trace metals. Anal Chem 51: 844-850

Tomsett AB, Sewell AK, Jones SJ, de Miranda J, Thurman DA (1992). Metal-binding proteins and metal-regulated gene expression in higher plants. In: Wray JL (ed) Society for experimental biology seminar series 49: inducible plant proteins, Cambridge University Press, Cambridge, pp 1-24

Trampczynska A, Gawronski SW, Kutrys S (2001) *Canna* x *generalis* as a plant for phytoextraction of heavy metals in urbanized area. Zeszyty Naukowe Politechniki Slaskiej 45(1487): 71-74

Vangronsveld J, Cunningham SD (1998) Metal-contaminated soils: *in-situ* inactivation and phytorestoration. Springer, Berlin Heidelberg New York

Varennes A, de Torres MO, Coutinho JF, Rocha MMGS, Neto MMPM, De-Varennes A (1996) Effects of heavy metals on the growth and mineral composition of a nickel hyperaccumulator. J Plant Nutr 19: 669-676

Vassil A, Kapulnik Y, Raskin I, Salt DE (1998) The role of EDTA in lead transport and accumulation by Indian mustard. Plant Physiol 117: 447-453

Vázquez MD, Barcelo J, Poschenrieder C, Madico J, Hatton P, Baker AJM, Cope GH (1992) Localization of zinc and cadmium in *Thlaspi caerulescens* (Brassicaceae), a metallophyte that can hyperaccumulate both metals. J Plant Physiol 140: 350-355

Vázquez MD, Poschenreider C, Barcelo J, Baker AJM, Hatton P, Cope GH (1994) Compartmentalization of zinc in roots and leaves of the zinc hyperaccumulator *Thlaspi caerulescens* J. & C. Presl. Bot Acta 107: 243-250

Vymazal J (1996) Constructed wetlands for wastewater treatment in the Czech Republic: the first 5 years experience. Water Sci Technol 34(11): 159-165

Watanabe ME (1997) Phytoremediation on the brink of commercialization. Environ Sci Technol 31: 182-186

Wenzel WW, Adriano DC, Alloway B, Doner HE, Keller C, Lepp NW, Mench M, Naidu R, Pierzynski GM (eds) (1999) Proceedings of the extended abstracts of the 5th international conference on biogeochemistry of trace elements, Vienna, vols 1 and 2, pp 1191

Wise, DL, Trantolo DJ, Cichon, EJ, Inyang HI, Stottmeister U (2000) Bioremediation of contaminated soils. Dekker, New York

Zhu YL, Pilon-Smits EAH, Tarun AS, Weber SU, Jouanin L, Terry N (1999b) Cadmium tolerance and accumulation in Indian mustard is enhanced by overexpressing glutamylcysteine synthase. Plant Physiol 121: 1169-1177

Zhu YL, Pilon-Smits EAH, Jouanin L, Terry N (1999a) Overexpression of glutathione synthetase in Indian mustard enhances cadmium accumulation and tolerance. Plant Physiol 119: 73-79

Zorpas AA, Constantinides T, Vlyssides AG, Aralambous I, Loizidou M (1999) Heavy metal uptake by natural zeolite and metals partitioning in sewage sludge compost. Bioresource Technol 71: 113-119

Chapter 15

Metal Removal from Sewage Sludge: Bioengineering and Biotechnological Applications

T.R. Sreekrishnan[1], R.D. Tyagi[2]

[1]Department of Biochemical Engineering and Biotechnology, Indian Institute of Technology, Hauz Khas, New Delhi 110 016, India
[2]Institut national de la recherche scientifique (INRS-Eau), Université du Québec, Complexe Scientifique, 2700 rue Einstein, C.P. 7500, Sainte-Foy, Québec, G1V4C7 Canada

15.1 INTRODUCTION

Heavy metals such as cadmium, chromium, copper, nickel, lead, zinc, etc., are normally found in varying concentrations in sewage. The metals present in domestic wastewaters come from a number of sources: feces, pharmaceutical products, cosmetics, washing and cleaning chemicals, paints and other surface coatings, etc. Since the source of these metals is non-point in nature, it is extremely difficult to practice source control. As a result, all these metals finally end up at the sewage treatment site. A significant part of these metals are removed during the primary decantation or primary treatment process as part of the primary sludge. During the secondary treatment (generally biological and most often activated sludge) process, a complex between metals and extracellular polymers (produced by the microorganisms used in the treatment) is produced, which results in a biofloc structure. The metals are either adsorbed on the cell surface or are simply entrapped in the bioflocs. These physical and biochemical mechanisms are responsible for the fate of 80-90% of the total metals originally present in the effluent. The metal concentrations observed in the sludge vary with the type of wastewater treatment, the presence of certain industries in the area, the habits of the population covered, the state of the wastewater transport system, etc.

The presence of high concentrations of toxic metals in sludge poses a constraint on the application of these sludges on agricultural land (Bruce and Davis 1989; Alloway and Jackson 1991; Korentajer 1991; USEPA 1993). A number of studies have been carried out in the last few years on the risks associated with the application of metal-charged sludge on agricultural land (Sterritt and Lester 1980; Coker and Matthews 1983; Mininni and Santori 1987). Copper, nickel and zinc are phytotoxic and high concentrations of these metals in soil can strongly affect the harvest yields (Sommers and Nelson 1981; Lester et al. 1983; Davis and Carlton-Smith 1984; Webber 1986; Mench et al. 1992). Metal accumulation in plants following sludge application on agricultural land was demonstrated for antimony, arsenic, cadmium, chromium, copper, iron, mercury, molybdenum, nickel, lead, selenium and zinc (Davis and Carlton-Smith 1980; Adamu et al. 1989; Levine et al. 1989; Cappon 1991; Granato et al. 1991; Hernandez et al. 1991; Jackson and Alloway 1991; Roca and Pomares 1991; Tadesse et al. 1991; Jing and Logan 1992; Obbard et al. 1993). Metals can also make it into the food chain by sticking to the surface of vegetation when metal-containing sludge is spread on agricultural land (Klessa and Desira-Buttigieg 1992).

The presence of heavy metals in the consumable parts of vegetation can pose a risk to human and animal health. For example, cadmium has phytotoxicity symptoms that are ten times more visible than its zootoxicity (Coker and Matthews 1983). In humans and animals, excessive absorption of cadmium leads to its accumulation in the kidney and liver, causing histological and functional damage. Biological effects of cadmium also include its interference with fundamental enzymatic functions, such as oxidative phosphorylation, by blocking thiol groups and nucleic acid synthesis (Doyle et al. 1978). Cadmium can also pose certain cardiological toxicity problems (NAS 1979; Carmignani et al. 1983). Toxic effects have been reported in bovines that had eaten from fields amended by sludge heavily polluted with lead (Webber 1986). Environmental exposure to weak lead contents is associated with a number of metabolic disorders and neuropsychological deficiencies in humans (NAS 1980; USEPA 1986): nuisible effects on red blood cells (RBC), metabolism levels, perturbation of calcium homeostases in the hepatocytes, bone and brain cells, neurological damage. A number of studies were also carried out confirming the ill effects of lead in human arterial hypertension (Nriagu 1988). Metals such as Al, As, Sb, Be, Bi, Cd, Hg, Cr, Co, Mn, Ni, Pb, Ti, V, Se and Zn could affect human reproduction or be an initiator or promotor of certain cancers by acting as an inhibitor in DNA and RNA synthesis, or as mutagenic agents (Jennette 1981; Martell 1981; Babich et al. 1985). The conventional sludge treatment processes, such as aerobic and anaerobic digestion, cannot remove these heavy metals from the sludge to any appreciable extent. In fact, the sludge digestion process, which causes a reduction in sludge solids, results in an effective increase in the concentration of potentially toxic heavy metals in the sludge.

15.2 REMOVAL OF METALS FROM SEWAGE SLUDGES

A number of methods have been explored for the purpose of removing the potentially toxic heavy metals from sewage sludges. These methods can be classified into three: (1) physical treatment, (2) chemical treatment, and (3) biological treatment. Physical treatment methods are based on the principle of extraction of the contaminants by physical or physicochemical means. These methods include leaching, flotation, classification, electric/electro-acoustic separation and ion exchange. Solubilization of metals from the sludge matrix can be achieved by means of some chemical reactions using reagents such as inorganic acids, organic acids, chelating agents, surfactants, oxidizing agents, bases and reducing agents. The considerable quantities of acid required to solubilize metals by chemical methods makes this technology uneconomical (Wong and Henry 1988). Also, it has been observed that prolonged periods of acid leaching increased metal solubilization efficiency only marginally.

Sludge samples containing significant amounts of heavy metals (Cr, Cu, Ni, Pb and Zn) were subjected to acid treatment using either hydrochloric, sulfuric, nitric or phosphoric acid by Naoum et al. (2001). The optimum combination, in terms of metal removal efficiency and environmental impacts, was sought through a variety of tests by applying a ratio of 1:5 of sludge quantity (g) per volume of acid (ml). The concentrations of the different acids used were in the range 5-20% and the contact times ranged between 15 and 60 min. The optimum combination was achieved when the sludge samples were in contact with H_2SO_4 (20% v/v) for 60 min. In order to estimate the metal leachability, the heavy metal content as well as the metal distribution in the residue were investigated and it was found that most of the heavy metal content was extracted while the remaining was removed from the initial mobile phases to the more stable ones.

Another approach is to effect the metal extraction using an organic acid, as reported by Veeken and Hamekers (1999). Organic acids could be attractive extracting agents because the extraction can be performed at mildly acidic conditions (pH 3-5) and they are biologically degradable. The extraction was studied for the heavy metals Cu and Zn and for competing metals Ca and Fe. The rate of extraction increases for increasing temperature and citric acid concentration. Cu can be extracted for 60-70% and Zn for 90-100% by citric acid at pH 3-4. A first economic valuation of the extraction and subsequent composting process showed that the total costs of the treatment process are below the costs of incineration.

Every study conducted has pointed out that it is possible to solubilize metals by acidification, but the quantities of acid required being very high can increase the treatment cost considerably. It can also cause severe damage to the soil structure due to excessive solubilization of soil solids. Use of chemical chelating agents such as ethylenediaminetetraacetic acid (EDTA) and nitrilotriacetic acid (NTA) to solubilize toxic metals present in sewage sludge has been studied (Jenkins et al. 1981; Campanella et al. 1985; Lo and Chen 1990). High operational costs, certain operational difficulties, and some times inefficient metal solubilization are some limitations of these techniques.

Ito et al. (2000) have reported a new chemical method for removal of heavy metals from anaerobically digested sewage sludge using ferric sulfate. The addition of ferric sulfate to the sludge caused the acidification of the sludge and the elution of heavy metals from the sludge. The pH of the sludge decreased with an increase in the amount of iron added and with a decrease in the sludge concentration. At a sludge solid concentration of 2% (w/w), the sludge pH dropped below 3 and the elution percentage of cadmium, copper and zinc was more than 80% when the added amount of ferric iron was more than 1.5 g/l of wet sludge. Furthermore, the method using ferric sulfate was compared with that using sulfuric acid at pH 3 in order to clarify the effect of ferric iron as an oxidation reagent on elution of heavy metals. Ferric iron eluted cadmium, copper and zinc more effectively than sulfuric acid. This effective elution of heavy metals was caused by the oxidation of the sludge solid by ferric iron added. From these results, it was concluded that ferric iron played a role in acidifying the sludge and oxidizing metallic compounds in the sludge and this new chemical method was useful for the removal of heavy metals from anaerobically digested sewage sludge.

15.2.1 Biological Leaching of Metals

The metabolic capability of certain microorganisms to solubilize metals is an interesting biological phenomenon. This phenomenon, which is catalyzed by microbial activity, principally bacterial, is called bioleaching and necessitates the presence of microorganisms capable of proliferating in extreme ecosystems (highly acidic pH, highly oxidizing conditions, high metal ion content) and can derive energy by the oxidation of mineral sulfur compounds. Bioleaching of metals can be achieved when the leaching bacteria oxidize the insoluble metal sulfides into soluble sulfates as follows:

$$MS + 2O_2 \rightarrow M^{2+} + SO_4^{2-} \tag{15.1}$$

The oxidation of these sulfides is catalyzed by the action of sulfide reductase or sulfide oxidase that allows the oxidation of sulfides to sulfates. The oxidation of sulfides to sulfate can be effected by cytochrome c oxidoreductase or from APS reductase.

Certain microorganisms derive their energy by the oxidation of ferrous ions into ferric, which contribute to the indirect mechanism for metal solubilization as follows:

$$2\ FeSO_4 + 0.5\ O_2 + H_2SO_4 \rightarrow Fe_2(SO_4)_3 + H_2O \tag{15.2}$$

$$Fe_2(SO_4)_3 + MS \rightarrow M^{2+} + SO_4^{2-} + 2\ FeSO_4 + S° \tag{15.3}$$

Reaction (15.2) is biological in nature, whereas reaction (15.3) is purely chemical in nature, with ferric ions acting as oxidizing agents in the acidic medium. The elemental sulfur naturally present or formed during the indirect oxidation of metal sulfides can be oxidized to sulfuric acid, leading to metal dissolution. This reaction is common to all species of thiobacilli and several other microorganisms (Kelly and Harrison 1988).

$$S° + 1.5\ O_2 + H_2O \rightarrow H_2SO_4 \tag{15.4}$$

The bacteria *Thiobacillus ferrooxidans* and *Thiobacillus thiooxidans* have long been considered as principal microorganisms involved in direct attack on metal sufides.

Removal of heavy metals (Cr, Cu, Zn, Ni and Pb) from anaerobically digested sludge has been studied in a batch system using isolated indigenous iron-oxidizing bacteria (Xiang et al. 2000). The inoculation of indigenous iron-oxidizing bacteria and the addition of $FeSO_4$ accelerated the solubilization of Cr, Cu, Zn, Ni and Pb from the sludge. The pH of the sludge decreased with an increase in Fe^{2+} concentrations and reached a low pH of 2-2.5 for treatments receiving both bacterial inoculation and $FeSO_4$. After 16 days of bioleaching, the heavy metal removal efficiencies were found to be Cr 55.3%, Cu 91.5%, Zn 83.3%, Ni 54.4%, and Pb 16.2%. In contrast, only 2.6% Cr, 42.9% Cu, 72.1% Zn, 22.8% Ni and 0.56% Pb were extracted from the control without the bacterial inoculation and addition of $FeSO_4$. The residual heavy metal content in the leached sludge was acceptable for unrestricted use for agriculture. The experimental results confirmed the effectiveness of using the isolated iron-oxidizing bacteria for the removal of heavy metals from sewage sludge.

Two thiobacilli groups (acidophilic and less acidophilic) are involved in bioleaching of metals using sulfur as an energy source for growth. The cooperation between these two groups, as well as the growth kinetics of these populations, have been measured during the metal-leaching process (Blais et al. 1993). The tests were carried out in five different types of sludges (one secondary, two aerobically digested, and two anaerobically digested sludges) with varying solid contents (7, 13, 14, 23, 41 g/l). During 5 days of bioleaching, the bacteria produced between 2.8 and 4.4 g/l of sulfate, corresponding to an oxidation of around 19-29% of added sulfur (Table 15.1).

Table 15.1 Variation of pH and sulfate production after 5 days of microbial leaching with elemental sulfur

| Sludges | pH | | Sulfate conc. | | Percentage | Sulfur |
	Initial	Final	Initial	Final	Oxidized
Beauceville (secondary)	6.04	2.05	509	3340	18.9
Black Lake (aerobically digested)	6.95	1.58	1510	5250	24.9
St. Claire (oxidation pond)	6.72	2.21	606	4080	23.2
St. Georges (secondary)	6.92	1.55	781	5200	29.4
Valcartier (anaerobically digested)	6.60	2.10	874	3930	20.3

Less acidophilic thiobacilli initially present in the sludge multiply and reduce the pH of the sludge to around 4.0. The acidophilic thiobacilli, added in the inoculum, then start proliferating and result in a significant reduction in pH. A fall in pH to values lower than 3.0 will cause a reduction in the number of less acidophilic thiobacilli to below the detection limit.

15.3 EFFECTS OF PROCESS PARAMETERS

Removal of heavy metals from sewage sludges by the bioleaching process is made possible by the growth and associated activities of the bacteria used for this purpose. This is true whether the leaching is carried out using *Thiobacillus ferrooxidans*, *T. thioparus* or *T. thiooxidans* or a combination of these species. Bacterial growth in any medium depends on a number of factors such as availability of carbon and nitrogen, availability of oxygen (except for strict anaerobes), availability of trace metals and elements in the growth medium, as well as on parameters such as temperature and pH of the medium. Consequently, bacterial growth in sewage sludge is a complex process and parameters such as temperature and pH do have considerable influence on the bacterial growth and product formation rates. This in turn will dictate the acid production rate as well as total acid concentration in the medium at any instant. The production of acid will result in a lower pH for the medium (sewage sludge). The magnitude of the change in pH will be governed by the nature and composition of the sludge and the sludge solids concentration. In short, there are many interrelated factors that will affect acid production, pH change and metal solubilization events. A proper identification of these relations, qualitatively as well as quantitatively, is essential for the optimization and scale-up of the process.

15.3.1 Solids Concentration

Total solids concentration of the sludge comprises the suspended solids and the dissolved solids present per unit volume of the sludge. The dissolved part of the total solids contains organic as well as inorganic salts that are water soluble and are generally small in quantity compared with the suspended solids. The suspended solids are comprised of insoluble compounds and cells. The nature and composition of the suspended solids varies considerably with the type of sludge. In primary sludges, the suspended solids are mainly water-insoluble compounds present in the sewage which are separated from the wastewater during the process of primary treatment. In general, primary sludges do not contain any appreciable amount of biomass (bacterial cells). Secondary sludges, on the other hand, contain a considerable amount of biomass produced during the wastewater treatment stage (secondary treatment). Again, the nature of the sludge biomass depends on the type of secondary treatment used, aerobic or anaerobic. The sludge solids concentration can vary, in general, anywhere between 0.5 and 4.0% on a weight per unit sludge volume basis. This variation in the quality and quantity of solids present in the sewage sludges affects the sludge bioleaching process considerably.

When a metal-bioleaching operation is carried out on sludges having different solids concentrations, but which are the same in all other aspects, the pH reduction in the sludge depends on the solids concentration. The decrease in pH with time is comparatively smoother and more gradual at higher sludge solids concentrations. This is due to the buffering action imparted by the sludge, which in turn is directly proportional to the sludge solids concentration.

In order to compare the acid production rates, one needs to analyze the sulfate accumulation with time during the metal-bioleaching process, in the sludges having different solids concentrations. It has been seen that during 120 h of incubation, acid accumulation is greater in the sludge having higher solids concentration. For a logical reasoning of the order of the chain of events, it is better to know how the sulfate production rates vary as a function of time and sludge solids concentration. The sulfate production rate at any instant is the slope of the tangent drawn to the time vs sulfate concentration curve, at that instant. Graphical tangent construction and slope measurement is tedious and prone to errors. Instead, one can resort to numerical curve fitting techniques that will give an equation relating time and sulfate concentration. The first differential of these equations with respect to time, at any instant, is the sulfate production rate for that instant.

Proceeding along these lines, it was observed that the highest sulfate production rate is recorded for the sludge having the highest solids concentration and the maximum sulfate production rate recorded is lower at lower solids concentrations. Also, sludges with lower solids content will reach their sulfate production rate maxima earlier than in those having higher solids concentration.

Experimental investigations have established that the growth rate of the bacteria is directly influenced by the pH of the medium. Growth of the thiobacilli species is associated with acid production, which brings down the pH of the medium. It is well known that bacterial growth is influenced by conditions of the growth environment such as temperature, pH of the medium, etc. Hence any change in the pH of the medium should normally reflect in the bacterial growth rate and subsequently the acid production rate. As we have already seen, the metal-bioleaching process is associated with a continuous change in the pH of the sludge. Under such circumstances, the bacterial growth and acid production rates also undergo a continuous change. The nature and magnitude of this change will depend on the pH optima for growth and acid production for the bacteria. We have already seen that the pH drop is much more abrupt in sludges with low solids concentration compared with those with high solids concentration. The drastic pH drop will likely have an adverse effect on bacterial growth and acid production. This would result in lower sulfate production rates and lower sulfate concentrations in sludges with lower solids concentration.

Detailed experimentation on the effect of solids concentration on metal solubilization from sewage sludges has shown that more of the metal is solubilized for the same amount of acid production in sludges with lower solids concentration. In other words, a greater quantity of acid is needed at higher solids concentrations to solubilize the same amount of metal. This is because metal solubilization depends not on the quantity of acid produced, but on the resulting pH of the medium.

15.3.2 Sludge pH

The sulfur-oxidizing, acid-producing bacteria consist of two distinct *Thiobacillus* species, the less acidophilic *Thiobacillus thioparus* having an optimum growth pH around 7.0 and the acidophilic *Thiobacillus thiooxidans* with an optimum pH for growth at around 4.0. The bacterial growth phenomenon is measured and quantitatively expressed in terms of its specific growth rate. Specific growth rate (μ) is defined as:

$$\mu = (1/N)\,(dN/dt) \tag{15.5}$$

where N is the number of bacterial cells (colony forming units per ml) and t is the time. It has been observed that, at any instant, the specific growth rate of the sulfur-oxidizing bacteria bears a linear relationship to the natural logarithm of the pH of the medium at that instant. In other words,

$$\mu = A + B\,\ln(pH) \tag{15.6}$$

where A and B are constants.

The less acidophilic species has an optimum pH of 7.0 and, consequently, its specific growth rate is maximum at a pH of 7.0. Denoting the maximum specific growth rate of the less acidophilic species as $\mu1^*$, the specific growth rate at any pH can be computed by the relationship:

$$\mu1 = \mu1 * \frac{C_1 + \ln(pH)}{C_1 + \ln(7.0)} \tag{15.7}$$

where C_1 ($=A_1/B_1$) is a constant, and its value has been computed as -1.33 ± 0.05. It is interesting to note that at a pH of 4.0, from Eq. (15.7),

$$\mu1 = 0.09\,\mu1^*.$$

This means that the specific growth rate of the less acidophilic species at pH 4.0 is very low. This is interesting since the acidophilic species has its optimum at this pH. In fact, the predominating organism in the bioleaching operation undergoes a shift from less acidophilic to acidophilic thiobacilli at a pH of around 4.0. Similarly, using the subscript '2' to denote the acidophilic bacteria, we have

$$\mu_2 = A_2 + B_2\,\ln(pH) \tag{15.8}$$

The acidophilic species of thiobacilli has a pH optimum, and consequently maximum specific growth rate, at a sludge pH of 4.0. If $\mu2^*$ represents the specific growth rate of the acidophilic species at a pH of 4.0,

$$\mu 2 = \mu 2 * \frac{C_2 + \ln(pH)}{C_2 + \ln(4.0)} \qquad (15.9)$$

where C_2 $(=A_2/B_2)$ is a constant with a value of -0.4 ± 0.05. Again, at a pH of 1.5, $\mu2=0.0055$ $\mu2*$. This means that the specific growth rate of the acidophilic species is very low at a pH of 1.5.

15.3.3 The Buffering Capacity Index (BCI)

Metal solubilization is closely linked to the pattern of pH decrease that, in turn, is dictated by the sulfate production rate as well as sludge solids concentration. Because of their dependency on each other, it is desirable that these three parameters be quantitatively related in a suitable manner. For this purpose, a new system parameter called the 'buffering capacity index' (or BCI) of the sludge has been introduced.

BCI for a sludge is defined as the quantity of sulfate required in mg to change the pH of a unit volume of that sludge by one unit at pH (4.0), and is calculated as:

$$BCI = d(SO_4^{2-})/d(pH) \text{ at pH} = 4.0 \qquad (15.10)$$

Based on experimental data using a number of sludges, it has been determined that the BCI-solids concentration relationship is linear, but varies depending on the type of sludge. That is, the BCI-solids concentration graphs for non-digested, anaerobically digested and aerobically digested sludges will be three different straight lines. This actually reflects the diversity among sewage sludges in terms of their chemical composition for each of the three different types of sludges. The non-digested sludge will have a high concentration of precipitated inorganic salts, whereas the anaerobically digested sludge will have a high concentration of anaerobic bacterial cell mass. The way in which each of these sludge constituents responds to acid accumulation in the sludge will dictate the pH-sulfate profile in each case.

The overall process taking place during the metal-bioleaching process can be visualized as the work of two distinct sulfur-oxidizing bacterial species, a less acidophilic species (pH optimum 7.0) and an acidophilic species (pH optimum 4.0). The less acidophilic species grows over the pH range 7.0-4.0, whereas growth of the acidophilic species occurs in the pH range 4.0-1.5. Sewage sludges normally have a pH of 6.0-7.0, under which conditions the less acidophilic species grows and the acidophilic species does not. The growth of the less acidophilic species in the presence of elemental sulfur is associated with the production of sulfuric acid, which brings down the medium pH. The lowering of medium pH has an adverse effect on the specific growth rate of the less acidophilic bacteria, but they continue to grow at a reduced rate, resulting in further acid production, until the sludge pH reaches a value of 4.0. At this point, the less acidophilic species ceases to grow and the acidophilic species takes over the task of acid production. The growth of the acidophilic species is also adversely affected by the progressively lower pH conditions encountered as a result of continued acid formation. Finally, at a pH of about 1.5, the growth of the acidophilic species also ceases. The process of metal solubilization proceeds in parallel with the acid production process. The quantity of metal(s) leached into the liquid phase depends on the sludge pH, sludge type (or, speciation of metal in the sludge) and the initial metal concentration present in the sludge.

15.3.4 Metal Solubilization

Of all the factors that are seen to affect the metal solubilization process, the most difficult ones to quantify are the concentrations, relative as well as absolute, of each metal present in the sludge, and the metal speciation in the sludge. This problem can be tackled by expressing the solubilities of various heavy metals in the form of a chart with metal concentration in the liquid phase, expressed as a percentage of the metal concentration initially present in the sludge, presented as a function of sludge pH. The effect of metal speciation on metal solubility can be accounted for by using three different solubility charts for each metal, depending on whether the sludge is non-digested, aerobically digested or anaerobically digested. Alternatively, metal solubilization could be quantified using artificial neural networks. A neural net consists of several "neurons" interconnected in a specific manner. The nature and strength of these connections varies from one neural net to another. The number of neurons, their configuration and the connecting strength between any two neurons are specific for a neural net constructed to serve a particular task. Neural nets work on the principle of "learning and reproducing". Given the constraints on writing mathematical expressions for the metal solubilization process, a neural-net-based metal solubilization model could serve the purpose satisfactorily. A neural net model could be developed for any given situation by deciding the critical input parameters, the interrelationships among them and then designing a neural net which satisfies those conditions. Alternatively, commercial neural net simulation software could be trained for the task. In a neural net model developed for predicting metal solubilization, a commercial product, BrainMaker Professional from California Scientific Software, Grass Valley, California, was used.

15.3.5 Effect of Temperature

Bacteria are highly susceptible to temperature variations and this will reflect in the rate of all biocatalyzed reactions that form part of the bioleaching operation. The difference in the rate of pH drop at different temperatures may be related to the growth rate of the less acidophilic and/or acidophilic bacteria at these temperatures. The growth of the less acidophilic bacteria is fastest at 42 °C and the growth rate decreases as the temperature is lowered. One should bear in mind the fact that the bacterial growth rate during the course of the leaching process is also a function of sludge pH as already explained.

The effect of temperature on the specific growth rate of the less acidophilic bacteria can be incorporated into the growth model (Eq. 15.7) by the following relationships:

$$C_1 = 1.2 + 1.73 \times 10^{-2}(T) - 3.0 \times 10^{-4} (T)^2 \tag{15.11}$$

and $$\mu 1^* = 1.18 \times 10^{-1} - 1.72 \times 10^{-2} (T) + 1.1 \times 10^{-3}(T)^2 - 1.6 \times 10^{-5}(T)^3 \tag{15.12}$$

where T is the temperature expressed in °C. These equations were derived from experimental data and are valid for the temperature range 7 °C $\leq$ T $\leq$ 42 °C. In the case of acidophilic species, cell proliferation is faster at higher temperatures and slows down with a reduction in temperature. However, this is not true at 42 °C, since the acidophilic species does not grow at this temperature. The effect of temperature on the specific growth rate of the less acidophilic bacteria can be incorporated into the growth model, Eq. (15.9), by the following relationships:

$$C_2 = 3.58 \times 10^{-1} - 3.5 \times 10^{-3}(T) \tag{15.13}$$

and $$\mu_2^* = 3.2 \times 10^{-2} - 1.15 \times 10^{-3}(T) + 2.37 \times 10^{-4}(T)^2 - 4.033 \times 10^{-6}(T)^3 \tag{15.14}$$

The percentage metal solubilization at a given sludge pH is not affected by the temperature. However, the leaching temperature indirectly affects the overall process by influencing the bacterial growth and sulfate production rates and hence the pH change, which in turn affects metal solubilization. The metal solubilization achieved at a particular sludge pH value is the same, regardless of any temperature variations.

15.4 SIMULTANEOUS SLUDGE DIGESTION AND METAL LEACHING

We have seen that metal leaching by the sulfur-oxidation process mainly involves two groups of bacteria, the less acidophilic and the acidophilic sulfur-oxidizing bacteria. These bacteria have optimum pH in the vicinity of 7.0 and 4.0, respectively. Since the metal-leaching process makes use of these two groups of bacteria in a serial manner with the second group (acidophilic sulfur oxidizers) taking up where the first group has let off, the overall process begins in the vicinity of neutral pH for the contaminated sludge. This, combined with the fact that the system is highly aerobic, opens up the possibility of achieving degradation of the sludge biomass fraction by an aerobic digestion process. In other words, metal leaching by the sulfur-oxidation process and aerobic sludge stabilization proceed simultaneously, resulting in a simultaneous sludge digestion and metal-leaching (SSDML) process.

During the SSDML process, elemental sulfur is oxidized into sulfuric acid by the sulfur-oxidizing bacteria, resulting in a lowering of the sludge pH; heavy metals present in the sludge are leached out into the liquid phase; the biodegradable fraction of the sludge volatile solids is aerobically digested; and pathogenic organisms present in the sewage sludge are eliminated. The SSDML process has great possibilities as a potential solution to the sludge disposal problem.

Sulfur is an insoluble substrate and only the outer surfaces of the sulfur particles are available for bacterial colonization and subsequent metabolism. Since the bacterial oxidation of sulfur depends on the surface area of sulfur, it is reflected in the quantity of sulfate produced. As bacteria grow and oxidize the sulfur, the surface area of sulfur available also undergoes changes. The exact reduction in surface area of sulfur will depend on the quantity of sulfate produced, which in turn will depend on the sludge pH.

The growth of sulfur-oxidizing bacteria could be limited by any of the many nutrients required to support it. Among these, the most important one is oxygen. Oxygen supplied to the process by way of air sparing can become available to the bacteria only when it is dissolved in the liquid phase. Since the solubility of oxygen is very low in the medium, almost always below 10 mg l^{-1}, demand for higher quantities of oxygen can be met only by increasing the rate at which it is transferred from the gas phase to the liquid phase. Transfer of oxygen to the medium can be improved by increasing the aeration and/or agitation rate and by modifying the reactor geometry (improved sparger design, addition of baffles, etc.) at a given temperature of operation. However, these methods have their own practical limitations and cannot increase the oxygen transfer rates indefinitely. In addition, the aerobic sludge digestion process that takes place in parallel with acid production also requires a considerable quantity of oxygen. Consequently, the system has to be studied with respect to its oxygen requirements and operated such that either there is no oxygen limitation taking place at any time or the kinetic expression used to design the system takes into account the effect of oxygen limitation on the system performance.

For a SSDML process carried out in the batch mode, the maximum oxygen uptake rate (OUR) is encountered at the beginning of the process and the value of OUR decreases drastically over the days of operation. This means that the possibility of encountering oxygen-limited kinetics is at the initial stage. Consequently, it is possible to operate the batch SSDML process with a variable aeration rate with the aeration rate decreasing over the days of operation.

During the SSDML process, the sludge pH decreases with time, resulting in a final value of around 2.0 or even lower. In the conventional aerobic sludge digestion process, the pH change is not so prominent. There, the drop in pH is caused by the nitrification reactions taking place and seldom decrease beyond pH 5.0. The steady reduction in the sludge pH during the SSDML process has a marked effect on the solids reduction rate.

The relation between MLVSS(B) concentration (M) and MLVSS(B) reduction rate (dM/dt) is influenced by (1) sludge pH, and (2) oxygen availability, and possibly by (3) sludge MLSS concentration. The sludge undergoing the SSDML process comes from the primary and/or secondary treatment stage clarifier(s) and is bound to reflect any disturbance to the wastewater treatment system. Such disturbance usually leads to variations in the quality of clarifier underflow. This means variations in the solids concentration of the sludge to be treated by the SSDML process.

The variation of the sludge pH during the SSDML process is very similar to the behavior of sludges having different solids concentrations during the bioleaching process using sulfur-oxidizing bacteria. There it was due to the higher buffering capacity of the sludge at higher solids concentrations, and it is true for the present situation also. The sulfate concentration increases faster in the sludge with higher solids concentration, but its effect on the sludge pH is buffered down by the higher buffering capacity of the sludge. Since the specific growth and product formation rates of the sulfur-oxidizing bacteria decrease with lowering of the sludge pH, the maximum sulfate production rate is observed for the sludge with the highest solids concentration. Thus, the effect of variation in solids concentration on the sludge acidification during the SSDML process is qualitatively similar to the effect observed in the case of sludge bioleaching.

A new, closed loop process for the disinfection, stabilization and removal of heavy metals from sewage sludge (consisting of a sludge/sulfuric acid reactor, hybrid H_2S generator and H_2S bioscrubber) has been described by Lowrie et al. (2002). Total solids (TS), volatile suspended solids (VSS), chemical oxygen demand (COD), acetate and propionate destruction in the hybrid H_2S generator have shown that digestion efficiency is not compromised in a hybrid reactor generating H_2S compared with a methanogenic reactor. 70% of the electron flow in the hybrid H_2S generator was diverted to methane at a COD/SO_4 ratio of 5.45:1. Enough H_2SO_4 could be generated from the H_2S emitted at this ratio to effect sufficient metal solubilization and pathogen removal from primary sludge.

15.5 TOXICITY REDUCTION THROUGH BIOLEACHING

Bioleaching of sewage sludge was successful in reducing the toxicity due to heavy metals (Renoux et al. 2001). Detoxification was demonstrated with five toxicological endpoints:

lettuce and barley germination (5 days), barley growth (14 days) and Microtox (15 min). Neither type of sludge was harmful in terms of earthworm body weight change (28 days) and genotoxicity as measured with the SOS Chromotest. However, given the high variability, no definitive conclusion could be drawn based on lettuce root elongation (5 days) or daphnid mortality (48 h), although the latter endpoint seemed to be more sensitive to the treated sludge elutriate.

Metal analyses showed the untreated sludge to be fairly contaminated with metals. Noticeable decreases were observed following treatment for Cd, Cu, Mn, Ni and Zn. In contrast, the treatment raised S levels in the sludge from 1 to 3.4%. However, the fertilizing value of the sludge (K and P) was preserved. Elutriates from both types of sludges also differed chemically. Among analyzed metals, the Zn concentration was higher in the bioleached sludge elutriate than in the untreated one, while the Mn concentration was lower. More C, P and S were also found in the treated sludge elutriate. Zn appeared noticeably more water soluble in the bioleached sludge, if one compares the proportion of metal recovered after elutriation. This increase in mobility was also observed, but to a lesser extent, with Cu and P.

Copper and Zn were in sufficiently high concentrations in both elutriates to account for *D. magna* mortality. At the LC_{50} dilutions observed, Cu concentrations were respectively 0.014 and 0.004 mg/l in the untreated and bioleached sludge elutriates. The presence of Zn in the elutriates (about 0.7 for the untreated and 1.6 for the bioleached sludge) may also have contributed to the toxic effect. Moreover, the sensitivity of *D. magna* to sludge elutriates may be due to organic matter, reflected by the total carbon content and conductivity, both higher in the bioleached sludge elutriate sample.

However, comparison of the other toxicological results with the corresponding metal analyses showed that individual metal concentrations could not explain the toxic effects observed. This was the case when the sludge elutriates were tested with *V. fischeri* bioluminescence. A study using the Microtox solid-phase procedure showed that the presence of Cu in the sludge, and to a lesser extent the presence of Pb and Zn, were responsible for bioluminescence inhibition. According to the study, Cu and Zn in solution affected *V. fischeri* bioluminescence differently. Fifty percent inhibition occurred at 1.7 and 0.45 mg/l for Cu, and at 1.9 and 2.7 mg/l for Zn. The Cu concentrations in the two elutriates (at least 10 times less) were too low to explain the inhibition observed (about 94% of inhibition with both elutriates half diluted). However, enhancement of the Cu-induced effect on *V. fischeri* by sewage water or sludge has already been observed. To obtain a better estimate of their effect on bioluminescence, Zn (sulfate form) and Mn (chloride form) were also tested in pure solutions. Microtox IC_{50} and toxic threshold for Zn were, respectively, 0.8 and 0.03 mg/l.

The combined action of all metals, including those not monitored with the ICP, was the most likely explanation for the higher impact of the untreated sludge elutriate, since it would appear that only a few organic pollutants are leachable from urban sludge.

As with the elutriates, the inhibiting action of the sludge applied to soil on barley growth cannot be directly attributed to the toxicity of one metal in particular. However, this inhibition only occurred at the higher untreated sludge application ratio, where the Cu concentration was 125 mg/kg. No effect was observed with the bioleached sludge or at lower ratios of application of untreated sludge, i.e. when Cu concentrations ranged between 12 and 75 mg/kg. In the present study, the Zn concentration in the sludge-soil mixture was higher than 100 mg/kg, regardless of mixture conditions, but subsequent barley uptake of Zn was unaffected by sludge application. In addition, leaf concentrations of Zn (around 70 mg/kg) were comparable to those reported in the literature for the same species, since 10-60 mg/kg were reported for barley background contents, 30-100 mg/kg in the presence of sludge, and leaf concentrations of Zn had to exceed 120 mg/kg to have a critical impact on yields of different crop species. In contrast to Zn, Cu bioaccumulated in barley sprouts increased up to 15 mg/kg of leaf after addition of the untreated sludge. Concentrations ranging between 14 and 25 mg/kg were judged to have a critical impact on barley growth. Normal leaf concentrations of 6-12 mg/kg of Cu were reported, corresponding to the values observed in the control of the present study. Hence, the increase in the internal Cu pool in the presence of the untreated sludge may have been directly responsible for the sprout growth inhibition. The presence of other metals may also have enhanced the effect. Sludge bioleaching not only eliminated inhibiting factors, but also restored the fertilizing value of the sludge.

Although the bioleaching treatment mostly led to a concomitant reduction in metal concentrations and in toxicity, their dependent relationship remains to be demonstrated. In order to strengthen this hypothesis, bioconcentration factors, which in the present study are ratios of single-element concentrations in barley sprout tissues and in the sludge-soil mixture, were calculated for Cu, Zn, and Mn. These factors varied with soil contamination, sludge-soil ratios, and type of sludge. The variations clearly showed that metal bioavailability differed in the soils amended with treated and untreated sludge. If the concentration factors are compared for Cu and Zn, at equal metal concentrations in the soil, they are higher for untreated than for bioleached sludge, indicating higher bioavailability. Moreover, the concentration factor tended to decrease as more contaminant or sludge was added to the soil. For these metals, the concentration factors were less than one, indicating that the organism limited net absorption or translocation to the leaves.

15.6 METAL RECOVERY FROM LEACHATE

Conditioning and dehydration of leached sludge are two very significant steps in the sludge treatment and decontamination chain. The conditions of acidity and redox potential prevailing in the decontaminated sludge are seldom (never) encountered during conventional sludge treatment processes. Use of a belt filter press is not preferred since it has many metallic parts that come into contact with the decontaminated sludge, leading to serious corrosion problems. Centrifugation immediately following the leaching stage is possible, but is extremely expensive. The plate and frame filter press is presently the favored means for dewatering of the sludge subjected to bioleaching.

Following the dehydration stage, the solubilized metals present in the filtrate must be recovered. The leachate is highly acidic and contains considerable quantities of dissolved and colloid organic matter and sulfates in addition to the solubilized metals. Use of biosorbants for a selective recovery of metals is one of the options being explored at present.

A number of technologies such as chemical precipitation, adsorption, membrane separation, solvent extraction, magnetic separation, electrolytic recovery, etc. can be used to recover the metals in solution. However, most of these technologies are specific to one metal and/or are too expensive to be used for the treatment of sludge leachate. Chemical precipitation using lime (calcium hydroxide) as neutralizing and precipitating agent is one of the most commonly used methods for the recovery of metals from a number of complex solutions.

All precipitation processes generate metal-contaminated sludge which causes problems during its disposal in the environment. A substantial part of the metals present in the acid leachate are in the form of free ions. The precipitation step consists of formation of flocs of metallic hydroxides which sediment easily, leading to a significant reduction in the metal contents of the leachate. Addition of polyelectrolytes to the leachate increases considerably the aggregation and sedimentation process.

The variation in optimum conditions for the removal of each metal is regulated by a number of factors. First, the minimum solubility of metallic hydroxides varies for each of the metals. There is also competition among the metals for the hydroxide ions (OH^-) available in the system. Because of this phenomenon, optimum metal removal can be measured at different pH values, permitting a minimal solubilization in an aqueous solution containing none but the given metal.

15.7   PROCESS ECONOMICS

The cost of carrying out the metal-leaching and removal process is extremely important and there will be competition among the various technological options available. Here we take a look at the cost of metal leaching by three different processes, viz. acid addition process, ferrous sulfate process and the process based on elemental sulfur addition. Of these three processes, the first is purely a chemical process; the second is chemical combined with a biological step; and the third purely biological.

15.7.1   Description of the Processes

In the chemical leaching process (acidification process) for metal removal from sludges, concentrated sulfuric acid is added to the sludge in a reactor vessel and mixed thoroughly. The acid addition is controlled by a pH monitor and controller system, and the sludge pH is brought down to 1.5-2.0, conditions under which many of the heavy metals present in the sludge will be solubilized. The dissolved metals are then separated from the sludge during the sludge-dewatering step. Sludge transfer from the sludge storage tank, controlled acid addition and the mixing operations altogether take 24 h. The treated sludge is then sent to the dewatering section. The process carried out like this is essentially a batch operation, one batch being completed every 24 h.

In the iron-oxidation process (ferrous sulfate process) for metal bioleaching from sewage sludge, raw sludge is pumped into the reactor and its pH is lowered to 4.0 by controlled addition of concentrated sulfuric acid. This is followed by addition of ferrous sulfate from an overhead bin. The inoculum for the bioleaching operation is provided by the previous batch of sludge, 10% of which is retained in the reactor. The sludge is aerated by compressed air sparged through a network of air diffusers placed at the bottom of the reactor. The reactor is equipped with a pH monitor/controller to monitor the pH drop during the initial acidification stage as well as during the bioleaching operation.

The total process, from the addition of sludge and acid to the completion of the leaching operation, takes 48 h. The treated sludge goes to the dewatering section.

In the sulfur-oxidation process for metal bioleaching, the sludge is pumped into the reactor and elemental sulfur added to it from an overhead bin. Inoculum is provided by the previous batch, 10% of which is retained for this purpose. Aeration is provided by a system of submerged air diffusers using compressed air. The batch time is 48 h and the treated sludge goes for dewatering.

The sludge dewatering and sludge and leachate disposal steps are common to all three processes and operate continuously. Dewatering is carried out using a centrifuge filter and the dewatered sludge, which is at a pH of about 2.0, is brought to neutral pH by adding lime, and is sent for land application. Metal compounds dissolved in the leachate are precipitated by lime addition and separated from the liquid phase by gravity settling followed by centrifugation. The liquid stream is recycled to the secondary wastewater treatment stage.

15.7.2 Cost Calculation

For the purpose of cost calculations, the metal-leaching and downstream processes associated with that operation were considered as battery-limit additions to an existing wastewater treatment facility. Consequently, the cost of land, building, stand-by power source (capital investment only) and lighting were excluded from these calculations.

At lower plant capacities (expressed in tonnes per day of dry sludge treated), the sulfur-oxidation process had the lowest computed cost figure (expressed as $ CAN. per tonne of dry sludge) and the iron-oxidation process the highest.

However, as the plant capacity is increased, the sulfur-oxidation process becomes costly compared with the acidification process. The cost computations also reveal that the cost of metal removal is related to the volume of sludge treated rather than the sludge dry weight.

For all three processes, increase in plant capacity is associated with decrease in equipment cost per unit amount of sludge treated. Irrespective of the process employed for the metal-leaching operation, sludges having lower solids concentration incur higher cost per tonne of dry sludge treated. This is due to the increase in volume of sludge handled while dealing with a sludge having low solids concentration. Another factor that has a considerable effect on the process economics is the aeration rate. Higher aeration rates call for higher capacity compressors and an associated rise in cost of utilities.

REFERENCES

Adamu CA, Bell PF, Mulchi C (1989) Residual metal concentrations in soils and leaf accumulations in tobacco a decade following farmland application of municipal sludge. Environ Pollut 56: 113-126
Alloway BJ, Jackson AP (1991) The behavior of heavy metals in sewage sludge-amended soils. Sci Total Environ 100: 151-176
Babich H, Devanas MA, Stotzky G (1985) The mediation of mutagenicity and clastogenicity of heavy metals by physicochemical factors. Environ Res 37: 253
Blais JF, Tyagi RD, Auclair JC (1993) Bioleaching of metals from sewage sludge: effects of temperature. Water Res 27: 111-120

Bruce AM, Davis RD (1989) Sewage sludge disposal: current and future options. Water Sci Technol 21: 1113-1128

Campanella L, Cardarelli E, Ferri T, Petronio BA, Pupella A (1985) Evaluation of toxic metals leaching from urban sludge. In: Pawlowski L, Alaerts G, Lacy WJ (eds) Chemistry for protection of the environment. Elsevier, Amsterdam, pp 151-161

Cappon CJ (1991) Sewage sludge as a source of environmental selenium. Sci Total Environ 100: 177-285

Carmignami M, Boscolo P, Ripanti G, Finelli VN (1983) Effects of chronic exposure to cadmium and/or lead on some neurohumoral mechanisms regulating cardiovascular function in the rat. Proceedings of the 4th international conference on heavy metals in the environment. CEP Consultants, Edinburgh, pp 557-558

Coker EG, Matthews PJ (1983) Metals in sewage sludge and their potential effects in agriculture. Water Sci Technol 15: 209-225

Davis RD, Carlton-Smith CH (1980) Crops as indicators of the significance of contamination of soil by heavy metals. Technical report 140. Water Research Center, Stevenage, Herts, UK

Davis RD, Carlton-Smith CH (1984) An investigation into the phytotoxicity of zinc, copper and nickel using sewage sludge of controlled metal content. Environ Pollut B 8: 163-185

Doyle PJ, Lester JN, Perry R (1978) Survey of literature and experience on the disposal of sewage sludge on land. Final report to the UK Department of the Environment

Granato TC, Richardson GR, Pietz RI, Lue-Hing C (1991) Prediction of phytotoxicity and uptake of metals by models in proposed USEPA 40 CFR part 503 sludge regulations: comparison with field data for corn and wheat. Water Air Soil Pollut 57/58: 891-902

Hernandez T, Moreno JI, Costa F (1991) Influence of sewage sludge application on crop yields and heavy metal availability. Soil Sci Plant Nutr 37: 201

Ito A, Umita T, Aizawa J, Takachi T, Morinaga K (2000) Removal of heavy metals from anaerobically digested sewage sludge by a new chemical method using ferric sulfate. Water Res 34: 751-758

Jackson AP, Alloway BJ (1991) The transfer of cadmium from sewage-sludge amended soils into the edible components of food crops. Water Air Soil Pollut 57/58: 873-881

Jenkins RL, Scheybeler BJ, Smith ML, Baird R, Lo MP, Haug RT (1981) Metals removal and recovery from municipal sludge. J Water Pollut Control Fed 53: 25-32

Jennette KW (1981) The role of metals in carcinogenesis: biochemistry and metabolism. In: Environmental health perspectives, vol 40. Role of metals in carcinogenesis

Jing J, Logan TJ (1992) Effects of sewage sludge cadmium concentration on chemical extractibility and plant uptake. J Environ Qual 21: 73-81

Kelly DP, Harrison AP (1988) Genus *Thiobacillus*. In: Holt JG, Staley JT, Bryant MP, Pfennig N (eds) Bergey's manual of determinative bacteriology, vol 3. Williams and Wilkins, Baltimore, MD, pp 1842-1858

Klessa DA, Desira-Buttigieg A (1992) The adhesion of leaf surfaces of heavy metals from sewage sludge applied to grassland. Soil Use Manage 8: 115-121

Korentajer L (1991) A review of the agricultural use of sewage sludge: benefits and potential hazards. Water South Afr 17: 189-196

Lester JN, Sterrit RM, Kirk PWW (1983) Significance and behavior of heavy metals in waste water treatment process. II. Sludge treatment and disposal. Sci Total Environ 30: 45-83

Levine MB, Hall AT, Barrett GW, Taylor DH (1989) Heavy metal concentrations during ten years of sludge treatment to an old-field community. J Environ Qual 18: 411-418

Lo KSL, Chen YH (1990) Extracting heavy metals from municipal and industrial sludges. Sci Total Environ 90: 99

Lowrie D, Hobson J, Stuckey DC (2002) Sulfate disinfection, stabilisation and heavy metal removal from sewage sludge-process: description and preliminary results. Water Sci Technol 45(10): 287-292

Martell AE (1981) Chemistry of carcinogenic metals. Environ Health Perspect 40: 207

Mench M, Juste C, Solda P (1992) Effets de l'utilization de boues urbaines en essai de longue duree:accumulation des metaux par les vegetaux superieurs. Bull Soc Bot Fr 139: 141

Mininni G, Santori M (1987) Problems and perspectives of sludge utilization in agriculture. Ecosystem Environ 18: 291-311

NAS (1979) Geochemistry of water in relation to cardiovascular disease. US National Academy of Sciences, Washington, DC

NAS (1980) Lead in the human environment. US National Academy of Sciences, Washington, DC

Nriagu JO (1988) A silent epidemic of environmental metal poisoning. Environ Pollut 50: 139-161

Naoum C, Fatta D, Haralambous KJ, Loizidou M (2001) Removal of heavy metals from sewage sludge by acid treatment. Journal of environmental science and health, part A, Toxic/hazardous substances and environmental engineering 36: 873-881

Obbard JP, Sauerbeck DR, Jones KC (1993) Rhizobium leguminosarum bv. trifolii in soils amended with heavy metal contaminated sewage sludges. Soil Biol Biochem 25: 227-231

Renoux AY, Tyagi RD, Samson R (2001) Assessment of toxicity reduction after metal removal in bioleached sewage sludge. Water Res 35: 1415-1424

Roca J, Pomares F (1991) Prediction of available heavy metals by six chemical extractants in a sewage sludge-amended soil. Commun Soil Sci Plant Anal 22: 2119-2136

Sommers LE, Nelson DW (1981) Monitoring the response of soils and crops to sludge applications. In: Bouchart JA, Jone WJ, Sprague GE, (eds) Sludge and its ultimate disposal. Ann Arbor Science, Ann Arbor, MI, 286 pp

Sterritt RM, Lester JN (1980) The value of sewage sludge to agriculture and effects of the agricultural use of sludges contaminated with toxic elements: a review. Sci Total Environ 16: 55-90

Tadesse W, Shuford JW, Taylor RW, Adriano DC, Sajwan KS (1991) Comparative availability to wheat of metals from sewage sludge and inorganic salts. Water Air Soil Pollut 55:397-408

USEPA (1986) Annual volume of sludge in the US standards for the disposal of sewage sludge; proposed rule. US EPA, Cincinnati, OH

USEPA (1993) Standards for the use and disposal of sewage sludge. 40 CFR Parts 257, 403 and 503. Final rule, US EPA, Cincinnati, OH

Veeken AHM, Hamelers HVM (1999). Removal of heavy metals from sewage sludge by extraction with organic acids. Water Sci Technol 40: 129-136

Webber MD (1986) Epandage des boues d'epuration sur les terres agricoles—une evaluation. Direction Generale de la Recherche, Agriculture Canada, 42 pp

Wong L, Henry JG (1988) Bacterial leaching of heavy metals from anaerobically digested sludge. In: Wise DL (ed) Biotreatment systems, vol 2. CRC Press, Boca Raton, FL, pp 125-169

Xiang L, Chan LC, Wong JW (2000) Removal of heavy metals from anaerobically digested sewage sludge by isolated indigenous iron-oxidizing bacteria. Chemosphere 41: 283-287

Chapter 16

Species-Selective Analysis for Metals and Metalloids in Plants

Dirk Schaumlöffel, Joanna Szpunar, Ryszard Łobiński

CNRS UMR 5034, Group of Bio-Inorganic Analytical Chemistry, 2, av. Pr. Angot, 64000 Pau, France

16.1 INTRODUCTION

Interactions of plants with metals and metalloids have attracted considerable attention in recent years (Merian 1991; Prasad 1996). This interest has been for a number of different reasons, reflecting the variety of areas of the importance of plants and the different roles played by the different elements. Indeed, whereas many elements (e.g. B, Se) are well-established essential nutrients for plant growth, others (e.g. Cd, Pb) are well-recognized stress factors. Plants produce edible fruits and vegetables that are important (and sometimes major) sources of trace elements in the human diet that makes legislation set limits regarding toxic metal concentrations. Tea, coffee and wine are important sources of trace elements in the human diet in some countries. Selenium-enriched plants are increasingly used in medicine and as food supplements. The use of specially engineered plants has been attracting growing interest in the remediation of polluted waters and soils (phytoremediation). For some elements, especially As, plants (e.g. algae, seaweeds, and phytoplankton) are an important link in food chains.

In the environment, plants are much more exposed to the pollution risk in comparison with other living organisms since they cannot move around. In particular, plants growing in urban areas and along busy motorways have suffered from automotive pollution from alkyllead compounds (antiknock additives to petrol) and their substituents: organomanganese compounds, and platinum released from the catalysts. The stress to plants from these sources would never have been noticed if not for the fact that this contamination affected vineyards of which the fruit is used to make wine.

The mechanisms of metal uptake by roots, metal translocation from roots to shoots, and plant tolerance to toxic metals are dependent on the molecular forms (speciation) of metals which can be further modified by the organism studied. Eukaryotic cells resist the cytotoxic effects of heavy metal ions by sequestering the ions in stable, intracellular macromolecular complexes. Different types of metal-chelating compounds have been developed by living organisms to regulate the intracellular metal ion concentration. They include: amino acids, citric acid, malic acid, phytochelatins and metallothioneins. The synthesis of some of these compounds, e.g. phytochelatins, is enzymatically mediated, i.e. requires a protein activated by the metal of interest. The identification of the complex of the metal with the enzyme and the characterization of the metal complexes with products of the enzyme-catalyzed reaction is an emerging field of research in environmental speciation analysis.

In order to be of concern in terms of essentiality and/or toxicity, trace elements in the human diet must be bioavailable, i.e. readily absorbable by the gut and further utilizable in the body. Since the bioavailability depends critically on the actual species of an element present, it is becoming more and more evident that information on the concentration of an element in a foodstuff tells us very little about how well the element will be assimilated. Precise information regarding the identity, nature and concentration of individual metal compounds present in a sample is therefore required. This concerns, in particular, food and feed supplements aimed at supplying a given oligoelement, e.g. selenium.

The success of using hyperaccumulating plants for phytoremediation of contaminated soils and waters requires a better understanding of the mechanisms of metal uptake, translocation and accumulation by plants. The prerequisite for the understanding of these mechanisms is the identification of the molecules involved, notably bioligands and metal complexes synthesized by a plant to be used for metal transport and suitable for bioaccumulation.

Whereas the total element concentration status has been a long-established parameter for assessing stress to plants, its nutritive value in terms of oligoelement supply or its toxicity, species-related information has often been neglected despite its crucial value. The reason for this has been the lack of analytical techniques able to deliver information about the concentration of several species in which each of the trace elements is present in a sample. The purpose of this chapter is to highlight the recent analytical techniques that allow a rapid, species-selective and sensitive analysis for compounds containing a metal or a metalloid present in plant tissues and foodstuffs of plant origin. The coverage of the chapter is limited to species that are thermodynamically stable and kinetically inert, i.e. that pass intact through a chromatographic column.

16.2 METAL AND METALLOID SPECIES IN PLANTS

In terms of chemistry, species of interest can be divided into organometallic species that contain a covalent bond between an atom of carbon and that of a metal or metalloid and coordination complexes in which metal is coordinated by a bioligand present or induced in plant cells. Organometallic species enter the plant as a result of external environmental contamination or can be synthesized in a plant (methylmercury, organoarsenic or organoselenium species). Metallocomplexes are synthesized by a plant itself from the metal ions that penetrate the plant cells. The species of interest can be divided into several categories as discussed below.

16.2.1 Species with a Covalent Metal (Metalloid)-Carbon Bond

Tetraethyllead, tetramethyllead and sometimes mixed methylethyllead compounds were commonly used in the past as antiknock agents. During combustion they undergo a variety of degradation processes leading to a number of mixed ionic organolead species: Me_3Pb^+, Me_2Pb^{2+}, Et_3Pb^+, Et_2MePb^+, Et_2Pb^{2+}, and others (van Cleuvenbergen and Adams 1991). Organolead contamination in plants may result either from direct intake of natural water through the roots or from absorption of the pollutants through the foliage. The contamination of fresh grass and tree leaves by organolead compounds has been studied (van Cleuvenbergen et al. 1990).

Other compounds with a metalloid-carbon bond that have attracted considerable interest include organoarsenic and organoselenium species (Cullen and Reimer 1989; Shibata et al. 1992). It was revealed that marine plants (algae and seaweeds) accumulate substantial amounts of arsenic distributed among a number of organoarsenic species. Besides arsenate (AsO_4^{3-}) and arsenite (AsO_3^{3-}), monomethyl- ($CH_3AsO_3^{2-}$) and dimethylarsenic (cacodylic) [$(CH_3)_2AsO_2^-$] acid, trimethylarsine oxide [$(CH_3)_3AsO$], tetramethylarsonium [$(CH_3)_4As^+$], arsenobetaine [$(CH_3)_3As^+CH_2COO^-$] and arsenocholine [$(CH_3)_3As^+CH_2CH_2OH$] have been identified. More complex organoarsenic compounds, so-called arsenosugars, were also isolated from a brown algae by Edmonds and Francesconi (1987), which were quantified in a subsequent work by the same group (Madsen et al. 2000). Recently, McSheehy et al. have identified a number of arsenosugars in algae (McSheehy et al. 2000a) and oysters (McSheehy et al. 2001a). Despite the fact that arsenic concentrations in terrestrial organisms are generally much lower than in the marine environment, Kuehnelt et al. (2000) were able to show that most of these arsenic compounds occur also in green plants and lichens.

Regarding selenium, besides the most common selenate (SeO_4^{2-}) and selenite (SeO_3^{2-}) species, a number of selenoamino acids have been identified in plants, the most ubiquitous predominant species being selenomethionine found in wheat, soybeans and selenium-enriched yeast (Beilstein et al. 1991; Shibata et al. 1992; Dauchy et al. 1994). The most widely studied plant is selenium-enriched yeast. The majority (>80%) of selenium in yeast is water-insoluble and present in a bound form, or more than 50% of the selenium is probably protein-bound or in other large organic species (Casiot et al. 1999). Gilon et al. (1995) reported the selenium in yeast to be in three forms: inorganic selenium, selenocysteine, and selenomethionine. More than 20 selenium compounds including selenocysteine, selenomethionine, methylselenocysteine, and inorganic forms were found to be present in selenium-enriched yeast (Ge et al. 1996; Bird et al. 1997), where, in a recent work, the structures of five selenium compounds have been identified (McSheehy et al. 2002b). Selenium-enriched garlic, onion and broccoli were studied to identify five selenium species and several unknown peaks (Cai et al. 1995). From Chinese garlic grown on naturally seleniferous soils, γ-glutamyl-Se-methylselenocysteine was isolated and characterized as the main selenium species (McSheehy et al. 2000b). Selenomethionine was the product of proteolytic enzyme digestion of the 7S globulin fraction of soya-bean protein (Yasumoto et al. 1988), and the same compound was also found as the primary selenium species in several nut varieties (Kannamkumarath et al. 2002). Several seleno analogs of sulfur-containing amino acids and their derivatives have been identified in terrestrial plants, especially so-called selenium-accumulator plants (Brown and Shrift 1982), which are also regarded as candidates for Se-phytoremediation. Three selenium species, Se-methylselenocysteine, Se-homocysteine and Se-cystathionine, were proposed to occur in the Se-accumulator *Brassica juncea* (Indian mustard), while, in genetically modified *Brassica juncea*, the overexpression of Se-cysteine methyltransferase leads to the formation of Se-methylselenocsteine (Montes-Bayón et al. 2002).

16.2.2 Metal Complexes with Phytochelatins and Metallothioneins

Phytochelatins (cadystins) are short metal-induced sulfhydryl-rich peptides possessing the general structure: $(\gamma\text{-GluCys})_n\text{-Gly}$ where n=2-11. They are synthesized from glutathione in plants and fungi exposed to metal ions: Cd^{2+}, Cu^{2+}, and Zn^{2+}. Metals are chelated through coordination with the sulfhydryl groups in cysteine. An intracellular complex formed by these thiol peptides is thought to detoxify the metal by sequestration in the vacuole (Rauser 1996; Zenk 1996). Two principal structures of phytochelatins have been reported in the literature: cadystin A $(\gamma\text{-EC})_3G$ and cadystin B $(\gamma\text{-EC})_2G$, where the γEC unit stands for $-\gamma$-Glu-Cys (Hayashi and Winge 1992).

Higher polypeptides, such as e.g. metallothioneins (MTs), have also been studied. MTs are a group of non-enzymatic polypeptides with low molecular mass (6-7 kDa), rich in cysteinyl residues and able to complex metal ions having sulfur affinity. This group of proteins is characterized by their resistance to thermocoagulation and acid precipitation, by the presence of ca. 60 non-aromatic amino acids, and by the absence of disulfide bonding (Stillman et al. 1992; Stillman 1995). MTs are most common in the animal kingdom; they have also been reported in several plants, e.g. tomato (Inouhe et al. 1991). The expression of metal-binding peptides and proteins such as MTs in genetically modified plants in order to enhance their metal accumulation properties for phytoremediation purposes was recently reviewed by Mejare and Bülow (2001).

Rapid differentiation between complex-bound and non-bound cadmium is essential to evaluate the content of toxic cadmium in edible resources and foodstuffs. Binding of heavy metals in excess with phytochelatins has been demonstrated for Cd and Cu by Leopold and Günther (1997) in cell cultures of plants as well as for Cd in rice, soybeans (Vacchina et al. 2000) and maize (Chassaigne et al. 2001). Individual and synergistic effects of Cu, Zn and Cd ions on the induction of metallothionein in cyanobacterium (the origin of trophic chains in an aquatic systems) were investigated (Takatera and Watanabe 1993). The in vivo selenium association with cyanobacterial metallothioneins following the coadministration of Zn and selenite or selenate was examined (Takatera et al. 1994).

16.2.3 Metal Complexes with Macrocyclic Chelate Ligands in Plants

Species of some trace elements, especially Co, Ni, and Mo, are essential for environmentally important processes, e.g. microbial production of methane and biomethylation of trace elements. They are components of coenzymes which show a very special functionality. A coenzyme (low molecular mass non-protein species) assists in enzymatic catalysis as a true substrate that determines the type of reaction while the corresponding apoenzyme (a protein) determines substrate specificity and the reaction rate. The most important class of metallocoenzymes are the tetrapyrroles that are at least partially unsaturated tetradentate macrocyclic ligands, which, in their deprotonated forms, can tightly bind divalent metal ions (for structures, see Kaim and Schwederski 1994).

Probably the best known are the analogs of cobalamin which are prevalent in nature as a consequence of bacterial synthesis and are responsible for the catalysis of methyl transfer reactions (Dolphin 1982; Hall 1983). Particular attention in environmental chemistry has been paid to coenzyme B_{12} (adenosylated form of cobalamin) which is responsible for the process of transfer of a CH_3^- ion to Hg, Pb and Sn (biomethylation). Cobalamins co-occur with other classes of corrinoid analogs devoid of enzymatic activity that lack the nucleotide moiety (5,6-dimethylbenzimidazole).

Another example of a complex of a tetradentate macrocycling with an essential element is the complex of the factor F_{430}, a nickel-containing prosthetic group of S-methyl coenzyme-M reductase found in methanogenic bacteria. Owing to the occurrence of this species, Ni appears as an essential element in the production of methane by the biological reduction of CO_2 by hydrogen to give methane by archaebacteria.

16.2.4 Metal Complexes with Polysaccharides

Polysaccharides and glycoproteins are important constituents of plants and contain numerous potential metal-binding sites. The detailed three-dimensional structures of carbohydrates, and the role of metal ions in determining and regulating these structures, constitute essentially unchartered territory in bioinorganic chemistry (Stillman et al. 1992) and one of the important research challenges in environmental speciation analysis.

Some recent studies have focused on a pectic structurally complex polysaccharide rhamnogalactorunan II (RG-II; Doco et al. 1993; Pellerin et al. 1996). RG-II isolated from cell walls represents a mixture of a monomer and a dimer, the latter being cross-linked by a 1:2 borate-diol ester (Ishii and Matsunaga 1996; Matsunaga et al. 1996, 1997; O'Neill et al. 1996). The fractions of the dimer isolated from plant cells were found to bind in vitro di- and trivalent metals with an ionic radius higher than 0.95 Å (Pb^{2+}, Sr^{2+}, Ba^{2+}, lanthanides; O'Neill et al. 1996). In contrast to the previously discussed metallocomplexes, polysaccharides represent a continuous polymorphism of the ligands involved that renders their characterization more difficult.

16.2.5 Metal Complexes with Organic Acids

A number of organic acids such as citric acid, oxalic acid or succinic acid as well as some amino acids and members of the mugineic acid family occur in plant tissues and are possible ligands for metal complexation (Łobiński and Potin-Gautier 1998). The thermodynamic stability of the metal complexes strongly depends on their equilibrium constants; thus, not every possible complex can be isolated with a chromatographic column. The toxicity of aluminum to plants is well known as a factor for the inhibition of root growth and the induction of other harmful effects. Bantan et al. (1999) studied the speciation of aluminum in plant sap of *Sempervivum tectorum* and *Sansevieria trifsciata*. Negatively charged low molecular weight Al species were found, predominant as Al-citrate and Al-aconitate complexes. Recently, it has been demonstrated that, in Chinese cabbage, aluminum is transported from the roots to the shoots as Al-malate and Al^{3+} (Bantan Polak et al. 2001).

Research on mechanisms of the resistance of plants is important in view of phytoremediation which is the use of specially selected and engineered metal-accumulated plants for the depollution of waters and soils. The following processes are of potential interest: phytoextraction—the use of metal-accumulating plants to remove toxic metals from soil, rhizofiltration—the use of plant roots to remove toxic metals from polluted waters, and phytostabilization—the use of plants to eliminate the bioavailability of toxic metals in soils (Raskin et al. 1997). Access to information on the molecular level by identifying and characterizing the metal complexes involved in the processes is the prerequisite to the understanding of the mechanisms involved and represents an important challenge to environmental analytical chemistry.

Remarkably little is known about the molecular forms of metals occurring in hyperaccumulating plants. The notable few exceptions include the correlation of free nickel and histidine levels in *Alyssum lesbiacum* suggesting a Ni-histidine complex being responsible for the xylem transport (Krämer et al. 1996) and the demonstration of citrate as one of the ligands binding nickel in the *Sebertia acuminata* latex (Sagner et al. 1998). Recent studies identified a nickel complex with nicotianamine in *Thlaspi caerulescens* (Vacchina et al. 2003) as well as in *Sebertia acuminata* (Schaumlöffel et al. 2003).

16.2.6 Miscellaneous

The introduction of automobile exhaust catalysts in 1974 in the US and later in Japan and Europe resulted in emissions of Pt into the environment that turned out to be considerably higher than had been expected. For Pt speciation in biotic materials a widespread forage plant (*Lolium multiflorum*) has preferably been used as a bioindicator, in particular for heavy metals (Peichl et al. 1992). It was shown that Pt was bound to a protein in the high molecular mass fraction of native grass (160-200 kDa; Messerschmidt et al. 1995). In Pt-treated cultures, however, about 90% of the total Pt is found in a fraction of low molecular mass (<10 kDa; Messerschmidt et al. 1994). The Pt-binding ligands could be identified only tentatively; some Pt coeluted with phytochelatin fractions and some with polygalacturonic acids (Klueppel et al. 1998).

The relatively high concentration of Al in tea has been the subject of concern, since for many people tea is the major source of Al in the diet. It was assumed that the metal-binding ligands in tea infusions are large polyphenolic compounds occurring widely in tea and other plants (Oedegard and Lund 1997).

In cocoa powder stable cadmium and lead species were found in the range between 6 and 48 kDa. These complexes were resistant to simulated gastrointestinal conditions suggesting possible ways of detoxifying Pb and Cd in cocoa (Mounicou et al. 2002b).

16.3 COUPLED TECHNIQUES FOR SPECIES-SELECTIVE ANALYSIS

The success of an analyst searching for a given organometallic moiety in a plant matrix depends on two factors. First, he or she must be sure of determining this and not another species (analytical selectivity). Secondly, the detection limit (sensitivity of the detector and noise level) of the instrumental setup should match the analyte's level in the sample.

Intrinsically species-selective techniques, such as Mössbauer spectroscopy, X-ray photoelectron spectroscopy (XPS), electron spin resonance spectroscopy (ESR), or mass or tandem mass spectrometry (MS or MS/MS), usually fail at trace levels in the presence of a real-sample matrix. ^{27}Al (70.4 MHz) NMR spectra were obtained on various intact samples of Al-accumulating plant tissues. None of the materials examined (*Hydrangea sepals* and leaves of three *Theaceae* species) contained detectable amounts of $Al(H_2O)_6^{3+}$, although the Al was present as hexacoordinated complexes, and in most instances in at least two forms (Nagata et al. 1991).

Selectivity in terms of species is typically achieved by on-line combination of a high-performance separation technique (chromatography or electrophoresis) with a parallel element-specific and molecule-specific detection. Non-specific detectors [UV, flame ionization detector (FID)] suffer from a large background noise and poor sensitivity. Analysis by a coupled technique is often preceded by a more or less complex wet chemical sample preparation. The latter is mandatory for the sample to meet the conditions (imposed by the separation technique) in terms of form to be presented to the instrumental system. There is a tendency to integrate this preliminary step into the whole experimental setup.

The abundance of coupled techniques available for elemental speciation analysis is shown in Fig. 16.1. The choice of the separation technique is determined by the physicochemical properties of the analyte (volatility, charge, polarity) whereas that of the detection technique is determined by the analyte's level in the sample. This is the sample

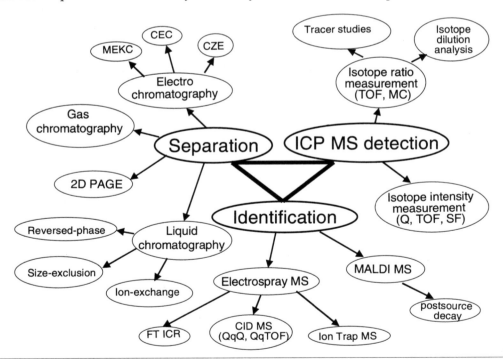

Fig. 16.1 Analytical coupled techniques in elemental speciation analysis

matrix (air, water, sediment, biomaterial) that dictates, in its turn, the choice of the sample preparation procedure (Łobiński 1997; Sturgeon 2000; Szpunar et al. 2000). Separation techniques for speciation analysis have been comprehensively reviewed (Szpunar 2000). For species which are volatile, or which are readily convertible to volatile ones by means of derivatization, gas chromatography (GC) is the method of choice (Łobiński 1994; Łobiński and Adams 1997). Species that do not fulfil the above requirement are separated by ion exchange or reversed-phase liquid chromatography. In particular, proteins and other biopolymers are separated by size-exclusion (gel-permeation) chromatography (Sutton et al. 1997; Barth et al. 1998; Makarov and Szpunar 1998; Zoorob et al. 1998; Gooding and Regnier 2002). The physicochemical similarity of many proteins stimulates the use of electrophoretic techniques for efficient separations (Richards and Beattie 1994; Hu and Dovichi 2002). In terms of detectors, plasma spectrometric techniques are favored over atomic absorption spectrometry (AAS) because of their much higher sensitivity. Microwave-induced plasma (MIP) atomic emission spectrometry is the choice for GC (Łobiński and Adams 1993), whereas the more energetic inductively coupled plasma (ICP) mass spectrometry is the choice for LC and capillary zone electrophoresis (CZE; Szpunar and Łobiński 1999).

Element-specific detectors do not allow for the identification of the species eluted. This drawback gains in importance as the wider availability of more efficient separation techniques and more sensitive detectors makes the number of unidentified species grow. Therefore, the prerequisite for progress in speciation analysis is a wider application of sensitive on-line techniques for compound identification, i.e. mass or tandem mass spectrometry with soft ionization [e.g. electrospray (ESI); Hofstadler et al. 1996] or matrix-assisted laser desorption/ionization (MALDI; Niessen and Tinke 1995).

16.3.1 Sample Preparation in Species-Selective Analysis of Plant Materials

Washing cells in a Tris-HCl buffer (pH 8) containing 1 M EDTA to remove metal ions reversibly bound to the cell wall is recommended (Behne et al. 1995). Metallothioneins and phytochelatins are prone to oxidation during isolation due to their high cysteine content. During oxidation disulfide bridges are formed and the MTs either copolymerize or combine with other proteins to move into the high molecular weight fraction. Since species of interest may be oxidized by, e.g. oxygen, Cu(I) or heme components, the homogenization of tissues and subsequent isolation of MTs should be normally performed in deoxygenated buffers and/or in the presence of a thiolic reducing agent, e.g. β-mercaptoethanol (Suzuki and Sato 1995).

Similar precautions are required during speciation of elemental redox states. Tervalent As is less stable than As(V) during storage in solution or in dried plant material; oxidation can be caused both by microorganisms and by, e.g., Fe(III). Caution in interpreting the results of As speciation analyses is thus advocated (Berghoff and von Willert 1988).

Approaches to speciation analysis in solid water-insoluble samples have been scarce because of the difficulties with a selective destruction of the solid matrix in a way that the metal complex of interest is preserved intact. To date, the studies have been limited to some organometallic species (organotin, methylmercury, alkyllead) of which the stability could readily be assured during the hydrolysis of proteinaceous matrix by tetramethylammonium hydroxide (van Cleuvenbergen et al. 1990) or proteolytic enzymes (Forsyth and Iyengar 1989).

16.3.1.1 Leaching with Water and Aqueous Buffers

In the most common procedures soluble extracts of tissues and cultured cells are prepared by homogenizing tissue samples in an appropriate buffer. Often water or water/methanol is used where no danger of affecting the complex acid-base stability occurs (organoselenium and organoarsenic speciation). Neutral buffers are usually used for extraction to avoid dissociation of the complexes. A 10-50 mM Tris-HCl buffer at pH 7.4-9 is the most common choice. In a recent study analyzing nickel species in hyperaccumulating plants, Ni-Tris complexes as artefacts were found when a Tris buffer was used for sample preparation (Vacchina et al. 2003). Therefore, in a subsequent study, ammonium acetate or simply water was used for extraction (Schaumlöffel et al. 2003). For cytosols containing Cd-induced MTs dilution factors up to 10 have been used whereas for those with natural MT levels equal amounts of tissue and buffer have been found suitable. Typical components of a homogenation mixture include: β-mercaptoethanol (antioxidant), NaN_3 (an antibacterial agent) and phenylmethylsulfonyl fluoride (protease inhibitor). The homogenization step is followed by centrifugation. The use of a refrigerated ultracentrifuge (100,000 g) is strongly recommended.

As a result two fractions, a soluble one (cell supernatant, cytosol) and a particulate one (cell membranes and organelles), are obtained. Only the supernatant is usually analyzed for biomacromolecules thus limiting the number of species of concern to those being cytosoluble. It is recommended that the supernatant be stored at -20 °C under nitrogen prior to analysis.

Filtration of the cytosol (0.2 μm filter) before introducing it onto the chromatographic column is strongly advised. A guard column should be inserted to protect the analytical column particularly from effects of lipids, that otherwise degrade the separation. Any organic species that adhere to the column can also bind inorganic species giving rise to anomalous peaks in subsequent runs. To avoid contamination of the analytical column by trace elements, buffers should be cleaned by cation exchange on Chelex-100.

The drawback of leaching procedures is their generally poor efficiency in recovering element species. The typical efficiencies are about 10-20% with the highest values reported of 50-85% and 30-57% in the case of Zn and Cd, respectively. Addition of sodium dodecyl sulfate (SDS) increases the yield for selenoproteins (Casiot et al. 1999). The increase in this recovery can be achieved by destruction (at least partly) of the sample matrix while the preservation of the original identity is ensured.

16.3.1.2 Total Alkaline or Acid Hydrolysis

Hydrolysis with aq. 25% tetramethylammonium hydroxide is a rapid and effective method to solubilize plant materials prior to chromatography (van Cleuvenbergen et al. 1990). Polypeptides and proteins undergo degradation under the reaction conditions and information on speciation of metals bound to or incorporated in phytochelatin, metallothioneins and higher proteins is lost. The method is thus limited to sample preparation of species with a covalent carbon-metal (metalloid) bond that is likely to survive, e.g. the case of alkyllead species. The method applied to selenium-enriched yeast solubilized the sample completely but the Se species present were entirely degraded to selenomethionine and inorganic selenium (Casiot et al. 1999). Inorganic alkali solutions or acids hydrolyze samples in the same way but the attack seems to be too aggressive even to preserve the carbon-metal bond.

16.3.1.3 Enzymic Solubilization of Solid Plant Tissues

In the case of metal complexes alkaline or acid hydrolysis cannot be accepted since it would affect the complexation equilibria. Selective degradation of the matrix at pH natural for a sample studied by enzymes is often the only possibility.

Cellulose and complex water-insoluble pectic polysaccharides are the main matrices of the water-insoluble residue after centrifugation of fruit and vegetable homogenates. The use of pectinolytic enzymes is therefore necessary to solubilize the solid sample. Pectinolysis is known to degrade efficiently large pectic polysaccharides but some of them, e.g. rhamnogalacturonan II, are considered enzyme-resistant (Szpunar et al. 1999). A mixture of commercial preparations, Rapidase LIQ and Pectinex Ultra-SPL, has been reported for release of metal complexes from the solid parts of edible plants, fruits and vegetables (Szpunar et al. 1999). Extraction of selenium compounds from selenium-enriched yeast with a mixture containing a proteolytic enzyme led to recoveries of Se species above 85%, the majority as selenomethionine (Casiot et al. 1999; Mounicou et al. 2002a).

16.3.2 GC with Element-Selective Detection

The commercial availability (Hewlett-Packard) of a gas chromatograph with microwave-induced plasma detection that allows detection limits at the picogram and sub-picogram levels for most metals and metalloids is the primary tool for speciation analysis of small molecules with a carbon-heteroelement bond. The majority of organometal or -metalloid compounds present in plant samples need to be derivatized, i.e. converted to thermally stable and sufficiently volatile species to be separated by GC. Alkyllead compounds can be converted to butyl derivatives without the loss of initial information leading to butylated methyl- and ethyllead species (van Cleuvenbergen et al. 1990; Łobiński et al. 1994). Selenoamino acids were derivatized with isopropyl chloroformate and bis(p-methoxyphenyl) selenoxide (Kataoka et al. 1994), with pyridine and ethyl chloroformate (Cai et al. 1995), or silylated with bis(trimethylsilyl)acetamide (Yasumoto et al. 1988). Selenomethionine forms volatile methylselenocyanide with CNBr (Zheng and Wu 1988; Ouyang et al. 1989). Methods for the conversion of arsenic compounds to volatile and stable derivatives are based on the reaction of monomethylarsonic acid and dimethylarsenic acid with thioglycolic acid methyl ester (TGM; Beckermann 1982; Dix et al. 1987; Haraguchi and Takatsu 1987). The determination of arsenite, arsenate and methanearsonic acid in aqueous samples by GC of their 2,3-dimercaptopropanol complexes has been reported (Fukui et al. 1983). Application of GC with element-selective detection to species analysis of plants and related samples are summarized in Table 16.1 (Yasumoto et al. 1988; Cai et al. 1994; Dodd et al. 1996).

Species-Selective Analysis for Metals and Metalloids in Plants

419

Table 16.1 Applications of GC with element-selective detection for speciation in plants and foodstuffs of plant origin

Sample	Sample preparation	Analytes	Derivatizing agent	Detection	Reference
Freshwater pondweed	Leaching with acetic acid	Mono-, di- and trimethylantimony Sb(III)	$NaBH_4$	PC GC - ICP MS	Dodd et al. (1996)
Grass and tree leaves		Alkyllead compounds	BuMgCl	PC GC - QF AAS	Van Cleuvenbergen et al. (1990)
Seaweed	Leaching with HCl, extraction into benzene, back-extraction into H_2O	Mono- and dimethylarsenic acids	Methyl mercaptoacetate	PC GC - MIP AES	Haraguchi and Takatsu (1987)
Garlic, onion, broccoli	Leaching with HCl, extraction into $CHCl_3$		Pyridine, ethyl chloroformate	MIP AES	Cai et al. (1995)
Elephant garlic	Leaching with aq. Na_2SO_4	Volatile Se-species		HS CGC - MIP AES	Cai et al. (1994)
Grapes, must	Extraction with DDTC into hexane	Me_3Pb^+, Me_2Pb^{2+}, Et_3Pb^+, Et_2Pb^{2+}	PrMgCl	CGC - MIP AES	Łobiński et al. (1994)
Soya bean protein hydrolysate	Tedious purification		Bis(trimethyl-silyl) acetamide	PC GC - MS	Yasumoto et al. (1988)

Table 16.2 Applications of size-exclusion chromatography with ICP-MS detection

Analyte and sample	Column	Eluent (flow rate)	Reference
Phytochelatins	Eurogel GFC-OH (50×7.5 mm) Eurogel GFC (300×7.5 mm)	0.01 M CH_3COONH_4 pH 7.0	Leopold and Günther (1997)
Pt species in grass	Bio-Gel SEC 20XL (30×7.8 mm)	0.025 M NaCl (0.7 ml min^{-1})	Klueppel et al. (1998)
Metals in tea	Superdex 75 HR (30 cm×10 mm)	0.1 M CH_3COONH_4 pH 5.5 (1 ml min^{-1})	Oedegard and Lund (1997)
B_{12} in radish roots	YMC-Pack Diol-120 (30 cm×8 mm)	0.2 M $HCOONH_4$ pH 6.5 (1 ml min^{-1})	Pellerin et al. (1996)
Pb, Ba, Sr in apples and carrots	Superdex 75 HR (30 cm×10 mm)	0.03 M $HCOONH_4$ pH 5.2 (0.6 ml min^{-1})	Szpunar et al. (1999)
Anacystis nidulans	Asahipak GFA-30F (30 cm×7.6 mm)	0.05 M Tris-HCl pH 7.5 and 0.2 M $(NH_4)_2SO_4$ with 0.1 mM EDTA (0.8 ml min^{-1})	Takatera et al. (1994)
Wine	Superdex 75 HR (30 cm×10 mm)	0.03 M formate buffer, pH 5.8	Raskin et al. (1997)
Cu, Zn, Cd in cyanobacterium	DuPont GF-250 (250×9.4 mm)	0.05 M Tris-HCl, pH 7.5, 0.2 M $(NH_4)_2SO_4$	Takatera and Watanabe (1993)
As in algae	Superdex peptide HR (30 cm×10 mm)	1% aq. acetic acid, pH 3 (0.6 ml min^{-1})	McSheehy et al. (2000a)
Cd phytochelatins in soybean and rice	Superdex peptide HR (30 cm×10 mm)	0.03 M Tris-HCl pH 7.5 (0.75 ml min^{-1})	Vacchina et al. (2000)
Ni in *Thlaspi caerulescens*	Superdex peptide HR (30 cm×10 mm)	0.005 M CH_3COONH_4 pH 7.0 (0.75 ml min^{-1})	Vacchina et al. (2002)
Ni in *Sebertia acuminata*	Superdex peptide HR (30 cm×10 mm)	0.005 M CH_3COONH_4 pH 6.8 (0.75 ml min^{-1})	Schaumlöffel et al. (2003)
Se in nuts	Superdex 30 PG (60×26 mm) Superdex peptide HR (30 cm×10 mm)	0.05 M CH_3COONH_4 + 0.02% SDS pH 4.5 (0.6 ml min^{-1})	Kannamkumarath et al. (2002)
Cd and Pb in cocoa	Superdex 75 HR (30 cm×10 mm)	0.03 M CH_3COONH_4 pH 7.2 (0.6 ml min^{-1})	Mounicou et al. (2002b)

16.3.3 Size-Exclusion Chromatography with ICP-MS

Size-exclusion chromatography (SEC) with on-line detection by ICP-MS is the primary technique that allows the detection of metals bound to macromolecular ligands in an unknown sample. It enables information on the distribution of the metal(s) present in the sample among the fractions of different molecular weight with detection limits down to 1 ng ml^{-1} to be obtained for most elements. Because the resolution of SEC is insufficient for the discrimination of the small amino acid heterogeneities, the coupling of SEC with ICP-MS is the most popular technique for the first screening of an unknown sample in view of the presence of macromolecular species of elements. Applications of SEC/ICP-MS to speciation analysis of plants and related materials are summarized in Table 16.2. In recent studies the most popular SEC column is the newly developed Superdex peptide column due to its separation range between 100 and 7000 kDa, which seems to be optimal for the separation of low molecular weight metal species in plant tissues. Figure 16.2 shows an example multielement chromatogram of a vegetable juice (carrot) sample. SEC using a short guard column with ICP-MS detection allows the rapid quantification of the bound metal fraction by comparison of the chromatographic signal with that, obtained in parallel, by flow-injection ICP-MS analysis of the sample (Shum and Houk 1993). As another example, Fig. 16.3 shows the SEC separation of nickel complexes in the extract of the latex from the hyperaccumulating plant *Sebertia acuminata* with subsequent mass spectrometric identification of the major peak.

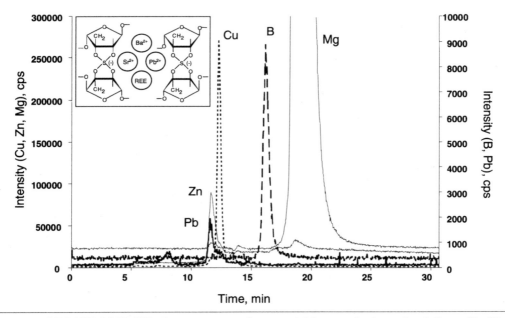

Fig. 16.2　SEC-HPLC/ICP-MS chromatogram of a vegetable sample (carrot), elution with 30 mM Tris-HCl buffer (pH 7.2) from the Superdex 75-HR (10×300 mm×13 μm) column. *Inset* Structure of the metal-polysaccharide complex

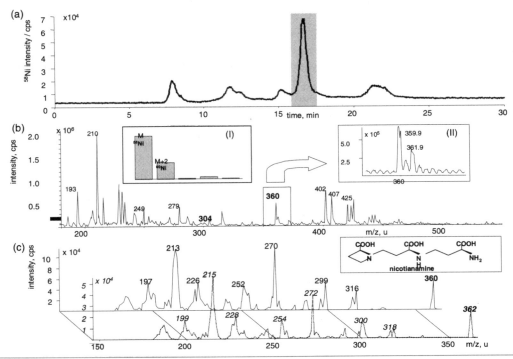

Fig. 16.3 Ni speciation analysis in the hyperaccumulating plant *Sebertia acuminata* (Schaumlöffel et al. 2003). **a** SE-HPLC/ICP-MS chromatogram of the latex extract, elution with 5 mM ammonium acetate buffer (pH 6.8) from the Superdex peptide HR (30 cm×10 mm) column; **b** ESI-MS spectrum of the fractions corresponding to the major peak (*shaded area*) in **a**, *inset I* theoretical Ni isotopic pattern, *inset II* observed Ni isotopic pattern; **c** ESI-MS/MS spectrum of the *m/z* 360 and 362 ions leads to the identification of a Ni-complex with nicotianamine, *inset* structure of the ligand

Advantages of ICP-MS detection, besides its high sensitivity, include the detection of several elements in one run and the possibility of the use of stable isotopes. For multielement SEC-HPLC/ICP-MS, in order to ensure acceptable sensitivity the elements of interest were divided into groups for which a separate chromatographic injection was made; within each group data acquisition conditions were optimized (Oedegard and Lund 1997). Eluents should not contain elements that give polyatomic interfering ions in the ICP step. The wide variety of buffers reported in the literature makes it relatively easy to choose one readily tolerated by ICP. Up to 30 mM Tris-HCl, formate or acetate buffer is the most common choice. Ammonium acetate buffer is preferred to Tris buffer if subsequent identification of the analytes by ESI-MS/MS is carried out due its tolerance in the electrospray source (absence of ionization suppression), even after preconcentration. The standard flow rate of 0.7-1.0 ml/min is compatible with most of the sample introduction systems of the ICP-MS instruments. An interface based on a low-flow-rate direct injection nebulizer has been reported (Shum and Houk 1993). In many applications SEC-HPLC/ICP-MS is used as a semiquantitative technique used to monitor relative changes in analytical signals in a well-defined series of samples, usually to follow metal uptake from a culture medium by a plant. Quantification of the signal obtained is usually performed by a peak area calibration either by changing the measurement system into the flow-injection mode after completing the chromatographic run or using a calibration graph if standards are available.

16.3.4 Ion-Exchange HPLC, Reversed-Phase HPLC and Capillary Zone Electrophoresis with ICP-MS Detection

Beside size-exclusion chromatography, the most common separation techniques used include ion-exchange and reversed-phase (RP) chromatography while ICP-MS is used to detect a metal (or metals) of interest. Anion- and cation-exchange chromatography is the preferred separation technique for arsenic (Gallagher et al. 2001; Kohlmeyer et al. 2002; van Hulle et al. 2002) and aluminum (Bantan Polak et al. 2001) speciation analysis while reversed-phase chromatography is mainly used for the separation of organoselenium compounds (Kotrebai et al. 2000; McSheehy et al. 2000b; Kannamkumarath et al. 2002; Montes-Bayón et al. 2002) and phytochelatins (Vacchina et al. 2000; Chassaigne et al. 2001). Figure 16.4 shows the analysis of selenium species in wild-type and genetically modified Indian mustard by RP-HPLC/ICP-MS.

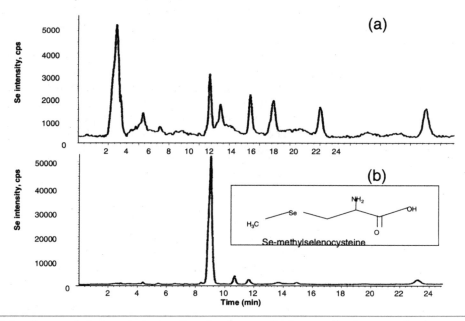

Fig. 16.4 Se speciation analysis in *Brassica juncea* (Indian mustard) grown in the presence of Na_2SeO_3 as Se source by reversed-phase chromatography, isocratic elution with 10% $MeOH/H_2O$ containing 0.1% HFBA from the Alltech C_8 (250×4.6 mm×5 *m*m) column. **a** RP-HPLC/ICP-MS chromatogram of the extract from a wild-type plant; **b** RP-HPLC/ICP-MS chromatogram of the extract from a genetically modified plant, which enlarged the production of the selenoamino acid Se-methylselenocysteine (*inset*; Montes-Bayón et al. 2002; reproduced by permission of The Royal Society of Chemistry)

A very promising technique for a highly resolved separation as well as the definitive verification of the purity of the target compounds appears to be capillary zone electrophoresis (CZE; Richards and Beattie 1994) because of the large number of theoretical plates available. The advantages of CZE, such as the possibility of analyzing for relatively labile species because of the absence of chromatographic packing and high resolution, are outweighed by the need for ultra-sensitive detectors, such as ICP sector field MS, because of the small sample amount injected. Although the CZE/ICP-MS interface developed by Schaumlöffel and Prange (1999) is now commercially available, applications in plant speciation analysis have been scarce and none really with element-selective detection and a sample from a plant tissue.

16.3.5 Electrospray-MS(/MS), HPLC/ESI-MS and CZE/ESI-MS in Speciation Analysis

Electrospray mass spectrometry that allows a precise (± 1 Da) determination of the molecular mass of a species is an invaluable tool for the identification and a prerequisite for further characterization of compounds in speciation analysis (Hofstadler et al. 1996). In particular, the potential of ESI-MS/MS was successfully demonstrated for the identification of aluminum complexes in plant sap (Bantan et al. 2001; Bantan Polak et al. 2001), organoselenium compounds in garlic (McSheehy et al. 2000b) and Indian mustard (Montes-Bayón et al. 2002), arsenosugars in algae (McSheehy et al. 2000a; van Hulle et al. 2002), and nickel complexes in hyperaccumulating plants (Schaumlöffel et al. 2003; Vacchina et al. 2003). The unmatched advantage of ESI-MS is that this measurement can also be realized on-line for an effluent of reversed-phase chromatography of a roughly purified analytical compound, as shown in Fig. 16.5. An additional dimension to ESI-MS in speciation analysis can be obtained by post-column demetallation of the metallocomplexes separated in the column in order to simplify the spectra and to identify the apoligand (Vacchina et al. 2000). The difference between the mass spectra of metallated and demetallated peaks can offer an attractive means for the determination of the stoichiometry of the separated compounds.

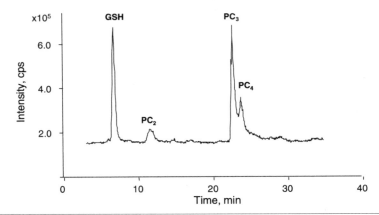

Fig. 16.5 RP-HPLC/ESI-MS chromatogram of phytochelatins, elution with 0.1% TFA with CH_3CN gradient of 0-50% (0-40 min) from the Brownlee Aquapore C_8 (150×1 mm×7 μm) column

Non-metallated peptides of limited molecular weight (up to 2500 Da) can be subjected to on-line sequence analysis by MS/MS using collision-induced dissociation. The advantage is the small samples necessary. An on-line coupling of capillary zone electrophoresis to ES-MS/MS has been recently demonstrated as an interesting tool for the separation and sequencing of phytochelatins in edible plants (Chassaigne et al. 2001; Mounicou et al. 2001). The CZE separation of phytochelatins in roots from a rice plant with subsequent on-line detection, identification and sequencing by ESI-MS/MS is shown in Fig. 16. 6. Other complementary techniques, e.g. NMR, can be necessary, as demonstrated for the analysis of citric acid as a ligand for nickel in *S. acuminata* (Sagner et al. 1998), but it should be noted that the amount of trace metal complexes being subjected to environmental speciation analysis is generally insufficient for their characterization by NMR unless tedious separation techniques are employed.

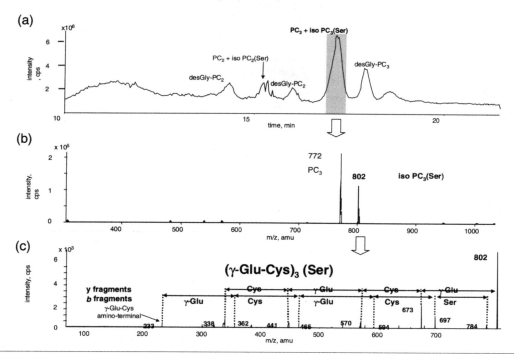

Fig. 16.6 Identification of phytochelatins in *Oryza sativa* (rice) exposed to Cd stress by CZE/ESI-MS/MS. **a** CZE/ESI-MS electropherogram (total ion current) of the purified extract of the rice roots; **b** ESI-MS spectrum of the major peak in (a) (*shaded area*); **c** ESI-MS/MS spectrum of the *m/z* 802 ion, sequencing of the peptide and identification of the phytochelatin

16.3.6 Multidimensional Separation Techniques with Parallel ICP-MS and ESI-MS/MS Detection

One of the major problems in the analysis of real samples is the unavailability of standards. Only a very restricted number of standards (e.g. some animal metallothioneins, cobalamin analogs) can be used for peak identification. In most applications, a multidimensional chromatographic technique should be developed where, after a preparative SEC separation of the plant extract, the metal-containing fractions are submitted to another separation mechanism such as ion exchange, reversed-phase chromatography or capillary zone electrophoresis. This enables verification of whether the SEC peak contains one or more metallocompounds and to remove the matrix that should allow a further analysis by molecular ion mass spectrometry. It should be noted that one subsequent separation technique after SEC is usually sufficient for the evaluation of the purity of a metal compound. In the case of an impure preparation the use of a second technique with another complementary separation mechanism should be considered which is, for example, reported for the analysis of arsenic species in algae (McSheehy et al. 2002a). Examples of multidimensional separation techniques for speciation analysis of plants and related materials are given in Table 16.3. Figure 16.7 demonstrates in an exemplary fashion the multidimensional approach for the identification of arsenosugar B in algae.

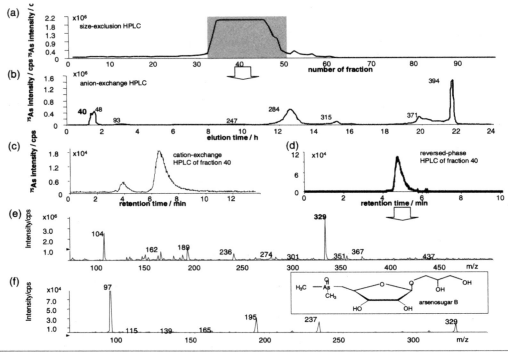

Fig. 16.7 Analysis of arsenosugars in algae by multidimensional (size exclusion, anion exchange, cation exchange, reversed phase) chromatography combined with ICP-MS and ESI-MS/MS (McSheehy et al. 2002a). **a** Purification of the algae extract by SE-HPLC/ICP-MS, fractions 32-50 (*shaded area*) were pooled and lyophilized; **b** separation of the pooled fractions from (a) by AE-HPLC/ICP-MS; **c** control of the purity of fraction 40 from **b** by CE-HPLC/ICP-MS; **d** further purification of fraction 40 from **b** by RP-HPLC/ICP-MS; **e** ESI-MS spectrum of the pooled fractions (*shaded area*) from (d); **f** ESI-MS/MS spectrum of the *m/z* 329 ion and identification as arsenosugar B (structure in the *inset*)

Table 16.3 Examples for the application of multidimensional separation techniques

Analyte and sample	First separation technique	Second separation technique	Third separation technique	Reference
Se in garlic	**SEC** Sephadex G-75 (70 cm × 16 mm) 1% acetic acid pH 3 (0.7 ml min^{-1})	**RP-HPLC** Sperisorb (25 cm × 4.6 mm) 0.3% acetic acid pH 3 (0.75 ml min^{-1})		McSheehy et al. (2000b)
Se in yeast	**SEC** Sephadex G-75 (70 cm × 16 mm) 1% acetic acid pH 3 (0.7 ml min^{-1})	**RP-HPLC** Sperisorb (25 cm × 4.6 mm) 0.3% acetic acid pH 3 (0.9 ml min^{-1})		McSheehy et al. (2001b)
Cd phytochelatins in soybean and rice	**SEC** Superdex peptide HR (30 cm × 10 mm) 0.03 M Tris-HCl pH 7.5 (0.75 ml min^{-1})	**RP-HPLC** Vydac C$_8$ (15 cm × 4.6 mm) 1% TFA pH 2.3, acetonitrile gradient		Vacchina et al. (2000)
Cd phytochelatins in maize seedlings	**RP-HPLC** Eurospher-C$_{18}$ (25 cm × 4 mm) 0.1% TFA, acetonitrile gradient (2.0 ml min^{-1})	**CZE/ESI-MS** Fused silica (95 cm × 75 mm) 5 mM ammonium acetate pH 4.0, 20.9 kV		Chassaigne et al. (2001)
As in algae	**SEC** Superdex peptide HR (30 cm × 10 mm) 1% acetic acid, pH 3 (0.6 ml min^{-1})	**AE-HPLC** Supelcosil (25 cm × 4.6 mm) 5 to 26 mM phosphate within 22 min, pH 6 (1.0 ml min^{-1})		McSheehy et al. (2000a)
As in algae	**SEC** Sephadex G-75 (70 cm × 16 mm) 1% acetic acid pH 3 (0.9 ml min^{-1})	**AE-HPLC** DEAE Sephadex A-25 (70 cm × 16 mm) 2 to 50 mM ammonium carbonate within 25 h, pH 8.9 (0.9 ml min^{-1})	**RP-HPLC** Intersil ODS-2 (25 cm × 4.6 mm) 4 mM malonic acid (0.75 ml min^{-1})	McSheehy et al. (2002a)

16.4 CONCLUSIONS

A large amount of fundamental data on metal speciation in plants nas been acquired without the sophisticated tools discussed above and without mentioning the term "speciation". This was possible by extremely laborious isolation and enrichment procedures to obtain the metal species to be characterized. Hyphenated techniques based on the on-line coupling of gas chromatography, high-performance liquid chromatography and capillary zone electrophoresis with sensitive, element-selective techniques such as atomic absorption, atomic emission and mass spectrometry with plasma and electrospray ionization sources offer an attractive instrumental approach to the identification, characterization and determination of the individual metal and metalloid species present in plants and in foodstuffs of plant origin. The high sensitivity, selectivity and, especially the rapidity with which the data can be acquired, are worth the wider implementation of these relatively new, however already well established, analytical techniques in research on the role of essential and toxic metals in plants.

REFERENCES

Bantan T, Milačič R, Mitrović B, Pihlar B (1999) Combination of various analytical techniques for speciation of low molecular weight aluminium complexes in plant sap. Fresenius J Anal Chem 365:545-552

Bantan Polak T, Milačič R, Pihlar B, Mitrović B (2001) The uptake and speciation of various Al species in the *Brassica rapa pekinensis*. Phytochemistry 57:189-198

Barth HG, Boyes BE, Jackson C (1998) Size exclusion chromatography and related separation techniques. Anal Chem 70:251R-278R

Beckermann B (1982) Determination of monomethylarsonic acid and dimethylarsenic acid [hydroxydimethylarsine oxide] by derivatization with thioglycolic acid methyl ester [methyl mercaptoacetate] and gas-liquid chromatographic separation. Anal Chim Acta 135:77-84

Behne D, Weiss-Nowak C, Kalcklosch M, Westphal C, Gessner H, Kyriakopoulos A (1995) Studies on the distribution and characteristics of new mammalian selenium containing proteins. Analyst 120:823-825

Beilstein MA, Whanger PD, Yang GQ (1991) Chemical forms of selenium in corn and rice grown in a high selenium area of China. Biomed Environ Sci 4:392-398

Berghoff RL, von Willert DJ (1988) Interference in the quantitative determination of arsenic(III) and arsenic(V) in nutrient solution and plant material. Fresenius Z Anal Chem 331:42-45

Bird SM, Ge H, Uden PC, Tyson JF, Block E, Denoyer E (1997) High-performance liquid chromatography of selenoamino acids and organoselenium compounds. Speciation by inductively coupled plasma mass spectrometry. J Chromatogr A 789:349-359

Brown TA, Shrift A (1982) Selenium: toxicity and tolerance in higher plants. Biol Rev 57:59

Cai XJ, Uden PC, Sullivan JJ, Quimby BD, Block E (1994) Headspace-gas chromatography with atomic-emission and mass selective detection for the determination of organoselenium compounds in elephant garlic. Anal Proc 31:325-327

Cai XJ, Block E, Uden PC, Zhang X, Quimby BD, Sullivan JJ (1995) *Allium* chemistry: identification of selenoamino acids in ordinary and selenium-enriched garlic, onion, and broccoli using gas chromatography with atomic emission detection. J Agric Food Chem 43:1754-1757

Casiot C, Szpunar J, Łobiński R, Potin-Gautier M (1999) Sample peparation and HPLC separation approaches to speciation analysis of selenium in yeast by ICP-MS. J Anal At Spectrom 14:645-650

Chassaigne H, Vacchina V, Kutchan TM, Zenk MH (2001) Identification of phytochelatin-related peptides in maize seedlings exposed to cadmium and obtained enzymatically *in vitro*. Phytochemistry 56:657-668

Cullen WR, Reimer KJ (1989) Arsenic speciation in the environment. Chem Rev 89:713-764

Dauchy X, Potin-Gautier M, Astruc A, Astruc M (1994) Analytical methods for the speciation of selenium compounds: a review. Fresenius J Anal Chem 348:792-805

Dix K, Cappon C-J, Toribara TY (1987) Arsenic speciation by capillary gas-liquid chromatography. J Chromatogr Sci 25:164-169

Doco T, Brillouet J-M, O'Neill MA (1993) Isolation and characterization of a rhamnogalacturonan II from red wine. Carbohydr Res 243:333-343

Dodd M, Pergantis SA, Cullen WR, Li H, Eigendorf GK, Reimer KJ (1996) Antimony speciation in freshwater plant extracts by using hydride generation-gas chromatography-mass spectrometry. Analyst 121:223-228

Dolphin D (ed) (1982) B-12 Biochemistry and medicine, vol 2. Wiley, New York

Edmonds JS, Francesconi KA (1987) Transformations of arsenic in the marine environment. Experientia 43:553-557

Forsyth DS, Iyengar JR (1989) Enzymatic hydrolysis of biological and environmental samples as pretreatment for analysis. J Assoc Off Anal Chem 72:997-1001

Fukui S, Hirayama T, Nohara M, Sakagami Y (1983) Determination of arsenite, arsenate and methanearsonic acid in aqueous samples by gas chromatography of their 2,3-dimercaptopropanol (BAL) complexes. Talanta 30:89-93

Gallagher PA, Shoemaker JA, Wei X, Brockhoff-Schwegel CA, Creed JT (2001) Extraction and detection of arsenicals in seaweed via accelerated solvent extraction with ion chromatographic separation and ICP-MS detection. Fresenius J Anal Chem 369:71-80

Ge H, Cai XJ, Tyson JF, Uden PC, Denoyer ER, Block E (1996) Identification of selenium species in selenium-enriched garlic, onion and broccoli using high-performance ion chromatography with inductively coupled plasma mass spectrometry detection. Anal Commun 33:279-281

Gilon N, Astruc A, Astruc M, Potin-Gautier M (1995) Selenoamino acid speciation using HPLC-ETAAS following an enzymic hydrolysis of selenoprotein. Appl Organomet Chem 9:623-628

Gooding KM, Regnier FE (2002) Size exclusion chromatography. Chromatographic Science Series 87 (HPLC of Biological Macromolecules, 2nd edn), pp 49-79

Hall CA (ed) (1983) The cobalamins. Churchill Livingstone, Edinburgh

Haraguchi H, Takatsu A (1987) Derivative gas chromatography of methylarsenic compounds with atmospheric-pressure helium microwave-induced plasma atomic-emission spectrometric detection. Spectrochim Acta 42B:235-241

Hayashi Y, Winge DR (1992) (γ-EC)$_n$G peptides. In: Stillman MJ, Shaw CF, Suzuki KT (eds) Metallothioneins. Synthesis, structure and properties of metallothioneins, phytochelatins and metalthiolate complexes. VCH, New York, pp 271-283

Hofstadler SA, Bakhtiar R, Smith RD (1996) Electrospray ionization mass spectrometry. I. Instrumentation and spectral interpretation. J Chem Educ 73:A82

Hu S, Dovichi NJ (2002) Capillary electrophoresis for the analysis of biopolymers. Anal Chem 74:2833-2850

Inouhe M, Inagawa A, Morita M, Tohoyama H, Joho M, Murayama T (1991) Native cadmium-metallothionein from the yeast *Saccharomyces cerevisine*: its primary structure and function in heavy-metal resistance. Plant Cell Physiol 32:475-482

Ishii T, Matsunaga T (1996) Isolation and characterization of a boron-rhamnogalacturonan-II complex from cell walls of sugar beet pulp. Carbohydr Res 284:1-9

Kaim W, Schwederski B (1994) Bioinorganic chemistry: inorganic elements in the chemistry of life. Wiley, Chichester, p 25

Kannamkumarath SS, Wrobel K, Wrobel K, Vonderheide A, Caruso JA (2002) HPLC-ICP-MS determination of selenium distribution and speciation in different types of nut. Anal Bioanal Chem 373:454-460

Kataoka H, Miyanaga Y, Makita M (1994) Determination of selenocyst(e)amine, selenocyst(e)ine and selenomethionine by gas chromatography with flame photometric detection. J Chromatogr 659:481-485

Klueppel D, Jakubowski N, Messerschmidt J, Stüwer D, Klockow D (1998) Speciation of platinum metabolites in plants by size-exclusion chromatography and inductively coupled plasma mass spectrometry. J Anal At Spectrom 13:255-262

Kohlmeyer U, Kuballa J, Jantzen E (2002) Simultaneous separation of 17 inorganic and organic arsenic compounds in marine biota by means of high-performance liquid chromatography/inductively coupled plasma mass spectrometry. Rapid Commun Mass Spectrom 16:965-974

Kotrebai M, Birringer M, Tyson JF, Block E, Uden PC (2000) Selenium speciation in enriched and natural samples by HPLC-ICP-MS and HPLC-ESI-MS with perfluorinated carboxylic acid ion-pairing agents. Analyst 125:71-78

Krämer U, Cotter-Howells JD, Charnock JM, Baker AJM, Smith JAC (1996) Free histidine as a metal chelator in plants that accumulate nickel. Nature 379:635-638

Kuehnelt D, Lintschinger J, Goessler W (2000) Arsenic compounds in terrestrial organisms. VI. Green plants and lichens from an old arsenic smelter site in Austria. Appl Organomet Chem 14:411-420

Leopold I, Günther D (1997) Investigation of the binding properties of heavy-metal-peptide complexes in plant cell cultures using HPLC-ICP-MS. Fresenius J Anal Chem 259:364-370

Łobiński R (1994) Gas chromatography with element selective detection in speciation analysis. Status and future prospects. Analusis 22:37-48

Łobiński R (1997) Elemental speciation and coupled techniques. Appl Spectrosc 51:260A-278A

Łobiński R, Adams FC (1993) Recent advances in speciation analysis by capillary gas chromatography-microwave induced plasma atomic emission spectrometry. Trends Anal Chem 12:41-49

Łobiński R, Adams FC (1997) Speciation analysis by gas chromatography with plasma source spectrometric detection. Spectrochim Acta Rev 52:1865-1903

Łobiński R, Potin-Gautier M (1998) Metals and biomolecules - bioinorganic analytical chemistry. Analusis 26:21-24

Łobiński R, Szpunar J, Adams FC, Teissedre PL, Cabanis JC (1994) Speciation analysis of organolead compounds in wine by capillary gas chromatography microwave induced plasma atomic emission spectrometry. J Assoc Off Anal Chem 76:1262-1267

Madsen AD, Goessler W, Pedersen SN, Francesconi KA (2000) Characterization of an algal extract by HPLC-ICP-MS and LC-electrospray MS for use in arsenosugar speciation studies. J Anal At Spectrom 15:657-662

Makarov A, Szpunar J (1998) The coupling of size-exclusion HPLC with ICP MS in bioinorganic analysis. Analusis 26(6):M26-M30

Matsunaga T, Ishii T, Watanabe H (1996) Speciation of water-soluble boron compounds in radish roots by size-exclusion HPLC/ICP-MS. Anal Sci 12:673-675

Matsunaga T, Ishii T, Watanabe-Oda H (1997) HPLC/ICP-MS study of metals bound to borate-rhamnogalacturonan-II from plant cell walls. In: Ando T, Fujita K, Mae T, Matsumoto H, Mori S, Sekiya J (eds) Plant nutrition - for suitable food production and environment. Kluwer, Dortrecht, pp 89-90

McSheehy S, Marcinek M, Chassaigne H, Szpunar J (2000a) Identification of dimethylarsinoyl-riboside derivatives in seaweed by pneumatically assisted electrospray tandem mass spectrometry. Anal Chim Acta 410:71-84

McSheehy S, Yang W, Pannier F, Szpunar J, Łobiński R, Auger J, Potin-Gautier M (2000b) Speciation analysis of selenium in garlic by two-dimensional high-performance liquid chromatography with parallel inductively coupled plasma mass spectrometric and electrospray tandem mass spectrometric detection. Anal Chim Acta 421:147-153

McSheehy S, Pohl P, Łobiński R, Szpunar J (2001a) Investigation of arsenic speciation in oyster test reference material by multidimensional HPLC-ICP-MS and electrospray tandem mass spectrometry (ES-MS-MS). Analyst 126:1055-1062

McSheehy S, Pohl P, Szpunar J, Potin-Gautier M, Łobiński R (2001b) Analysis for selenium speciation in selenized yeast extracts by two-dimensional liquid chromatography with ICP-MS and electrospray MS-MS detection. J Anal At Spectrom 16:68-73

McSheehy S, Pohl P, Vélez D, Szpunar J (2002a) Multidimensional liquid chromatography with parallel ICP MS and electrospray MS/MS detection as a tool for the characterization of arsenic species in algae. Anal Bioanal Chem 372:457-466

McSheehy S, Szpunar J, Haldys V, Tortajada J (2002b) Identification of selenocompounds in yeast by electrospray quadrupole-time of flight mass spectrometry. J Anal At Spectrom 17:507-514

Mejare M, Bülow L (2001) Metal-binding proteins and peptides in bioremediation and phytoremediation of heavy metals. Trends Biotechnol 19:67-73

Merian E (ed) (1991) Metals and their compounds in the environment. VCH, Weinheim

Messerschmidt J, Alt F, Tölg G (1994) Platinum species analysis in plant material by gel permeation chromatography. Anal Chim Acta 291:161-167

Messerschmidt J, Alt F, Tölg G (1995) Detection of platinum species in plant material. Electrophoresis 16:800-803

Montes-Bayón M, LeDuc DL, Terry N, Caruso JA (2002) Selenium speciation in wild-type and genetically modified Se accumulating plants with HPLC separation and ICP-MS/ES-MS detection. J Anal At Spectrom 17:872-879

Mounicou S, Vacchina V, Szpunar J, Potin-Gautier, Łobiński R (2001) Determination of phytochelatins by capillary zone electrophoresis with electrospray tandem mass spectrometry detection (CZE-ES MS/MS). Analyst 126:624-632

Mounicou S, McSheehy S, Szpunar J, Potin-Gautier M, Łobiński R (2002a) Analysis of selenized yeast for selenium speciation by size-exclusion chromatography and capillary zone electrophoresis with inductively coupled plasma mass spectrometric detection (SEC-CZE-ICP-MS). J Anal At Spectrom 17:15-20

Mounicou S, Szpunar J, Łobiński R, Andrey D, Blake C-J (2002b) Bioavailability of cadmium and lead in cocoa: comparison of extraction procedures prior to size-exclusion fast-flow liquid chromatography with inductively coupled plasma mass spectrometric detection (SEC-ICP-MS). J Anal At Spectrom 17:880-886

Nagata T, Hayatsu M, Kosuge N (1991) Direct observation of aluminium in plants by nuclear magnetic resonance. Anal Sci 7:213-215

Niessen WMA, Tinke AP (1995) Liquid chromatography-mass spectrometry. General principles and instrumentation. J Chromatogr A 703:37-57

O'Neill MA, Warrenfeltz D, Kates K, Pellerin P, Doco T, Darvill AG, Albersheim P (1996) Rhamnogalacturonan-II, a pectic polysaccaharide in the walls of growing plant cell, forms a dimer that is covalently cross-linked by a borate ester. J Biol Chem 271:22923-22930

Oedegard KE, Lund W (1997) Multielement speciation of tea infusion using cation-exchange separation and size-exclusion chromatography in combination with inductively coupled plasma mass spectrometry. J Anal At Spectrom 12:403-408

Ouyang Z, Wu J, Xie L (1989) Method for indirect determination of trace bound selenomethionine in plants and some biological materials. Anal Biochem 178:77-81

Peichl L, Wäber M, Reifenhäuser W (1992) Schwermetallmonitoring mit der Standardisierten UWSF. Z Umweltchem Ökotoxikol 6(1994):63-69

Pellerin P, Doco T, Vidal S, Williams P, Brillouet J-M, O'Neill MA (1996) Structural characterization of red wine rhamnogalacturonan II. Carbohydr Res 290:183-197

Prasad MNV (ed) (1996) Plant ecophysiology. Wiley, New York

Raskin I, Smith RD, Salt DE (1997) Phytoremediation of metals: using plants to remove pollutants from the environment. Curr Opin Biotechnol 8:221-226

Rauser WE (1996) Phytochelatins and related peptides. Structure, biosynthesis, and function. Plant Physiol 109:1141-1149

Richards MP, Beattie JH (1994) Analysis of metalloproteins and metal-binding peptides by capillary electrophoresis. J Cap Electrophoresis 1:196-207

Sagner S, Kneer R, Wanner G, Cosson J-P, Deus-Neumann B, Zenk MH (1998) Hyperaccumulation, complexation and distribution of nickel in *Sebertia acuminata*. Phytochemistry 47:339-347

Schaumlöffel D, Prange A (1999) A new interface for combining capillary electrophoresis with inductively coupled plasma-mass spectrometry. Fresenius J Anal Chem 364:452-456

Schaumlöffel D, Ouerdane L, Łobiński R (2003) Speciation analysis of nickel in the latex of the hyperaccumulating tree *Sebertia acuminata* by HPLC with ICP MS and electrospray MS/MS detection. J Anal At Spectrom 18:120-127

Shibata Y, Morita M, Fuwa K (1992) Selenium and arsenic in biology: their chemical forms and biological functions. Adv Biophys 28:31-80

Shum SCK, Houk RS (1993) Elemental speciation by anion-exchange and size exclusion chromatography with detection by inductively coupled plasma mass spectrometry with direct injection nebulization. Anal Chem 65:2972-2976

Stillman MJ (1995) Metallothioneins. Coord Chem Rev 144:461-511

Stillman MJ, Shaw CF, Suzuki KT (eds) (1992) Metallothioneins. Synthesis, structure and properties of metallothioneins, phytochelatins and metalthiolate complexes. VCH, New York

Sturgeon RE (2000) Current practice and recent developments in analytical methodology for trace element analysis of soils, plants, and water. Commun Soil Sci Plant Anal 31:1479-1512

Sutton K, Sutton RMC, Caruso JA (1997) Inductively coupled plasma mass spectrometric detection for chromatography and capillary electrophoresis. J Chromatogr 789:85-126

Suzuki KT, Sato M (1995) Preparation of biological samples for quantification of metallothionein with care against oxidation. Biomed Res Trace Elem 6:51

Szpunar J (2000) Bio-inorganic speciation analysis by hyphenated techniques. Analyst 125:963-988

Szpunar J, Łobiński R (1999) Species-selective analysis for metal-biomacromolecular complexes using hyphenated techniques. Pure Appl Chem 71:899-918

Szpunar J, Pellerin P, Makarov A, Doco T, Williams P, Łobiński R (1999) Speciation of metal-carbohydrate complexes in fruit and vegetable samples by size-exclusion HPLC-ICP-MS. J Anal At Spectrom 14:639-644

Szpunar J, Bouyssiere B, Łobiński R (2000) Sample preparation techniques for elemental speciation studies. Compr Anal Chem 33:7-40

Takatera K, Watanabe T (1993) Individual and synergistic effects of heavy-metal ions on the induction of cyanobacterial metallothionein examined by high-performance liquid chromatography/inductively coupled plasma mass spectrometry. Anal Sci 9:19-23

Takatera K, Osaki N, Yamaguchi H, Watanabe T (1994) Characterization and quantitation of metallothionein isoforms induced an a cyanobacterium using reversed-phase HPLC/ICP mass spectrometry. Anal Sci 10:567-572

Vacchina V, Łobiński R, Oven M, Zenk MH (2000) Signal identification in size-exclusion HPLC-ICP-MS chromatograms of plant extracts by electrospray tandem mass spectrometry (ES MS/MS). J Anal At Spectrom 15:529-534

Vacchina V, Mari S, Czernic P, Margues L, Pianelli K, Schaumlöffel D, Lebrun M, Łobiński R (2003) Speciation of nickel in a hyperaccumumating plant by HPLC-ICP MS and electrospray MS/MS assisted by cloning using yeast complementation. Anal Chem (in press)

Van Cleuvenbergen RJA, Adams FC (1991) Organolead compounds. In: Hutzinger O (ed) Handbook of environmental chemistry, vol 3E. Springer, Berlin Heidelberg New York

Van Cleuvenbergen R, Chakraborti D, Adams F (1990) Speciation of ionic alkyllead in grass and tree leaves. Anal Chim Acta 228:77-84

Van Hulle M, Zhang C, Zhang X, Cornelis R (2002) Arsenic speciation in Chinese seaweeds using HPLC-ICP-MS and HPLC-ES-MS. Analyst 127: 634-640

Yasumoto K, Suzuki T, M Yoshida (1988) Identification of selenomethionine in soya-bean protein. J Agric Food Chem 36:463-467

Zenk MH (1996) Heavy metal detoxification in higher plants - a review. Gene 179:21-30

Zheng O, Wu J (1988) Determination of selenomethionine in selenium yeast by cyanogen bromide gas chromatography. Biomed Chromatogr 2:258-259

Zoorob GK, McKiernan JW, Caruso JA (1998) ICP-MS for elemental speciation studies. Mikrochim Acta 128:145

Chapter 17

Experimental Characterisation of Metal Tolerance

Karin I. Köhl[1], Rainer Lösch[2]

[1]Max Planck Institut für Molekulare Pflanzenphysiologie, Am Mühlenberg 1, 14476 Golm, Germany
[2]Abt. Geobotanik, H. Heine-Universität, Universitätsstr. 1, 40225 Düsseldorf, Germany

17.1 INTRODUCTION

The capacity of some plants to survive on soils that contain high concentrations of certain (heavy) metals has fascinated ecologists for decades. Much work (reviewed by Antonovics et al. 1971; Ernst 1974; Woolhouse 1983; Baker 1987; Baker and Walker 1989, 1990; Schat and Ten Bookum 1992a; Macnair 1993) has been dedicated to find differences in plant metal tolerance and decipher the underlying physiological and genetic basis for these differences. The topic has its applied aspects as well. Detrimental effects of high aluminium concentration on crop production prompted the screening for aluminium tolerance in commercially important germplasms (Aniol and Gustafson 1990).*

The aim of tolerance tests is the prediction of metal effects on fitness or final yield, which are difficult or laborious to measure. Therefore, short-term methods measuring the effect of the metal on more easily observable characters, such as growth, have been devised (Macnair 1993). As even growth measurements are laborious in large-scale screening experiments, the toxicological concept of 'biomarkers' (see Sect. 17.6) has been introduced into plant metal tolerance research (Sneller et al. 1999a). The quality of any test parameter will depend on how closely the effect of metal on these parameters resembles the effect of metal on the relevant character, e.g. yield of a cultivar in an agricultural breeding program or fitness for an ecological research project.

Statisticians constantly make pleas aimed at other scientists to think about parameters and statistics before embarking on their experimental work. Therefore, this review will

*The scientifically interesting aspect of metal-accumulating plants, that might be used as tools for the extraction of metals from contaminated soils (phytoremediation), has been recently reviewed (Raskin and Ensley 2000; McGrath et al. 2001) and is not covered in this review.

introduce parameters and statistics before presenting some assay methods and conclude by discussing the relative merits of different approaches to characterising tolerance for different objectives.

17.2 PARAMETERS

The effect of a metal on a plant can be described by a dose-response curve (Fig. 17.1), in which the response of the organism is plotted against the concentration of the metal in the medium. The response is an observable character of the phenotype, like root length, growth rate or survival. In principle, this variable can be used directly to quantify tolerance. However, when comparing genotypes with different root growth rates, it is not possible to distinguish whether differences in root growth under metal treatment result from differences in tolerance genes, or from differences in genes determining root growth rate. Therefore, different parameters have been defined that are assumed to be exclusively dependent on tolerance and independent of genetic or environmental effects on the response character.

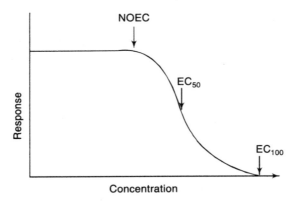

Fig. 17.1 Cardinal points of a dose-response curve. *NOEC* (no observed effect concentration) is the highest concentration that does not result in a response different from the control, EC_{50} (effective concentration 50) is the concentration that results in 50% of the response, and EC_{100} is the lowest concentration that results in 100% inhibition (=zero response)

The first, and still most common, parameter to characterise metal tolerance is the tolerance index, TI, which is calculated as:

TI = Response at elevated test metal concentration/Response at control conditions

where response is a measurable character, e.g. increase in root length in the classical root-elongation test (Wilkins 1978). Alternatively, the effect index (EI) can be calculated as EI = 1-TI (Sneller et al. 1999a). The response to control conditions and to the elevated metal concentration can be measured either sequentially or in parallel (Wilkins 1978; Bannister and Woodman 1992).

In a sequential test, the plant is first cultivated under control conditions and then cultivated at elevated metal concentrations. The *sequential tolerance index* is the quotient from the response after the second cultivation period divided by the response after the first (control) cultivation period (Bannister and Woodman 1992). The sequential tolerance index can be calculated for individual plants and is an important prerequisite for genetic work (Wilkins 1978), allowing repeated measurements to remove variability between individuals from the error term. However, the sequential determination is very sensitive to time effects on the response, resulting from temporal variation in the environmental or developmental effects (Wilkins 1978; Bannister and Woodman 1992). Growth measurements therefore need to be performed in the time of linear or log-linear growth (Wilkins 1978; Baker 1987; Bannister and Woodman 1992), with plant cultivation in controlled environments to ensure reproducibility (Wilkins 1978; Schat and Ten Bookum 1992b).

For the determination of the parallel tolerance index, responses to control conditions and elevated metal concentrations are measured at the same time in different plants. The parallel tolerance index is less sensitive to developmental or environmental effects on the response (Wilkins 1978; Bannister and Woodman 1992). However, unless cloned material is used, variation in the TI will reflect variance in the measured response (e.g. growth rate) plus variance in the underlying tolerance (Macnair 1990).

Although the parallel tolerance index is less sensitive than the sequential index to changes in the growth rate over time, the result of long-term experiments may be affected by the cultivation time. If the growth rate decreases with the size of the plant due to developmental effects or space constraints, this effect will affect the growth rate of control plants earlier in the experiment than the growth rate of growth-inhibited plants. This will result in a seeming increase in the tolerance index. This restriction to long-term metal exposure experiments can be overcome by the use of weight-corrected relative growth rates (RGR; Sneller et al. 1999a). In this approach, RGR is fitted as a function of fresh weight (FW) using the Von Bertalanffy (Sneller et al. 1999a) growth model to data from control plants:

$$RGR = a * FW^{(-1/3)} - b$$

where a and b are estimated parameters.

For the treatment plants, an expected RGR can be calculated for each FW and related to the RGR calculated from consecutive FW measurements to calculate the tolerance index or the effect index.

The control measurements are performed at a concentration that is supposed to allow optimal growth. As low metal concentrations can enhance root growth (Wilkins 1978), control measurements are better performed at defined low metal concentration than without added metals; in the latter case, metal concentrations vary greatly from haphazard contaminations. The ideal elevated test metal concentration for a quantitative comparison of genotypes will result in a TI of more than 0 in the most sensitive genotype, and less than 100 in the most resistant ecotype (Wilkins 1978; Schat and Ten Bookum 1992b). Concentrations fulfilling this requirement may not exist when plants of widely different tolerance are compared (Schat and Ten Bookum 1992b). Even if such a concentration exists, the differences between the TI values of genotypes with different dose-response curves will still depend on the choice of the test concentration (Fig. 17.2).

In multiple-concentration tests, organisms are treated in parallel or serially with an arithmetic or geometric series of metal concentrations (Schat and Ten Bookum 1992b; Schat and Vooijs 1997). The resulting dose-response curve can be described by the three toxicological cardinal points: the NOEC, the EC_{50} and the EC_{100} (Fig. 17.1). The NOEC (no

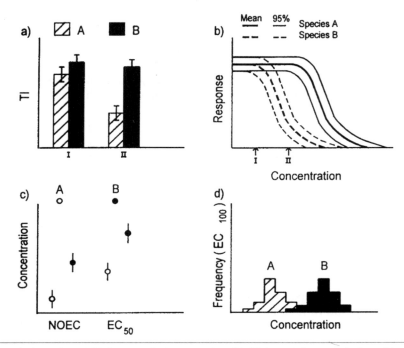

Fig. 17.2 Description of metal tolerance in two species (*A*, *B*) by **a** tolerance index at test concentrations *I* and , **b** dose-response curve with 95% confidence interval, **c** NOEC and EC_{50}, and **d** frequency distribution of the EC_{100}

observed effect concentration) is the highest concentration that does not result in a response different from the control, the EC_{50} (effective concentration 50) results in 50% of the response and the EC_{100} is the lowest concentration that results in zero response (=100% inhibition of the response). The tolerance of a population can be described by any of these three points. In parallel experiments, variance of NOEC or EC_{50} will reflect variance in tolerance and variance in measured response (e.g. growth rate). Observation of individual plant response in sequential experiments yields a distribution of the EC_{100} as a measure for the tolerance distribution in a population (Schat and Ten Bookum 1992b).

Small concentration steps (i.e. testing many different concentrations) allow the identification of small differences in tolerance (Schat and Ten Bookum 1992b). However, the longer time required for multiple-concentration tests has often to be compensated for by testing a lower number of replicates or a lower number of genotypes, which may jeopardise the precision of the test (Macnair 1993).

17.3 STATISTICS

Statistical treatment of the response data depends on the type of variable. Growth rate and biomass are continuous variables, for which differing states can be expressed in numbers, which theoretically can assume any value between two fixed values. For ranked variables, the relative order of observations is expressed by assigning a rank or order, for example I = healthy, II = slight toxicity symptoms, III = heavy toxicity symptoms, IV = dead. The difference in magnitude between ranks is not constant or proportional, which means that ranks are non-additive. Attributes or categorical variables express qualities, like 'dead' or 'alive', and can be treated statistically when combined with frequency data. Continuous variables can be converted into categorical variables, by defining a limit for the change from one state to the other (e.g. complete growth inhibition if growth <10% of control; Sokal and Rohlf 1995).

The effect of treatments or species on continuous variables like the TI can be tested by analysis of variance (ANOVA; van Frenckell-Insam and Hutchinson 1993; Tilstone and Macnair 1997). ANOVA, however, assumes normal distribution and homogenous variances of the data. Homogeneity of variance is achieved by log-transformation of absolute biomass data (Dueck et al. 1987; Harmens et al. 1993a; Pollard and Baker 1996), whereas relative biomass data (% of control) are arcsine-transformed (Hickey et al. 1991; van Frenckell-Insam and Hutchinson 1993) or log-transformed (Cox and Hutchinson 1979; Humphrey and Nicholls 1984). Significant differences between means are established by the t-test (Harrington et al. 1996) or by multiple comparison of means (van Frenckell-Insam and Hutchinson 1993). Cox and Hutchinson (1979) avoided the potential problems arising from the assumption of normal data distribution by performing Kruskal-Wallis analysis of variance by ranks and employing the non-parametric Mann and Whitney U statistic for comparison of means.

Dueck et al. (1987) evaluated tests in which several response parameters are measured at a single treatment, by multiway factorial analysis of variance (MANOVA; for details, see Scheiner 1993). Liu et al. (1996) ranked TIs based on five different growth parameters for each treatment system and analysed the rank values as a randomised complete block design with each treatment system as a block. Fairbrother et al. (1998) compared ANOVA with NCAA (non-metric cluster and association analysis) for their power to detect patterns in potential biomarkers in response to chemical exposure and found the latter method to be more sensitive.

Dose-response curves describe the relationship between the (independent variable) concentration of a substance to which an organism has been exposed and the dependent response variable. Clinical and toxicological dose-response studies are mainly based on frequency data of dichotomous variables (dead/alive) or time-to-occurrence data (Crump 1979; Salsburg 1986). In contrast, response variables in dose-response curves for metal tolerance in plants are mainly continuous variables like growth or TI. These curves (see Fig. 17.3) are generally fitted by linear regression of the relative biomass data on log-transformed treatment concentration, in a concentration range that results neither in complete inhibition nor in no effect (e.g. TI = 15-85%; Hickey et al. 1991; Schat and Ten Bookum 1992b; Harmens et al. 1993a; Sneller et al. 1999b). To compare dose-response curves, regression parameters or values like NOEC, EC_{50} or EC_{100}, calculated from the regression parameters, are compared by parametric or non-parametric tests (Hickey et al. 1991; Schat and Ten Bookum 1992a; Johnson et al. 1997a).

As the complete dose-response curve (Fig. 17.1) obviously cannot be described by linear regression, non-linear regression is used to fit curves for graphic display and data comparison (Hickey et al. 1991; Johnson et al. 1997a). Probit analysis may be more suitable for the interpretation of dose-response curves (Wilkins 1978). Probit analysis

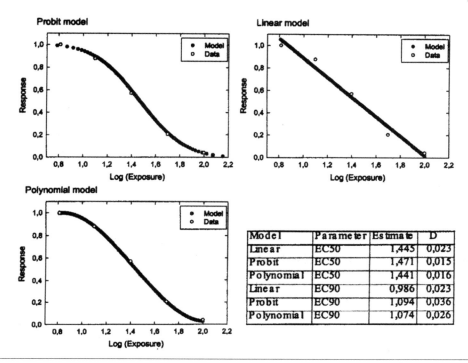

Fig. 17.3 Probit, linear and polynomial models for dose-response curve fitting (data from Sneller et al. 1999b) and EC_{50} and EC_{90} calculated from estimated model parameters. Notice increasing differences between models for EC_{90} estimates compared with EC_{50} estimates and larger deviation (D) for EC_{92}-EC_{90} estimates compared with the EC_{52}-EC_{50} estimates for the non-linear models. The polynomial model overfits the data, as four parameters are estimated from five data points

(Fig. 17.3) was originally designed for the interpretation of frequency data (e.g. percentage of dead organisms; Mead et al. 1993). The basic assumption of normal distribution of the concentration tolerated by each unit is generally met when concentrations are transformed to a log scale. The regression relation is written:

$$\text{probit}(p) = \alpha + \beta \log(c)$$

where α and β are regression parameters

$\quad\quad c$ = treatment concentration

$\quad\quad p$ = proportional response variable (e.g. % plants without root growth).

For each proportion p, the value of probit(p) can be derived from the relationship between the proportion of the normal distribution and the normal deviate z. For the EC_{50}, probit(0.5) equals 0, which means that $EC_{50}=\alpha/\beta$. EC_{100} and NOEC can only be approximated by this method, as probit(0) and probit(100) are infinitesimal (Forbes and Forbes 1994). Angle transformation allows the calculation of EC_{100} and NOEC. However, the confidence limits for estimated EC values close to EC_{100} and NOEC become very wide in both models, so as to render the estimated values useless for biological interpretation (Salsburg 1986). Crump (1979) compared dose-response models based on probit and on polynomial function for the purpose of estimating NOECs.

The methods mentioned above are not suitable for the description of segregation data from crossing experiments, as these data are not derived from a priori known groups of normally distributed data. Methods to describe frequency distribution of individual EC_{100} values from breeding populations and the evaluation of crossing experiments can be found in Humphrey and Nicholls (1984), Macnair (1990, 1993) and Schat and Vooijs (1997). The sources for variation in metal tolerance and metal accumulation capacity within a species (*Arabidopsis halleri*)—within and between population variation—have been analysed using ANOVA on the continuous variable metal content (Bert et al. 2002; Macnair 2002) and G statistic on frequency of non-tolerant individuals (Bert et al. 2000).

An additional approach for the evaluation of multiple-concentration tests on several species is the Finlay-Wilkinson plot (Nicholls and McNeilly 1979). In this plot, the mean performance of a species at a given concentration is depicted against the mean performance of *all* species at this concentration and the linear regression is calculated. Species are compared by comparison of the regression coefficients.

17.4 SHORT-TERM ROOT-ELONGATION TEST

The estimation of metal tolerance by the root-elongation test is based on the observation that toxic concentrations of many metals inhibit root growth (Woolhouse 1983). Bradshaw (1952; from Baker 1987) found that root growth on lead mine soils is more inhibited in *Agrostis tenuis* populations from normal soils than in those from lead-rich soils. In the first test, grass tillers were subjected to sequential series of increasing Pb (and Zn) concentrations in hydroponic systems (Wilkins 1957). After 1-day growth at the respective metal concentration, increase in root length of the longest root was measured as a response variable. This *root-elongation test* has been widely used to quantify metal tolerance for several different metals, like Pb (Simon and Lefèbvre 1977; Bradshaw and McNeilly 1981), Cu (Macnair 1983; Schat and Ten Bookum 1992a; Tilstone and Macnair 1997), Zn (Simon and Lefèbvre 1977), Ni (Gabbrielli et al. 1991; Tilstone and Macnair 1997), Co (Gabbrielli et al. 1991; Schat and Ten Bookum 1992a; Murphy and Taiz 1995a) and Cd (Schat and Ten Bookum 1992a; Murphy and Taiz 1995a). Aniol (1984) developed a similar system to measure Al effects on root meristems by monitoring root growth during a recuperation period after the application of Al stress.

The composition of the hydroponic solution strongly affects the toxicity of a given metal concentration. Wilkins (1957) discovered that Ca alleviates Pb toxicity. Reduction in metal availability, by the precipitation of insoluble metal sulphates and phosphates in the nutrient solution, can reduce toxicity, especially for Pb or Cu (Baker 1987). Similarly, chelators such as EDTA, which are used to keep iron in a plant-available state in nutrient solutions, can form metal chelates that alter metal availability. Many authors have, therefore, performed root-elongation tests in Ca-nitrate solution (Wilkins 1957; Tilstone and Macnair 1997), which maintains membrane integrity, and is thought to be suitable for a short-term test, of up to 2 weeks duration (Wilkins 1978). However, the use of Ca-nitrate as a matrix solution may mask specific interaction of the toxic metal with the uptake of other nutrient ions and may lead to nutrient deficiencies in long-term tests. Other authors (Schat and Ten Bookum 1992a; Baier et al. 1995; Murphy and Taiz 1995a,b) used complete nutrient solutions with or without Fe chelators or phosphate. Additionally, nutrient solutions are often pH-buffered (Schat and Ten Bookum 1992a), as the formation of insoluble metal hydroxides at higher pH can affect metal availability. The metal concentration that results in 50% growth inhibition is generally higher in complete nutrient solution than in nutrient solution without phosphate or in Ca-nitrate solution (Wilkins 1978; Baker 1987; Wu 1990). The background solution may also affect the relative tolerance ranking of different populations (Baker 1987; Bannister and Woodman 1992).

Metals are added to the hydroponic solution mostly as sulphates (Wilkins 1957; Aniol 1984; Humphrey and Nicholls 1984; Schat and Ten Bookum 1992a; Tilstone and Macnair 1997), or occasionally as chloride or nitrate (Wilkins 1957; Cox and Hutchinson 1979; Murphy and Taiz 1995a; Parker 1995). Fixed standard concentrations were suggested for single concentration tests for different metals (Baumeister and Ernst 1978; Bradshaw and McNeilly 1981). However, the ideal concentration (see Sect. 17.1), which depends on the studied species and the question asked (see Sect. 17.5), is generally chosen based on the results of preliminary multiple-concentration assays (Baker 1978; Macnair 1983; Baker et al. 1994; Murphy and Taiz 1995a,b; Tilstone and Macnair 1997). Table 17.1 compiles some data on the concentration ranges employed in multiple-concentration root-elongation tests.

Table 17.1 Concentration range for different metals in root elongation assays

Metal	Range (μM)	Species	Source
Al	0-1500	*Arabidopsis thaliana*	Murphy and Taiz (1995a)
Al	0-720	*Triticum aestivum*	Johnson et al. (1997a)
Cd	0-45	*Chloris barbata*	Patra et al. (1994)
Cd	0-110	*Arabidopsis thaliana*	Murphy and Taiz (1995a)
Cd	0-440	*Silene vulgaris*	Schat et al. (1996)
Cd	0-400	*Silene vulgaris*	Schat and Vooijs (1997)
Cd	0-50	*Thlaspi caerulescens*	Ebbs et al. (2002)
Co	0-360	*Silene vulgaris*	Schat and Vooijs (1997)
Co	0-500	*Alyssum bertolonii*	Gabbrielli et al. (1991)
Cr	0-420	*Arabidopsis thaliana*	Murphy and Taiz (1995a)
Cu	0.1-400	*Silene vulgaris*	Schat and Ten Bookum (1992b)
Cu	0-115	*Arabidopsis thaliana*	Murphy and Taiz (1995a)
Cu	0-110	*Silene vulgaris*	Schat et al. (1996)
Hg	0-2.5	*Chloris barbata*	Patra et al. (1994)
Ni	0-500	*Alyssum bertolonii*	Gabbrielli et al. (1991)
Ni	0-800	*Arabidopsis thaliana*	Murphy and Taiz (1995a)
Ni	0-65	*Silene vulgaris*	Schat and Vooijs (1997)
Zn	0-500	*Alyssum bertolonii*	Gabbrielli et al. (1991)
Zn	10-1000	*Arabidopsis halleri*	Macnair et al. (1999)
Zn	100-2000	*Arabidopsis halleri*	Bert et al. (2000)
Zn	0-1400	*Arabidopsis thaliana*	Murphy and Taiz (1995a)
Zn	0-3250	*Silene vulgaris*	Schat et al. (1996)

The effect of metal treatment on root growth is monitored by measuring the increase in root length of the longest root, the main root (for dicots) or all roots (Simon and Lefèbvre

1977; Baker 1978; Shaw 1984, cited from Baker 1987; Schat and Ten Bookum 1992a,b; Baker et al. 1994; Parker 1995). Alternatively, the absolute length of the longest root or all roots after the metal treatment is measured (Wu and Antonovics 1975; Cox and Hutchinson 1979; Baier et al. 1995; Wundram et al. 1996; Johnson et al. 1997a). De Koe et al. (1992) showed that maximum root growth (MRG=the highest increment in root length for an individual root out of all roots on the plant, for a given time span) is a better parameter for sequential tests than root growth in a defined root, as MRG is not correlated with cultivation time and shows a lower variance than the latter variable.

In an alternative root-elongation test, tolerance is measured by the rooting capacity at a high concentration that prevents root growth in non-tolerant plants (Macnair 1983). However, this method yields only a qualitative estimate of tolerance and, in some cases, no concentration can be found that inhibits root growth in all non-tolerant plants and allows growth in all tolerant plant (Schat and Ten Bookum 1992b). Schat and Ten Bookum (1992b) therefore suggest a sequential multiple-concentration test, in which the lowest metal concentration that inhibits root growth (EC_{100}) is determined for each individual, and populations are characterised by their EC_{100} frequency distribution. The root bending and extension assay (Murphy and Taiz 1995a), developed to screen for metal tolerance in *Arabidopsis thaliana* populations, is based on the same principle. Murphy and Taiz (1995a) replaced the traditional hydroponic system by a newly developed vertical mesh transfer (VMT) technique that allows fast transfer and screening of a large number of plants in species with fast germinating, small seeds.

A number of techniques have been developed to facilitate the detection of changes in root length, namely staining of the roots with dyes (Aniol 1984) or charcoal (Schat and Ten Bookum 1992a) or changing the root orientation to induce geotropic root bending (Murphy and Taiz 1995a). Computerised image analysis has been employed to monitor and evaluate root growth continuously or after the treatment (Murphy and Taiz 1995b; Parker 1995).

Root-elongation tests are short-term tests, with treatment times at a single metal concentration of ≤2 days in the majority of tests, and maximum treatment times of about 8 days (Wilkins 1957; Simon and Levèbvre 1977; Schat and Ten Bookum 1992a; Murphy and Taiz 1995a; Schat et al. 1996; Schat and Vooijs 1997; Tilstone and Macnair 1997; Bert et al. 2000; van Hoof et al. 2001). The method is more suitable to test for tolerance to metals that have a fast effect on root growth (e.g. Cu) than for those that affect root growth more slowly (e.g. Zn and Cd; Schat and Ten Bookum 1992b). In the latter case, results are less reproducible and EC_{100} concentrations are about 20 times higher in a 2-day assay than for long-term growth (Schat and Ten Bookum 1992b). The discrepancy between short-term tests and long-term tests may result from acclimation response, or the accumulation of toxic effects. The latter results in a decrease in the EC_{100} with the time of exposure to the metal (Schat and Ten Bookum 1992b). Thus, lower tolerance levels will be overestimated relative to higher tolerance levels, as in a sequential test the overall metal exposure time is shorter for low metal treatments than for high metal treatments. In contrast, pre-treatment with low metal concentrations (Aniol 1984; Outridge and Hutchinson 1991; Murphy and Taiz 1995a) was found to increase Cu, Al or Cd tolerance in several species. In some cases, an acclimation response was found only in some populations, whereas other populations were constitutively tolerant or non-tolerant (Murphy and Taiz 1995a). The interaction between populations and treatment time may lead to discrepancies between the results of short-term and long-term assays (Köhl 1997).

17.5 LONG-TERM GROWTH TEST

Chronic effects of metals are assessed by treating plants with single or multiple metal concentrations for several weeks or months. As environmental and developmental effects necessarily result in a change in growth rate over this period of time, long-term growth assays can only be performed as parallel assays. The growth response is generally assessed by measuring the final biomass or related parameters (see below) at harvest. However, as the biomass is more than zero at the beginning of the experiment, zero tolerance (no growth at all) will not lead to a TI of 0, thus violating one of the assumptions of the TI concepts (Wilkins 1978). The problem can be solved by determining the change in the biomass during the metal treatment by at least measuring the biomass at the beginning and the end of the treatment. The initial biomass data can be used to account for effects of different initial sizes on the growth rate and thus the final biomass (e.g. weight-corrected growth rate, see Sect. 17.2). This reduces the necessity to normalise the material to plants of similar age and size.

For long-term metal treatments, the cultivation system must contain the metal under investigation in a form available to the plant, while supplying the essential nutrients for growth. The necessity to meet the specific demands of the studied species makes cultivation conditions for long-term growth tests less uniform than those for short-term tests.

The growth responses of plants to elevated concentrations of Cd, Ni, Pb, and Zn have been measured by cultivating plants for several weeks in hydroponics with complete nutrients solution (Baumeister and Burghardt 1956; Mathys 1973; Bernal and McGrath 1994; Harrington et al. 1996; Liu et al. 1996; Ebbs and Kochian 1997; Krämer et al. 1996). The major advantages of hydroponic systems are the good reproducibility of the supply of nutrients and potentially toxic metals, the accessibility of the roots for observation during the experiment, and the ease of harvesting the complete plant. However, the iron chelator EDTA has a high affinity for Cu, Zn, Ni, and Cd, too. The interaction between EDTA and these metals can decrease the free metal concentration and thereby result in a lower apparent toxicity of these metals (Ernst 1972a) and reduced Fe availability (Chaney and Bell 1987). The computer program GEOCHEM PC (Parker et al. 1995) can be used to calculate the activity of free metal ions and complexed metal species assuming thermodynamic equilibrium. In contrast to soils, nutrient solutions lack ion-buffering capacities, which results in changes of ion concentrations with time unless nutrient solutions are exchanged frequently. Changes in pH are critical especially in studies with metals that change toxicity or solubility with pH (Al, Pb). pH-buffering systems like MES have been used in more recently reported metal tolerance assays (Schat and Vooijs 1997; Sneller et al. 1999a), whereas chelators to buffer the concentrations of metals other than Fe are less widely used (Ibekwe et al. 1998). Moreover, plants in nutrient solution lack an exodermis (Zimmermann and Steudle 1998) and cannot establish a rhizosphere or mycorrhiza (Rorison and Robinson 1986), thus differing from soil-grown plants in factors that may affect metal tolerance.

Solid media-cultivation systems allow the establishment of a rhizosphere and mycorrhiza, but roots cannot be observed during cultivation unless special root observation chambers are used (Kahle 1993; Whiting et al. 2000). The complete harvest of the belowground biomass is difficult. When plants are grown on soils that are 'naturally' rich in metals, like mine soils or sewage-sludge amended soils, treatment concentration can be varied by amending these soils with material of low 'metal' concentration (peat, compost or garden soil; Walley et al. 1974; Bernal et al. 1994; Meerts and van Isacker 1997; McGrath et al. 1997). Dall'Agnol et al. (1996) varied Al availability by liming. In these systems, metal treatments are unavoidably varied together with other soil parameters like nutrient availability or water capacity. Alternatively, variable amounts of the metal are added to a soil with low content of this metal (Dueck et al. 1987; Homer et al. 1991; Hagemeyer et al. 1994; Hagemeyer and Lohrie 1995). The treatment metal should be added to the soil well in advance of planting the plants, as the availability of added heavy metals changes after application due to rapid ion-exchange processes or more slowly through metal diffusion into soil minerals (Brümmer et al. 1986).

The major advantage of growth tests on soil is the close resemblance of the cultivation conditions to those in the natural habitat, especially the buffering of the ion concentrations by the ion-exchange capacity of the soil. The major problem of this system is the very poor reproducibility of the treatment conditions. The availability and toxicity of the metal strongly depend on soil parameters like ion-exchange capacity, pH and concentrations of other ions (e.g. Ca; Brümmer et al. 1986; Blamey et al. 1992; Metwally et al. 1993) that are not identical in any two soils and can be very heterogeneous even in a single batch of soil.

Sand culture provides a compromise between soil cultures and hydroponics, with reproducible growing conditions, while allowing the establishment of rhizosphere and mycorrhiza. The very low ion content of pure-grade sand can be supplemented either by soaking in nutrient solution or mixing with solid nutrient sources, like ion-laden ion-exchange resins (Köhl 1996, 1997). The root harvest from these systems is more laborious than from hydroponics, although easier than from soils with organic material. In sand-nutrient solution cultures, nutrient concentrations and pH are not buffered and even more likely to change with time than in hydroponics, as the solution volume is smaller and the exchange more difficult than in hydroponics. In sand/ion-exchange resin systems, however, the concentrations of nutrients and toxic metals are buffered by the ion-exchange capacity of the resins.

The tolerance of plants to long-term metal treatment is mostly quantified by measuring the final biomass of the plants after the treatment period. Most authors determine the final dry mass of the shoot (Baker et al. 1983; Dueck et al. 1987; Bernal and McGrath 1994; Pollard and Baker 1996), or of shoot and roots (Baumeister and Burghardt 1956; Ernst 1972a; Mathys 1973; Bernal and McGrath 1994; Liu et al. 1996; Ebbs and Kochian 1997; Köhl 1997; Whiting et al. 2000). Plant growth under metal treatment was also quantified by measuring the final fresh weight (Harrington et al. 1996; Sunkar et al. 2000; Dominguez-Solis et al. 2001), the total leaf length (Westerbergh 1994), the height of the shoot and the length of roots (Denny and Wilkins 1987; Harrington et al. 1996; Liu et al. 1996; Ebbs and Kochian 1997; Whiting et al. 2000), the lateral root density along the tap root (Ebbs and Kochian 1997), or the total root volume (Kahle 1993). Root length seems to respond more sensitively to metal treatment than root dry weight (Brown and Wilkins 1985; Denny and Wilkins 1987; Baker and Walker 1990). Whiting et al. (2000) additionally employed

[14]C-marker techniques to measure the relative distribution of current assimilates to the roots in soil compartments that differed in Zn supply and found a good agreement between the partitioning of harvested root mass and ^{14}C activity.

17.6 BIOMARKERS

Measurement of metal tolerance based on the previously outlined growth assays is both time-consuming and laborious with a low degree of automation, thus rendering the approach unsuitable for large-scale (screening) studies. Thus, the 'biomarkers' concept has been adopted into the metal-tolerance studies (Sneller et al. 1999a). Biomarkers were originally defined as biochemical, physiological or morphological changes as a consequence of chemical exposure (Huggett et al. 1992). Ideally, the biomarker must be widely distributed in the taxonomic group of interest (e.g. plants) and the change in the biomarker must be measured quickly and cheaply and closely correlated to the effect of interest, e.g. the effect of the chemical on growth, yield or fitness effect (see Sect. 17.1). In recent years, biochemical changes, especially phytochelatin production, have been tested for their potential as biomarkers (Sneller et al. 1999a). However, a number of morphological and physiological features have been tested and used in this respect, namely, plasmolysis capacity, pollen viability, seed germination, photosynthesis and respiration.

Phytochelatins (PCs) are small peptides that are rapidly produced in many plant species upon exposure to heavy metals (Grill et al. 1987). PC concentration correlates with short-term toxicity of Cd in *Zea mays* and *Silene vulgaris* (Keltjens and Vanbeusichem 1998; Sneller et al. 1999a) and As in *Silene vulgaris* (Sneller et a. 1999b). In long-term experiments, the growth (measured as effect index based on weight-corrected growth rates) correlated with PC concentration only at Cd concentrations up to 2 μM, whereas higher Cd concentration resulted in low PC concentrations in spite of strong growth inhibition and high Cd concentration in the plant (Sneller et al. 1999a). In chronically As-exposed plants, PC concentrations correlated with As exposure and As concentration in the plant, although the applied As concentration had no effect on growth (Sneller et al. 1999b). In *Holcus lanatus* populations that differed in As tolerance, PC concentrations generally increased with As exposure. However, the PC concentration at a given As exposure did not correlate with the effect of this As exposure on growth. Thus, PC measurements seem to yield good information about Cd, Cu and As exposure in a limited range of exposure concentrations and are more specific to metal stress than any of the effects mentioned below. However, PC concentration does not allow prediction of the growth effects of this exposure as the relationship between PC concentrations and growth is not unambiguous.

Another biochemical marker that has been used for the rapid detection of toxic metal effects is ATP concentration. ATP concentration measurements have similar shortcomings in the prediction of long-term consequences (Hickey et al. 1991) and are much less specific to metal stress than PC concentrations. Fatty acid composition (Verdoni et al. 2001) and peroxidase activity (MacFarlane 2002) have been suggested as biochemical markers for metal exposure. In maize, peroxidase activity responded more sensitively to Cd exposure than growth parameters (Lagriffoul et al. 1998). Stress proteins (heatshock proteins) have been discussed as biomarkers for a range of stresses (Lewis et al. 1999). In the alga *Enteromorpha intestinalis*, the induction of stress-70 responded to Cu exposure as sensitively as the change in growth (Lewis et al. 1998). A general proteomics approach to identify protein candidates whose concentrations change upon metal exposure has so far only been reported for bacteria (Dilworth et al. 2001).

The methods of 'comparative protoplasmology' (Iljin 1935, cited from Ernst 1974) test for the metal tolerance of a specific cell or tissue type. Isolated fragments of tissue are exposed to heavy metals for up to 48 h and assessed for survival by measuring the capacity to recover from sugar-induced plasmolysis. Higher tolerance in these assays corresponds to provenience of the material from metal-rich sites (Repp 1963; Gries 1966; Ernst 1972a,b, 1976) and to higher tolerance in growth assays (Wu and Antonovics 1978).

The effects of Zn or Cu on pollen development, germination and the growth rate of pollen tubes have been measured in artificial media and in situ in metal-treated plants (Searcy and Mulcahy 1985; Dueck et al. 1987; Harper et al. 1998). Searcy and Mulcahy (1985) found effective selection for pollen with metal tolerance genes during pollen development, but no difference in pollen germination or pollen tube growth.

Seed germination and seedling survival have been monitored to compare metal tolerance of different species. The NOEC for the inhibition of seed germination was found to be higher in tolerant populations than in non-tolerant populations, tolerance ranking being based on measurements in adult plants (Ernst 1968, 1974; Walley et al. 1974), whereas other authors found no such differences (Cox 1979, from Baker and Walker 1989). However, metal concentrations that inhibit seed germination are considerably higher than those that decrease radicle growth or survival rates of seedlings (Ernst 1965; Walley et al. 1974; Baker et al. 1983; Zeid 2001), suggesting that seed germination is not a sensitive parameter for assessing metal tolerance.

Photosynthesis, chlorophyll fluorescence, respiration and transpiration can be measured on living plants and thus allow monitoring of rapid effects or time-resolved measurements (Lu et al. 2000). Chlorophyll fluorescence and photosynthesis may be more sensitive markers for toxicity of Cu, Pb and Zn, which interact directly with the photosynthetic apparatus, than for Cd and Ni (Szalontai et al. 1999). Chlorophyll fluorescence is indeed affected by lower Cu concentration than root elongation in *Alyssum montanum* (Ouzounidou 1994). However, Homer et al. (1980) found no difference in Pb and Cd sensitivity between root elongation and chlorophyll fluorescence. Davies et al. (2001) monitored the toxicity of different Cr species and the interaction between Cr and mycorrhiza by photosynthesis measurements and growth measurements. Rivera-Becerril et al. (2002) used chlorophyll fluorescence and growth measurements to investigate the interaction between Cd and arbuscular mycorrhiza. Respiration and photosynthesis measurements have been further employed to quantify Cu tolerance in slow-growing organisms like lichens and bryophytes (Nash 1990; Wells and Brown 1995).

17.7 CONCLUSIONS: THE IMPORTANCE OF TOLERANCE TESTS

The comparison of related populations that differ in metal tolerance is an important tool to find out whether a physiological trait is causally related to metal tolerance. Tolerance testing is necessary to show that test populations from sites that differ in metal content differ in tolerance (Baker 1987; Westerbergh 1994; Köhl 1997). For comparisons of mutants, tolerance tests need to establish whether a mutation results in hypersensitivity or resistance compared to the wild type. When physiological responses are observed in response to a metal treatment, it is essential to establish whether the experimental treatment is tolerated by the plant; otherwise, the physiological reaction may be a mere stress response. Phytochelatin production, for example, has been discussed as an important mechanism for metal tolerance, based on the induction of phytochelatin synthesis by Zn, Cd and Cu treatment (Steffens 1990). However, phytochelatin concentration increases only at Cu or Zn treatments above the NOEC of the plant (Schat and Kalff 1992; Harmens et al. 1993b), and phytochelatin production in sensitive populations is higher than or equal to that in tolerant plants, when the populations are treated with the metal concentration of their respective NOEC or EC_{50}. Similarly, accumulation of organic acids, which were discussed as metal chelators, is now suspected to reflect metabolic disturbance (Woolhouse 1983; Verkleij and Schat 1990).

The determination of the effect of the metal treatment on a plant is also important when plants are experimentally screened for metal hyperaccumulation (i.e. their capacity to accumulate high metal concentrations in the shoot). When plants are grown at Cu, Zn or Pb concentrations that result in growth reduction or toxicity symptoms, shoot concentrations of these metals increase abruptly in comparison with concentrations at lower metal treatments (Antonovics et al. 1971; Baker et al. 1983; Denny and Wilkins 1987; Köhl et al. 1997). Metal hyperaccumulators, in contrast, show high metal concentrations at nontoxic metal treatments. To distinguish between metal hyperaccumulation as a physiological trait and metal accumulation as a toxicity effect, shoot metal concentration needs to be measured under physiological metal treatment, i.e. below the NOEC. Thus, in physiological studies, tolerance tests are needed to establish that the treatment is in the physiological range of the plant.

Genetic analysis of metal-tolerant plants has demonstrated the heritability of metal tolerance and enumerated the genes involved (reviewed by Macnair 1993), elucidated the question of co-tolerance (Schat and Vooijs 1997; Tilstone and Macnair 1997), and will in future help to establish the causal link between tolerance and physiological function by co-segregation analysis and gene identification. Originally, tolerance determination in genetic analysis was based on the TI, which, however, has some serious disadvantages (Macnair 1990). First, the TI has a large error variance, which may lead to a continuous TI distribution masking a discontinuous Mendelian segregation. Second, it is not possible to distinguish between the interaction of the root growth genes with the environment and the tolerance genes with the environment, which makes the interpretation of small differences in TI very difficult, if control root lengths differ significantly. Additionally, conclusions about the direction of dominance will depend on the test concentration in single concentration tests based on the TI (Schat and Ten Bookum 1992b). These problems were overcome by developing tolerance parameters that are independent of root length growth. Macnair (1983) measured tolerance as the ability to grow roots at a metal concentration that inhibited root growth in 100% of the non-tolerant plants, whereas Schat and Ten Bookum (1992b) quantified tolerance by the determination of the EC_{100} in sequential assays.

Ecological studies and their applied agricultural branches try to predict fitness or yield of a genotype in a real environment based on the results of standardised short-term tests. The test conditions, e.g. concentrations in hydroponics, cannot be directly compared with those in the field. Therefore, the test predictions are relative assuming that the genotype that performs best in the short-term test will do likewise in the long-term field situation. Most work to test this assumption has been done on Al resistance. The haematoxylin assay for Al resistance was shown to be an effective tool to select for dominant Al resistance genes (Johnson et al. 1997a), whereas selection for Al resistance in callus tissue culture was ineffective with regard to short-term performance on soil (Dall'Agnol et al. 1996). When plants that were selected for Al tolerance in solution culture were compared in field trials, the agronomic benefit of Al tolerance was dependent on the edaphic environment and the genetic background (Johnson et al. 1997b).

In the future, prediction of field performance from short-term tests will be of increasing importance to select promising candidate lines from transgenic or TILLING (McCallum et al. 2000) populations ahead of costly field trials.

REFERENCES

Aniol A (1984) Induction of aluminium tolerance in wheat seedlings by low doses of aluminium in the nutrient solution. Plant Physiol 75: 551-555

Aniol A, Gustafson JP (1990) Genetics of tolerance in agronomic plants. In Shaw AJ (ed) Heavy metal tolerance in plants: evolutionary aspects. CRC Press, Boca Raton, pp 255-267

Antonovics J, Bradshaw AD, Turner RG (1971) Heavy metal tolerance in plants. Adv Ecol Res 7: 1-85

Baier AC, Somers DJ, Gustafson JP (1995) Aluminium tolerance in wheat: correlating hydroponic evaluations with field and soil performance. Plant Breeding 114: 291-296

Baker AJM (1978) Ecophysiological aspects of zinc tolerance in *Silene maritima* With. New Phytol 80: 635-642

Baker AJM (1987) Metal tolerance. New Phytol 106 [Suppl]: 93-111

Baker AJM, Walker PI (1989) Physiological responses of plants to heavy metals and the quantification of tolerance and toxicity. Chem Spec Bioavailab 1: 9-17

Baker AJM, Walker PI (1990) Ecophysiology of metal uptake by tolerant plants. In: Shaw AJ (ed) Heavy metal tolerance in plants: evolutionary aspects. CRC Press, Boca Raton, pp 155-177

Baker AJM, Brooks RR, Pease AJ, Malaisse F (1983) Studies on copper and cobalt tolerance in three closely related taxa within the genus *Silene* L (Caryophyllaceae) from Zaïre. Plant Soil 73: 377-385

Baker AJM, Reeves RD, Hajar ASM (1994) Heavy metal accumulation and tolerance in British populations of the metallophyte *Thlaspi caerulescens* J. and C. Presl. (Brassicaceae). New Phytol 127: 61-68

Bannister P, Woodman RF (1992) The influence of tolerance indices and growth on metal tolerance of pasture legumes and serpentine plants. In: Baker AJM, Proctor J, Reeves RD (eds) The vegetation of ultramafic (serpentine) soils. Intercept, Andover, pp 375-390

Baumeister W, Burghardt H (1956) Über den Einfluß des Zinks bei *Silene inflata* Smith. Ber Deutsch Bot Ges 69: 161-168

Baumeister W, Ernst W (1978) Mineralstoffe und Pflanzenwachstum, 3rd edn. Fischer, Stuttgart

Bernal MP, McGrath SP (1994) Effects of pH and heavy metal concentrations in solution culture on the proton release, growth and elemental composition of *Alyssum murale* and *Raphanus sativus* L. Plant Soil 166: 83-92

Bernal MP, McGrath SP, Miller AJ, Baker AJM (1994) Comparison of the chemical changes in the rhizosphere of the nickel hyperaccumulator *Alyssum murale* with the non-accumulator *Raphanus sativus*. Plant Soil 164: 251-259

Bert V, Macnair MR, de Laguerie P, Saumitou-Laprade P, Petit D (2000) Zinc tolerance and accumulation in metallicolous and nonmetallicolous populations of *Arabidopsis halleri* (Brassicaceae). New Phytol 146: 225-233

Bert V, Bonnin I, Saumitou-Laprade P, de Laguerie P, Petit D (2002) Do *Arabidopsis halleri* from nonmetallicolous populations accumulate zinc and cadmium more effectively than those from metallicolous populations? New Phytol 155: 47-57

Blamey FPC, Edemeades DC, Wheeler DM (1992) Empirical models to approximate the calcium and magnesium ameliorative effects and genetic differences in aluminium tolerance in wheat. Plant Soil 144: 281-287

Bradshaw AD, McNeilly T (1981) Evolution and pollution. Studies in biology 130. Arnold, London

Brown MT, Wilkins DA (1985) Zinc tolerance in *Betula*. New Phytol 99: 91-100

Brümmer GW, Gerth J, Herms U (1986) Heavy metal species, mobility and availability in soils. Z Pflanzenernähr Bodenkd 149: 382-398

Chaney RL, Bell PF (1987) Complexity of iron nutrition: lessons for plant-soil interaction research. J Plant Nutr 10: 963-994

Crump KS (1979) Dose response problems in carcinogenesis. Biometrics 35: 157-167

Cox RM, Hutchinson TC (1979) Metal co-tolerance in the grass *Deschampsia cespitosa*. Nature 279: 231-233

Davis FT, Puryear JD, Newton RJ, Egilla JN, Grossi JAS (2001) Mycorrhizal fungi enhance accumulation and tolerance of chromium in sunflower (*Helianthus annuus*). J Plant Physiol 158: 777-786

Dall'Agnol M, Bouton JH, Parrott WA (1996) Screening methods to develop alfalfa germplasms tolerant of acid, aluminum toxic soils. Crop Sci 36: 64-70

De Koe T, Geldmeyer K, Jaques NMM (1992) Measuring maximum root growth instead of longest root elongation in metal tolerance tests for grasses (*Agrostis capillaris*, *Agrostis delicatula* and *Agrostis castellana*). Plant Soil 144: 305-308

Denny HJ, Wilkins DA (1987) Zinc tolerance in *Betula* ssp. I. Effect of external concentration of zinc on growth and uptake. New Phytol 106: 517-524

Dilworth MJ, Howieson JG, Reeve WG, Tiwari RP, Glenn AR (2001) Acid tolerance in legume root nodule bacteria and selecting for it. Aust J Exp Agric 41: 435-446

Dominguez-Solis JR, Gutierrze-Alcala G, Romero LC, Gotor C (2001) The cytosolic O-acetylserine(thiol)lyase gene is regulated by heavy metals and can function in cadmium tolerance. J Biol Chem 276: 9297-9302

Dueck TA, Wolting HG, Moet DR, Pasman FJM (1987) Growth and reproduction of *Silene cucubalus* Wib. intermittently exposed to low concentrations of air pollutants, zinc and copper. New Phytol 105: 633-645

Ebbs SD, Kochian LV (1997) Toxicity of zinc and copper to *Brassica* species: implications for phytoremediation. J Environ Qual 26: 776-781

Ebbs S, Lau I, Ahner B, Kochian L (2002) Phytochelatin synthesis is not responsible for Cd tolerance in the Zn/Cd hyperaccumulator *Thlaspi caerulescens* (J. and C. Presl). Planta 214: 635-640

Ernst W (1965) Über den Einfluss des Zinks auf die Keimung von Schwermetallpflanzen und auf die Entwicklung der Schwermetallpflanzengesellschaft. Ber Deutsch Bot Ges 78: 205-212

Ernst W (1968) Das Violetum calaminariae westfalicum, eine Schwermetallpflanzengesellschaft bei Blankenrode in Westfalen. Mitt Florist-Soz Arbeitsgem NF 13: 263-268

Ernst W (1972a) Schwermetallresistenz und Mineralstoffhaushalt. Forschungsberichte des Landes NRW no 2251, Westdeutscher Verlag, Opladen

Ernst W (1972b) Ecophysiological studies on heavy metal plants in south central Africa. Kirkia 8: 125-145

Ernst W (1974) Schwermetallvegetation der Erde. Fischer, Stuttgart

Ernst W (1976) Ökologische Grenze zwischen Violetum calaminariae und Gentiano-Koelerietum. Ber Deutsch Bot Ges 89: 381-390

Fairbrother A, Landes WG, Dominques S, Shiroyama T, Buchholz P, Roze MJ, Matthews GB (1998) A novel nonmetric multivariate approach to the evaluation of biomarkers in field studies. Ecotoxicology 7: 1-10

Forbes VE, Forbes TL (1994) Ecotoxicology in theory and practice. Chapmann and Hall, London

Gabbrielli R, Mattioni C, Vergnano O (1991) Accumulation mechanisms and heavy metal tolerance of a nickel hyperaccumulator. J Plant Nutr 14: 1067-1080

Gries B (1966) Zellphysiologische Untersuchungen über die Zinkresistenz bei Galmeiökotypen und Normalformen von *Silene cucubalus* Wib. Flora B 156: 271-290

Grill E, Winnacker EL, Zenk MH (1987) Phytochelatins, a class of heavy-metal-binding peptides from plants are functionally analogous to metallothioneins. PNAS 84: 439-443

Hagemeyer J, Lohrie K (1995) Distribution of Cd and Zn in annual xylem rings of young spruce trees (*Picea abies* (L.) Karst.) grown in contaminated soil. Trees 9: 195-199

Hagemeyer J, Heppel T, Breckle S-W (1994) Effects of Cd and Zn on the development of annual xylem rings of young Norway spruce (*Picea abies*) plants. Trees 8: 223-227

Harmens H, Gusmao NGCPB, den Hartog PR, Verkleij JAC, Ernst WHO (1993a) Uptake and transport of zinc in zinc-sensitive and zinc-tolerant *Silene vulgaris*. J Plant Physiol 141: 309-315

Harmens H, den Hartog PR, Ten Bookum WM, Verkleij JAC (1993b) Increased zinc tolerance in *Silene vulgaris* (Moench) Garcke is not due to increased production of phytochelatins. Plant Physiol 103: 1305-1309

Harper FA, Smith SE, Macnair MR (1998) Can an increased copper requirement in copper tolerant *Mimulus guttatus* explain the cost of tolerance? II. Reproductive phase. New Phytol 140: 637-654

Harrington CF, Roberts DJ, Nickless G (1996) The effect of cadmium, zinc and copper on the growth, tolerance index, metal uptake and production of malic acid in two strains of the grass *Festuca rubra*. Can J Bot 74: 1742-1752

Hickey CW, Blaise C, Costan G (1991) Microtesting appraisal of ATP and cell recovery toxicity end points after acute exposure of *Selenastrum capricornutum* to selected chemicals. Environ Toxicol Water Qual 6: 383-403

Homer JR, Cotton R, Evans EH (1980) Whole leaf fluorescence as a technique for measurement of tolerance of plants to heavy metals. Oecologia 45: 88-89

Homer FA, Morrison RS, Brooks RR, Clemens J, Reeves RD (1991) Comparative studies of nickel, cobalt, and copper uptake by some nickel hyperaccumulators of the genus *Alyssum*. Plant Soil 138: 195-205

Huggett RJ, Kimerle RA, Mehrle PM, Bergman JR (eds) (1992) Biochemical, physiological and histological markers of anthropogenic stress. Lewis Publ, Boca Raton

Humphrey MO, Nicholls MK (1984) Relationships between tolerance to heavy metals in *Agrostis capillaris* L. (*A. tenuis* Sibth.). New Phytol 98: 177-190

Ibekwe AM, Angle JS, Chaney RL, Van Berkum P (1998) Zinc and cadmium effects on rhizobia and white clover using chelator-buffered nutrient solution. Soil Sci Soc Am J 62: 204-211

Johnson JP Jr, Carver BF, Baligar VC (1997a) Expression of aluminum tolerance transferred from atlas 66 to hard winter wheat. Crop Sci 37: 103-108

Johnson JP Jr, Carver BF, Baligar VC (1997b) Productivity in Great Plains acid soil of wheat genotypes selected for aluminium tolerance. Plant Soil 188: 101-106

Kahle H (1993) Response of roots of trees to heavy metals. Exp Environ Bot 33: 99-119

Keltjens WG, Vanbeusichem ML (1998) Phytochelatins as biomarkers for heavy metals stress in maize (*Zea mays* L.) and wheat (*Triticum aestivum* L.)—combined effects of copper and cadmium. Plant Soil 203: 119-129

Köhl K. (1996) Population-specific traits and their implication for the evolution of a drought-adapted ecotype in *Armeria maritima*. Bot Acta 109: 206-215

Köhl KI (1997) Do *Armeria maritima* (Mill.)Willd. ecotypes from metalliferous soils and non-metalliferous soils differ in growth response under Zn stress? A comparison by a new artificial soil method. J Exp Bot 48: 1959-1967

Köhl KI, Harper FA, Baker AJM, Smith JAC (1997) Defining a metal-hyperaccumulator plant: the relationship between metal uptake, allocation and tolerance. Plant Physiol 114 [Suppl]: 124

Krämer U, Cotter-Howells JD, Charnock JM, Baker AJM, Smith JAC (1996) Free histidine as a metal chelator in plants that accumulate nickel. Nature 379: 635-638

Lagriffoul A, Mocquot B, Mench M, Vangronsveld J (1998) Cadmium toxicity effects on growth, minerals and chlorophyll contents and activities of stress related enzymes in young maize plants (*Zea mays* L.). Plant Soil 200: 241-250

Lefèbvre C (1975) Evolutionary problems in heavy metal tolerant *Armeria maritima*. International conference on heavy metals in the environment. Toronto, Ontario, pp 155-168

Lewis S, May S, Donkin ME, Depledge MH (1998) The influence of copper and heatshock on the physiology and cellular stress response of *Enteromorpha intestinalis*. Mar Environ Res 46: 421-424

Lewis S, Handy RD, Cordi B, Billinghurst Z, Deplege MH (1999) Stress proteins (HSP's): methods of detection and their use as an environmental biomarker. Ecotoxicology 8: 351-368

Liu H, Heckman JR, Murphy JA (1996) Screening fine fescues for aluminum tolerance. J Plant Nutr 19: 677-688

Lu CM, Chau CW, Zhang JH (2000) Acute toxicity of excess mercury on the photosynthetic performance of cyanobacterium, *S. platensis*—assessment by chlorophyll fluorescence analysis. Chemosphere 41: 191-196

MacFarlane GR (2002) Leaf biochemical parameters in *Avicennia marina* (Forsk.) Vierh. as potential biomarkers of heavy metal stress in estuarine ecosystems. Mar Pollut Bull 44: 244-256

Macnair MR (1983) The genetic control of copper tolerance in the yellow monkey flower *Mimulus guttatus*. Heredity 50: 283-293

Macnair MR (1990) The genetics of metal tolerance in natural populations. In: Shaw AJ (ed) Heavy metal tolerance in plants: evolutionary aspects. CRC Press, Boca Raton, pp 235-253

Macnair MR (1993) Tansley review no 49. The genetics of metal tolerance in vascular plants. New Phytol 124: 541-559

Macnair MR (2002) Within and between population genetic variation for zinc accumulation in *Arabidopsis halleri*. New Phytol 155: 59-66

Macnair MR, Bert V, Huitson SB, Saumitou-Laprade P, Petit D (1999) Zinc tolerance and hyperaccumulation are genetically independent characters. Proc R Soc Lond Ser B Biol Sci 266: 2175-2179

Mathys W (1973) Vergleichende Untersuchungen der Zinkaufnahme von resistenten und sensitiven Populationen von *Agrostis tenuis* Sibth. Flora 162: 492-499

McCallum CM, Comai L, Greene EA, Henikoff S (2000) Targeted screening for induced mutations. Nat Biotechnol 18: 455-457

McGrath SP, Shen ZG, Zhao FJ (1997) Heavy metal uptake and chemical changes in the rhizosphere of *Thlaspi caerulescens* and *Thlaspi ochroleucum* grown in contaminated soils. Plant Soil 188: 153-159

McGrath SP, Zhao FJ, Lombi E (2001) Plant and rhizosphere processes involved in phytoremediation of metal-contaminated soils. Plant Soil 232: 207-214

Mead R, Curnow RN, Hasted AM (1993) Statistical methods in agriculture and experimental biology, 2nd edn. Chapman and Hall, London

Meerts P, van Isacker N (1997) Heavy metal tolerance and accumulation in the metallicolous and non-metallicolous populations of *Thlaspi caerulescens* from continental Europe. Plant Ecol 133: 221-231

Metwally AI, Mashhady AS, Falatah AM, Reda M (1993) Effect of pH on Zn adsorption and solubility in different clays and soils. Z Pflanzenernähr Bodenkd 156: 131-135

Murphy A, Taiz L (1995a) A new vertical mesh transfer technique for metal-tolerance studies in *Arabidopsis*. Plant Physiol 108: 29-38

Murphy A, Taiz L (1995b) Comparison of metallothionein gene expression and nonprotein thiols in ten *Arabidopsis* ecotypes. Plant Physiol 109: 945-954

Nash TH III (1990) Metal tolerance in lichens. In: Shaw AJ (ed) Heavy metal tolerance in plants: evolutionary aspects. CRC Press, Boca Raton, pp 119-131

Nicholls MK, McNeilly T (1979) Sensitivity of rooting and tolerance to copper in *Agrostis tenuis* Sibth. New Phytol 83: 653-664

Outridge PM, Hutchinson TC (1991) Induction of cadmium tolerance by acclimation transferred between ramets of the clonal fern *Salvina minima* Baker. New Phytol 117: 597-605

Ouzounidou G (1994) Copper-induced changes on growth, metal content and photosynthetic function of *Alyssum montanum* L. plants. Environ Exp Bot 34: 165-172

Parker DR (1995) Root growth analysis: an underutilised approach to understanding aluminium rhizotoxicity. Plant Soil 171:151-157

Parker DR, Norvell WA, Chaney RL (1995) GEOCHEM-PC—a chemical speciation program for IBM and compatible personal computers. In: Loeppert RH, Schwab P, Goldberg S (eds) Chemical equilibrium and reaction models. SSSA special publication 42. SSA, Madison, WI

Patra J, Lenka M, Panda BB (1994) Tolerance and co-tolerance of the grass *Chloris barbata* Sw to mercury, cadmium and zinc. New Phytol 128: 165-171

Pollard AJ, Baker AJM (1996) Quantitative genetics of zinc hyperaccumulation in *Thlaspi caerulescens*. New Phytol 132: 113-118

Raskin I, Ensley BD (2000) Phytoremediation of toxic metals. Using plants to clean up the environment. Wiley, New York

Rivera-Becerril F, Calantzis C, Turnau K, Caussanel JP, Belimov AA, Gianinazzi S, Strasser RJ, Gianinazzi-Pearson V (2002) Cadmium accumulation and buffering of cadmium-induced stress by arbuscular mycorrhiza in three *Pisum sativum* L. genotypes. J Exp Bot 53: 1177-1185

Repp G (1963) Die Kupferresistenz des Protoplasmas höherer Pflanzen der Kupfererzböden. Protoplasma 57: 643-659

Rorison LH, Robinson D (1986) Mineral nutrition. In: Moore PD (ed) Methods in plant ecology, 2nd edn. Blackwell, Oxford, pp 145-211

Salsburg DS (1986) Statistics for toxicologists. Dekker, New York

Schat H, Kalff MMA (1992) Are phytochelatins involved in differential metal tolerance or do they merely reflect metal-imposed strain? Plant Physiol 99: 1475-1480

Schat H, Ten Bookum WM (1992a) Metal-specificity of metal tolerance syndromes in higher plants. In: Baker AJM, Proctor J, Reeves RD (eds) The vegetation of ultramafic (serpentine) soils. Intercept, Andover, pp 337-351

Schat H, Ten Bookum WM (1992b) Genetic control of copper tolerance in *Silene vulgaris*. Heredity 68: 219-229

Schat H, Vooijs R (1997) Multiple tolerance and co-tolerance to heavy metals in *Silene vulgaris*: a co-segregation analysis. New Phytol 136: 489-496

Schat H, Vooijs R, Kuiper E (1996) Identical major gene loci for heavy metal tolerance that have independently evolved in different local populations and subspecies of *Silene vulgaris*. Evolution 50: 1888-1895

Scheiner SM (1993) MANOVA: multiple response variables and multispecies interactions. In: Scheiner SM, Gurevitch J (eds) Dosing and analysis of ecological experiments. Chapman and Hall, London, pp 94-112

Searcy KB, Mulcahy DL (1985) Pollen selection and the gametophytic expression on metal tolerance in *Silene dioica* (Caryophyllaceae) and *Mimulus guttatus* (Scrophulariaceae). Am J Bot 72: 1700-1706

Simon E, Lefèbvre C (1977) Aspects de la tolerance aux metaux lourds chez *Agrostis tenuis* Sibth., *Festuca ovina* L. et *Armeria maritima* (Mill.)Willd. Oecol Plant 12: 95-110

Sneller FEC, Noordover ECM, Ten Bookum WM, Schat H, Bedaux JJM, Verkleij JAC (1999a) Quantitative relationship between phytochelatin accumulation and growth inhibition during prolonged exposure to cadmium in *Silene vulgaris*. Ecotoxicology 8: 167-175

Sneller FEC, van Heerwaarden LM, Kraaijeveld-Smit FJL, Ten Bookum WM, Koevoets PLM, Schat H, Verkleij JAC (1999b) Toxicity of arsenate in *Silene vulgaris*, accumulation and degradation of arsenate-induced phytochelatins. New Phytol 1444: 223-232

Sokal RR, Rohlf FJ (1995) Biometry. The principles and practice of statistics in biological research, 3rd edn. Freeman, San Francisco

Steffens JC (1990) The heavy-metal binding peptides of plants. Annu Rev Plant Physiol Plant Mol Biol 41: 553-575

Sunkar R, Kaplan B, Bouche N, Arazi T, Dolev D, Talke IN, Maathuis FJM, Sanders D, Bouchez D, Fromm H (2000) Expression of a truncated tobacco NtCBP4 channel in transgenic plants and disruption of the homologous *Arabidopsis* CNGC1 gene confer Pb^{2+} tolerance. Plant J 24: 533-542

Szalontai B, Horvath LI, Debreczeny M, Droppa M, Horvath G (1999) Molecular rearrangements of thylakoids after heavy metal poisoning, as seen by Fourier transform infrared (FTIR) and electron spin resonance (ESR) spectroscopy. Photosynth Res 61: 241-252

Tilstone GH, Macnair MR (1997) Nickel tolerance and copper-nickel co-tolerance in *Mimulus guttatus* from copper mine and serpentine habitats. Plant Soil 191: 173-180

Van Frenckell-Insam BAK, Hutchinson TC (1993) Nickel and zinc tolerance and co-tolerance in populations of *Deschampsia cespitosa* (L.) Beauv. subject to artificial selection. New Phytol 125: 547-553

Van Hoof NALM, Hassinen VH, Hakvoort HWJ, Ballintijn KBF, Schat H, Verkleij JAC, Ernst WHO, Karenlampi SO, Tervahouta AI (2001) Enhanced copper tolerance in *Silene vulgaris* (Moench) Garcke populations from copper mines is associated with increased transcript levels of a 2b-type metallothionein gene. Plant Physiol 126: 1519-1526

Verkleij JAC, Schat H (1990) Mechanisms of metal tolerance in higher plants. In: Shaw AJ (ed) Heavy metal tolerance in plants: evolutionary aspects. CRC Press, Boca Raton, pp 179-193

Verdoni N, Mench M, Cassagne C, Bessoule JJ (2001) Fatty acid composition of tomato leaves as biomarkers of metal-contaminated soils. Environ Toxicol Chem 20: 382-388

Walley KA, Khan MSI, Bradshaw AD (1974) The potential for evolution of heavy metal tolerance in plants. I. Copper and zinc tolerance in *Agrostis tenuis*. Heredity 32: 309-319

Wells JM, Brown DH (1995) Cadmium tolerance in a metal-contaminated population of the grassland moss *Rhytidiadelphus squarrosus*. Ann Bot 75: 21-29

Westerbergh A (1994) Serpentine and non-serpentine *Silene dioica* plants do not differ in nickel tolerance. Plant Soil 167: 297-303

Whiting SN, Leake JR, McGrath S, Baker AJM (2000) Positive responses to Zn and Cd by roots of the Zn and Cd hyperaccumulator *Thlaspi caerulescens*. New Phytol 145: 199-210

Wilkins DA (1957) A technique for the measurement of lead tolerance in plants. Nature 4575: 37-38

Wilkins DA (1978) The measurement of tolerance to edaphic factors by means of root growth. New Phytol 80: 623-633

Woolhouse HW (1983) Toxicity and tolerance in the response of plants to metals. In: Lange OL, Nobel PS, Osmond CB, Ziegler H (eds) Physiological plant ecology. II. Responses to the chemical and biological environment. Encyclopedia of plant physiology, new series vol 12C. Springer, Berlin Heidelberg New York, pp 245-300

Wu L (1990) Colonization and establishment of plants in contaminated sites. In: Shaw AJ (ed) Heavy metal tolerance in plants: evolutionary aspects. CRC Press, Boca Raton, pp 269-285

Wu L, Antonovics J (1975) Zinc and copper uptake by *Agrostis stolonifera*, tolerant to both zinc and copper. New Phytol 75: 231-237

Wu L, Antonovics J (1978) Zinc and copper tolerance of *Agrostis stolonifera* in tissue culture. Am J Bot 65: 268-271

Wundram M, Selmar D, Bahadir M (1996) The *Chlamydomonas* test: a new phytotoxicity test based on the inhibition of algal photosynthesis enables the assessment of hazardous leachates from waste disposals in salt mines. Chemosphere 32: 1623-1631

Zeid IM (2001) Responses of *Phaseolus vulgaris* to chromium and cobalt treatments. Biol Plant 44: 111-115

Zimmermann U, Steudle E (1998) Transport across young maize roots: effect of apoplastic barriers. Bulgarian J Plant Physiol Special Issue 207

Subject Index